Lecture Notes in Artificial Intelligence 12275

Subseries of Lecture Notes in Computer Science

Series Editors

Randy Goebel
University of Alberta, Edmonton, Canada
Yuzuru Tanaka
Hokkaido University, Sapporo, Japan
Wolfgang Wahlster
DFKI and Saarland University, Saarbrücken, Germany

Founding Editor

Jörg Siekmann
DFKI and Saarland University, Saarbrücken, Germany

More information about this series at http://www.springer.com/series/1244

Gang Li · Heng Tao Shen ·
Ye Yuan · Xiaoyang Wang ·
Huawen Liu · Xiang Zhao (Eds.)

Knowledge Science, Engineering and Management

13th International Conference, KSEM 2020
Hangzhou, China, August 28–30, 2020
Proceedings, Part II

 Springer

Editors
Gang Li (ID)
Deakin University
Geelong, VIC, Australia

Ye Yuan
Beijing Institute of Technology
Beijing, China

Huawen Liu (ID)
Zhejiang Normal University
Jinhua, China

Heng Tao Shen (ID)
University of Electronic Science
and Technology of China
Chengdu, China

Xiaoyang Wang (ID)
Zhejiang Gongshang University
Hangzhou, China

Xiang Zhao (ID)
National University of Defense Technology
Changsha, China

ISSN 0302-9743   ISSN 1611-3349   (electronic)
Lecture Notes in Artificial Intelligence
ISBN 978-3-030-55392-0   ISBN 978-3-030-55393-7   (eBook)
https://doi.org/10.1007/978-3-030-55393-7

LNCS Sublibrary: SL7 – Artificial Intelligence

This Springer imprint is published by the registered company Springer Nature Switzerland AG
The registered company address is: Gewerbestrasse 11, 6330 Cham, Switzerland

Preface

The International Conference on Knowledge Science, Engineering and Management (KSEM) provides a forum for researchers in the broad areas of knowledge science, knowledge engineering, and knowledge management to exchange ideas and to report state-of-the-art research results. KSEM 2020 is the 13th in this series, which builds on the success of 12 previous events in Guilin, China (KSEM 2006); Melbourne, Australia (KSEM 2007); Vienna, Austria (KSEM 2009); Belfast, UK (KSEM 2010); Irvine, USA (KSEM 2011); Dalian, China (KSEM 2013); Sibiu, Romania (KSEM 2014); Chongqing, China (KSEM 2015); Passau, Germany (KSEM 2016); Melbourne, Australia (KSEM 2017); Changchun, China (KSEM 2018); and Athens, Greece (KSEM 2019).

The selection process this year was, as always, competitive. We received received 291 submissions, and each submitted paper was reviewed by at least three members of the Program Committee (PC) (including thorough evaluations by the PC co-chairs). Following this independent review, there were discussions between reviewers and PC chairs. A total of 58 papers were selected as full papers (19.9%), and 27 papers as short papers (9.3%), yielding a combined acceptance rate of 29.2%.

We were honoured to have three prestigious scholars giving keynote speeches at the conference: Prof. Zhi Jin (Peking University, China), Prof. Fei Wu (Zhejiang University, China), and Prof. Feifei Li (Alibaba Group, China). The abstracts of Prof. Jin's and Prof Wu's talks are included in this volume.

We would like to thank everyone who participated in the development of the KSEM 2020 program. In particular, we would give special thanks to the PC for their diligence and concern for the quality of the program, and also for their detailed feedback to the authors. The general organization of the conference also relies on the efforts of KSEM 2020 Organizing Committee.

Moreover, we would like to express our gratitude to the KSEM Steering Committee honorary chair, Prof. Ruqian Lu (Chinese Academy of Sciences, China), the KSEM Steering Committee chair, Prof. Dimitris Karagiannis (University of Vienna, Austria), Prof. Chengqi Zhang (University of Technology Sydney, Australia), who provided insight and support during all the stages of this effort, and the members of the Steering Committee, who followed the progress of the conference very closely with sharp comments and helpful suggestions. We also really appreciate the KSEM 2020 general co-chairs, Prof. Hai Jin (Huazhong University of Science and Technology, China), Prof. Xuemin Lin (University of New South Wales, Australia), and Prof. Xun Wang (Zhejiang Gongshang University, China), who were extremely supportive in our efforts and in the general success of the conference.

We would like to thank the members of all the other committees and, in particular, those of the Local Organizing Committee, who worked diligently for more than a year to provide a wonderful experience to the KSEM participants. We are also grateful to Springer for the publication of this volume, who worked very efficiently and effectively.

Finally and most importantly, we thank all the authors, who are the primary reason why KSEM 2020 is so exciting, and why it will be the premier forum for presentation and discussion of innovative ideas, research results, and experience from around the world as well as highlight activities in the related areas.

June 2020

Gang Li
Heng Tao Shen
Ye Yuan

Organization

Steering Committee

Ruqian Lu (Honorary Chair)	Chinese Academy of Sciences, China
Dimitris Karagiannis (Chair)	University of Vienna, Austria
Yaxin Bi	Ulster University, UK
Christos Douligeris	University of Piraeus, Greece
Zhi Jin	Peking University, China
Claudiu Kifor	University of Sibiu, Romania
Gang Li	Deakin University, Australia
Yoshiteru Nakamori	Japan Advanced Institute of Science and Technology, Japan
Jorg Siekmann	German Research Centre of Artificial Intelligence, Germany
Martin Wirsing	Ludwig-Maximilians-Universität München, Germany
Hui Xiong	Rutgers University, USA
Bo Yang	Jilin University, China
Chengqi Zhang	University of Technology Sydney, Australia
Zili Zhang	Southwest University, China

Organizing Committee

Honorary Co-chairs

Ruqian Lu	Chinese Academy of Sciences, China
Chengqi Zhang	University of Technology Sydney, Australia

General Co-chairs

Hai Jin	Huazhong University of Science and Technology, China
Xuemin Lin	University of New South Wales, Australia
Xun Wang	Zhejiang Gongshang University, China

Program Committee Co-chairs

Gang Li	Deakin University, Australia
Hengtao Shen	University of Electronic Science and Technology of China, China
Ye Yuan	Beijing Institute of Technology, China

Keynote, Special Sessions, and Tutorial Chair

Zili Zhang   Southwest University, China

Publication Committee Co-chairs

Huawen Liu   Zhejiang Normal University, China
Xiang Zhao   National University of Defense Technology, China

Publicity Chair

Xiaoqin Zhang   Wenzhou University, China

Local Organizing Committee Co-chairs

Xiaoyang Wang   Zhejiang Gongshang University, China
Zhenguang Liu   Zhejiang Gongshang University, China
Zhihai Wang   Zhejiang Gongshang University, China
Xijuan Liu   Zhejiang Gongshang University, China

Program Committee

Klaus-Dieter Althoff   DFKI and University of Hildesheim, Germany
Serge Autexier   DFKI, Germany
Massimo Benerecetti   Università di Napoli Federico II, Italy
Salem Benferhat   Université d'Artois, France
Xin Bi   Northeastern University, China
Robert Andrei Buchmann   Babes-Bolyai University of Cluj Napoca, Romania
Chen Chen   Zhejiang Gongshang University, China
Hechang Chen   Jilin University, China
Lifei Chen   Fujian Normal Univeristy, China
Dawei Cheng   Shanghai Jiao Tong University, China
Yurong Cheng   Beijing Institute of Technology, China
Yong Deng   Southwest University, China
Linlin Ding   Liaoning University, China
Shuai Ding   Hefei University of Technology, China
Christos Douligeris   University of Piraeus, Greece
Xiaoliang Fan   Xiamen University, China
Knut Hinkelmann   FHNW University of Applied Sciences and Arts
                                     Northwestern Switzerland, Switzerland
Guangyan Huang   Deakin University, Australia
Hong Huang   UGOE, Germany
Zhisheng Huang   Vrije Universiteit Amsterdam, The Netherlands
Frank Jiang   Deakin University, Australia
Jiaojiao Jiang   RMIT University, Australia
Wang Jinlong   Qingdao University of Technology, China
Mouna Kamel   IRIT, Université Toulouse III - Paul Sabatier, France
Krzysztof Kluza   AGH University of Science and Technology, Poland

Longbin Lai	Alibaba Group, China
Yong Lai	Jilin University, China
Qiujun Lan	Hunan University, China
Cheng Li	National University of Singapore, Singapore
Ge Li	Peking University, China
Jianxin Li	Deakin University, Australia
Li Li	Southwest University, China
Qian Li	Chinese Academy of Sciences, China
Shu Li	Chinese Academy of Sciences, China
Ximing Li	Jilin University, China
Xinyi Li	National University of Defense Technology, China
Yanhui Li	Northeastern University, China
Yuan Li	North China University of Technology, China
Shizhong Liao	Tianjin University, China
Huawen Liu	Zhejiang Normal University, China
Shaowu Liu	University of Technology Sydney, Australia
Zhenguang Liu	Zhejiang Gongshang University, China
Wei Luo	Deakin University, Australia
Xudong Luo	Guangxi Normal University, China
Bo Ma	Chinese Academy of Sciences, China
Yuliang Ma	Northeastern University, China
Stewart Massie	Robert Gordon University, UK
Maheswari N	VIT University, India
Myunghwan Na	Chonnam National University, South Korea
Bo Ning	Dalian Maritime University, China
Oleg Okun	Cognizant Technology Solutions GmbH, China
Jun-Jie Peng	Shanghai University, China
Guilin Qi	Southeast University, China
Ulrich Reimer	University of Applied Sciences St. Gallen, Switzerland
Wei Ren	Southwest University, China
Zhitao Shen	Ant Financial Services Group, China
Leilei Sun	Beihang University, China
Jianlong Tan	Chinese Academy of Sciences, China
Zhen Tan	National University of Defense Technology, China
Yongxin Tong	Beihang University, China
Daniel Volovici	ULB Sibiu, Romania
Quan Vu	Deakin University, Australia
Hongtao Wang	North China Electric Power University, China
Jing Wang	The University of Tokyo, Japan
Kewen Wang	Griffith University, Australia
Xiaoyang Wang	Zhejiang Gongshang University, China
Zhichao Wang	Tsinghua University, China
Le Wu	Hefei University of Technology, China
Jia Xu	Guangxi University, China
Tong Xu	University of Science and Technology of China, China
Ziqi Yan	Beijing Jiaotong University, China

Bo Yang	Jilin University, China
Jianye Yang	Hunan University, China
Shiyu Yang	East China Normal University, China
Shuiqiao Yang	University of Technology Sydney, Australia
Yating Yang	Chinese Academy of Sciences, China
Feng Yi	UESTC: Zhongshan College, China
Min Yu	Chinese Academy of Sciences, China
Long Yuan	Nanjing University of Science and Technology, China
Qingtian Zeng	Shandong University of Science and Technology, China
Chengyuan Zhang	Central South University, China
Chris Zhang	Chinese Science Academy, China
Chunxia Zhang	Beijing Institute of Technology, China
Fan Zhang	Guangzhou University, China
Songmao Zhang	Chinese Academy of Sciences, China
Zili Zhang	Deakin University, Australia
Xiang Zhao	National University of Defense Technology, China
Ye Zhu	Monash University, Australia
Yi Zhuang	Zhejiang Gongshang University, China
Jiali Zuo	Jiangxi Normal University, China

Additional Reviewers

Weronika T. Adrian	Piotr Wiśniewski
Taotao Cai	Yanping Wu
Xiaojuan Cheng	Zhiwei Yang
Viktor Eisenstadt	Xuan Zang
Glenn Forbes	Yunke Zhang
Nur Haldar	Qianru Zhou
Kongzhang Hao	Borui Cai
Sili Huang	Hui Chen
Francesco Isgro	Shangfu Duan
Gongjin Lan	Uno Fang
Enhui Li	Huan Gao
Shuxia Lin	Xin Han
Patryk Orzechowski	Xin He
Roberto Prevete	Xuqian Huang
Najmeh Samadiani	Krzysztof Kutt
Bi Sheng	Boyang Li
Beat Tödtli	Jiwen Lin
Beibei Wang	Yuxin Liu
Yixuan Wang	Ning Pang

Contents – Part II

Social Knowledge Analysis and Management

Text Mining and Document Analysis

Deep Learning

Contents – Part I

Knowledge Management for Education

Knowledge-Based Systems

Machine Learning

MA-TREX: Mutli-agent Trajectory-Ranked Reward Extrapolation via Inverse Reinforcement Learning

Sili Huang[1,2], Bo Yang[1,2(✉)], Hechang Chen[2,3], Haiyin Piao[4], Zhixiao Sun[5], and Yi Chang[2,3]

[1] College of Computer Science and Technology, Jilin University, Changchun, China
ybo@jlu.edu.cn
[2] Key Laboratory of Symbolic Computation and Knowledge Engineering, Ministry of Education, Beijing, China
[3] College of Artificial Intelligence, Jilin University, Changchun, China
[4] School of Electronics and Information, Northwestern Polytechnical University, Xi'an, China
[5] Unmanned System Research Institute, Northwestern Polytechnical University, Xi'an, China

Abstract. Trajectory-ranked reward extrapolation (T-REX) provides a general framework to infer users' intentions from sub-optimal demonstrations. However, it becomes inflexible when encountering multi-agent scenarios, due to its high complexity caused by rational behaviors, e.g., cooperation and communication. In this paper, we propose a novel Multi-Agent Trajectory-ranked Reward EXtrapolation framework (MA-TREX), which adopts inverse reinforcement learning to infer demonstrators' cooperative intention in the environment with high-dimensional state-action space. Specifically, to reduce the dependence on demonstrators, the MA-TREX uses self-generated demonstrations to iteratively extrapolate the reward function. Moreover, a knowledge transfer method is adopted in the iteration process, by which the self-generated data required subsequently is only one third of the initial demonstrations. Experimental results on several multi-agent collaborative tasks demonstrate that the MA-TREX can effectively surpass the demonstrators and obtain the same level reward as the ground truth quickly and stably.

Keywords: Mutli-agent system · Inverse reinforcement learning · Reward extrapolation · Iterative extrapolation · Knowledge transfer

1 Introduction

Existing multi-agent reinforcement learning can effectively deal with multi-agent tasks with reasonable reward design [14]. However, in many complex scenarios, it

This work was supported in part by National Natural Science Foundation of China under grants 61876069, 61572226, 61902145, Jilin Province Key Scientific and Technological Research and Development project under grants 20180201067GX and 20180201044GX.

G. Li et al. (Eds.): KSEM 2020, LNAI 12275, pp. 3–14, 2020.
https://doi.org/10.1007/978-3-030-55393-7_1

is difficult for experts to design reasonable rewards and goals, and agents cannot learn the behaviors people expect [2,16]. If the agent cannot obtain the reward signal, inverse reinforcement learning can find a reasonable reward function from demonstrations, which are provided by the demonstrator [1]. It has been confirmed that in a complex multi-agent environment when the agent can obtain the high-performance expert trajectory, the reward function highly related to the basic facts can be restored [1]. Unfortunately, various complex tasks cannot provide high-quality expert demonstrations [18,21], and the problem is more serious in the multi-agent field.

If a demonstrator is sub-optimal and can inform their intentions, the agent can use these intents to learn performance beyond the demonstrator [5,20]. But most of the existing inverse reinforcement learning algorithms cannot do this, and usually look for reward functions that make the demonstration look close to the best [8,9,17,22]. Therefore, when the demonstrator is sub-optimal, IRL will also lead to sub-optimal behavior such as behavior cloning [19]. Imitation learning method [3] directly imitates behavior without reward inference, which also has the same disadvantage. Brown proposed an algorithm learned from the sub-optimal demonstrator [5], but it is only effective for single-agent problems, and reward inference is limited to the demonstrator. Different from the single agent, multi-agent problems usually use Nash equilibrium [11] as the optimal solution, which makes the algorithm more demanding on the demonstrator and more difficult for reward inference.

In view of this, inspired by the trajectory-ranked reward extrapolation (T-REX) algorithm [5], we propose a novel multi-agent trajectory-ranked reward extrapolation (MA-TREX) framework, and give an iterative form of reward extrapolation using self-generated demonstrations. Specifically, through the ranked team trajectories, the reward function learns to allocate higher team rewards for better trajectories based on the global state, and guides the agent to achieve performance beyond the demonstrator. In order to break through the demonstrators' restrictions on reward reasoning, collect new trajectories generated during the agents' learning process, and add ranking labels as a new training set. The new reward function uses the new ranked demonstrations to reason about higher returns, and is then used to train agents with higher performance. In the learning process of the new reward function, a knowledge transfer method is adopted, which takes only a small amount of demonstrations to complete the learning after inheriting the parameters of the previous round of reward function. Our contributions can be summarized as following:

- A novel multi-agent trajectory-ranked reward extrapolation (MA-TREX) framework is proposed. To the best of our knowledge, this is the first framework for MA-IRL, which only uses a few ranked sub-optimal demonstrations to infer the users' intentions in multi-agent tasks.
- Learning from the trajectory generated during the agent training process further reduces the dependence on the demonstrator, and the reward function learning from generated trajectories can achieve the same level reward as the ground-truth quickly and stably.

- By combining the idea of knowledge transfer in the iterative process, the self-generated trajectories required to learn the reward function subsequently is only one-third of the initial trajectories, thereby reducing the cost of adding preference labels to pairwise trajectories.
- The effectiveness of our proposed MA-TREX is validated by using several simulated particle environments in that simulated particle environments are representative and most of the cutting-edge MA-IRL algorithms are validated based on them.

2 Preliminaries

In this section, we introduce Markov game concepts and existing algorithms involved in the experiment, and give definitions of commonly used symbols.

2.1 Markov Games

Markov games [13] are generalizations of Markov decision processes to the case of N interacting agents, which can be represented as a tuple (N, S, A, P, η, r). In a Markov game with N agents, where S represents the global state and $\{A_i\}_{i=1}^N$ represents the set of actions taken by agents, $P : S \times A_1 \times ... \times A_n$ is the state transition probability of the environment. At time t, the agents are in the state s^t, chooses the action $(a_1...a_N)$, and the probability of the state transitioning to s^{t+1} is $P\left(s^{t+1}|s^t, a_1, ..., a_N\right)$. The agent can get a reward through the function $r_i : S \times A_1 \times ... \times A_N \to R$. η represents the distribution of the initial environmental state. We use π without subscript i to represent the agent's joint policy, a_i represents the action of agent i, and a_{-i} represents the set of actions of all agents except for i. The goal of each agent i is to maximize their expected returns $E_\pi\left[\sum_{t=1}^T \gamma^t r_{i,t}\right]$, where γ is the discount factor and $r_{i,t}$ is the reward obtained by agent i at step t in the future.

2.2 Trajectory-Ranked Reward Extrapolation

Suppose agent cannot obtain the ground-truth reward signal r, but there are some demonstrations D provided by demonstrator. D is the set of trajectories $\{\tau_i\}_{i=1}^m$, which is obtained by sampling after expert π_E interacts in the environment. Unlike traditional inverse reinforcement learning, when the demonstrator is sub-optimal, but experts can rank these trajectories without using ground-truth rewards, the goal of trajectory-ranked reward extrapolation (TREX) is to infer the users potential intention through the ranked demonstrations. Utilizing this intention allows agents to learn policies beyond the demonstrator.

More specifically, given a sequence of m ranked trajectories τ_t for $t = 1...m$, where $\tau_i \prec \tau_j$ if $i < j$. The goal of TREX is to predict the cumulative return $J(\tau)$ of the trajectory, and classify the pairwise trajectories (τ_i, τ_j) in order to learn the potential optimization goals of experts. The objective function of the classifier is defined in the form of cross entropy:

$$L\left(\theta\right) = -\sum_{\tau_i \prec \tau_j} \log \frac{exp \sum_{s \in \tau_j} r_\theta\left(s\right)}{exp \sum_{s \in \tau_i} r_\theta\left(s\right) + exp \sum_{s \in \tau_j} r_\theta\left(s\right)} \tag{1}$$

where r_θ is the evaluation of the state s by the reward function.

3   Methodology

In this section, we first describe our MA-TREX algorithm, which is a multi-agent version of TREX. Then, we will introduce the iterative form MA-TREX, which is an improved version of MA-TREX.

3.1   Multi-agent Trajectory Ranked Reward Extrapolation

Similar to the TREX assumption, we use expert knowledge to rank demonstrations without ground-truth rewards [4,15]. MA-TREX infers the cooperation intention of the demonstrator based on the ranking. As is shown in Fig. 1, given T demonstrations, from the worst to the best $(\tau_{11}, ..., \tau_{1N}), ..., (\tau_{T1}, ..., \tau_{TN})$. MA-TREX has two main steps: (1) joint reward inference and (2) policy optimization.

Given the ranked demonstrations, the MA-TREX uses a neural network to predict the team return $r_\theta\left(S\right)$ for the global state $S : (s_1, s_2, ..., s_N)$, and performs reward inference such that $\sum_{S \in (\tau_{i1}, ..., \tau_{iN})} r_\theta(S) < \sum_{S \in (\tau_{j1}, ..., \tau_{jN})} r_\theta(S)$, when $(\tau_{i1}, ..., \tau_{iN}) \prec (\tau_{j1}, ..., \tau_{jN})$. The reward function r_θ can be trained with ranked demonstrations using the generalized loss function:

$$L\left(\theta\right) = E_{(\tau_{i1}, ..., \tau_{iN}),(\tau_{j1}, ..., \tau_{jN}) \sim \pi} [\xi\left(P\left(J_\theta(\tau_{i1}, ..., \tau_{iN}) < J_\theta(\tau_{j1}, ..., \tau_{jN})\right)\right),$$
$$(\tau_{i1}, ..., \tau_{iN}) \prec (\tau_{j1}, ..., \tau_{jN})] \tag{2}$$

where π represents the joint distribution of the team demonstration, $\prec$ represents the preference relationship between the pairwise trajectories, ξ corresponds to the binary classification loss function, and J_θ is the cumulative return to the team trajectory τ calculated using the reward function.

Specifically, we use cross entropy as the classification loss function. The cumulative return J_θ is used to calculate the softmax normalized probability distribution P. We can derive the pairwise trajectories classification probability and loss function:

$$P(J_\theta(\tau_{i\tau}) < J_\theta(\tau_{j\tau})) \approx \frac{exp \sum_{S \in \tau_{j\tau}} r_\theta(S)}{exp \sum_{S \in \tau_{i\tau}} r_\theta(S) + exp \sum_{S \in \tau_{j\tau}} r_\theta(S)} \tag{3}$$

$$L(\theta) = -\sum_{\tau_{i\tau} \prec \tau_{j\tau}} \log \frac{exp \sum_{S \in \tau_{j\tau}} r_\theta(S)}{exp \sum_{S \in \tau_{i\tau}} r_\theta(S) + exp \sum_{S \in \tau_{j\tau}} r_\theta(S)} \tag{4}$$

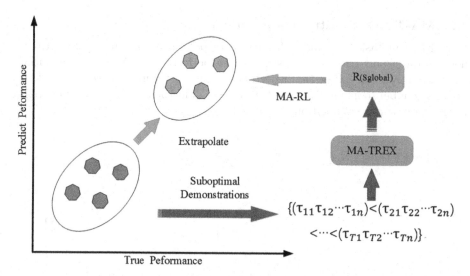

Fig. 1. The MA-TREX obtains some ranked demonstrations and learns a joint reward function from these rankings. Through multi-agent reinforcement learning, the learned joint reward function can be used to train joint strategies better than demonstrator.

where $\tau_{i\tau} = (\tau_{i1}, ..., \tau_{iN})$. Through the above loss function, a classifier can be trained, and the classifier calculates which trajectory is better based on the cumulative return of the team. This form of loss function follows from the classic Bradley-Terry and Luce-Shephard models of preferences [4,15] and has been shown to be effective for training neural networks from preferences [6,12]. To increase the number of training samples, we use data augmentation to obtain pairwise preferences from partial trajectories, which can reduce the cost of generating demonstrations. The specific scheme is to randomly select pairwise team trajectories from demonstrations and extract partial state sets, respectively. By predicting the return of the state, the cumulative return of the trajectory is calculated as the logit value in the cross entropy.

Based on the above method, the MA-TREX can obtain the team's cumulative return $r_\theta(S)$ from the demonstrations. We use multi-agent reinforcement learning to train the joint policy π through $r_\theta(S)$. The optimization goal of agent i is:

$$J(\pi_i) = E[\sum_{t=0}^{\infty} \gamma^t r_\theta(S) | \pi_i, \pi_{-i}] \tag{5}$$

where the reward function in the formula is not a ground-truth reward, but the return value predicted by the neural network for the global state S. Using the predicted reward function, agents with better performance than experts can be obtained.

3.2   MA-TREX Iterative Optimization

In multi-agent tasks, our algorithm can extrapolate rewards from sub-optimal demonstrator and train agents with better performance. As with the initial assumption, we can collect the trajectory during the training process and add a preference label to generate a new training set, and then use the above method to train a new reward function.

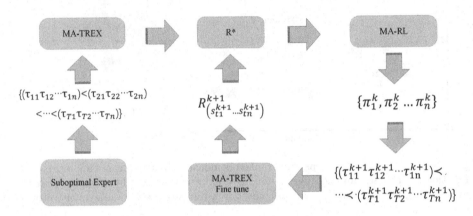

Fig. 2. An iterative form MA-TREX. After the first round of reward function learning, the new demonstrations generated during each multi-agent reinforcement learning process is combined with the fine tune method to train a new reward function.

The iterative training process is shown in Fig. 2. Unlike the initial iteration, the training uses demonstrations that are not provided by human experts, but are generated independently by the model. Humans only need to provide ranking labels for new demonstrations. In addition, although the demonstrations used in each iteration are different, the tasks are the same, so we combined the idea of knowledge transfer in meta-learning. Similar to the application of fine-tune technology in image classification, the problem of pairwise trajectories classification can be viewed as two steps: the first step is to extract the features of the global state, and the second step is to evaluate the features. Intuitively, the second step should be re-learned, so that in subsequent iterations, the new reward function inherits the first half of the previous reward function network. The advantage of using parameter inheritance is that subsequent training of the new reward function only needs to generate one-third of the initial demonstrations, and the reward can be extrapolated again.

For the new demonstrations, we still use cross entropy as the loss function, and calculate the softmax normalized probability distribution p by predicting the cumulative return of the new trajectory. We can derive the classification probability and loss function in iterative form:

$$P(J_{\theta_k}(\tau_{i\tau}^{k-1}) < J_{\theta_k}(\tau_{j\tau}^{k-1}k)) \approx \frac{exp\sum_{S\in\tau_{j\tau}^{k-1}} r_{\theta_k}(S)}{exp\sum_{S\in\tau_{i\tau}^{k-1}} r_{\theta_k}(S) + exp\sum_{S\in\tau_{j\tau}^{k-1}} r_{\theta_k}(S)} \quad (6)$$

$$L(\theta_k) = - \sum_{\tau_{i\tau}^{k-1} \prec \tau_{j\tau}^{k-1}} \log \frac{exp \sum_{S \in \tau_{j\tau}^{k-1}} r_{\theta_k}(S)}{exp \sum_{S \in \tau_{i\tau}^{k-1}} r_{\theta_k}(S) + exp \sum_{S \in \tau_{j\tau}^{k-1}} r_{\theta_k}(S)} \qquad (7)$$

where the demonstration for the k-th network is generated by the training policy at the k-1th iteration. When $k = 1$, the formula is the same as the MA-TREX without iteration.

Each iteration of the MA-TREX can obtain a new reward function $r_{\theta_k}(S)$. Combining multi-agent reinforcement learning, we fuse the reward function learned in multiple iterations to train new joint policy π_k. The iterative multi-agent reinforcement learning objective function is:

$$J(\pi_{k,i}) = E[\sum_{t=0}^{\infty} \sum_{j=1}^{k} \gamma^t w_j r_{\theta_j}(S) | \pi_{k,i}, \pi_{k,-i}] \qquad (8)$$

where w_j represents the weight of the reward function r_{θ_j} in the fusion reward function, k represents the $k - th$ iteration. In the experiment, specify that the latest round of reward function has a weight of 0.5, because the demonstrations used in the new round of iterative training is usually better. We summarize the MA-TREX iterative training process in Algorithm 1.

4   Experiments

In this section, we evaluate our MA-TREX algorithm on a series of simulated particle environments. Specifically, consider the following scenarios: 1) Cooperative navigation, three agents need to reach three target points by cooperating with each other while maintaining no collision; 2) Cooperative communication, two agents, a speaker and a listener, navigate to the target location by cooperating with each other; 3) Cooperative reference, similar to cooperative communication, but each agent acts as both speaker and listener.

4.1   Experiment Demonstrations

To generate expert trajectories, we use ground-truth rewards and the standard multi-agent deep deterministic policy gradient (MADDPG) algorithm to train sub-optimal demonstrator models. In order to investigate the reward extrapolation ability of the MA-TREX under different performance demonstrations, three demonstrator models with different performances were trained for different tasks. Specifically, for each task, we train 500 steps, 1000 steps, and 1500 steps, and collect 1500 pairwise trajectories, respectively. In iterative optimization, new policies are trained using predicted rewards, and 500 pairwise trajectories are collected from the training process.

Algorithm 1: MA-TREX Iterative Optimization Algorithm.

Input: Expert policy $\pi^1 = \pi^E$;Ranked sub-optimal Expert trajectories
$D_{\pi^1} = \{\tau_j^1\}$; Markov game as a black box with parameters
(N, S, A, P, η); Maximum number of iterations K; The set of joint
policies $\{\pi^k\}_{k=1}^K$; The set of reward functions $\{r_{\theta_k}\}_{k=1}^K$

Output: The set of reward functions $\{r_{\theta_k}\}_{k=2}^K$; The last joint policy π^K

1 Initialize Joint policies $\{\pi^k\}_{k=1}^K$; Reward functions $\{r_{\theta_k}\}_{k=1}^K$.

2 **for** $k = 2, ..., K$ **do**

3   Inherit the first half network of θ_{k-1} to θ_k.

4   Add preference labels to new trajectories $\{\tau_j^{k-1}\}$.

5   Sample pairwise trajectories with preference labels:

6      $(\tau_{i1}^{k-1}, \tau_{i2}^{k-1}, ..., \tau_{iN}^{k-1}) \prec (\tau_{j1}^{k-1}, \tau_{j2}^{k-1}, ..., \tau_{jN}^{k-1}) \sim D_{\pi_{k-1}}$

7   Update θ_k to optimization objective function in EP. 7:

8      $L(\theta_k) = -\sum_{\tau_{i\tau}^{k-1} \prec \tau_{j\tau}^{k-1}} \log \frac{exp \sum_{S \in \tau_{j\tau}^{k-1}} r_{\theta_k}(S)}{exp \sum_{S \in \tau_{i\tau}^{k-1}} r_{\theta_k}(S) + exp \sum_{S \in \tau_{j\tau}^{k-1}} r_{\theta_k}(S)}$

9   **for** $i = 1, 2, ..., N$ **do**

10   Update $\pi_{k,i}$ with fusion reward to increase the objective in EP. 8:

11         $J(\pi_{k,i}) = E[\sum_{t=0}^{\infty} \sum_{j=1}^{k} \gamma^t w_j r_{\theta_j}(S) | \pi_{k,i}, \pi_{k,-i}]$

12   **end**

13 **end**

14 **Return** $\{r_{\theta k}\}_{k=1}^K$; π^K.

4.2   Experiment Setup

We use 1500 random pairwise trajectories with preference labels based on trajectory ranking instead of ground-truth rewards to train the first reward network. In the iterative training phase of the new reward function, the fine tune technology is used to inherit the first half of the parameters of the previous reward network, and only 500 new random trajectories are used for training.

In order to evaluate the quality of predicted rewards, the policy is trained through the multi-agent deep deterministic policy gradient (MADDPG) algorithm to maximize the reward function. For the iterative training process, with the latest reward accounting for half of the fusion reward standard, the learning rate is fixed at 0.01 and the batch size is fixed at 100, to ensure that the non-reward function external factors have minimal impact on performance. After training is completed, the ground-truth reward signal is utilized to evaluate the performance of the joint policy.

4.3   Result and Analysis

We compared against two multi-agent inverse reinforcement learning (MA-IRL) algorithms that have achieved remarkable results in the task: multi-agent generative adversarial imitation learning (MA-GAIL) [21] and multi-agent adversarial inverse reinforcement learning (MA-AIRL) [18]. MA-GAIL and MA-AIRL are

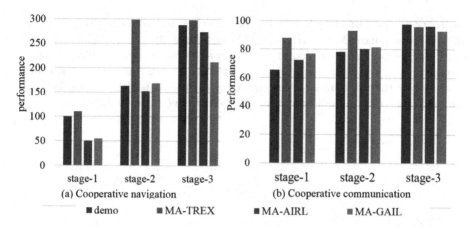

Fig. 3. Performance comparison of the sub-optimal expert trajectory in three stages of two collaborative tasks. Performance is obtained by calculating the average value of ground-truth rewards for 500 tasks.

the multi-agent versions of the famous algorithms generative adversarial imitation learning [10] and adversarial inverse reinforcement learning [7], which very representative in multi-agent inverse reinforcement learning.

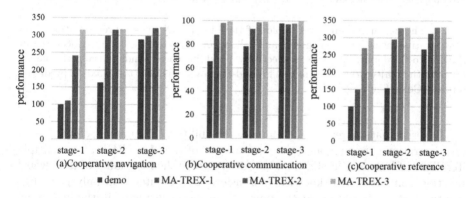

Fig. 4. Performance comparison of the MA-TREX performing multiple iteration training in three collaborative tasks. For all phase tasks, our MA-TREX algorithm basically achieves very high performance within 3 iterations of learning.

Without using iterative optimization, we tested the learning ability of the sub-optimal demonstrator in three different stages. As is shown in Fig. 3, in the second stage of Cooperative navigation and the first and second stages of Cooperative navigation, the performance of the strategy learned by the MA-TREX is significantly better than that of the demonstrator. Since the demonstrator is close to optimal in the third stage, there is no significant improvement in performance. In the first stage of Cooperative navigation, provided trajectories are so poor that all algorithms are not effective, but our algorithm still has obvious advantages. MA-GAIL and MA-AIRL usually cannot achieve significantly higher performance than demonstrators, because they are just imitating demonstrations rather than inferring the intent of demonstrations. Experimental results show that in a multi-agent environment, the MA-TREX can effectively use preference information to surpass the performance of demonstrator.

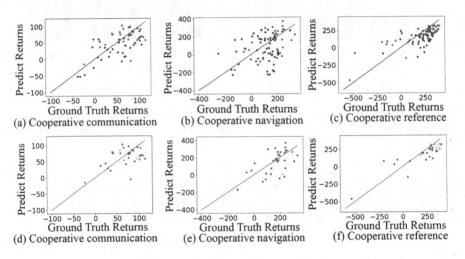

Fig. 5. The extrapolation rewards plot of the MA-TREX under three iterations in 3 collaborative tasks. The blue, green, and yellow points correspond to the trajectories of the first to third extrapolations, respectively. Solid line corresponds to the performance range of the trajectory. The x axis is ground-truth returns, and the y-axis is predicted return. (Color figure online)

To verify that the reward function has the ability to learn using self-generated demonstrations, we compared the performance of the MA-TREX after multiple iterations of learning. In addition, in order to prove the rationality of the reward function combined with knowledge transfer, the new iteration only generates one-third of the initial demonstrations. As is shown in Fig. 4, in the first stage of Cooperative navigation and Cooperative reference, although the performance of the MA-TREX after the first reward extrapolation has improved, it is still far from the level of ground-truth reward. From the above experimental results, conclusions can be drawn as follows: 1) the ability to infer rewards is limited by the demonstrator; 2) the MA-TREX achieves high performance in all stages after iterative training through self-generated demonstrations; 3) the MA-TREX can effectively inherit the knowledge it has learned through iterative training and is no longer limited to the initial demonstrator.

In order to investigate the ability of the MA-TREX to extrapolate the trajectory of experts, we compared the ground-truth returns and predicted returns of the trajectory from the demonstration generated in the iteration. Figure 5 shows the demonstrations generated by the MA-TREX at different iterations. It can be seen from the figure that with the iterative learning of the reward function, the positive correlation between the predicted reward and the ground truth reward gradually increases, which corresponds to the previous performance comparison experiment. For example, in Fig. 5 (b), the reward function after the second iteration (green dots) has a significant improvement in the positive correlation with the ground-truth reward compared to the first iteration (blue dots). It is consistent with the phenomenon of greatly improving performance occurring in Fig. 4 (a). In summary, the experimental results show that the reward function is learning in a more reasonable direction, and the performance gradually approaches the ground-truth reward level with iteration.

5   Conclusion

In this paper, we present a novel reward learning framework, MA-TREX, which uses sub-optimal ranked demonstrations to extrapolate agent intentions in multi-agent tasks. After combining the reward function with multi-agent deep reinforcement learning, it achieves better performance than the demonstrator, and it is also superior to the MA-AIRL and MA-GAIL methods. Furthermore, combining the knowledge transfer idea and using the model's self-generated demonstrations, the iterative optimization form of the MA-TREX is realized. And the reward function can reach the same level as the ground truth within three iterations by using self-generated demonstrations. In the future, one direction is to complete subsequent iterative learning without adding new labels.

References

1. Abbeel, P., Ng, A.Y.: Apprenticeship learning via inverse reinforcement learning. In: Proceedings of the Twenty-first International Conference, Machine Learning, Canada, vol. 69 (2004)
2. Amodei, D., Olah, C., Steinhardt, J., Christiano, P.F., Schulman, J., Mané, D.: Concrete problems in AI safety. CoRR abs/1606.06565 (2016)
3. Argall, B.D., Chernova, S., Veloso, M.M., Browning, B.: A survey of robot learning from demonstration. Robot. Auton. Syst. **57**(5), 469–483 (2009)
4. Bradley, R.A., Terry, M.E.: Rank analysis of incomplete block designs: I. The method of paired comparisons. Biometrika **39**(3–4), 324–345 (1952)
5. Brown, D.S., Goo, W., Nagarajan, P., Niekum, S.: Extrapolating beyond sub-optimal demonstrations via inverse reinforcement learning from observations. In: Proceedings of the 36th International Conference on Machine Learning, USA, vol. 97, pp. 783–792 (2019)
6. Christiano, P.F., Leike, J., Brown, T.B., Martic, M., Legg, S., Amodei, D.: Deep reinforcement learning from human preferences. In: Advances in Neural Information Processing Systems 2017, USA, pp. 4299–4307 (2017)
7. Finn, C., Christiano, P.F., Abbeel, P., Levine, S.: A connection between generative adversarial networks, inverse reinforcement learning, and energy-based models. CoRR abs/1611.03852 (2016)

8. Finn, C., Levine, S., Abbeel, P.: Guided cost learning: deep inverse optimal control via policy optimization. In: Proceedings of the 33nd International Conference on Machine Learning, USA, vol. 48, pp. 49–58 (2016)

9. Henderson, P., Chang, W., Bacon, P., Meger, D., Pineau, J., Precup, D.: Option-GAN: learning joint reward-policy options using generative adversarial inverse reinforcement learning. In: Proceedings of the Thirty-Second AAAI Conference on Artificial Intelligence, USA, pp. 3199–3206 (2018)

10. Ho, J., Ermon, S.: Generative adversarial imitation learning. In: Advances in Neural Information Processing Systems 2016, Spain, pp. 4565–4573 (2016)

11. Hu, J., Wellman, M.P.: Multiagent reinforcement learning: theoretical framework and an algorithm. In: Proceedings of the Fifteenth International Conference on Machine Learning USA, pp. 242–250 (1998)

12. Ibarz, B., Leike, J., Pohlen, T., Irving, G., Legg, S., Amodei, D.: Reward learning from human preferences and demonstrations in Atari. In: Advances in Neural Information Processing Systems 2018, Canada, pp. 8022–8034 (2018)

13. Littman, M.L.: Markov games as a framework for multi-agent reinforcement learning (1994)

14. Lowe, R., Wu, Y., Tamar, A., Harb, J., Abbeel, P., Mordatch, I.: Multi-agent actor-critic for mixed cooperative-competitive environments. In: Advances in Neural Information Processing Systems 2017, USA, pp. 6379–6390 (2017)

15. Luce, D.R.: Individual choice behavior: a theoretical analysis. J. Am. Stat. Assoc. **115**(293), 1–15 (2005)

16. Ng, A.Y., Harada, D., Russell, S.J.: Policy invariance under reward transformations: theory and application to reward shaping. In: Proceedings of the Sixteenth International Conference on Machine Learning, Slovenia, pp. 278–287 (1999)

17. Ramachandran, D., Amir, E.: Bayesian inverse reinforcement learning. In: Proceedings of the 20th International Joint Conference on Artificial Intelligence, India, pp. 2586–2591 (2007)

18. Song, J., Ren, H., Sadigh, D., Ermon, S.: Multi-agent generative adversarial imitation learning. In: Advances in Neural Information Processing Systems 2018, Canada, pp. 7472–7483 (2018)

19. Torabi, F., Warnell, G., Stone, P.: Behavioral cloning from observation. In: Proceedings of the Twenty-Seventh International Joint Conference on Artificial Intelligence, Sweden, pp. 4950–4957 (2018)

20. Wang, S., Hu, X., Yu, P.S., Li, Z.: MMRate: inferring multi-aspect diffusion networks with multi-pattern cascades. In: The 20th ACM SIGKDD International Conference on Knowledge Discovery and Data Mining, USA, pp. 1246–1255 (2014)

21. Yu, L., Song, J., Ermon, S.: Multi-agent adversarial inverse reinforcement learning. In: Proceedings of the 36th International Conference on Machine Learning, California, vol. 97, pp. 7194–7201 (2019)

22. Ziebart, B.D., Maas, A.L., Bagnell, J.A., Dey, A.K.: Maximum entropy inverse reinforcement learning. In: Proceedings of the Twenty-Third AAAI Conference on Artificial Intelligence, USA, pp. 1433–1438 (2008)

An Incremental Learning Network Model Based on Random Sample Distribution Fitting

Wencong Wang[1], Lan Huang[1,2], Hao Liu[1], Jia Zeng[1], Shiqi Sun[1], Kainuo Li[3], and Kangping Wang[1,2(✉)]

[1] College of Computer Science and Technology, Jilin University,
Changchun 130012, China
`wangkp@jlu.edu.cn`
[2] Key Laboratory of Symbolic Computation and Knowledge Engineering,
Jilin University, Changchun 130012, China
[3] Information Center of Political and Law Committee of Jilin Provincial Committee,
Changchun 130001, China

Abstract. The training of the classification network has tough requirements of data distribution. The more the training data did not consistent with the distribution of the real target function, the higher error rate the network will produce. In the context of incremental learning, the data distribution of the subsequent training tasks may not consistent with the data distribution of the previous tasks. To handle this problem, lots of learning methods were introduced, most of these methods are complicated and heavy computing. In this paper, a novel method which is faster and simpler is proposed to uniform subsequent training data. Artificial training samples are produced from random inputs in current trained network. In subsequent task, these artificial samples were mixed with new real data as training data. The experiments with proper parameters show that new features from new real data can be learned as well as the old features are not forgot catastrophically.

Keywords: Incremental learning · Sample distribution fitting · Classification network

1 Introduction

1.1 Relevant Background

Incremental learning is a type of learning mode in which the system can incessantly obtain new knowledge from persistent training samples. In some cases, since the training data cannot be obtained in the same time or stored completely,

Supported by the Jilin Province Science and Technology Development Plan Project under Grant 20190201273JC, and by the Jilin Provincial Key Laboratory of Big Data Intelligent Computing.

the training process may be spilt into many batches, each batch is called as a task. Transfer learning [16] is often used to establish the transformation from different tasks in multi-task incremental learning. If the knowledge that task A and task B going to learn are similar, the model of task A can be used while training task B, we can also use a part or all of the parameters in the network of task A to initialize the training network of task B. The transfer info can be the sample data of the original training set [2], the parameters of the model, or the features that the network extracted [18]. In this way, we can build a process of learning tasks from A to B and the process of learning task B to C, and so on. Although transfer learning can handle some problems in multi-task incremental learning, there are still some problems left. One problem is we can only obtain the trained network model and network weights instead of training data of task A when training task B. If the data of task A and task B do not have the same distribution, the ability of solving task A will gradually decline in the process of training task B. This sharp decline in the performance of previous learning tasks when new knowledge was added is called "catastrophic forgetting" [3], which is exactly the key issue that incremental learning needs to handle.

1.2   Related Work

The main idea of incremental learning is retaining the relevant information of previous tasks while training new tasks. Since convolutional neural network do not have a good interpretability [20], it has been found that incremental learning becomes difficult after convolutional networks begin to be applied to machine learning. In order to deal with this problem, the concept of knowledge distillation is first proposed in [5], which information of the old network model is transmitted to the new network through the redefined softmax function, in this way the new network is updated without catastrophic forgetting. Exploiting the relevant ideas of knowledge distillation, the LwF (Learning without Forgetting) method was proposed in [7]. LwF works better on both new tasks and old tasks comparing with Feature Extraction and Finetune-FC. It only needs information from the new tasks, that is more flexible than Joint Training method. The new loss function of the update network is defined in [7], which is the sum of the distillation loss value from the old model and the loss value of the new model. Using this new loss function, [14] and [15] have applied incremental learning in the field of target detection and semantic segmentation. The knowledge from early learning can still be forgotten using the LwF method after multiple tasks of learning. In order to improve performance, the iCaRL (Incremental classifier and Representation Learning) [12] method is proposed, which uses a part of representative data feature of the old task along with the data of the new task. iCaRL needs the data of the old task, which may consume an extra amount of storage. It's obviously that a network's parameters consume less storage than a big training data set. Based on this idea, [13] used the GAN model to train an experience playback of the old tasks' data while training the new tasks. This method will train a GAN [4] network which generate data to simulate old tasks data. For a classification network, we can also train a CGAN network [9,19].

GAN is hard to train and the data that generate by GAN is very likely to be quite different distribution with the real data. In the other way, comparing the weight difference or gradient difference between tasks in the process of learning can also achieve good performance. By comparing the weight difference, an Elastic Weight Consolidation (EWC) method is proposed in [6], which defines the importance of each weight in the network affects the training results by Fisher matrix [10], which guides the network to update the unimportant weights of the network while training and minimally affecting the previous results. The methods of comparing gradient difference such as GEM (Gradient Episodic Memory) [8] and its improved version A-GEM [1] constrain the gradient update direction of the new tasks. This can avoid catastrophic forgetting caused by too much critical deviation from the old tasks' gradient.

2 Random Sample Fitting Distribution Model (RFD)

2.1 Network Structure

Our classification network contains convolutional and fully connected layers, which is modified from LeNet [17] includes five layers which are input layer, convolutional layer, activation layer, pooling layer, and fully connected layer. This structure is inspired by an incremental learning model which uses GAN network to assist the incremental [13]. In the referred model, the tasks are training both on the classification network and on the GAN network in same time. After training, the generator of the GAN network can generate samples with the same distribution as the training set, and the classification network can also classify the samples well. The subsequent tasks use samples generated by the GAN network and real samples to train a new classification network. The training time of the GAN network is very long and its discriminator network does not outperform our model. We try to erase the GAN network and prove that the GAN network is unnecessary for preventing forgetting. Our incremental learning network consists of two parts, the Frozen network and the Update network. Frozen network is not changing during a training task while Update network is training. In the start of next task, the new Frozen network is copied from Update network. While training in a new task, the supervise data are mixing from new real data and outputs of the Frozen network generated by random inputs. These data are flow into the Update network together for typical gradient descent. This proposed network architecture is shown in Fig. 1.

2.2 Formula Derivation

Adding noise to training samples is a way of data augmentation [11], so can all samples be able to represent network information? From the Frozen Network, lots of supervised samples can be generated that reflect the parameters of previous Update Network. It's not easy to ensure the distribution of random input is consistent with previous training data. But our goal is rather preventing forgetting than keeping consistency. So, input distribution is uniform and proportions

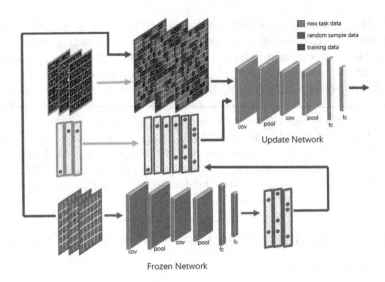

Fig. 1. A depiction of Random sample Fitting Distribution incremental learning network (RFD). Red color represents the training data composing by the new task data (green color) and the random sample data (blue color) generate by Frozen Network (grey color). These data finally flow into the Update Network (yellow color). (Color figure online)

of each category are same as previous training data. We hope to some extent that the random samples and their classification-generated by the old network can prevent forgetting effectively. Assuming that the previous training data is represented by D_{old}, the new training data is represented by D_{new} and the randomly generated data is represented by D_{random}. There are $|D_{old}|$ number of data in the previous training data set, $|D_{new}|$ in the new training data, and $|D_{random}|$ in the randomly generated data. Let n be the number of categories in all training data. The vector $P_{old} = (p_1, p_2, p_3, p_4, \cdots p_n)$ is used to represent the data classification of each category in the previous training data. A component p_a in $|P_{old}|$ represents the probability distribution of the old training data set on a category. It can be calculated that there are a total of $p_a \cdot |D_{old}|$ previous training data on a category. Similarly, the vector $P_{new} = (p'_1, p'_2, p'_3, p'_4, \cdots p'_n)$ is used to represent the data classification of each category of the new data. $P_{random} = (p''_1, p''_2, p''_3, p''_4, \cdots p''_n)$ represents the data classification of each category in generated data. The randomly generated data is generated by the Frozen network, which represents the data distribution of the previous training model, so we suppose $P_{random} = (p''_1, p''_2, p''_3, p''_4, \cdots p''_n) = (p_1, p_2, p_3, p_4, \cdots p_n)$. The distribution of the exact training data set in next tasks as follows.

$$P'_{new} = \frac{p_1 \cdot |D_{old}| + p_1 \cdot |D_{random}| + p'_1 \cdot |D_{new}|}{|D_{old}| + |D_{random}| + |D_{new}|} + \frac{p_2 \cdot |D_{old}| + p_2 \cdot |D_{random}| + p'_2 \cdot |D_{new}|}{|D_{old}| + |D_{random}| + |D_{new}|}$$
$$+ \cdots + \frac{p_n \cdot |D_{old}| + p_n \cdot |D_{random}| + p'_n \cdot |D_{new}|}{|D_{old}| + |D_{random}| + |D_{new}|}$$

$$(1)$$

When the number of training data $|D_{new}| \ll |D_{random}|$, $P'_{new} = P_{old}$, which will not lead to catastrophic forgetting, but the network will never learn new information. If the number of new training data is $|D_{new}| \gg |D_{random}|$, $P'_{new} \neq P_{old}$, the network is prone to forgetting. Let $|D_{new}| = \phi|D_{random}|$, set ϕ to a larger number, and the network can be updated at a slower speed without catastrophic forgetting. In this way, the total Loss function L_{total} of the network can be defined as follows.

$$L_{total} = L_{new} + \phi L_{random} \tag{2}$$

The function L_{total} is derived from the distribution formula (see formula 1). When the network is updated, the loss value consists of two parts. The loss value generated by the random sample is recorded as L_{random}, and the loss value generated by the post-training data is recorded as L_{new}. The Frozen network has the frozen loss value of 0 does not participate in the update of the gradient. It can also be concluded from the above-mentioned distribution definition (see formula 2) that changes in ϕ will affect the update speed of the network. In neural networks, the direction of network updating is generated by the gradient of data. We can change the loss function formula (see formula 2) to a gradient formula (see formula 3) for a single update of the network. The total gradient G_{total} consists of gradient G_{new} from new data and gradient G_{random} of random generated data.

$$G_{total} = G_{new} + \phi G_{random} \tag{3}$$

In more extensive application scenarios, we can apply our model to different distribution data which have the same gradients. Keeping the random generated data to have the same gradient as the old data will have the same influence to network, which is the original intention of our model design.

3   Experiments

The verification experimental data in this paper are the MNIST handwritten dataset, which contains a training set and a test set. There are 60,000 samples in the training set and 10,000 samples in the validation set. Each sample is a 28×28 single-channel picture sequence and corresponding label.

3.1   Comparison with Non-incremental Learning Method

Failure Caused by Inconsistent Distribution of Training Data. When two tasks in incremental learning have obeyed different distributions, the training of the second task will affect the knowledge already learned in the first task. To verify this opinion, 800 samples in the data set are selected for the first task. These samples are uniformly distributed, and the probability of occurrence of 0–9 labels is 1/10. The process of training on LeNet is shown in Fig. 2. After 40 rounds of training, the accuracy of the MNIST test set reached 93%. As the number of trainings increases, the accuracy of the training set and val set gradually rises to 98% and 92%, and the loss value gradually reduced to 0.06 and 0.03, which is an acceptable value. There are another 800 samples with different distribution selected for second task. Among these samples, only the data labeled 0–4 appeared, and the probability of each type is 1/5. The data labeled 5–9 are not included, the probability is 0. Under this situation, the difference distribution will result in catastrophic forgetting. The training process for the second task is shown in Fig. 3. After 40 rounds of training, the training accuracy in the train set reached nearly 99%, and the loss value in the train set also dropped to a very low level. But after 17 epochs of training the accuracy of val set gradually decreases, and the loss value is not getting smaller, which indicates that the network training has been overfitted. Notice that the test set still contains 5 categories that were not included in training data for second task. After 40 epochs, the model has an accuracy of 50.77% on the test set of MINST. The model has forgotten lots of knowledge obtained from the first training when learning new knowledge. The remaining knowledge of the model is only the data labeled 0–4, and the data labeled 5–9 is forgotten, so the total accuracy is close to 50%.

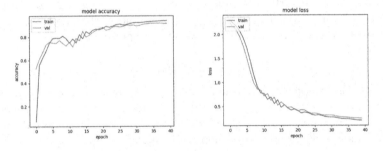

Fig. 2. The process of training in first task.

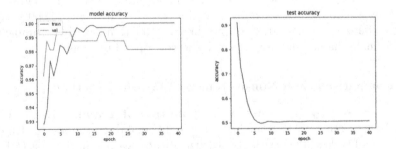

Fig. 3. Failure due to inconsistent distribution of training data, the training accuracy and training loss perform good during the second stage training, but the test loss and accuracy break up which means a catastrophic forgetting has happened.

Jitter Caused by Small Training Step. In some situation, the network can only get a small number of samples once, and the samples cannot be stored. In order to show this process, a network similar to the previous experiment (see Sect. 3.1) is trained, There are 10000 samples which are split into 100 tasks equally, and the epoch size for each task is 10. The total epochs size is 10000(100 × 10). This experiment is different with mini-batch method whose batch size is 10. The mini-batch method only performs one gradient descent, while 10 gradient descents are performed for the same 10 samples in this experiment. The training process is shown in Fig. 4. Because the whole training data have same distribution with test data, the accuracy increases when training data accumulate. But there are lots of jitters caused by multi-times training with a small part of training data which pull down final accuracy. In incremental learning, the training data were input like stream data which not available in next task. It's necessary to train multi times for new knowledge. There is contradiction between multi-times training and jitters avoiding. Some improvement should be introduced to handle this problem. The jitters which are representation of partly forgetting is harmful and should be eliminated as much as possible.

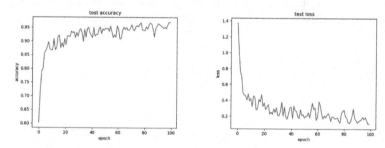

Fig. 4. Jitters caused by small training step. The jitter occurred not only on the training set but also on test set.

3.2 RFD Method

Suppressing Catastrophic Forgetting. First, the experiments show the inhibition of our method between two tasks which have different distributions, like the first situation of Sect. 3.1. Task 1 uses 800 10-label data to train to an accuracy of 93%, then Task 2 uses 800 5-label data for incremental learning. The results on the test set of MNIST are shown in Fig. 5. A significant difference can be seen after using our method in the training process of Task 2. In Task 2, the accuracy of a typical back propagation algorithm on the test set drops rapidly to less than 10% and the loss quickly increases. Exploiting our method, the accuracy only drops to 83%, and the speed of network forgetting is significantly slowed down.

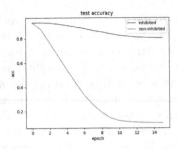
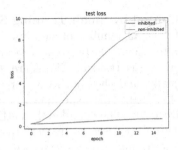

Fig. 5. Inhibition of forgetting by using model, the accuracy value drops slowly when using our model and the loss value rise only a little.

Continuous Learning with RFD. We have shown a harmful jitter in the second experiment of Sect. 3.1 which is mainly caused by partial forgetting. As a comparison, we made another experiment with same training dataset as the second experiment in Sect. 3.1. The results are shown in Fig. 6. Compared with Fig. 4, Fig. 6 has smoother lines, which is consistent with our expectation. RFD method can also suppress partial forgetting caused by too small step size in continuous learning.

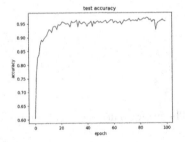
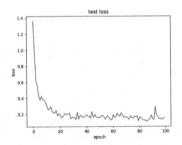

Fig. 6. Suppressing Jitter in Continuous learning. There are 100 data in a single task with 10 epochs for each task is performed. The ϕ in RFD method is set to 60.

3.3   Network Feature Verification

We want to make a comparison with non-incremental models with low learning rates to prove some characteristics of our model. Let the current network weights be W and the updated network weights is W', the learning rate is α, and the gradient is G, then the formula for a single gradient descent can be written as $W' = W + \alpha G$. Our Loss function is composed of two parts, L_{random} and L_{new}. At each gradient descent, Gradients come from both new task samples and random samples. Let the gradient provided by the new task samples be G_{new} and the gradient provided by the random samples be G_{random}. The gradient G

is a function of loss, it can be written as a function $G = f(loss)$. The formula for the single gradient descent of our model can be written as formula 4).

$$W' = W + \alpha \cdot (\phi G_{random} + G_{new}) = W + \alpha \cdot f(\phi L_{random} + L_{new}) \qquad (4)$$

It can be guessed that if the random sample does not provide useful knowledge information, the gradient brought by the random samples is also random. The total gradient of all random samples should be close to 0, that is, $G_{random} \approx 0$. At this time, the loss of the model is only related to L_{new}. In this way, the learning rate of the model is equivalent to the original $\alpha/(\phi + 1)$, which is equivalent to a model with a low learning rate. We simultaneously train non-incremental models with various learning rates. A comparison of 100 data in a single task with 10 epochs for each task is performed. This comparison results are shown in Fig. 7. The loss curve and the accuracy curve of the non-incremental model have severely jittered, and this phenomenon is as same as the previous experiments that the small training step size causes jitters (see Fig. 4). And our model has both a fast descent rate and a good accuracy rate. The knowledge of the previous task is indeed contained in the artificial samples. In the definition of the Loss method (see formula 2), the ϕ coefficient is a very important parameter. According to the previous discussion, it can be known that when the ϕ is large, the network tends not to update, and when the ϕ is small, the network tends to forget the past knowledge, so the proper ϕ value has become a key issue. We compare the various values of ϕ respectively. The comparison results are as follows (see Fig. 8). In order to measure the degree of this forgetting, we use a smooth filtered curve to fit the points on the current polyline. The distance between this smooth curve and the accuracy polyline is the degree of forgetting during training. Manhattan Distance (L1) is selected to measure the distance between the two curves (red and blue). The bigger the distance between the two curves are, the greater the curve jitter is and the more forgetting is. The distances are shown in Table 1. L1 tends to become smaller when ϕ value increases, that is,

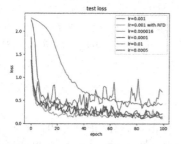
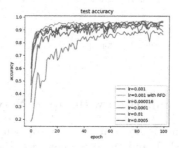

Fig. 7. Comparison with low learning rate training, we select learning rate of 0.01, 0.001, 0.0005, 0.0001 and a special value 0.000016. The learning rate α of our RFD model is 0.01, and the ratio of sample numbers ϕ is 60:1. If random samples do not provide useful knowledge information, the equivalent learning rate is $\alpha/(\phi + 1) = 0.000016$ (Color figure online)

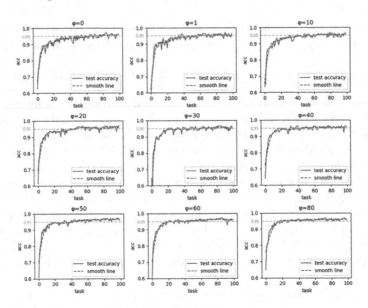

Fig. 8. Test set accuracy under various values of ϕ. This time we choose 60 data in a single task, so after 100 tasks of training all MNIST training data were trained. The blue line represents the accuracy changes when new tasks were added and the red line fitting the blue line. (Color figure online)

the degree of forgetting of the network is becoming smaller, which is consistent with our expectation. When ϕ is large enough, the training cost of the network becomes high because the training data increases. If the training effect does not improve significantly, we should not continue to increase the value of ϕ. This value is approximately 60 in previous experiments.

Table 1. L1 results under various values of ϕ. These values are calculated from Fig. 8.

ϕ	$\phi = 0$	$\phi = 1$	$\phi = 10$	$\phi = 20$	$\phi = 30$	$\phi = 40$	$\phi = 50$	$\phi = 60$	$\phi = 80$
L1	1.098	1.0554	1.004	0.9041	0.917	0.9142	0.8164	0.7455	0.7321

3.4 Comparison with Other Networks

Experiments are run on a laptop with an Intel i7-6700HQ CPU, a NVIDIA 960M GPU, and 16GB memory. LwF method, GAN-based incremental learning method, and EWC method are tested for comparison. To ensure that these methods are comparable, same data and related proportions are provided. The core of the LwF method is loss distillation. The loss value of the method is expressed as $L = L_{new} + \sigma L_{distillation}$. Using the same proportionality coefficient $\sigma = \phi$, RFD and LWF can be compared. The incremental training model based on GAN consists of fake data and new task data. It can be considered that the gradient is composed of new data and fake data. The proportion of new data and fake data is as same as our model. The loss function of the EWC method refers to the weights of the previous task and is defined as $L'(\theta) = L(\theta) + \lambda(W_i - W_0)$. Similarly, the proportionality coefficient λ and ϕ can ensure the comparability of the model. The test accuracy of the four models are comparable, and the accuracy on the test set of the incremental model on the MNIST is also very close. But the time costs are very different. The comparison results are shown in Table 2. The training time of the GAN method consists of a classification network and a GAN network. It is very time-consuming, so the total time is longest. The EWC method needs to repeatedly calculate the fisher matrix, which will become a large time overhead when the amount of data is large. Our method is faster in total time because it is easier to generate data than LwF (Fig. 9).

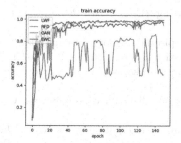 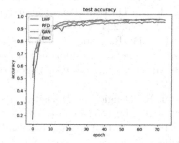

Fig. 9. Comparison results, typical three increasement learning methods and our method (RFD).

Table 2. Comparison of training time and accuracy.

Method	Total time	Model train time	Accuracy
LWF	186.93 s	**79.64 s**	97.59%
GAN	3886.02 s	93.71 s + 3434.74 s	97.19%
EWC	1674.700	1647.96	95.65%
Ours (RFD)	**169.77 s**	79.74 s	**97.66%**

4   Conclusion

In this paper, We propose an incremental learning network model based on a random sample fitting distribution. This method mainly simplifies the generator of a GAN-based incremental learning model. It consists of two parts: The Frozen network responsible for generating random samples and the New network used for training. In the experiments, our new model has the same accuracy as the traditional method while learning faster.

References

1. Chaudhry, A., Ranzato, M., Rohrbach, M., Elhoseiny, M.: Efficient lifelong learning with a-gem. In: International Conference on Learning Representations (ICLR) (2019)
2. Dai, W., Yang, Q., Xue, G.R., Yu, Y.: Boosting for transfer learning. In: Proceedings of the 24th International Conference on Machine Learning (2007)
3. French, R.M.: Catastrophic forgetting in connectionist networks. Trends Cogn. Sci. **3**(4), 128–135 (1999)
4. Goodfellow, I.J., et al.: Generative adversarial nets. In: Proceedings of the 27th International Conference on Neural Information Processing Systems, vol. 2, pp. 2672–2680 (2014)
5. Hinton, G., Vinyals, O., Dean, J.: Distilling the knowledge in a neural network. Comput. Sci. **14**(7), 38–39 (2015)
6. Kirkpatrick, J., et al.: Overcoming catastrophic forgetting in neural networks. Proc. Natl. Acad. Sci. U.S.A. **114**(13), 3521–3526 (2017)
7. Li, Z., Hoiem, D.: Learning without forgetting. IEEE Trans. Pattern Anal. Mach. Intell. **40**, 2935–2947 (2017)
8. Lopez-Paz, D., Ranzato, M.A.: Gradient episodic memory for continual learning. In: Advances in Neural Information Processing Systems 30, pp. 6467–6476 (2017)
9. Mehdi Mirza, S.O.: Conditional generative adversarial nets. Comput. Sci. 2672–2680 (2014)
10. Pascanu, R., Bengio, Y.: Revisiting natural gradient for deep networks. Comput. Sci. **37**(s 10–11), 1655–1658 (2013)
11. Perez, L., Wang, J.: The effectiveness of data augmentation in image classification using deep learning. In: The IEEE Conference on Computer Vision and Pattern Recognition (CVPR) (2017)
12. Rebuffi, S.A., Kolesnikov, A., Sperl, G., Lampert, C.H.: ICARL: incremental classifier and representation learning. In: CVPR (2016)
13. Shin, H., Lee, J.K., Kim, J., Kim, J.: Continual learning with deep generative replay. In: Advances in Neural Information Processing Systems 30 (2017)
14. Shmelkov, K., Schmid, C., Alahari, K.: Incremental learning of object detectors without catastrophic forgetting. In: IEEE International Conference on Computer Vision (ICCV), vol. 1, pp. 3420–3429 (2017)
15. Tasar, O., Tarabalka, Y., Alliez, P.: Incremental learning for semantic segmentation of large-scale remote sensing data. IEEE J. Sel. Top. Appl. Earth Observ. Remote Sens. **PP**(99), 1–14 (2019)
16. Tzeng, E., Hoffman, J., Saenko, K., Darrell, T.: Adversarial discriminative domain adaptation. In: The IEEE Conference on Computer Vision and Pattern Recognition (CVPR), July 2017
17. Yann, L., Leon, B., Bengio, Y., Patrick, H.: Gradient-based learning applied to document recognition. Proc. IEEE **86**, 2278–2324 (1998)
18. Yosinski, J., Clune, J., Bengio, Y., Lipson, H.: How transferable are features in deep neural networks? In: Proceedings of the 27th International Conference on Neural Information Processing Systems, vol. 2, pp. 3320–3328 (2014)
19. Zhai, M., Chen, L., Fred Tung, J.H., Nawhal, M., Mori, G.: Lifelong GAN: continual learning for conditional image generation. In: ICCV (2019)
20. Zhou, B., Khosla, A., Lapedriza, A., Oliva, A., Torralba, A.: Learning deep features for discriminative localization. In: CVPR, June 2016

Parameter Optimization and Weights Assessment for Evidential Artificial Immune Recognition System

Rihab Abdelkhalek[1][(✉)] and Zied Elouedi[2]

[1] LARODEC, Institut Supérieur de Gestion de Tunis, Tunis, Tunisia
rihab.abdelkhalek@gmail.com
[2] Université de Tunis, Tunis, Tunisia
zied.elouedi@gmx.fr

Abstract. Nowadays, the Artificial Immune Recognition Systems (AIRS) are considered as powerful supervised learning approaches inspired by the natural immunological elements. They achieved a big success in the area of machine learning. Nevertheless, the majority of AIRS versions does not take into account the effect of uncertainty related to the classification process. Managing uncertainty is undoubtedly among the fundamental challenges in real-world classification problems. That is why, in this research work, a novel AIRS approach is proposed under the belief function theory where the number of training antigens represented by each memory cell is considered as the derived weight used in the classification process. Furthermore, a parameter optimization strategy is proposed under uncertainty to find out optimal or near-optimal parameter values through the minimization of a given error function. The performance of the new weighted evidential method with parameter optimization is shown based on a comparative evaluation with the other traditional AIRS versions.

Keywords: Machine learning · Artificial Immune Recognition Systems · Uncertain reasoning · Belief function theory · Parameter optimization

1 Introduction

During the last decade, the field of Artificial Immune Systems (AIS) [2] has become among the important branches of the computational intelligence. Inspired by the incorporation of natural immunological metaphors such as affinity maturation, clonal selection, and memory cell retention, these systems have shown impressive and intriguing performance in tackling complicated computer engineering problems. Artificial Immune Recognition Systems (AIRS) [1] are among the most powerful supervised classification techniques. They have achieved a considerable success on various machine learning issues by reaching good and competitive classification results. Yet, several approaches of AIRS

© Springer Nature Switzerland AG 2020
G. Li et al. (Eds.): KSEM 2020, LNAI 12275, pp. 27–38, 2020.
https://doi.org/10.1007/978-3-030-55393-7_3

have been conceived, such as AIRS2 [14] and AIRS3 [15]. While these versions
have been proposed to solve decision-making problems under a certain environment, they suffer from some inefficiency while working in an uncertain context. To overcome this limitation, which has attracted a great interest of many
researchers, new AIRS approaches have been proposed based on uncertainty
theories, like the possibility [10] and the fuzzy set theories [6–9]. Otherwise, the
belief function theory (BFT) [3,4], also called evidence theory, is considered as
a robust and rich environment for handling uncertainty and dealing with the
total ignorance. Its success owes much to its flexibility and its ability to aggregate different sources of information. Recently, an extension of AIRS2 has been
suggested under such theory named Evidential AIRS2 (EAIRS2) [11]. Within
such method, all memory cells have been treated with equal importance during
the classification process. However, it is obvious that training antigens should be
weighted differently so as to further improve the classification performance. For
this purpose, a new weighted AIRS approach under the uncertain framework of
the belief function, called WE-AIRS, has been proposed in [16] where the weight
output of AIRS3 assigned during the learning phase has been considered. These
evidential classification approaches are considered as efficient methods handling
imperfection within AIRS. Nevertheless, their effectiveness is influenced by the
values of their corresponding parameters. In fact, the challenge of tuning the
parameters integrated in the classification process was ignored. That is why, our
aim in this paper is to improve the classification performance of both EAIRS2
and Weighted Evidential AIRS3 by determining the optimal or near-optimal
values of the assigned parameters. Actually, optimization is considered among
the fundamental problems of analysis scientific and engineering design. In our
approach, the parameter optimization procedure is carried out by minimizing
the error rate of the classification. More specifically, we opt for one of the most
popular optimization strategies, which is the gradient descent, under the belief
function theory.

The rest of the paper is divided as follows: Sect. 2 recalls the main concepts
of the belief function theory. The related work of Artificial Immune Recognition System is presented in Sect. 3. Our approach, named Optimized Evidential
Weighted AIRS, is presented in Sect. 4. Then, the experimental results are presented in Sect. 5. In the end, Sect. 6 concludes this paper.

2   Belief Function Theory

Using the belief function framework [3–5], we represent a given problem by a
finite set of events named the frame of discernment, that we denote by Θ. It
includes the hypotheses related to the problem [5] and it is defined as follows:

$$\Theta = \{\theta_1, \theta_2, \cdots, \theta_n\} \tag{1}$$

The potentially values that can be taken by each subset of Θ is referred to as
the power set of Θ, 2^Θ, where $2^\Theta = \{A : A \subseteq \Theta\}$.

A basic belief assignment (*bba*) can be defined as the belief related to the different elements of Θ [4] such that:

$$m : 2^{\Theta} \rightarrow [0, 1] \quad and \quad \sum_{A \subseteq \Theta} m(A) = 1 \tag{2}$$

Each mass $m(A)$, called a basic belief mass (*bbm*), measures the amount of belief allocated to an event A of Θ.

Considering two *bba*'s m_1 and m_2 corresponding to reliable and independent information sources, the evidence can be combined using Dempster's combination rule which is defined as follows:

$$(m_1 \oplus m_2)(A) = k. \sum_{B,C \subseteq \Theta : B \cap C = A} m_1(B) \cdot m_2(C) \tag{3}$$

$$where \quad (m_1 \oplus m_2)(\varnothing) = 0 \quad and \quad k^{-1} = 1 - \sum_{B,C \subseteq \Theta : B \cap C = \varnothing} m_1(B) \cdot m_2(C)$$

For the decision making process, beliefs can be transformed into pignistic probabilities as following:

$$BetP(A) = \sum_{B \subseteq \Theta} \frac{|A \cap B|}{|B|} \frac{m(B)}{(1 - m(\varnothing))} \; for \; all \; A \in \Theta \tag{4}$$

3   The Artificial Immune Recognition System

The Artificial Immune Recognition Systems, referred to as AIRS, are considered as powerful supervised learning approaches inspired by the natural immunological elements [1]. Given a training set T, each instance corresponds to an antigen having a similar representation as an antibody. AIRS2 [14] and AIRS3 [15] have been proposed as two improved versions of AIRS. The classification procedure of AIRS2 can be outlined as following: First, a pre-processing stage is performed where all numerical features of T are normalized and the distances between all the antigens in the system are computed. The memory cell pool (*MC* pool) and the Artificial Recognition Balls pool (*ARB* pool) are then initialized. A computation of the affinity measure is performed for each antigen in the *MC* pool belonging to the same class as the training set. Based on this similarity, we pick up the cell having the best match memory cell, denoted by *mc_match*. After the selection of *mc_match*, this memory cell is employed to produce a set of mutated clones which are integrated into the *ARB* pool. Finally, the antigen representing the highest similarity will be considered as the candidate memory cell (*mc_candidate*). The next step corresponds to the introduction of memory cells where the memory cell pool is revised. Subsequently, the obtained *MC* pool is employed in the decision making process. To do so, the k-nearest neighbors classifier is applied and the class of the test antigen is assigned using the majority vote of the k-nearest neighbors. Further, AIRS3 has been proposed in [15], where the main goal is to integrate a new component in order to hold the number of represented antigens for each cell in the MC pool, denoted by $numRepAg$. The obtained values of $numRepAg$ are maintained in the training stage for the classification task. In such approach, the value of k turns into $numRepAg$ rather than the whole neighbors. In such way, the total of $numRepAg$ corresponding to the selected cells would be equivalent to k. Then, the total of $numRepAg$ of all the chosen antigens with the same label is derived leading to the classification of the test antigen.

On the other hand, managing uncertainty is an important challenge when dealing with AIRS. For instance, an AIRS approach based on Fuzzy-KNN has been introduced in [9]. Otherwise, authors in [10] have proposed a possibilistic AIRS, where they represented the training set via possibilistic values. Lately, an evidential AIRS approach called EAIRS2 has been developed in [11]. In such approach, the authors considered the k-nearest cells as different information sources contributing in the classification. Despite its good performance, all memory cells in the EAIRS2 are treated equally throughout the decision making which may alter the classification results. That is why, authors in [16] proposed a weighted evidential AIRS (WE-AIRS) considering the weight of each resulting memory cell in the MC pool. Nevertheless, the effectiveness of both EAIRS2 and WE-AIRS3 is influenced by the values of their corresponding parameters. Obviously, determining the optimal or near-optimal values of the assigned parameters would rather improve the classification performance. Thus, tuning the parameters involved in the classification process should not be ignored.

In that regard, the aim of this paper is not only to handle uncertainty in AIRS but also to propose a weighted evidential AIRS method with parameter optimization.

4   Optimized Weighted Evidential AIRS

The main goal of the proposed approach, that we denote by Optimized WE-AIRS, is to enhance the accuracy of the classification results. Given a new antigen to be classified, each obtained memory cell in the MC pool is assigned a weight related to $numRepAg$. This weight is computed during the first phase corresponding to the learning-reduction procedure. Once the $numRepAg$ corresponding to each memory cell is derived, it will be used in the creation of the R-MC pool leading to the cells selection. Based on the obtained R-MC pool, an evidential optimization process is performed using the gradient descent technique aiming to find the parameter values minimizing a given cost function. Finally, the evidential classification and the decision making process are then performed using the evidential k-nearest neighbors [12,13] weighted by the extracted $numRepAg$. The process of our Optimized Weighted Evidential AIRS is described in Fig. 1.

In the following, we detail the five steps of our Optimized WE-AIRS namely, (1) the learning-reduction procedure, (2) the creation of the R-MC pool, (3) the evidential optimization process, (4) the evidential classification and (5) the decision making.

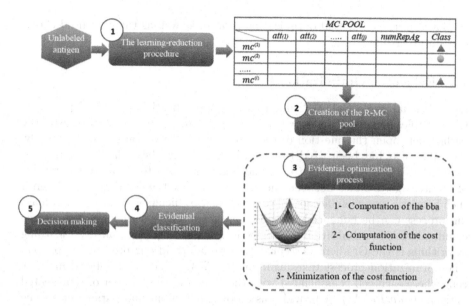

Fig. 1. The Optimized Weighted Evidential AIRS process

4.1   The Learning-Reduction Procedure

In order to enhance the classification results of AIRS, we introduce during this phase a new component named *numRepAg*. Such weight is assigned by preserving the number of original training cells represented by each antigen in the final *MC* Pool [15]. The result of the learning-reduction process, is a new *MC* pool where all the antigens are weighted by their corresponding *numRepAg*. In the next phase, the derived *numRepAg* of each memory cell will be considered during the selection process.

4.2   Creation of the R-MC Pool

Let us define $\Theta = \{c_1, c_2, \ldots, c_p\}$ which contains a finite set of p classes. This phase takes as input the set of antigens in the *MC* pool matched with their respective weights (*numRepAg*) computed during the previous step. We tend here to select the best stimulated antigens by computing the similarity of the test cell and the other antigens in the memory cell pool. In our approach, we do not pick up the k-nearest neighbors like the other traditional AIRS algorithms, but we rather select the s nearest cells where s is the sum of *numRepAg* of all samples. In the end of this phase, we obtain the *R-MC* pool which corresponds to the final reduced memory cell pool. The obtained *R-MC* pool contains the most stimulated cells to the unlabeled antigen getting: $\sum numRepAg = k$. The *R-MC* pool includes the s selected memory cells as well as their corresponding classes c_i such that: $R\text{-}MC = \{(mc^{(1)}, c_1), \cdots, (mc^{(s)}, c_s)\}$. It will be further employed in the evidential optimization process as well as the evidential classification procedure.

4.3 Evidential Optimization Process

In traditional evidential AIRS methods, the classification process depends on two variables namely, α and $\gamma = (\gamma_1, \ldots, \gamma_q)$, where $q \in \{1, \ldots, p\}$. However, the authors left open the question of the choice of these two parameters. Actually, the value of α has been fixed to 0.95 which gives the best classification accuracy [12,13]. When it comes to γ_p, one suggestion was to compute the inverse of the average distance between the antigens having the same class c_q. Even if such heuristic may yield acceptable outcomes, the efficiency of the decision making process can be enhanced if a given optimization is performed. One of the commonly used optimization strategies in machine learning and deep learning algorithms is the Gradient descent, which we adopt in our proposed approach. The optimization process takes as input the $R\text{-}MC$ pool with labeled memory cells where each cell is associated with the class and the number of represented antigens ($numRepAg$). It provides as an output the optimal parameter γ. The Evidential optimization process can be divided into three fundamental steps, namely (1) Computation of the bba, (2) Computation of the cost function and (3) Minimization of the cost function.

1. **Computation of the bba**

   Let $mc^{(l)}$ be a memory cell of the established $R\text{-}MC$ pool belonging to class c_q and its class membership is encoded as a vector $t^{(l)} = (t_1^{(l)}, \ldots, t_p^{(l)})$ of p binary indicator variables $t_j^{(l)}$ presented by $t_1^{(l)} = 1$ if $j = q$ and $t_1^{(l)} = 0$ otherwise. We recall that the $R\text{-}MC$ pool is the reduced pool containing the most stimulated antigens of $mc^{(l)}$ belonging to class c_q and s is the number of the selected cells. Inspired by [13], we generate a bba $m^{(l)}$ presenting the evidence related to the class of $m^{(l)}$ using the following equations:

$$m^{(l)}(\{c_q\}) = \frac{1}{N} \left(1 - \prod_{i \in I_{s,q}^{(l)}} (1 - \alpha\phi_q(\omega d^{(l,i)})) \right) \prod_{r \neq q} \prod_{i \in I_{s,r}^{(l)}} (1 - \alpha\phi_q(\omega d^{(l,i)}))$$

$$\forall q \in \{1..p\} \tag{5}$$

$$m^{(l)}(\Theta) = \frac{1}{N} \prod_{r=1}^{p} \prod_{i \in I_{s,r}^{(l)}} (1 - \alpha\phi_q(\omega d^{(l,i)})) \tag{6}$$

where $I_{s,q}^{(l)}$ denotes the set of the indexes of the neighbors of $mc^{(l)}$ selected in the $R\text{-}MC$ pool having class c_q, $I_{s,r}^{(l)}$ is the set of the indexes of the neighbors of $mc^{(l)}$ in the $R\text{-}MC$ pool having class c_r, $\omega d^{(l,i)}$ is the weighted euclidean distance between $mc^{(l)}$ and $mc^{(i)}$ where ω corresponds to the inverse of

$numRepAg(i)$ related to $mc_{(i)}$, N is a normalizing factor [12] and $\phi_q(\omega d^{(l,i)})$ is defined as:

$$\phi_q(\omega d^{(l,i)}) = e^{(-\gamma_q \omega d^{(l,i)2})} \tag{7}$$

2. **Computation of the cost function**

In this step, the pignistic probability function ($BetP$) is computed in order to get an output vector containing a pignistic probability distribution from the generated bba $m^{(l)}$. This distribution $BetP^{(l)}$ associated to $m^{(l)}$ is defined as: $BetP^{(l)} = (BetP^{(l)}(\{c_1\}), \dots, BetP^{(l)}(\{c_p\}))$.

Ideally, the vector $BetP^{(l)}$ should be close as possible to vector $t^{(l)}$ in order to have a short distance between the predicted class and the expected class. This closeness will be our cost function which is defined by the squared error computed as following:

$$E(mc^{(l)}) = \sum_{q=1}^{p}(BetP_q^{(l)} - t_q^{(l)})^2 \tag{8}$$

The Mean Squared Error (MSE) in the R-MC pool is finally equal to:

$$E = \frac{1}{s}\sum_{l=1}^{s} E(mc^{(l)}) \tag{9}$$

E denotes the average value, over all possible memory cells, of the squared difference between the output and the actual class. It is considered as a cost function used in order to adjust the parameter vector γ and obtain better classification accuracy. The optimization process corresponds to the derivation of the mathematical equation for the gradient of $E(mc^{(l)})$ related to γ. Such iterative derivation allows then the determination of γ_q through a gradient search procedure.

3. **Minimization of the cost function**

To minimize the cost function, we used the gradient descent over the parameter γ. In fact, the gradient descent is an iterative algorithm which uses the derivative of $E(mc^{(l)})$) with respect to each parameter γ_q such as:

$$\frac{\partial E(mc^{(l)})}{\partial \gamma_q} = \sum_{i \in I_{s,q}^{(l)}} \frac{\partial E(mc^{(l)})}{\partial \phi_q(\omega d^{(l,i)})} \frac{\partial \phi_q(\omega d^{(l,i)})}{\partial \gamma_q} \tag{10}$$

with

$$\frac{\partial E(mc^{(l)})}{\partial \phi_q(\omega d^{(l,i)})} = \sum_{r=1}^{p} \frac{\partial E(mc^{(l)})}{\partial BetP_r^{(l)}} \frac{\partial BetP_r^{(l)}}{\partial \phi_q(\omega d^{(l,i)})}$$

$$= \sum_{r=1}^{p} 2(BetP_r^{(l)} - t_r^{(l)}) \left[\frac{\partial m^{(l)}(\{c_r\})}{\partial \phi_q(\omega d^{(l,i)})} + \frac{1}{p}\frac{\partial m^{(l)}(\Theta)}{\partial \phi_q(\omega d^{(l,i)})} \right] \tag{11}$$

and

$$\frac{\partial \phi_q(\omega d^{(l,i)})}{\partial \gamma_q} = -\omega d^{(l,i)2}\phi_q(\omega d^{(l,i)}) \tag{12}$$

The optimal value of parameter γ after the tuning is the one provided with the minimum cost function.

4.4   Evidential Classification Process

This step is devoted to the classification process under the belief function theory. It is divided into two principal parts: (1) Evidence representation and (2) Evidence combination, which will be detailed below.

1. **Evidence representation**

   Classical approaches in this phase induce k basic belief assignments (bba's) for the selected nearest neighbors. However, in some situations where the k value is high, this protocol seems to be complicated and may decrease the efficiency of the decision making. In our work, we intent to resolve such issue by inducing just the basic belief assignments of the resulting antigens in the reduced memory cell pool. Therefore, the obtained memory cells will be taken as sources of evidence. Based on these pieces of evidences, the assignment of the right class will be processed. We first initialize α to the value 0.95. Then, unlike the classical AIRS methods [11], we do not use the traditional γ_p as the reciprocal of the average distance between two antigens having the same class c_q but we employ the optimal value obtained during the optimization process that we denote by γ_p^*. Consequently, the generation of bba's from the picked cells in $R\text{-}MC$ pool is achieved as follow:

$$m(.|mc^{(i)}) = \begin{cases} m(\{c_p\}|mc^{(i)}) = \alpha \ e^{-(\gamma_p^* \cdot (\omega d^{(l,i)})^2)} \\ m(\Theta|mc^{(i)}) = 1 - (\alpha \ e^{-(\gamma_p^* \cdot (\omega d^{(l,i)})^2)}) \end{cases} \tag{13}$$

2. **Evidence combination**

   The final stage of the evidential classification process is the fusion of the obtained weighted bba's produced for each memory cell in the $R\text{-}MC$ pool. This combination is reached through the Dempster rule of combination (Eq. 3).

4.5   Decision Making

Once all the previous steps are accomplished, we move to the next stage which is the decision making. In this phase, we have to assign to the unlabeled antigen the adequate class. Therefore, we use the pignistic probability ($BetP$) and according to the resulting $BetP$ values, we assign to the test pattern the class getting the highest value.

5   Experimentation

In order to emphasize the efficiency of our approach, that we denoted by Optimized WE-AIRS, an experimental study has been carried out based on four real world data sets collected from the UCI machine learning repository [17]. A comparison of our approach with the standard versions of AIRS has also been investigated. The compared methods are respectively, AIRS2, AIRS3, Fuzzy AIRS2, Evidential AIRS2 (EAIRS2) and Weighted Evidential AIRS (WE-AIRS).

5.1 Framework

In our experiments, we draw on four real data sets which are described in Table 1. We note that $InstancesNb$ corresponds to the antigens number. The attributes number is denoted by $AttributesNb$ while $ClassNb$ is the number of classes.

Table 1. Description of the data sets

Databases	$InstancesNb$	$AttributesNb$	$ClassNb$
Cryotherapy (C)	90	6	2
Wine (W)	178	13	3
Fertility (F)	100	9	2
Somerville Happiness Survey (SHS)	143	6	2

During our several experiments, the parameter values have been set as follows: Clonal rate (Cr) = 10, Mutation rate (Mr) = 0.4, HyperClonal rate (Hr) = 2, Number of resources (Nr) = 200, Stimulation threshold (St) = 0.3, Affinity threshold scalar (Ats) = 0.2. Furthermore, different values of k have been used such as: $k = \{3, 5, 7, 8, 9, 10\}$. All experiments have been performed using MATLAB R2018a.

5.2 Evaluation Metric

During our tests, the evaluation of our approach was based on the Percent Correctly Classified denoted by: PCC. This evaluation metric allows us to measure the classification efficiency. It is based on the following equation:

$$PCC = \frac{Number\ of\ correctly\ classified\ instances}{Total\ number\ of\ classified\ instances} \tag{14}$$

In addition, the Cross-Validation (CV) has been performed in order to measure the performance of our approach. Particularly, we used the 10-fold CV where the average of the accuracies related to the 10 repetitions has been computed.

5.3 Experimental Results

To confirm the achievement of our proposed approach above the other traditional versions of AIRS which are AIRS2, AIRS3, Fuzzy AIRS2, EAIRS2 and WE-AIRS, we did the comparison relying on the Percent Correctly Classified criterion of the diverse values of k. Figure 2 below illustrates the obtained results of these tests. For instance, for the Wine database having $k = 5$, the PCC attains 90.04% with our approach Optimized WE-AIRS, whereas, it is equal to 60.67% for AIRS2, 86.47% for AIRS3, 76.00% for Fuzzy AIRS2, 65.00% with EAIRS2 and 87.64% with WE-AIRS.

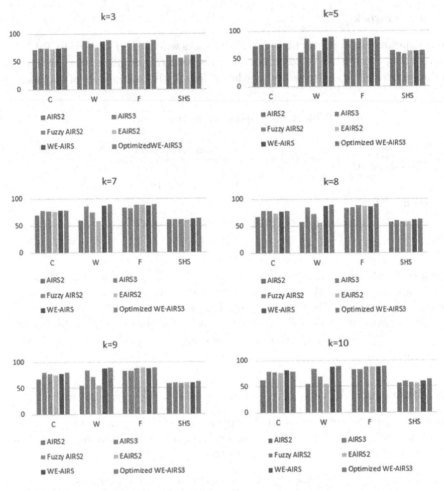

Fig. 2. *PCC* of used databases for various values of k

In the following table, the mean *PCC* obtained using the different values of k is reported for each database.

Overall, the experimental results reported in Table 2 prove that our approach beats the traditional AIRS releases for all the given data sets. In fact, we notice that Optimized WE-AIRS reaches the highest classification efficiency with a value of PCC equal to 89.21% for the data set W compared to 61.72% for AIRS2, 85.4% for AIRS3, 76.18% for Fuzzy AIRS2, 64.55% for EAIRS2 and 86.94% for WE-AIRS. The obtained results emphasize the high efficiency of our approach against the classical AIRS methods.

Table 2. The mean PCC (%)

Data sets	AIRS2	AIRS3	Fuzzy AIRS2	EAIRS2	WE-AIRS	Optimized WE-AIRS
C	68.02	75.78	75.82	74.04	76.11	**76.82**
W	61.72	85.40	76.18	64.55	86.94	**89.21**
F	82.79	84.07	85.53	84.79	86.22	**88.58**
SHS	59.35	60.51	58.75	59.10	61.17	**62.35**

6   Conclusion

In this paper, we employed the evidential gradient descent as a powerful optimization algorithm combined with the AIRS approach to enhance the classification accuracy under an uncertain environment. The conjunction of the belief theory with the optimization process as described in our paper, has led to a more effective classification AIRS method. This efficiency was proved through the different experimental results comparing our method with the standard AIRS approaches.

References

1. Watkins, A.: A resource limited artificial immune classifier. In: The 2002 Congress on Evolutionary Computation, pp. 926–931 (2002)
2. Castro, L.N., De Castro, L.N., Timmis, J.: Artificial Immune Systems: A New Computational Intelligence Approach. Springer, Heidelberg (2002)
3. Dempster, A.P.: A generalization of Bayesian inference. J. R. Stat. Soc. Ser. B (Methodol.) **30**, 205–247 (1968)
4. Shafer, G.: A Mathematical Theory of Evidence. Princeton University Press, Princeton (1976)
5. Smets, P.: The transferable belief model for quantified belief representation. In: Smets, P. (ed.) Quantified Representation of Uncertainty and Imprecision. HDRUMS, vol. 1, pp. 267–301. Springer, Dordrecht (1998). https://doi.org/10.1007/978-94-017-1735-9_9
6. Polat, K., Gunes, S.: Automated identification of diseases related to lymph system from lymphography data using artificial immune recognition system with fuzzy resource allocation mechanism (fuzzy-airs). Biomed. Signal Process. Control **1**(4), 253–260 (2006)
7. Golzari, S., Doraisamy, S., Sulaiman, M.N., Udzir, N.I.: Effect of fuzzy resource allocation method on airs classifier accuracy. J. Theor. Appl. Inf. Technol. **5**, 18–24 (2005)
8. Polat, K., Gunes, S.: Principles component analysis, fuzzy weighting pre-processing and artificial immune recognition system based diagnostic system for diagnosis of lung cancer. Expert Syst. Appl. **34**(1), 214–221 (2008)
9. Chikh, M.A., Saidi, M., Settouti, N.: Diagnosis of diabetes diseases using an artificial immune recognition system2 (AIRS2) with fuzzy k-nearest neighbor. J. Med. Syst. **36**(5), 2721–2729 (2012)

10. Hentech, R., Jenhani, I., Elouedi, Z.: Possibilistic AIRS induction from uncertain data. Soft. Comput. **20**(1), 3–17 (2015). https://doi.org/10.1007/s00500-015-1627-3

11. Lahsoumi, A., Elouedi, Z.: Evidential artificial immune recognition system. In: Douligeris, C., Karagiannis, D., Apostolou, D. (eds.) KSEM 2019. LNCS (LNAI), vol. 11775, pp. 643–654. Springer, Cham (2019). https://doi.org/10.1007/978-3-030-29551-6_57

12. Denoeux, T.: A k-nearest neighbor classification rule based on dempster-shafer theory. IEEE Trans. Syst. Man Cybern. **25**, 804–813 (1995)

13. Zouhal, L.M., Denoeux, T.: An evidence theoretic kNN rule with parameter optimization. IEEE Trans. Syst. Man Cybern. **28**(2), 263–271 (1998)

14. Watkins, A., Timmis, J.: Artificial immune recognition system (AIRS): revisions and refinements. In: AISB 2004 Convention, p. 18 (2004)

15. Jenhani, I., Elouedi, Z.: Re-visiting the artificial immune recognition system: a survey and an improved version. Artif. Intell. Rev. **42**(4), 821–833 (2012). https://doi.org/10.1007/s10462-012-9360-0

16. Abdelkhalek, R., Elouedi, Z.: WE-AIRS: a new weighted evidential artificial immune recognition system. In: International Conference on Robotics and Artificial Intelligence (to appear)

17. Asuncion, A., Newman, D.: UCI machine learning repository (2007). https://archive.ics.uci.edu/ml/index.php

Improving Policy Generalization for Teacher-Student Reinforcement Learning

Gong Xudong[1] , Jia Hongda[1], Zhou Xing[1], Feng Dawei[1], Ding Bo[1(✉)], and Xu Jie[2]

[1] National University of Defense Technology, Changsha, China
dingbo@nudt.edu.cn
[2] University of Leeds, Leeds, UK

Abstract. Teacher-student reinforcement learning is a popular approach that aims to accelerate the learning of new agents with advice from trained agents. In these methods, budgets are introduces to limit the amount of advice to prevent over-advising. However, existing budget-based methods tend to use up budgets in the early training stage to help students learn initial policies fast. As a result, initial policies are some kind solidified, which is not beneficial for improving policy generalization. In this paper, to overcome advising intensively in the early training stage, we enable advising in the entire training stage in a decreasing way. Specifically, we integrate advice into reward signals and propose an advice-based extra reward method, and integrate advice into exploration strategies and propose an advice-based modified epsilon method. Experimental results show that the proposed methods can effectively improve the policy performance on general tasks, without loss of learning speed.

Keywords: Reinforcement learning · Agent training · Advising strategy · Policy generalization

1 Introduction

Multi-agent reinforcement learning (MARL) [3] has been widely used in dynamic learning problems in multi-agent systems (MAS) and has gained considerable success in real time strategy games, e.g. DOTA2 [10]. In the application of MARL, new agents should be deployed to extend system capability or to replace failed agents. In these situations, the system ability to resume is determined by how fast newly entering agents can learn their policies. Thus, researches on how to speed up the learning of newly entering agents are vital challenges in MAS.

Teacher-student reinforcement learning [14] was proposed to meet the above challenges. In these methods, an experienced "teacher" agent helps accelerate the "student" agents learning by providing advice on which action to take next. [1]. Besides, helping students learn policies with strong generalization [11] should

© Springer Nature Switzerland AG 2020
G. Li et al. (Eds.): KSEM 2020, LNAI 12275, pp. 39–47, 2020.
https://doi.org/10.1007/978-3-030-55393-7_4

also be considered, otherwise, students can copy teacher policies immediately. Therefore, budgets are introduced to constrain the amount of advice [12,13]. However, existing budget-based methods tend to use up budgets in the early training stage, which means advising imposes an intensive impact to students' exploration in the early training stage. This leads to that students learn relatively solid initial policies fast, which, however, are not beneficial for learning policies with strong generalization. Similar inspiration can be found in pedagogy: if students follow the guidance too much in the early learning stage, they may lack the motivation for change and innovation in the future [2].

The main idea of this paper highlights that advising should be enabled in the entire training stage in a decreasing way, so that advice can provide students continuous reference to learn better policies. Based on this idea, we investigate the framework of reinforcement learning and find that reward signals and exploration strategies are two functional units that take effect in the entire training stage. Thus, we propose the advice-based extra reward (ER) method where we extend reward signals by providing a student with an extra reward if he selects an action that is similar to advice. And propose the advice-based modified epsilon (ME) method where we modify exploration strategies by asking for advice with a descending probability when a student decides to explore the environment.

We test the two proposed methods on the coordinated multi-agent object transportation problem (CMOTP) [3] and a variation of the CMOTP, that is, the r-CMOTP. Comparisons conducted with state-of-the-art advising strategies show that the two proposed methods can improve policy performance on general tasks effectively, without loss of learning speed.

The remainder of this paper is organized as follows. Section 2 presents the necessary background and related works. Section 3 introduces the advice-based ER and ME method. Section 4 compares the proposed methods with state-of-the-art methods on the CMOTP and r-CMOTP. Section 5 concludes the paper.

2   Background and Related Work

2.1   Motivated Scenario

As over advising hinders student learning [13], existing advising methods are generally designed with budgets [8,13] to limit the amount of advice. We apply existing budget-based methods to the CMOTP [3] (detailed in Sect. 4.1). For convenience, a CMOTP **task** refers to two agents allocated to certain specific positions aiming to transport goods to a home area. Different tasks are marked by different initial positions of the two agents.

Figure 1 shows the results when initial positions of the two agents are fixed in the training and testing. Figure 1(a) indicates that budgets are used in the early training stage, as students can finish a training episode quickly in the early stage and slowly in middle and late stages, while Fig. 1(b) illustrates that students can perform well in this specific task even when budgets have not been depleted. This process indicates that advice in budget-based methods takes effect in the early training stage to help students learn initial policies.

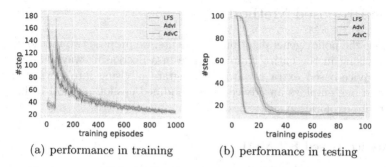

(a) performance in training   (b) performance in testing

Fig. 1. LFS means learning from scratch without advice, AdvI is the abbreviation of the advice importance strategy [13], and AdvC is the abbreviation of the advice correct strategy [13], both are budget-based methods. The y-axis represents the mean step that agents take to finish the task in testing, while the x-axis represents the amount of training episodes that have been used to train the policy. We terminate an episode if the episode length exceeds 100 in the tests. The faster a curve drops, the faster a student learns its policy.

In further experiments (detailed in Sect. 4.2), we find that training after budgets are depleted worsens policy performance on specific tasks, but is necessary to enhance policy performance on general tasks. However, as our experimental results demonstrate, policy performance on specific tasks and policy performance on general tasks obtained by budget-based methods are both worse than method without advising when training finished. These results suggest that although it is able to accelerate the learning of students, budget-based methods suffers from a low policy generalization problem.

2.2   Teacher-Student Reinforcement Learning

Teacher-student reinforcement learning was proposed by Clouse et al. [4]. The two roles in these methods are teachers that have learned their policies, and students that do not have their policies yet. The thought of these methods is that teachers can give students some advice basing on some heuristics (e.g. ask uncertain heuristic [4], advice importance heuristic [13] etc.) to accelerate their learning [5–7,13], but the amount of advice should be limited [13]. Over-advising is a major trouble in this setting, since it may hinder students' learning and consume too many communication resources [13]. When students receive advice, they execute these advice immediately and evolve their policies based on the reward signal from the environment. Generally, teachers and students can take different representing types of environment states and different learning algorithms, but share a common action set [13].

3   The Proposed Method

To improve the policy generalization of students, we propose two methods which enable advising in the entire training stage in a decreasing way. The first one uses an advice-based extra reward to integrate advising into reward signals. The second one employs an advice-based modified epsilon method to integrate advising into exploration strategies.

3.1   Advice-Based ER Method

In reinforcement learning framework, agents evolve their policies with direct feedback from reward function, which plays a vital role in the entire training stage. To distribute advice in the entire training stage in a decreasing way, we extend the reward function by considering information from advice.

Equation 1 shows that when the action chosen either by the exploration strategy or by the evolving action policy equals the advice from a teacher, the teacher provides the student with an extra reward, which is calculated by Eq. 2,

$$r'_{st}(s, \mathbf{a}) = r_{st}(s, \mathbf{a}) + \varphi(s, \mathbf{a}, t), \tag{1}$$

$$\varphi(s, \mathbf{a}, t) = \begin{cases} \omega + \mu e^{-\nu t}, if\ a_{st} = \pi_{tc}(s), \\ 0, else \end{cases} \tag{2}$$

where s is the state of the environment, $\mathbf{a}$ is the joint action of all the agents in the environment, t is the iteration that the policy has been trained, $r_{st}(s, \mathbf{a})$ is the reward function of the student, subscript st denotes the student, and subscript tc denotes the teacher, $\omega \in \mathbb{R}, \mu \in [0, +\infty), \nu \in [0, +\infty)$.

3.2   Advice-Based ME Method

In reinforcement learning framework, exploration strategies also play an important role in the entire training stage in helping agents learn an optimal and strong-generalization policy. To distribute advice in the entire training stage in a decreasing way, we let students ask for advice with a decreasing probability when he uses exploration strategy.

Equation 3 shows that when a student uses exploration strategies to interact with the environment, he asks for advice from a teacher with a specified probability,

$$a_{st} = \begin{cases} \pi_{st}(s), if\ x \in [\varepsilon(t), 1] \\ \pi_{tc}(s), if\ x \in [0, \varepsilon(t))\ and\ x' \in [0, g(t)], \\ \pi_\varepsilon(s), else \end{cases} \tag{3}$$

where t is the iteration that the policy has been trained, $x \sim U(0, 1), x' \sim U(0, 1)$ are two random variables, $\varepsilon(t)$ is the exploring probability at iteration t, and $g(t) = \omega + \mu e^{-\nu t}$, where $\omega \in [0, 1], \mu \in [0, 1 - \omega], \nu \in [0, +\infty)$ is the asking-for-advice probability at iteration t when the student explores.

4   Evaluation

4.1   CMOTP Environment and Experimental Settings

We evaluate the proposed methods on the CMOTP [3], and a variation of the CMOTP (the initial positions of the two agents are randomized, denoted as r-CMOTP below). As the main purpose of our work is to study the advising methods, for convenience, we implement independent learners [3,9] with Q-learning algorithm as our test algorithm. Available actions for each agent, state representation, network output, and environment restrictions are all same as [3]. First, we train the two agents in the r-CMOTP. Next, we set one of the trained agents as the teammate and the other as the teacher. For the proposed methods, we set $\omega = 0.01, \mu = 0.09, \nu = 0.0001$ for the ER and $\omega = 0.1, \mu = 0.1, \nu = 0.001$ for the ME. We conduct experiments with these settings.

4.2   Evaluation on the CMOTP

We train the student with a specific task, in which the initial positions of the two agents are fixed. We use 100 random seeds in the training to obtain 100 different policies for each method. First, we test the student agent with the same specific task to demonstrate policy quality in this specific task. In the training, we train policy with 10,000 episodes and conduct a test every 10 episodes and terminate a test if the episode length exceeds 100. Figure 2(a) and Table 1 exhibit the corresponding results. Next, we test each policy with 100 different tasks, record the number of tests in which the policy performs optimally. Average value on the 100 different tests is shown to demonstrate the policy generalization. Figure 2(b) and Table 2 demonstrate the corresponding results.

Table 1 exhibits that the policy quality in the specific task is high by the time-point when budgets are depleted for AdvI, AdvC, AskI, and AdvC-AskI. However, this finding does not mean that we can terminate training at this time-point, because Table 2 shows that policy generalization is low for all budget-based methods at this time-point. This result indicates that initial policies learned by budget-based methods are poor in generalization. Consequently, further training is required to enhance policy generalization. However, when the training is complete, the highest policy generalization obtained by the budget-based methods (13.08 by AdvC-AskU) is lower than 13.29 of the LFS method. This finding indicates existing budget-based methods are not beneficial to learn policies with better generalization than LFS.

4.3   Evaluation on the R-CMOTP

We further conduct experiments on a general case, that is, the r-CMOTP, to show differences in policy generalization. In this section, the initial agent position is randomized in the training and testing. We train policy with 55,000 episodes and measure the policy quality by testing each policy with 100 fixed tasks and record the average steps the policy need to complete these tasks. Figure 3(a) and Table 3

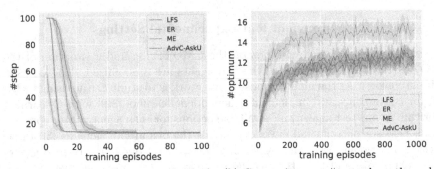

(a) Comparison on the steps to finish the task

(b) Comparison on #test where the policy performs optimally out of 100 test

Fig. 2. Comparison among LFS, ER, ME, and AdvC-AskI on the CMOTP. For (a), the y-axis represents the mean step that agents take to finish the task, while the x-axis represents the amount of training episode that has been used to train the policy. For (b), the y-axis represents the number of test in which the policy performs optimally out of 100 test, while the x-axis is same as (a). For both figures, the solid line is the result averaged on 100 different random seeds, and the shaded area is the standard deviation.

Table 1. Comparison on the policy quality on the specific task among different methods[a]

Methods	LFS	AdvI [13]	AdvC [13]	AskI [1]	AskU [4]	AdvC-AskI [1]	AdvC-AskU [1]	ER	ME
#OB[b]	–	89	90	97	14	92	37	–	–
#OT[c]	100	33	32	83	98	90	88	100	100
#EB[d]	–	62.00	69.11	55.41	55.65	75.66	–	–	–

[a]The initial position of the two agents are fixed in both the training and the testing.
[b]#policy that performs optimally when budgets are depleted (the AdvC-AskU method is tested at 80th episode).
[c]#policy that performs optimally when training is complete.
[d]#episode that has been used to train the policy when budgets are depleted (the AdvC-AskU does not use up budgets by the end of training but rarely uses them since 80 episodes).

Table 2. Comparison on the policy generalization among different methods[a]

Methods	LFS	AdvI	AdvC	AskI	AskU	AdvC-AskI	AdvC-AskU	ER	ME
#OB[b]	–	4.47	5.09	7.21	8.49	6.04	9.04	–	–
#OT[c]	13.29	8.56	8.75	11.50	11.17	12.34	13.08	15.51	11.98

[a]The initial positions of agents are fixed in training and random in testing.
[b]#test where the policy performs optimally when budgets are depleted
[c]#test where the policy performs optimally when training complete.

exhibit the corresponding results. Meanwhile, we measure policy generalization by testing each policy with 100 different tasks and record the number of tasks in which the policy performs optimally (the final value is averaged on 100 random seeds). Figure 3(b) and Table 3 show the corresponding results.

Figure 3(a) shows that ER and ME achieve faster rates in improving policy quality compared with AdvC-AskU. As can be observed in the first row of Table 3, the final average number of steps to finish a task is 9.39 for ER and 9.47 for ME, respectively, both of which are lower than those of budget-based methods. This finding suggests that ER and ME can improve policy quality. Figure 3(b) demonstrates that ER and ME have faster speeds in improving policy generalization compared with AdvC-AskU. In addition, the second row of Table 3 shows that the number of optima a policy obtains in 100 different tasks is 76.06 for ER and 75.77 for ME. Both are higher than LFS and budget-based methods, which indicates that ER and ME can improve policy generalization.

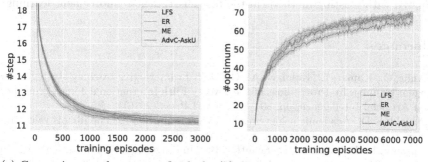

(a) Comparison on the steps to finish the task

(b) Comparison on #test where the policy performs optimally out of 100 test

Fig. 3. Comparison among LFS, ER, ME, and AdvC-AskI on the r-CMOTP. The illustration of (a) is same as Fig. 2(a) while the illustration of (b) is same as Fig. 2(b).

Table 1 demonstrates that the two proposed methods perform best on the specific task of the CMOTP. Meanwhile, Table 2 illustrates that ER achieves the best policy generalization (i.e., 15.51, which is higher than the LFS and budget-based methods). This finding suggests that ER can effectively improve policy generalization on the specific task of the CMOTP. Nevertheless, ME does not perform better than all budget-based methods, the reason may be that for the specific task, exploration is significantly important for students to improve policy generalization, however, asking for advice takes a certain proportion in exploration strategies.

Table 3. Comparison on the policy quality and generalization among different methods[a]

Methods	Optimal	Teacher	LFS	AdvI	AdvC	AskI	AskU	AdvC-AskI	AdvC-AskU	ER	ME
#St[b]	8.88	10.44	9.43	9.60	9.66	9.64	9.54	9.59	9.50	**9.39**	9.47
#Opt[c]	–	73.01	74.85	70.01	72.69	73.66	73.53	74.68	73.81	**76.06**	75.77

[a]The initial position of the two agents are random in both the training and the testing.
[b]The average steps for each policy on 100 specific tasks.
[c]The average number of test in which the policy performs optimally when training complete.

5   Conclusions

This study investigates methods of advising agents to accelerate learning and improve policy generalization. We propose the advice-based extra reward method and the advice-based modified epsilon method and conduct experiments on the coordinated multi-agent object transportation problem. Experimental results show that the proposed methods can effectively improve policy generalization compared with existing methods in the teacher-student reinforcement learning.

References

1. Amir, O., Kamar, E., Kolobov, A., Grosz, B.J.: Interactive teaching strategies for agent training. In: Proceedings of the Twenty-Fifth International Joint Conference on Artificial Intelligence, pp. 804–811. AAAI Press (2016)
2. Arnold, K.D.: Academic achievement - a view from the top. The Illinois valedictorian project. Academic Achievement, p. 76 (1993)
3. Buşoniu, L., Babuška, R., De Schutter, B.: Multi-agent reinforcement learning: an overview. In: Srinivasan, D., Jain, L.C. (eds.) Innovations in Multi-agent Systems and Applications - 1. SCI, vol. 310, pp. 183–221. Springer, Heidelberg (2010). https://doi.org/10.1007/978-3-642-14435-6_7
4. Clouse, J.A.: On integrating apprentice learning and reinforcement learning (1997)
5. Cruz, F., Magg, S., Weber, C., Wermter, S.: Improving reinforcement learning with interactive feedback and affordances. In: 4th International Conference on Development and Learning and on Epigenetic Robotics, pp. 165–170. IEEE (2014)
6. Cruz, F., Magg, S., Weber, C., Wermter, S.: Training agents with interactive reinforcement learning and contextual affordances. IEEE Trans. Cogn. Dev. Syst. 8(4), 271–284 (2016)
7. Griffith, S., Subramanian, K., Scholz, J., Isbell, C.L., Thomaz, A.L.: Policy shaping: integrating human feedback with reinforcement learning. In: Advances in Neural Information Processing Systems, pp. 2625–2633 (2013)
8. Ilhan, E., Gow, J., Perez-Liebana, D.: Teaching on a budget in multi-agent deep reinforcement learning. In: 2019 IEEE Conference on Games, pp. 1–8. IEEE (2019)
9. Matignon, L., Laurent, G.J., Le Fort-Piat, N.: Independent reinforcement learners in cooperative Markov games: a survey regarding coordination problems. Knowl. Eng. Rev. 27(1), 1–31 (2012)
10. OpenAI: Openai five (2018). https://blog.openai.com/openai-five/
11. Sutton, R.S., Barto, A.G.: Reinforcement Learning: An Introduction. MIT Press, Cambridge (2018)

12. Taylor, M.E., Carboni, N., Fachantidis, A., Vlahavas, I., Torrey, L.: Reinforcement learning agents providing advice in complex video games. Connect. Sci. **26**(1), 45–63 (2014)
13. Torrey, L., Taylor, M.: Teaching on a budget: agents advising agents in reinforcement learning. In: Proceedings of the 2013 International Conference on Autonomous Agents and Multi-agent Systems, pp. 1053–1060 (2013)
14. Zimmer, M., Viappiani, P., Weng, P.: Teacher-student framework: a reinforcement learning approach. In: AAMAS Workshop Autonomous Robots and Multirobot Systems (2014)

Recommendation Algorithms and Systems

Towards Effective Top-k Location Recommendation for Business Facility Placement

Pu Wang, Wei Chen, and Lei Zhao$^{(\boxtimes)}$ (iD)

School of Computer Science and Technology,
Soochow University, Suzhou 215006, China
{wangpu,robertchen,zhaol}@suda.edu.cn

Abstract. In the process of selecting locations for establishing new business facilities, location recommendation offers the optimal candidates, which maximizes the number of customers served to bring the maximum profits. In most existing work, only spatial positions of customers are considered, where social relationships and temporary activities, which are significant factors for candidate locations, are ignored. Additionally, current studies fail to take the capacity of service facilities into consideration. To overcome the drawbacks of them, we introduce a novel model MITLR (Multi-characteristic Information based Top-k Location Recommendation) to recommend locations with respect to capacity constraints. The model captures the spatio-temporal behaviors of customers based on historical trajectory and employs social relationships simultaneously, to determine the optimal candidate locations. Subsequently, by taking advantage of feature evaluating and parameter learning, MITLR is implemented through a hybrid B-tree-liked framework called CLTC-forest (tree). Finally, the extensive experiments conducted on real-world datasets demonstrate the better effectiveness of proposed MITLR.

Keywords: Top-k location recommendation · Spatio-temporal trajectory · Social relationship · Capacity constraint

1 Introduction

The study of top-k facility locations selection aims to identify the appropriate k locations for new business facilities from a range of available candidate locations. In this context, location is defined as a special site in a road network with facility on it in terms of a given service for customers, the selection is based on factors such as the number of customers served or the returns on facility investments. This kind of query has been widely applied in a variety of recommendation applications, such as planing to establish new electric vehicle charging stations, mobile toilets, or retail stores in a city.

As GPS and mobile devices are developed in recent years, daily trajectories have been recorded and utilized widely, as well as an increasing number of

© Springer Nature Switzerland AG 2020
G. Li et al. (Eds.): KSEM 2020, LNAI 12275, pp. 51–63, 2020.
https://doi.org/10.1007/978-3-030-55393-7_5

studies with location recommendation come to focus on trajectories [1,4–6,10]. Furthermore, advances in social network technology are facilitating interpersonal communication, friend-based recommendation becomes a growingly significant and relevant factor in recommendation system. For instance, customers will receive electronic red envelopes or coupons occasionally while consuming, and then share them with friends in some instant messaging Apps like Wechat, QQ, Alipay, etc. With shared electronic red envelopes or coupons, their friends could get a discount when they are consuming afterwards [11].

However, existing studies of trajectory-based location recommendation evaluate the correlation between customers and facility locations by their spatial distances solely, they fail to evaluate the timestamp of trajectories and the effects of customer social relationships on facility locations querying, which will render the recommendation results inaccurate or uneconomic. To illustrate the necessity of considering the friend-based recommendation, a straightforward example is demonstrated as below.

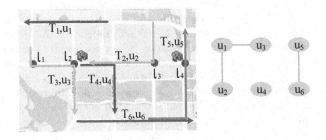

Fig. 1. A toy example with candidate locations, trajectories and social relationships.

Example 1. As shown in Fig. 1, there are four candidate locations l_1 to l_4 with electric vehicle charging stations, and six vehicle trajectories T_1 to T_6 corresponding to six distinct customers u_1 to u_6. To select the optimal facility location for u_6, as the shortest spatial distances from T_6 to l_2, l_3 and l_4 are all equal, a random one between l_2, l_3 and l_4 would be chosen by several existing algorithms such as NetClus [6], while spatial distances with candidate locations are considered merely. However, Fig. 1 shows that u_5 and u_6 could readily share electronic red envelopes or coupons since they are close friends. Therefore, if u_5 has been served by l_4 (as u_5 passes l_4 directly), there is a great probability that u_6 will also be served by l_4 due to the reciprocal recommendation of u_5.

To overcome the aforementioned deficiencies of earlier work, we formalize the problem of constrained top-k facility location recommendation into a novel model MITLR. To determine whether a candidate location is recommended or not, the total service utilities of candidate location are predicted. Unlike the previous work that only considers the road spatial distance, this model examines the importance of spatial and temporal features of trajectories, as well as the

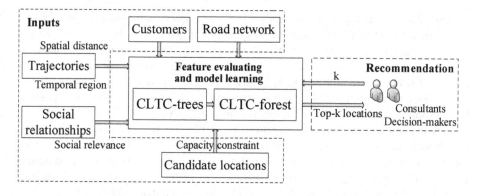

Fig. 2. The framework of MITLR.

social relationships of customers at the same time. As a result, the evaluation of service utilities is rendered more precisely by these significant characteristics and the new objective function. In addition, we take a formal approach to consider the capacity of all candidate locations with regards to the real living conditions of customers. We then develop a new hybrid index using service utilities and location capacities, which is referred to CLTC (Constrained Location and social-Trajectory Clusters) forest, to incorporate spatio-temporal trajectories, social relationships and candidate locations into CLTC-trees that ultimately form the final CLTC-forest. Based on this hybrid index, this study presents an efficient query algorithm that exploits a simple greedy strategy to obtain the final top-k results for recommending.

The overall framework of the proposed model is outlined in Fig. 2, and our key contributions are summarized as follows.

1. This study proposes the top-k location recommendation problem and a novel model MITLR, and defines the service utility function to predict the correlation while capturing the spatio-social and temporal behaviors.
2. We have developed a new index structure of the CLTC-forest (tree) which combines both candidate locations and trajectories, as well as present the process of parameter learning and efficient query approach.
3. Extensive experiments are performed on real datasets to offer insight into the effectiveness and feature weights of the proposed model.

2   Related Work

We cover the existing work on next optimal location prediction and the optimal k locations prediction in turn.

Several studies [1–3,7,9] focus on the next optimal location prediction problem by taking advantage of various metrics and objective functions. Sun et al. [7] acknowledge the service capacity of each location facility, they suggest that there is a certain limitation on the number of served customers and the metric is only examined by spatial distances. Li et al. [3] query a location for establishing facility based the optimal segment query of trajectories, they assign a score on each segment as in [1] but without recognizing the candidate location as a specific position on a road network. Yao et al. [9] take advantage of a recurrent model SERM for the next optimal location prediction in semantic trajectories, where both spatio-temporal transitions and relevant textual information posted by customers are considered to improve the precision.

Recent researches have concentrated on exploring the problem of the optimal k location recommendation [4–6,10]. In more details, Li et al. [4] mine the most influential k-location, from this point of view, they evaluate the maximum number of unique trajectories that traverse a location in a given spatial region, therefore, the common practicability of this work is greatly restricted by the traverse limitation. Mitra et al. [5,6] focus on the top-k location query problem with respect to trajectory merely, they [6] propose an index framework of NetClus for TOPS query by covering a wide range of objective functions, their work assumes that each of candidate locations has a radius parameter τ, as a result, the construction of NetClus leads to lots of index instances which are calculated and stored with different values of τ and cluster radii. They further extend TOPS to TIPS [5] in an attempt to minimize either the maximum customer inconvenience or the average inconvenience.

3   Problem Statement

In this section, we formalize the problem and the model MITLR, Table 1 summarizes the frequently used notations throughout this paper.

Table 1. Notation and corresponding description.

Notation	Description
L, C	Set of candidate locations and capacities
Γ	Set of spatio-temporal trajectories
U, S	Set of customers and social relationships
$f_d(l_i, u_j)$	Spatial distance of l_i and u_j
$f_s(l_i, u_j)$	Social relevance of l_i and u_j
$f_t(l_i, u_j)$	Temporal region of l_i and u_j
$F(u_j)$	Friend set of u_j
$SU(l_i, u_j)$	Service utility of l_i and u_j
$\Psi(l_i)$	Service utilities of l_i
$\Im$	Set of optimal k locations

Considering a setting where candidate locations and trajectories are located in a road network and social relationships of customers in trajectories can be captured. The road network is defined as a directed weighted graph $G = \{V_g, E_g\}$, where V_g denotes the set of vertices of road intersections and E_g denotes the set of directed edges of road segments, the weight of directed edge denotes its spatial distance. The candidate location l is a place for establishing a certain facility or service like electric vehicle charging station or mobile toilet, and the service capacity of l_i is defined as c_{l_i}, which means that it cannot be exceeded in real serving applications. A trajectory T is represented in the sequential form of $T = \{(v_1, t_1), ..., (v_\zeta, t_\zeta)\}, v_\zeta \in V_g$, where t_ζ denotes the timestamp when T crosses v_ζ. The social relationships of customers are simply modeled as an undirected and unweighted graph $S = \{V_u, E_u\}$, V_u is the set of nodes representing the customers, and E_u is the set of edges, where an edge denotes that there is a friend relationship between two corresponding customers. Besides, we suppose $|L| = m$ and $|\Gamma| = |U| = n$.

First of all, three significant characteristics of spatial distance, social relevance and temporal region in service utility are provided in detail.

Spatial Distance. The formula of the shortest spatial distance between location and trajectory is adopted, as presented in [6]. Therefore, the spatial distance characteristic of customer and location is outlined:

$$f_d(l_i, u_j) = \frac{\min_{\forall v_{jk}, v_{j\iota} \in T_j}\{d(v_{jk}, l_i) + d(l_i, v_{j\iota}) - d_a(v_{jk}, v_{j\iota})\} - \rho_{\min}}{\rho_{\max} - \rho_{\min}} \quad (1)$$

where $\rho_{\min}$ and $\rho_{\max}$ are normalization factors, $d_a(v_{jk}, v_{j\iota})$ denotes the shortest spatial distance from v_{jk} to $v_{j\iota}$ by going along T_j. To illustrate, a customer u_j deviates from v_{jk} of T_j to l_i, and then returns to $v_{j\iota}$ on G in her/his usual trajectory, note that the additional distance on T_j is not included.

Social Relevance. For assessing the social relevance of l_i and u_j, we assume that $F(u_j)$ is set of customers who have friend relationships with u_j, and $CU(l_i)$ represents set of customers that have already been evaluated to l_i for being served. Then the social relevance between u_j and l_i is defined as follows:

$$f_s(l_i, u_j) = \frac{|\{u_k | u_k \in F(u_j) \wedge u_k \in CU(l_i)\}|}{|CU(l_i)| + \lambda_s} \quad (2)$$

where $|\cdot|$ denotes the number of elements and λ_s is the Laplace smoothing coefficient. The intuition behind the social relevance feature is the friend-based recommendation through shared electronic red envelopes or coupons, and only the direct friendships between customers are concerned here.

Temporal Region. Supposing that u_j departs from her/his usual trajectory to one location at timestamp t_{jk} in Eq. (1), and arrives at location l_i at timestamp $AT(u_j, l_i) = t_{jk} + \Delta t$, where Δt is a constant timestamp value representing the duration from departure to arrival. For simplicity, Δt of each customer is set to equal and one day is divided into 24 equal segments. As a consequence, the temporal region characteristic is given:

$$f_t(l_i, u_j) = \frac{\sum\limits_{u_k \in CU(l_i)} \min\{|I(u_j, u_k, l_i)|, 24 - |I(u_j, u_k, l_i)|\}/24}{|CU(l_i)| + \lambda_t} \quad (3)$$

where $I(u_j, u_k, l_i) = AT(u_k, l_i) - AT(u_j, l_i)$, and λ_t is also the Laplace smoothing coefficient. The ground truth of temporal region is that, if a customer intends to stagger her/his arrival time with others who have already been evaluated to be served in the same facility, then the customer will get more guaranteed service.

Next, the service utility function that is raised to evaluate the correlation between l_i and u_j can be presented:

$$SU(l_i, u_j) = \begin{cases} 1 & f_d(l_i, u_j) = 0 \\ -\alpha_\vartheta * f_d(l_i, u_j) + \beta_\vartheta * f_s(l_i, u_j) + \gamma_\vartheta * f_t(l_i, u_j) & otherwise \end{cases} \tag{4}$$

where α_ϑ, β_ϑ and γ_ϑ are feature weights and $\alpha_\vartheta + \beta_\vartheta + \gamma_\vartheta = 1$, if $f_d(l_i, u_j) = 0$, it demonstrates that T_j of u_j just traverses l_i straightforward. The greater value of $SU(l_i, u_j)$ indicates there is a closer connection between l_i and u_j, and there is also a higher probability of l_i serving u_j, vice versa. Moreover, from the perspective of a candidate location, the total service utilities (also as service revenue) for all the served customers are defined as follows:

$$\Psi(l_i) = \sum_{u_j \in U'} SU(l_i, u_j), U' \in U, |U'| \leq C_i \tag{5}$$

where $\forall u_v \in U', \forall u_l \in U - U', SU(l_i, u_v) \geq SU(l_i, u_l)$. Consequently, the model of MITLR is formally stated.

Problem Definition. Given a query with parameter k, a set of spatio-temporal trajectories Γ with corresponding customers U on G, a set of social friend relationships S, and a set of candidate locations L with capacities C, the MITLR seeks to select the optimal location set $\Im$ ($\Im \in L, |\Im| = k$), which maximizes the sum of total service utilities Υ without exceeding the capacity limitations of each selected location, where $\Upsilon = \arg \max \sum_{i=1}^{k} \Psi(l_i), l_i \in \Im$.

4   Model Implementation

It is recognized that candidate locations in close proximity are prone to serve a great number of identical customers, and when $k \ll m$ and $c_i \ll n, i \in [1, m]$, the locations selected in query results are all keeping a certain distance from each other, therefore, the CLTC-tree (forest) is designed for MITLR in terms of model implementing and model learning. For specified k, the final k locations are returned from CLTC-forest by adopting a simple greedy manner.

4.1   CLTC-forest

The CLTC-forest is composed of a series of CLTC-trees, and the CLTC-tree is a B-tree-liked hybrid index structure that integrates candidate locations, trajectories, and customers with social relationships according to their service utilities and location capacities. Each tree node links three pieces of additional information, which include a candidate location l_i as its representation (label), a set of

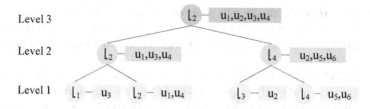

Level 3

Level 2

Level 1

Fig. 3. A simple CLTC-tree.

customers that will be served by $U(l_i)$ with their corresponding SU limited by facility capacity, as well as a total service utility of Ψ.

However, the none-leaf nodes of CLTC-tree are quite different from leaf nodes, as the representation of a none-leaf node is one of the labels of its two children nodes, the customers are a subset of a union which is constituted by customers of two children nodes that do not exceed the capacities. A simple example of one CLTC-tree that represents candidate locations, trajectories and customers with corresponding social relationships of Fig. 1 is denoted as in Fig. 3, where Ψ is omitted in each of specific tree node.

4.2 Generation Processes

The CLTC-forest construction includes two steps of clustering and combining.

Clustering. In order to cluster candidate locations with their most relevant customers into leaf nodes, a constrained k-medoid-liked cluster algorithm is proposed, in which all of candidate locations are served as 'medoids' and service utilities calculated by Eq. (4) are served as the metric. It can be seen that the final clusters are constrained by service capacities of each location on the scales.

However, if a number of customers that pass one candidate location simultaneously happen to overwhelm the capacity of the location, those customers whose initiating position or ending position is in the proximity of this location are clustered firstly. The is because that a customer will not be visibly disturbed or interrupted if she/he chooses one facility near to the initiating position or the ending position on the trace of her/his trajectory.

Combining. Since each leaf node is also a simple CLTC-tree which has merely one root node, a series of merging approaches can be adopted to combine two CLTC-trees into one while utilizing their root node information. By repeating this process, the entire CLTC-trees from leaf nodes to root nodes could be constructed, and then, CLTC-forest is formed finally. Before introducing the CLTC-trees-merging, the definition of the coherence of two CLTC-trees Ct_i and Ct_j is proposed as $CO(Ct_i, Ct_j) = \max\{\Psi(Lrn(Ct_i)), \Psi(Lrn(Ct_j))\}$, where $Lrn(Ct_i)$ is the representation of root node in Ct_i. $CO(Ct_i, Ct_j)$ indicates the service utilities that combines Ct_i and Ct_j into one by taking advantage of the additional information of two separate former root nodes. Greater value of $CO(Ct_i, Ct_j)$

represents that there is closer relevance of the two CLTC-trees, such as inti-mate relationships between the two customers set, or adjacent spatial distances between candidate locations. Therefore, CLTC-trees can be combined according to their coherence.

If two CLTC-trees is merged into one, a new tree node will be created to represent the root node, where two sub-trees are the two former CLTC-trees. For the newly root node, the service utility is equal to CO, the set of evaluated customers is U', where $U' \in Urn(Ct_i) \cup Urn(Ct_j)$ by Eq. (5), Urn is the corresponding customers in tree node, and its representation is the candidate location with larger value of CO in two children CLTC-trees.

4.3   Model Learning

With the help of Eq. (4), it can be observed that the service utility is just referred as a linear combination of the inputs, accordingly, several different kinds of regression algorithms could be deployed to learn these parameters. In this study, a linear regression with regularization is utilized, which the goal is to minimize the error e_M between the ground-truth location facilities and the recommended location results returned. Supposing that the overall parameters are denoted as $\theta_M(\alpha_\vartheta, \beta_\vartheta, \gamma_\vartheta)$, then the corresponding optimization function is defined as:

$$\min_{\theta_M} \sum_{i=1}^{k} d_E^2(l_{pi}, l_{ri}) + \gamma_e \|\theta_M\|^2 \tag{6}$$

where γ_e is the regularization parameter and set equal to 10^{-8} as demonstrated in [2], l_{pi} is the predicted location and l_{ri} is the corresponding ground truth location, $d_E(l_{pi}, l_{ri})$ indicates the Euclidean distance between l_{pi} and l_{ri}.

4.4   Location Recommending

In querying, the tree levels of CLTC-forest are marked in a top-down fashion firstly, we and assume that the highest tree level is $\hbar$, where $\hbar < \lg(\lfloor m \rfloor + 1)$. All of the root nodes of CLTC-trees are marked with $ls_\hbar$, and the children of root nodes (in $ls_\hbar$, if have) are marked as $ls_{\hbar-1}$, by repeating the process until there is no node left to be marked (until to ls_1), then the entire tree level marks with the corresponding tree nodes are inserted into a set LS. Within the specified k, the mark of tree level is selected while $|ls_i| = k$, if exists, the k candidate locations in the nodes of ls_i are the querying results. However, if the mark does not exist, two marks of ls_i and ls_{i-1} are chosen where $|ls_i| < k < |ls_{i-1}|$. Subsequently, by utilizing a simple greedy manner, the k distinct candidate locations, which boast the maximum total service utilities, are selected to the recommendation results from the nodes of ls_i and ls_{i-1}. It is noticed that node in ls_i and two of its children nodes in ls_{i-1} in one CLTC-tree share one label, so they could not be chosen into the result together.

5 Experimental Evaluation

5.1 Datasets

The most widely used urban datasets of Beijing and Shanghai are employed in this study, where the intersections of road network are utilized to represent candidate locations. To simulate and generate customers with the corresponding trajectories, algorithm of discovering *Popular Routes* in [8] has been adopted with two real urban datasets. We extract the customer check-in and following data from Sina WeiBo[1], where there is a timestamp in each check-in point that is accurate to seconds, as well as a part of the trajectory traces of automobiles are collected. The social relationships are also extracted from Sina WeiBo, where two customers who are following with each other show that they are close friends. The statistics of datasets are listed in Table 2.

Table 2. Statistics of the datasets.

Categories	Beijing	Shanghai
# of intersections	171,186	333,766
# of road segments	226,237	440,922
# of customers (trajectories)	412,032	230,303
# of candidate locations	171,186	333,766
# of social relationships	26,139,861	13,687,459
# of CR	300	313
# of VCS	854	868

Two categories of popular existed facilities are prepared for model training and model testing in two cities, which are chained restaurants (CR) including KFC, MacDonald and Pazza Hut, as well as fast vehicle charging stations (VCS). The number of two existed facilities is also presented in Table 2 respectively. Furthermore, their geographic coordinates are obtained from AutoNavi[2].

5.2 Evaluation Plans

Competitive Approaches. To the best of our knowledge, no existing studies have been committed to the top-k candidate recommendation by exploiting customer trajectories and social relationships in a city-scale road network so far, therefore, we compare our model with a series of competitive methods, which are k-Medoids, SERM [9] and NetClus [6] by means of slight modifications.

Evaluation Metrics. A couple of the metrics of *Precision* and *Root Mean Square Error* (RMSE) are designed carefully in effectiveness evaluating. On one

---

[1] https://www.weibo.com/.
[2] https://www.amap.com/.

hand, we suppose that $\Im$ is the top-k querying results obtained from testing data, as well as L' is the corresponding existed facilities with the same category, then the precision is given as $\mathrm{P}_k = \frac{\sum_{i=1}^{k} hit(\Im_i, L'_i)}{k}$, where $hit(\Im_i, L'_i) = 1$ indicates there is a corresponding facility L_i^{ϑ} that satisfies $d_E(\Im_i, L'_i) \leq \tau_E, \Im_i \in \Im$ and $L'_i \in L^{\vartheta}$. On the other hand, RMSE is also adopted to measure the deviations between the recommended locations and the ground-truth facilities, the definition of RMSE is $RMSE_k = \sqrt{\frac{\sum_{i=1}^{k} \min{(d_E(\Im_i, L'_i))^2}}{k}}$. Note that each pair of $\Im_i$ and L'_i is evaluated only once in two metrics.

Basic Settings. The corresponding datasets are divided into training part and testing part, which consist of 70% and 30% of the whole datasets selected randomly, each experiment is evaluated by 10 times and the average results are returned. The default values of $\alpha_{\vartheta}, \beta_{\vartheta}$ and γ_{ϑ} are all equal to $1/3$ at the beginning of model learning, λ_s and λ_t are set equal to 1. Meanwhile, τ_E is set to 200 metres, k is initialized as 20, 50, 80, and 100 separately. During the method practices, the multi-process programming is utilized to accelerate the whole evaluations while a total of 20 CPU cores are handled.

5.3   Experimental Results

Figure 4 and Fig. 5 have illustrated the precision and RMSE of varying k in two facility categories that are operated on Beijing and Shanghai respectively. It can be seen that the proposed MITLR significantly outperforms the other methods

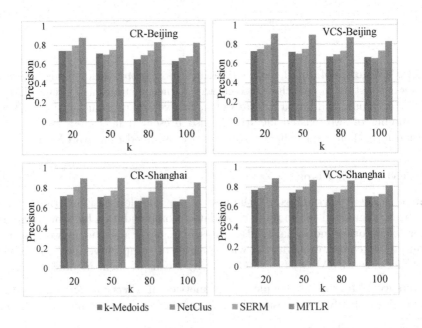

Fig. 4. Performance in terms of precision

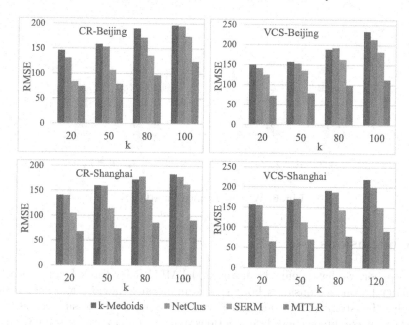

Fig. 5. Performance in terms of RMSE

under all circumstances in precision, the reasons would be analyzed in several important aspects, as we not only consider the effects of customer historical trajectories on facility placing, but also take advantage of the friend relationship based reciprocal recommendation, besides, the serious acknowledgements of service capacity improves the accuracies of prediction as well. Furthermore, the precision declines softly along with the raise of k, for the larger value of k, the hitting accuracy will experience diminishing returns.

Subsequently, we can see that our proposed model has better achievements compared with all competitors in RMSE, and the results has a reverse manner comparing with the precision, the reason could by analyzed from their definitions directly. In other words, if the value of precision is larger, the candidate locations recommended will be better represented by the corresponding facilities in road network, it also demonstrates that the larger RMSE will result in a worse performance on predicting contrarily.

We further investigate three feature weights in Fig. 6, when referring to $\alpha_\vartheta, \beta_\vartheta$ and γ_ϑ, the characteristic of social relevance is a principal factor on the evaluation of service utility especially in VCS, this is because that the majority of customers are more prone to be influenced by red envelopes or positive comments posted by their close friends when they are going to have consumptions at VCS, and vice versa.

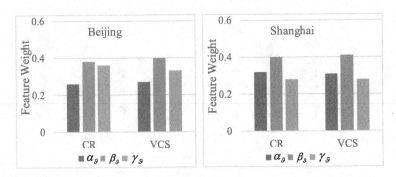

Fig. 6. Weights learned by MITLR

6   Conclusions

In this paper, we have defined a novel model MITLR for top-k facility location recommendation, it considers both spatio-temporal behaviors and social relationships of customers. In order to achieve effective query processing, CLTC-tree (forest) that combines candidate locations and customers are presented, and a query algorithm is also examined to obtain the results. Finally, extensive experiments with real datasets are performed to offer insights into the effectiveness.

References

1. Bao, J., He, T., Ruan, S., Li, Y., Zheng, Y.: Planning bike lanes based on sharing-bikes' trajectories. In: Proceedings of the 23rd ACM SIGKDD International Conference on Knowledge Discovery and Data Mining, pp. 1377–1386. ACM (2017)
2. Karamshuk, D., Noulas, A., Scellato, S., Nicosia, V., Mascolo, C.: Geo-spotting: mining online location-based services for optimal retail store placement. In: Proceedings of the 19th ACM SIGKDD International Conference on Knowledge Discovery and Data Mining, pp. 793–801. ACM (2013)
3. Li, X., Čeikute, V., Jensen, C.S., Tan, K.L.: Trajectory based optimal segment computation in road network databases. In: Proceedings of the 21st ACM SIGSPATIAL International Conference on Advances in Geographic Information Systems, pp. 396–399. ACM (2013)
4. Li, Y., Bao, J., Li, Y., Wu, Y., Gong, Z., Zheng, Y.: Mining the most influential k-location set from massive trajectories. In: Proceedings of the 24th ACM SIGSPATIAL International Conference on Advances in Geographic Information Systems, pp. 1–4. ACM (2016)
5. Mitra, S., Saraf, P., Bhattacharya, A.: Tips: mining top-k locations to minimize user-inconvenience for trajectory-aware services. IEEE Trans. Knowl. Data Eng. (2019)
6. Mitra, S., Saraf, P., Sharma, R., Bhattacharya, A., Ranuy, S., Bhandari, H.: NetClus: a scalable framework for locating top-k sites for placement of trajectory-aware services. In: 2017 IEEE 33rd International Conference on Data Engineering (ICDE), pp. 87–90. IEEE (2017)

7. Sun, Y., Huang, J., Chen, Y., Zhang, R., Du, X.: Location selection for utility maximization with capacity constraints. In: Proceedings of the 21st ACM International Conference on Information and Knowledge Management, pp. 2154–2158. ACM (2012)
8. Wei, L.Y., Zheng, Y., Peng, W.C.: Constructing popular routes from uncertain trajectories. In: Proceedings of the 18th ACM SIGKDD International Conference on Knowledge Discovery and Data Mining, pp. 195–203. ACM (2012)
9. Yao, D., Zhang, C., Huang, J., Bi, J.: SERM: a recurrent model for next location prediction in semantic trajectories. In: Proceedings of the 2017 ACM on Conference on Information and Knowledge Management, pp. 2411–2414. ACM (2017)
10. Yin, H., Cui, B., Chen, L., Hu, Z., Zhang, C.: Modeling location-based user rating profiles for personalized recommendation. ACM Trans. Knowl. Discov. Data (TKDD) **9**(3), 1–41 (2015)
11. Yin, M., Li, Q., Liu, Y.: Study on herd behavior and its influence on consumer impulse buying behavior during online shopping festival: an empirical study based on social influence theory and herd behavior. China Bus. Mark. **33**(8), 99–107 (2019)

Pairwise-Based Hierarchical Gating Networks for Sequential Recommendation

Kexin Huang[1], Ye Du[1], Li Li[1(✉)], Jun Shen[2], and Geng Sun[2]

[1] School of Computer and Information Science, Southwest University,
Chongqing, China
{huangkexin,duye99}@email.swu.edu.cn, lily@swu.edu.cn
[2] University of Wollongong, Wollongong, Australia
{jshen,jsun}@uow.edu.au

Abstract. The sequential pattern behind users' behaviors indicates the importance of exploring the transition relationships among adjacent items in next-item recommendation task. Most existing methods based on Markov Chains or deep learning architecture have demonstrated their superiority in sequential recommendation scenario, but they have not been well-studied at a range of problems: First, the influence strength of items that the user just access might be different since not all items are equally important for modeling user's preferences. Second, the user might assign various interests to certain parts of items, as what often attracts users is a specific feature or aspect of an item. Third, many methods ignore the complex item relations in user's previous actions. In this paper, we present a novel recommendation approach with gating mechanism and encoding module to address above problems. Specifically, the pair-wise encoding layer is first introduced to build 3-way tensor for modeling the relationships among items in user interact histories. We also apply two gating layers to filter useful information and capture user's short-term preference from aspect-level and item-level. We also follow the similar sprits to model user's long-term preference by integrating user latent embeddings. Empirical results on three public datasets show that our method achieves effective improvements over the state-of-the-art sequence-based models.

Keywords: Sequential recommendation · Collaborative filtering · Gating mechanism · Pairwise encoding

1 Introduction

Recommender system has become a popular way to alleviate information overload issue. Among the various recommendation methods, Collaborative filtering (CF) [1] is a most essential model to capture users' general preferences owing to its effectiveness and interpretability, but it fails to model the sequential dynamics in recommendation task. Leveraging users' behavior history sequences instead of ratings to predict their future behaviors has become increasingly popular in

© Springer Nature Switzerland AG 2020
G. Li et al. (Eds.): KSEM 2020, LNAI 12275, pp. 64–75, 2020.
https://doi.org/10.1007/978-3-030-55393-7_6

recent years [2,3]. This is because users access items in chronological order and the items that user will consume may be related to the items that he has just visited. To facilitates this task, a line of works convert users' historical actions into an action sequence order by operating timestamps [3–5].

Different from the conventional recommendation models, sequential recommendation methods usually based on Markov Chains (MCs) [2], which is a classic model that assume next action depends on previous actions and model the transition relationships between adjacent items for predicting user preferences. Although MCs-based models perform well in sparse scenarios, yet they cannot capture complex sequential dynamics. Another line of researches make use of deep neural networks (DNNs) to model both personalization and transitions based on item sequences, which outperform the MCs-based baselines. For example, Convolutional Neural Networks (CNNs) [6,7] have been introduced to capture user's short-term preferences, it adopts convolutional feature detectors to extract local patterns from item sequences by various sliding windows.

However, common neural methods regard sequence as a whole to calculate the impact on next item, which are difficult to gather relation features of different positions. Since a user may focus on one specific aspect of an item and pay different attention to various aspects of the same item. Furthermore, the item influence strength based on users' behaviors is diverse and dynamic, yet DNNs-based models fail to consider the specific aspect or feature of different items and ignore the item importance based on users' sequential actions.

In this paper, we also take user's sequences of recent interactions into account for sequential recommendation and follow the similar spirits [8] to apply Gate Convolutional Networks for modeling sequential dynamics for better recommendation. Specifically, we propose Hierarchical Pairwise Gating Model (HPGM) to effectively capture the sequential pattern then applying two gate linear units to model transition relationships and represent high-level features. For better relation extraction, we further devise pairwise encoding layer with concatenation which learn more meaningful and comprehensive representation of item sequence. We also conduct a series of experiments on serval benchmarks datasets. More importantly, experimental results show our model achieves better improvement over strong baselines.

2   Related Work

2.1   General Recommendation

General recommendation always focus on user's long-term and static preferences by modeling user's explicit feedbacks (e.g., ratings). Matrix Factorization (MF) [9] is the basis of many state-of-the-art methods such as [10], it seeks to uncover latent factors for representing users' preferences and items' properties from user-item rating matrix through the inner product operation. MF relies on user's explicit feedbacks but user's preferences also can be mined from implicit feedbacks (e.g., clicks, purchases, comments). The pair-wise methods [11] based on MF have been proposed to mining users' implicit actions and assume that user's observed feedbacks should only be 'more preferable' than unobserved feedbacks then optimize the pairwise rankings of pairs.

Neighborhood-based and model-based methods also have been extended to tackle implicit feedbacks, a line of works are based on Item Similarity Matrix such as SLIM [12], FISM [13]. These methods calculate preference scores for a new item by measuring its similarities with previous items. Recently, various deep learning techniques have been introduced to extract item features from the description of items such as images and texts by neural network in recommendation task [14].

2.2 Sequential Recommendation

For sequential recommendation task, Markov Chains (MCs) is an effective method to model sequential dynamics from successive items. Factorized Personalized Markov Chains (FPMC) [2] is a classic sequential recommendation model that combines MF and factorized MCs to model user preference and sequential patterns simultaneously. Hierarchical Representation Model (HRM) [4] extends the FPMC by introducing aggregation operations like max-pooling to model more complex. He et al's method (TransRec) [3] models the third-order interactions between sequential items by combining with metric embedding approaches.

Besides, another line of works model user sequences via deep learning techniques and show effective performance in sequential recommendation task [15]. Convolutional Sequence embedding (Caser) [6] captures sequential patterns and transitions from previous item sequence by convolutional operation with various filters. Another popular method RNN is also used to model user's sequential interactions because RNN is good at capturing transition patterns in sequence [16,17]. Attention Mechanisms have been incorporated into next item recommendation to model complex transitions for better recommendation [18]. Self-attention based sequential model (SASRec) [19] relies on Transformer instead of any recurrent and convolutional operations, it models the entire user sequence to capture user's long-term and short-term preferences then make predictions on few actions.

3 Proposed Methodology

The objective of our task is to predict next behaviors of users depending on previous chronological actions. We use $\mathcal{U}$ and $\mathcal{I}$ to present user set and item set (respectively) in sequential recommendation scenario. Given user u's action sequence $\mathcal{S}^u = (\mathcal{S}_1^u, \mathcal{S}_2^u, \cdots, \mathcal{S}_{|\mathcal{S}^u|}^u)$, where $\mathcal{S}_t^u \in \mathcal{I}$ denotes user u ever interacted with the item at time step t. To train the network, we extract every L successive items $(\mathcal{S}_1^u, \mathcal{S}_2^u, \cdots, \mathcal{S}_L^u)$ of each user $u \in \mathcal{U}$ as the input, its expected output as the next T items from the same sequence: $(\mathcal{S}_{L+1}^u, \mathcal{S}_{L+2}^u, \cdots, \mathcal{S}_{L+T}^u)$. In this section, we introduce our model via an embedding layer, pairwise encoding layer, hierarchical gating layer and prediction layer. The detailed network architecture is showed in Fig. 1.

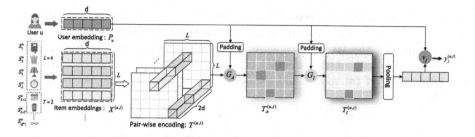

Fig. 1. The detail network architecture of **HPGM**. Previous successive item embeddings are transmitted into the pairwise encoding layer, the output and user embedding after pooling are passed into the hierarchical gating layer (G_A, G_I) and then predict the next item by combining the original user embedding and the sequence embedding after pooling operation.

3.1 Embedding Layer

Let $E_i \in \mathbb{R}^d$ be the item embedding corresponding to the i-th item in the item sequence, where d is the latent dimensionality. The embedding look-up operation retrieves previous L items' embeddings and stack them together to form the input matrix $X^{(u,t)} \in \mathbb{R}^{L \times d}$ for user u at time step t. Along with item embedding, we also represent user features in latent space with user embedding $P_u \in \mathbb{R}^d$.

3.2 Pairwise Encoding Layer

In order to capture intricate item relations among a specific item sequence and improve the flexibility of the model, we use pair-wise encoding layer to build a sequential tensor $T^{(u,t)} \in \mathbb{R}^{L \times L \times 2d}$ to store various item relationships. $T^{(u,t)}$ is composed by the item pair(i, j) of item subsequence, which concatenate the embedding of item i and j. The encoded 3-way tensor is similar to "image feature map" in the CNN-based model for computer vision tasks, so $T^{(u,t)}$ can replace the sequential item embedding $X^{(u,t)}$ as the input to downstream layers. Note we padding the user embedding with "1" and generate a user tensor $\hat{P}_u$ with same dimensions of $T^{(u,t)}$ for feeding the user embedding into the subsequent layers.

3.3 Hierarchical Gating Layer

Original GLU integrate convolution operation and simplified gating mechanism to make predictions [20], motivated by the gated linear unit (GLU) utilized on recommendation task [8], we also adopt similar spirits to model sequence dynamics. GLU control what information should be propagated for predicting next item, so we can select specific aspect/feature of item and particular item that is related to future items.

Aspect-Level Gating Layer. A user generally decides whether to interact with the item by looking for the specific attractive aspects of the item. Therefore, we modify the GLU to capture sequence pattern based on user-specific preference. The convolution operation is replaced by inner product to reduce the parameters of the model and the user's aspect-level interest can be generated by:

$$T_A^{(u,t)} = T^{(u,t)} * \sigma(W_1 \cdot T^{(u,t)} + W_2 \cdot \hat{P}_u) \tag{1}$$

where $*$ is the element-wise multiplication, $\cdot$ represents inner product operation, $W_1, W_2 \in \mathbb{R}^{1 \times 2d \times 2d}$ and $b \in \mathbb{R}^{1 \times 1 \times 2d}$ are the corresponding 3-way weight terms and the bias term, $\sigma(\cdot)$ denotes the sigmoid function. And the aspect-specific information can be propagated to the next layer by the aspect-level gating layer.

Item-Level Gating Layer. Users will assign higher weight attention to a particular item in real life. Exiting models ignore the item importance in modeling users' short-term preferences and attention mechanism is a success way to capture item-level interest. In this paper, we also adopt an item-level gating layer to achieve the same or even better performance. And the results after this layer can be calculated as:

$$T_I^{(u,t)} = T_I^{(u,t)} * \sigma(W_3 \cdot T_A^{(u,t)} + W_4 \cdot \hat{P}_u) \tag{2}$$

where $W_3 \in \mathbb{R}^{1 \times 1 \times 2d}$, $W_4 \in \mathbb{R}^{1 \times L \times 2d}$ are learnable parameters. By performing aspect-level and item-level gating module operations on item embedding, our model selects informational items and their specific aspects, meanwhile eliminates irrelevant features and items. Then we apply average pooling on the sequence embedding after item-level gating layer to make aggregation by accumulating the informative parts:

$$\hat{E}^{(u,t)} = average\left\{T_I^{(u,t)}\right\} \tag{3}$$

3.4   Prediction Layer

After computing user's short-term preference by preceding operation, we induce an implicit user embedding P_u to capture user's general preferences then we employ the conventional latent factor model (matrix factorization) to generate prediction score as follows:

$$y_j^{(u,t)} = \hat{E}^{(u,t)} v_j + P_u v_j \tag{4}$$

where $y_j^{(u,t)}$ can be interpreted as the probability of how likely user u will interact with item j at time step t and v_j denotes the item embedding. Note we adopt the full-connected layer to reduce the high-dimension before prediction.

3.5   Network Training

To train the network, we adopt the binary Bayesian Personalized Ranking loss [11] as the objective function:

$$L = \sum_{(u,i,j)\in\mathcal{D}} -ln\sigma(y_i^u - y_j^u) + \lambda_\Theta(\|\Theta\|^2) \tag{5}$$

where $\Theta = \{X, P_u, W_1, W_2, W_3, W_4, b\}$ denotes the model parameters, which are learned by minimizing the objective function on training set. Note we use some tricks to learn these 3-way parameters by PyTorch and their dimensions are derived from experiments. λ_Θ is the regularization parameter and $\sigma(x) = 1/(1 + e^{-x})$, $\mathcal{D}$ is the set of training triplets:

$$\left\{(u, i, j) \,|\, u \in \mathcal{U} \wedge i \in \mathcal{I}_u^+ \wedge j \in \mathcal{I}_u^- \right\} \tag{6}$$

we also randomly generate one negative item j from a candidate set of each user in each time step t, the candidate set of each user is defined by $\{j \in \mathcal{I}^- | \mathcal{I}^- = \mathcal{I} - \mathcal{S}^u\}$ and the Adam Optimizer [21] is used to optimize the network.

4   Experiments

In order to evaluate our model, we experiment with various baselines on three large-scale real-world datasets. The datasets cover different domains and sparsity. All the datasets and code we used are available online.

4.1   Datasets

We evaluate our model on three real-world dataset and these datasets vary greatly in domain, variability, sparsity and platform:

Amazon[1]. This dataset is collected from *Amazon.com* that contains large corpora of products ratings, reviews, timestamps as well as multiple types of related items. In this work, we choose the "CDs" category to evaluate the quantitative performance of the proposed model.

MovieLens[2]. MovieLens is created by the Grouplens research group from *Movielen.com*, which allows users to submit ratings and reviews for movies they have watched.

GoodReads[3]. A new dataset introduced in [22], comprising a large number of users, books and reviews with various genres. This dataset is crawled from

---

[1] http://jmcauley.ucsd.edu/data/amazon/.
[2] https://grouplens.org/datasets/movielens/.
[3] https://sites.google.com/eng.ucsd.edu/ucsdbookgraph/home.

Goodreads, a large online book review website. In this paper, we adopt the genres of Comics to evaluate the proposed model.

For each of the above datasets, we follow the same preprocessing procedure from [6]. We converted star-ratings to implicit feedback and use timestamps to determine the sequence order of users' actions. In addition, we discard users and items with less than 5 related actions. We also partition the sequence S^u for each user u into three parts: (1) the first 70% of actions in S^u as the training set. (2) the second 10% of actions for validation. (3) the remaining 20% of actions are used as a test set to evaluate performance of the model. Statistics of each dataset after pre-processing are shown in Table 1.

Table 1. Dataset statistics.

Dataset	#users	#items	#actions	Avg. #actions /user	Avg. #actions /item
Amazon CDs	17.0K	35.1K	0.47M	27.69	13.44
MovieLens	129.9K	13.6K	9.9M	76.43	726.89
GoodReads Comics	34.4K	33.1K	2.4M	70.00	72.80

4.2 Comparison Methods

We contain three groups of recommendation baselines to show the effective of HPGM. The first group are general recommendation models which only take user feedbacks into account instead of considering user's sequential behaviors.

– **PopRec**: PopRec ranks items according to the order of their overall popularity which decided by the number of the interactions.
– **Bayesian Personalized Ranking (BPR-MF)** [11]: This model combines matrix factorization and learning personalized ranking from implicit feedback by Bayesian Personalized Ranking.

The next group of the methods models the sequence of user actions to explore user's preference in sequential recommendation:

– **Factorized Markov Chains (FMC)** [2]: FMC factorizes the first-order Markov transition matrix to capture 'global' sequential pattern but it ignores the personalized user interaction.
– **Factorized Personalized Markov Chains (FPMC)** [2]: FPMC combines the matrix factorization and factorized Markov Chains as its recommender and it captures item-to-item transition and users' long-term preference simultaneously.

The final group includes methods which consider serval previously visited items to make predictions by deep-learning technique.

- **Convolutional Sequence Embeddings (Caser)** [6]: Caser captures sequential dynamic by convolutional operations on embedding matrix with length L.
- **GRU4Rec** [23]: This model treats users' action sequence as a session and utilizes RNNs to model user feedback sequences for session-based recommendation.
- **GRU4Rec$^+$** [24]: GRU4Rec$^+$ extends the GRU4Rec method by applying a different loss function and sampling strategy and achieve great sequential recommendation performance.

4.3 Evaluation Metrics

In order to evaluate performance of sequential recommendation, we adopt two common Top-N metrics Recall@N and NDCG@N. Recall@N measure Top-N recommendation performance by counting the proportion of times that the ground-truth next item is among the top N items and NDCG@N is a position-aware metric that distribute high weights on the higher positions. Here N is set from $\{5, 10, 15, 20\}$.

4.4 Implementation Details

The parameters of baselines are initialized as corresponding number in original paper. The latent dimension d is tested in $\{10, 20, 30, 40, 50\}$ and the learning rate for all models are tuned amongst $\{0.001, 0.005, 0.01, 0.02, 0.05\}$. We tune the batch size in $\{16, 32, 64, 128\}$ and margin λ_Θ is tuned in $\{0.001, 0.005, 0.01, 0.02\}$. After tuning processing on validation set, the learning rate is set to 0.001, $d = 50$, $\lambda_\Theta = 0.001$ and the batch size is 256. We also follow the same setting in: the Markov order L is 5 and predict the future $T = 3$ items. All experiments are implemented with PyTorch[4].

4.5 Recommendation Performance

Overall performance results of HPGM and baselines are summarized in Table 2 and Fig. 2, which clearly illustrate that our model obtains promising performance in terms of Recall and NDCG for all reported values in sequential recommendation task. We can gain the following observations:

The performance of BPR-MF is better than PopRec but is not as good as FMC, which demonstrates that local adjacent sequential information plays an vital role under the typical sequential recommendation setting. Compared to conventional sequential-based models (FMC and FPMC), we find that item-to-item relations is necessary to comprehend user's sequential actions. Furthermore, the performance results show that our proposed model can effectively capture item relationships and sequential dynamics in real-world datasets.

---

[4] https://pytorch.org/.

Table 2. Performance comparison with baselines on three datasets and the best results highlight in bold (Higher is better). The improvement is calculated by the best performance of baselines and our method.

Dataset	Amazon-CDs		MovieLens		GoodReads-Comics	
Metrics	Recall@10	NDCG@10	Recall@10	NDCG@10	Recall@10	NDCG@10
PopRec	0.0181	0.0095	0.0560	0.0487	0.0426	0.0503
BPR-MF	0.0233	0.0145	0.0774	0.0685	0.0688	0.0613
FMC	0.0240	0.0149	0.0819	0.0724	0.0745	0.778
FPMC	0.0245	0.0151	0.0847	0.0751	0.0813	0.0833
GRU4Rec	0.0302	0.0164	0.0924	0.0815	0.0958	0.0912
GRU4Rec+	0.0336	0.0171	0.1004	0.0946	0.1088	0.1128
Caser	0.0297	0.0163	0.1139	0.1016	0.1273	0.1329
HEPG	**0.0347**	**0.0181**	**0.1150**	**0.1087**	**0.1320**	**0.1430**
%Improv.	3.36	6.08	0.97	7.01	3.67	7.56

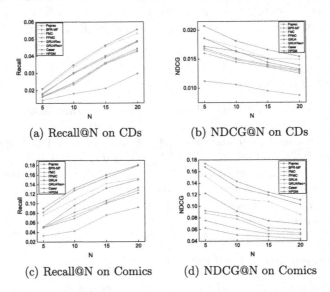

(a) Recall@N on CDs   (b) NDCG@N on CDs

(c) Recall@N on Comics   (d) NDCG@N on Comics

Fig. 2. Ranking performance (NDCG and Recall) with baselines on Amazon-CDs and GoodReads-Comics.

Another observation is sequential methods GRU4Rec and Caser based on neural network achieve better performance than conventional sequential recommendation model such as FPMC. We can conclude that neural network is suitable to model the complex transition between previous feedbacks and future behaviors of the user. Since baseline models have a lot of limitation, Caser only considers group-level influence by adopting CNN with horizontal and vertical filters but ignores the specific aspect-level influence of successive items.

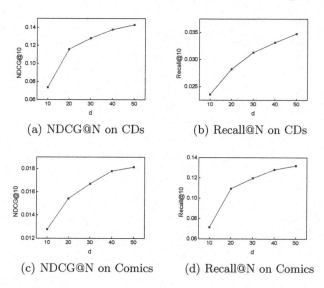

(a) NDCG@N on CDs   (b) Recall@N on CDs

(c) NDCG@N on Comics   (d) Recall@N on Comics

Fig. 3. Performance change with different dimension of embeddings d on Amazon-CDs and GoodReads-Comics.

In a word, our method can beat baselines with ground-truth ranking and shows effectiveness of our model on item relation, sequential dynamics and user's general preferences.

4.6 Influence of Hyper-parameters

In this subsection, we also analyze the effect of two key hyper-parameters: the latent dimensionality d and the length of successive items L. Figure 3 shows the effect of dimension d by evaluating with NDCG@10 and Recall@10 of all methods varying from 10 to 50 on Amazons-CDs and GoodReads-Comics. We also can conclude that our model typically benefits from lager dimension of item embeddings. Since small latent dimension cannot express the latent feature completely and with the increase of d, the model can achieve better performance on the real-world datasets.

Previous analysis can demonstrate that modeling sequence patterns are crucial for next-item recommendation, hence the length of sequence is a significant factor to determine model's performance. We also study the influence of different length of successive items L and Fig. 4 shows that the model does not consistently benefit from increasing L and a large L may lead to worse results since higher L may introduce more useless information. In most cases, $L = 5$ achieve better performance on the two datasets.

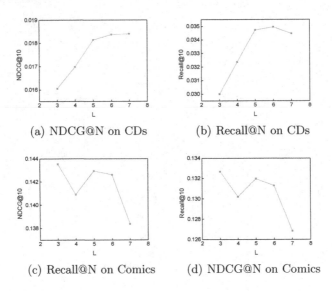

(a) NDCG@N on CDs   (b) Recall@N on CDs

(c) Recall@N on Comics   (d) NDCG@N on Comics

Fig. 4. The performance of HPGM with varying L on Amazon-CDs and GoodReads-Comics.

5   Conclusion

In this paper, we present a novel recommendation approach with gating mechanism to learn personalized user and item representations from user's sequential actions and generating prediction score by aggregating user's long-term and short-term preferences. Specifically, in order to model item relations in user behaviors, we apply pair-wise encoding layer to encode a sequence of item embedding into a pairwise tensor. Moreover, we build a hierarchical gating layer to model aspect-level and item-level influence among items to capture latent preferences of the user. We also conduct extensive experiments on multiple large-scale datasets and the empirical results show that our model outperforms state-of-the-art baselines. In the future, we plan to extend the model by exploring sequential patterns and make predictions from various types of context information.

Acknowledgement. This research was supported by NSFC (Grants No. 61877051), and Natural Science Foundation Project of CQ, China (Grants No. cstc2018jscx-msyb1042, and cstc2018jscx-msybX0273). Li Li is the corresponding author for the paper.

References

1. Ekstrand, M.D., Riedl, J.T., Konstan, J.A., et al.: Collaborative filtering recommender systems. Found. Trends® Hum.-Comput. Interact. **4**(2), 81–173 (2011)
2. Rendle, S., Freudenthaler, C., Schmidt-Thieme, L.: Factorizing personalized Markov chains for next-basket recommendation. In: Proceedings of the WWW, pp. 811–820. ACM (2010)

3. He, R., Kang, W.-C., McAuley, J.: Translation-based recommendation. In: Proceedings of the RecSys, pp. 161–169. ACM (2017)
4. Wang, P., Guo, J., Lan, Y., Xu, J., Wan, S., Cheng, X.: Learning hierarchical representation model for next basket recommendation. In: Proceedings of the SIGIR, pp. 403–412. ACM (2015)
5. Yu, L., Zhang, C., Liang, S., Zhang, X.: Multi-order attentive ranking model for sequential recommendation. In: Proceedings of the AAAI (2019)
6. Tang, J., Wang, K.: Personalized top-n sequential recommendation via convolutional sequence embedding. In: Proceedings of the WSDM, pp. 565–573. ACM (2018)
7. Yuan, F., Karatzoglou, A., Arapakis, I., Jose, J.M., He, X.: A simple convolutional generative network for next item recommendation. In: Proceedings of the WSDM, pp. 582–590. ACM (2019)
8. Ma, C., Kang, P., Liu, X.: Hierarchical gating networks for sequential recommendation. In: Proceedings of the KDD, pp. 825–833 (2019)
9. Koren, Y., Bell, R., Volinsky, C.: Matrix factorization techniques for recommender systems. Computer 8, 30–37 (2009)
10. He, X., Liao, L., Zhang, H., Nie, L., Hu, X., Chua, T.-S.: Neural collaborative filtering. In: Proceedings of the WWW, pp. 173–182. ACM (2017)
11. Rendle, S., Freudenthaler, C., Gantner, Z.: BPR: Bayesian personalized ranking from implicit feedback. In: Proceedings of the Twenty-Fifth Conference on Uncertainty in Artificial Intelligence, pp. 452–461. AUAI Press (2009)
12. Ning, X., Karypis, G.: Slim: sparse linear methods for top-n recommender systems. In: Proceedings of the ICDM, pp. 497–506. IEEE (2011)
13. Kabbur, S., Ning, X., Karypis, G.: FISM: factored item similarity models for top-n recommender systems. In: Proceedings of the KDD, pp. 659–667. ACM (2013)
14. Kang, W.-C., Fang, C., Wang, Z., McAuley, J.: Visually-aware fashion recommendation and design with generative image models. In: Proceedings of the ICDM, pp. 207–216. IEEE (2017)
15. Yan, A., Cheng, S., Kang, W.-C., Wan, M., McAuley, J.: CosRec: 2D convolutional neural networks for sequential recommendation. arXiv preprint arXiv:1908.09972 (2019)
16. Xu, C., Zhao, P., Liu, Y.: Recurrent convolutional neural network for sequential recommendation. In: Proceedings of the WWW, pp. 3398–3404. ACM (2019)
17. Yu, Z., Lian, J., Mahmoody, A., Liu, G., Xie, X.: Adaptive user modeling with long and short-term preferences for personalized recommendation. In: Proceedings of the IJCAI, pp. 4213–4219, July 2019
18. Zhou, C., et al.: ATRank: an attention-based user behavior modeling framework for recommendation. In: Proceedings of the AAAI (2018)
19. Kang, W.-C., McAuley, J.: Self-attentive sequential recommendation. In: Proceedings of the ICDM, pp. 197–206. IEEE (2018)
20. Dauphin, Y.N., Fan, A., Auli, M., Grangier, D.: Language modeling with gated convolutional networks. In: Proceedings of the ICML, vol. 70, pp. 933–941. JMLR.org (2017)
21. Kingma, D.P., Ba, J.: Adam: a method for stochastic optimization. In: Proceedings of the ICLR (2015)
22. Wan, M., McAuley, J.: Item recommendation on monotonic behavior chains. In: Proceedings of the RecSys, pp. 86–94. ACM (2018)
23. Hidasi, B., Karatzoglou, A., Baltrunas, L., Tikk, D.: Session-based recommendations with recurrent neural networks. arXiv preprint arXiv:1511.06939 (2015)
24. Hidasi, B., Karatzoglou, A.: Recurrent neural networks with top-k gains for session-based recommendations. In: Proceedings of the CIKM, pp. 843–852. ACM (2018)

Time-Aware Attentive Neural Network for News Recommendation with Long- and Short-Term User Representation

Yitong Pang[1], Yiming Zhang[1], Jianing Tong[2], and Zhihua Wei[1(✉)]

[1] Department of Computer Science and Technology,
Tongji University, Shanghai, China
{1930796,1652325,zhihua_wei}@tongji.edu.cn
[2] DellEmc, Shanghai, China
johnny_tong@dell.com

Abstract. News recommendation is very critical to help users quickly find news satisfying their preferences. Modeling user interests with accurate user representations is a challenging task in news recommendation. Existing methods usually utilize recurrent neural networks to capture the short-term user interests, and have achieved promising performance. However, existing methods ignore the user interest drifts caused by time interval in the short session. Thus they always assume the short-term user interests are stable, which might lead to suboptimal performance. To address this issue, we propose the novel model named Time-aware Attentive Neural Network with Long-term and Short-term User Representation (TANN). Specifically, to reduce the influence of interest drifts, we propose the Time-aware Self-Attention (T-SA) which considers the time interval information about user browsing history. We learn the short-term user representations from their recently browsing news through the T-SA. In addition, we learn more informative news representations from the historical readers and the contents of news articles. Moreover, we adopt the latent factor model to build the long-term user representations from the entire browsing history. We combine the short-term and long-term user representations to capture more accurate user interests. Extensive experiments on two public datasets show that our model outperforms several state-of-the-art methods.

Keywords: News recommendation · Self attention · Time-aware · Long-term interest · Short-time interest · Representation learning

1 Introduction

Nowadays, online news platforms have become popular with people to acquire daily information, such as MSN News and Google News. However, with the explosion of news contents and services, users are overwhelmed by tremendous news. News recommendation can find news that satisfies the personalized interests of

G. Li et al. (Eds.): KSEM 2020, LNAI 12275, pp. 76–87, 2020.
https://doi.org/10.1007/978-3-030-55393-7_7

users, and is an important method to alleviate the information overload [7]. Therefore, news recommendation has received the increasing attention on both academics fields and industry fields.

In news recommendation, the key task is learning the accurate user representations to reflect the user interests. Because of the uncertain user behavior and limited information, it is difficult to capture appropriate user interests. Traditional methods, like collaborative-filtering (CF) based methods ignore the sequence information about user browsing history. They can not learn the current interest of users exactly. Recently, some novel models based on deep learning were proposed for personalized news recommendation. Some deep learning based methods utilize the recurrent neural networks (RNN) and attention mechanism to capture the short-term user interests from recently viewed news [14,20,22,23]. Besides, some methods exploit to utilize the long-term and short-term interest of users together for more accurate representation [1,7]. For example, [1] proposed to learn long-term interests from user IDs and capture short-term user interests from recently viewed news by GRU model.

Although these deep learning based methods have achieved encouraging results, they still have two potential limitations. Firstly, they ignore the short-term user interest drifts, and consider that user preferences are stable in the short term. Short-term user interest is usually dynamic, but it is critical for making the recommendation decisions. Short-term user interests usually are dynamic but important to making news recommendation decisions. For example, the current interest of a user may have changed when he resumes browsing news after a short interval. Secondly, they generally recommend news by matching user interests with news content, but overlook the effective collaborative information. It is worth noting that users with similar interests may also read similar news, e.g., if Peter has similar browsing history to Bob and has browsed the news C, then Bob will probably also read C, even if the content of C is different from the news that Bob recently browsed. These mentioned methods fail to take into account this collaborative information.

For addressing above issues, this paper proposes a **T**ime-aware **A**ttentive **N**eural **N**etwork with long- and short-term user representation for news recommendation (TANN). In order to reduce the influence of user interest drifts, we propose the time-aware self-attention (T-SA) which considers the time interval information between two browsing records. We learn the short-term representations of users from recently browsed news via the T-SA. In addition, to integrate the collaborative information, we apply attention mechanism to learn informative news representations from the historical readers and the contents of news articles. Furthermore, we utilize latent factor model to model the long-term interests from entire browsing history of users, and combine the long-term interests with the short-term user interests for better representations. The experimental results on two real-world datasets show our method has achieved the promising performance and is superior to several state-of-the-art methods.

2   Related Work

Today, personalized news recommendations have received extensive attention from academia and industry, and have been extensively studied. A variety of news recommendation methods have been proposed, including conventional methods and methods based on deep learning [18,23].

Conventional methods include the CF-based methods, content-based methods and hybrid methods. CF-based methods assume that users with similar behavior will show similar preferences for news [3,6,7]. They usually learn the latent representations from entire historical interactions to get the general user interests. For example, latent factor model [9] directly models the user-item interaction with inner product. But CF-based methods have to face the problem of cold start and lack the ability to capture the short-term interests of users.

Content-based and hybrid recommendation can alleviate the cold-start problem. Content-based methods analyze the actual content or attributes of the news articles for recommendation Content-based methods analyze the content of historical news articles browsed by users for recommendation [12,16,17]. For example, [16] adopt vector space model similarity to evaluate the relevance of different news and recommend relevant news to users. Hybrid methods usually combine several different methods [10,11,13], such as SCENE [10] proposed a two-stage recommendation framework to blend the CF-based method and content-based method. However, these methods neglect the sequence information in user browsing history and can not effectively learn the short-term user interests. In addition, they fail to consider the contexts and semantic information in news which are important to learning representations of news and users.

Deep learning based methods are proposed for personalized news recommendation recently. These methods utilize the neural network models such as recurrent neural network and attention mechanisms, to learn short-term user interests from the sequence of the recently browsed news. Besides, these methods learn deep news features from news contents via neural networks [1,7,14,18,21–23]. For example, [14] adopt denoising auto-encoder to obtain effective news feature from news attributes, and learn users representations from browsed news via GRU network. Moreover, some works exploit to integrate the long-term and short-term interest of users for more accurate representations [1,7,24], such as [24] proposed to the time-LSTM which uses the time gate to control the impact of time, and introduces time interval information to obtain more reasonable user representations.

However, these deep learning based methods ignore the short-term user interest drifts and assume that user interests are stable in the short term. In addition, they usually take the idea of content-based methods, but ignore the effective collaborative information. Inspired by [2,24], we propose a time-aware attentive neural network, and consider the time interval information to alleviate the effects of user interest drifts. Furthermore, we exploit to learn the news representations from the content and the historical reader of news to integrate the collaborative information.

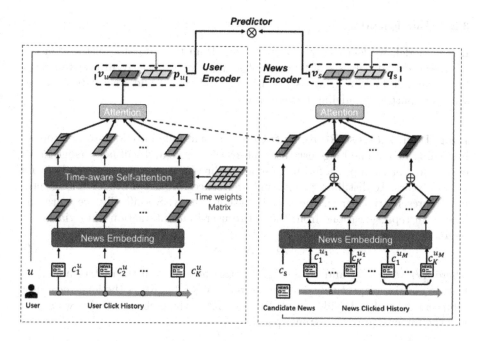

Fig. 1. The architecture of the proposed *TANN* for news recommendation.

3 Methodology

In this part, we present the TANN for news recommendation. The architecture of our proposed model is shown in Fig. 1. There are three major modules, i.e., the *User Encoder* with time-aware self-attention and latent factor model, the *News Encoder* with attention mechanism, and the *Predictor* to predict the probability that users will click on candidate news articles.

3.1 Problem Formulation

Assume that there are m users $U = \{u_1, u_2, ..., u_m\}$ and n news articles $V = \{v_1, v_2, ..., v_n\}$. According to the implicit feedback of users, the entire browsing data can be defined as matrix $\mathbf{M} \in \mathbb{R}^{m \times n}$, where $\mathbf{M}_{u,s} = 1$ indicates the user u browsed the news s, otherwise $\mathbf{M}_{u,s} = 0$. For each user, we chronologically organize the browsing history as a sequence of tuples $O^u = \{(v_j^u, t_j^u)\}_{j=1}^{l_u}$, where $t_1^u \leq t_2^u \leq ... \leq t_{l_u}^u$, v_j^u is the news in V, l_u indicates the number of articles viewed by user u, and (v_j^u, t_j^u) means that user u browsed news v_j^u at time t_j^u. In this paper, we exploit to build a prediction function $\hat{y}_{u,s} = F(u, s; \Theta) \in [0, 1]$ from the user-item matrix $\mathbf{M}$ and the click sequences O^u. Prediction function $F(u, s; \Theta)$ is used to calculate the probability that user u will click on a candidate news s.

3.2   User Encoder

In our model, we learn user representations from the browsing history via the *User Encoder*. It consists of two modules, namely, the long-term user representation module for capturing user consistent preferences and a short-term representation part for modeling temporal interests.

Long-Term Interest Representation. Users usually have long-term interests in reading news. The long-term user interest is stable and will not change much over time. For example, a football fan might browse many sports news about La Liga for years. To distill user consistent preferences, we exploit to learn long-term user representations from the user browsing history. Specifically, we learn the long-term representations from the entire user-item interaction data via latent factor model [9]:

$$\mathbf{M} = \mathbf{W}_U \cdot \mathbf{W}_V^T \tag{1}$$

where $\mathbf{W}_U \in \mathbb{R}^{m \times k}$ and $\mathbf{W}_V \in \mathbb{R}^{n \times k}$ are the user feature matrix and item feature matrix, respectively. We denote $\mathbf{p}_u = \mathbf{W}_U[u] \in \mathbb{R}^k$ as the long-term user representation of u. Besides, $\mathbf{q}_s = \mathbf{W}_V[s] \in \mathbb{R}^k$ is the embedding of news s, and k is the dimension of the latent space.

Short-Term Interest Representation. Short-term user interest reflects a common situation: users tend to be attracted to something temporarily. For example, the temporal interest of a user may have changed when he resumes browsing news after a short interval. To reduce the influence of interest drifts, we propose **T**ime-aware **S**elf-**A**ttention (T-SA) mechanism to learn the short-term user interests. T-SA considers two aspects of time factor: time order information and time interval information. Given the recent browsing history of user u: $C_u = \{(v_1, t_1), (v_2, t_2), ..., (v_K, t_K)\} \subseteq O^u$, where $t_i < t_{i+1}$, and the corresponding news embedding sequence $\{\mathbf{e}_1, ..., \mathbf{e}_K\}$ through the News Embedding Layer. We adopt the pre-trained word emebedding model to obtain the news embeddings.

For the time order information, we encode it into the news embedding. The dimension size of the news embedding is h, and the calculation of each dimension in the time order vector is as follows:

$$PE(pos, 2i) = sin(pos/10000^{2i/h}) \tag{2}$$

$$PE(pos, 2i + 1) = cos(pos/10000^{(2i+1)/h}) \tag{3}$$

where $pos \in 1, 2, ..., K$ is the time order. The time order code is PE_{pos}, and the new vector $\_\mathbf{e}_i$ which contains time order information as follows:

$$\_\mathbf{e}_i = \mathbf{e}_i + PE_i \tag{4}$$

For the time interval information, we obtain the time interval sequence $T_\Delta = \{\Delta_1, \Delta_2, ...\Delta_K\}$, where $\Delta_i = t_i - t_{i-1}$ and $\Delta_1 = 0$. Based on the attention mechanism, we propose a time weight matrix to control the attention weight between two news embeddings. Through the time weight, we can achieve that the bigger the time interval is, the smaller the attention weights between the two news are. Each embedding attention weight in the sequence is computed as follows:

$$\alpha_{i,j} = \frac{w_{i,j}^{\Delta} exp(-\mathbf{e}_i^T \mathbf{Q}^w \_ \mathbf{e}_j)}{\sum_{k=1}^{K} w_{i,k}^{\Delta} exp(-\mathbf{e}_i^T \mathbf{Q}^w \_ \mathbf{e}_k)} \tag{5}$$

$$\mathbf{e}_i = \sum_{k=1}^{K} \alpha_{i,k} \_ \mathbf{e}_k \tag{6}$$

where $\mathbf{Q}^w \in \mathbb{R}^{h*h}$ is the trainable projection parameter. $w_{i,j}^{\Delta}$ is the time interval weight between i-th news and j-th news. The principle of time weight setting is as follows:

$$w_{i,i}^{\Delta} = 1 \tag{7}$$

$$w_{i,j}^{\Delta} = w_{j,i}^{\Delta} = w_{i,j-1}^{\Delta} * p_{j-1,j}, i < j \tag{8}$$

$$p_{j-1,j} = \begin{cases} q < 1, \ if \Delta_j \geq \Delta_{threshold} \\ 1, \quad otherwise \end{cases} \tag{9}$$

where $\Delta_{threshold}$ is the time threshold and $q \in (0,1)$ control the magnitude of the change in time weight. For example, the time interval sequence is $\{0, 100, 20, 50\}$, and the $\Delta_{threshold} = 30$. Then the time interval weight matrix is shown as follows:

$$\begin{pmatrix} 1, \ q, \ q, \ q^2 \\ q, \ 1, \ 1, \ q \\ q, \ 1, \ 1, \ q \\ q^2, \ q, \ q, \ 1 \end{pmatrix} \tag{10}$$

We can get a new news embedding sequence $\{\mathbf{e}_1', \mathbf{e}_2', ..., \mathbf{e}_K'\}$ through the T-SA. This sequence considers the influence of time factor on short-term user interest. T-SA can learn the sequence correlation of each news article, and detect weak sequence correlation because of taking into account the time interval. Therefore, T-SA can alleviate the effects of short-term user interest drifts.

Moreover, we apply the attention mechanism to learn relevance between the news recently clicked by the user and the candidate news v_s. The specific calculation method is as follows:

$$\mathbf{d}_i = \tanh(\mathbf{W}_u \mathbf{e}_i' + \mathbf{w}_u) \tag{11}$$

$$\mathbf{d}^s = \tanh(\mathbf{W}_u \mathbf{e}_s + \mathbf{w}_u) \tag{12}$$

$$\alpha_i^u = \frac{exp(\mathbf{q}^u(\mathbf{d}_i + \mathbf{d}^s))}{\sum_{j=1}^{K} exp(\mathbf{q}^u(\mathbf{d}_j + \mathbf{d}^s))} \tag{13}$$

$$\mathbf{v}_u = \sum_{i=1}^{K} \alpha_i^u \mathbf{e}_i' \tag{14}$$

where $\mathbf{e}_s$ is the embedding of candidate news, $\mathbf{W}_u \in \mathbb{R}^{g*h}, \mathbf{q}^u \in \mathbb{R}^g$ are trainable parameters, and $\mathbf{w}_u \in \mathbb{R}^g$ is bias parameter. $\mathbf{v}_u$ denotes the short term interest representation of user u.

3.3   News Encoder

The *News Encoder* is designed to extract the abstract features of candidate news. We can learn the collaborative feature from the historical readers information. To leverage the collaborative information, we learn representations of a news article from the article content and users who recently browsed this article. For a candidate news v_s, $U_s = \{u_1, u_2, ..., u_M\}$ denotes the users who have recently browsed v_s. $C_i = \{v_1^{u_i}, v_2^{u_i}, ..., v_K^{u_i}\}$ denotes the news recently browsed by user u_i. As mentioned above, We can obtain the embedding of each news through the News Embedding Layer. The embedding set corresponding to the C_i is $E_i = \{\mathbf{e}_1^{u_i}, \mathbf{e}_2^{u_i}, ..., \mathbf{e}_K^{u_i}\}$. For the set of users who have clicked v_s, we apply the average sum of news embedding to get the content embedding of each user:

$$\mathbf{e}^{u_k} = \sum_{i=1}^{K} \frac{\mathbf{e}_i^{u_k}}{K} \tag{15}$$

where $\mathbf{e}_i^{u_k} \in \mathbb{R}^h$ represents the vector of the i-th news that user u_k has browsed. The set $E_s^u = \{\mathbf{e}^{u_1}, \mathbf{e}^{u_2}, ..., \mathbf{e}^{u_M}\}$ represents the textual feature of the news recently clicked by the each user. It indicates the reading features of current readers of news v_s and indirectly reflects the context of news v_s. For the each embedding $\mathbf{e}_i^{s'}$ in the sequence $\{\mathbf{e}_s, \mathbf{e}^{u_1}, ..., \mathbf{e}^{u_M}\}$, we utilize the attention mechanism as follows:

$$\alpha_i^s = \frac{exp(\mathbf{q}^w \tanh(\mathbf{W}_c \mathbf{e}_i^{s'} + \mathbf{w}_b))}{\sum_{j=1}^{M+1} exp(\mathbf{q}^w \tanh(\mathbf{W}_c \mathbf{e}_j^{s'} + \mathbf{w}_b))} \tag{16}$$

where $\mathbf{W}_c \in \mathbb{R}^{g*h}, \mathbf{q}^w \in \mathbb{R}^g$ are trainable parameters, and $\mathbf{w}_b \in \mathbb{R}^g$ is bias parameter. α_i^s represents the weight of the i-th embedding in the sequence. The final news representation $\mathbf{v}_s$ which contains text information and collaborative information is denoted as follows:

$$\mathbf{v}_s = \sum_{i=1}^{M+1} \alpha_i^s \mathbf{e}_i^{s'} \tag{17}$$

Besides, for the candidate news v_s, we also get the latent news embedding $\mathbf{q}_s$ by latent factor model.

3.4   Predictor

To obtain more accurate representations, we combine the long- and short-term user representation together. The final user and news representation are as followed:

$$\mathbf{v}_u^U = [\mathbf{v}_u; \mathbf{p}_u] \tag{18}$$

Table 1. Datasets Statistics.

Dataset	#user	#news	#interaction	Avg. #articles seen per user
Globo	322,897	46,033	2,988,181	9.25
Adressa	640,503	20,428	2,817,881	4.40

$$\mathbf{v}_s^I = [\mathbf{v}_s; \mathbf{q}_s] \tag{19}$$

where $[\cdot; \cdot]$ is the concatenation operation.

To predict the probability of the user u clicking the candidate news article v_s, we utilize the cosine function as the *Predictor*: $\hat{r}_{u,s} = cosine(\mathbf{v}_u^U, \mathbf{v}_s^I)$. $\hat{r}_{u,s}$ denotes the clicking probability.

3.5 Loss Function

We apply the pairwise learning method to train our proposed model. For the input triple $< u, p, n >$, where u, p, n respectively denote users, positive sample and negative sample, we minimize objection function as follows:

$$\arg\min_{\Theta} \sum_{(u,p,n) \in D} max\{0, m - (\hat{r}_{u,p} - \hat{r}_{u,n})\} + \lambda \Omega(\Theta) \tag{20}$$

where m is the margin between positive and negative sample, Θ is the trainable parameters of the model, $\Omega(\cdot)$ denotes the L2 regularization and λ is the penalty weight.

Since not all users and news articles can participate in the model training, just like the new users or new articles, we cannot obtain the long-term representation of each user during the prediction phase of our model. In order to solve this problem, we use a random masking strategy followed [1] during model training. Specifically, we randomly mask the long-term user representations $\mathbf{p}_u$ and the latent representation of news $\mathbf{q}_s$ with a certain probability p_m. The mask operation sets all dimensions of the vector to zero. Thus, the representations can be reformulated as:

$$\mathbf{p}_u = p \cdot \mathbf{W}_u[u]$$
$$\mathbf{q}_s = p \cdot \mathbf{W}_s[s] \tag{21}$$
$$p \sim M(1, 1 - p_m)$$

where M denotes the binomial distribution This design is in line with the actual situation and can reduce the impact of cold start on the model prediction [1].

4 Experiments

4.1 Experiments Setup

Dataset and Evaluation Protocol. We conduct experiments on two public real-world datasets: Globo[1] and Adressa[2]. Data statistics is shown in Table 1.

---

[1] https://www.kaggle.com/gspmoreira/news-portal-user-interactions-by-globocom.
[2] http://reclab.idi.ntnu.no/dataset/.

Globo dataset comes from a popular news platform in Brazil, which provides the pre-trained content embeddings of news articles. Adressa dataset [4] comes from a social news platform in Norwegian. We use one-week version of Adressa and adopt the pre-trained Glove embeddings [15] to build the news embeddings.

For each user in each dataset, the interaction history is split into the training set, validation set and test set with the ratio of 80%, 10% and 10% respectively. We use negative sampling to construct training data, and treat news articles that have not been browsed by users as negative samples. During model training, the validation set is used to adjust the hyper-parameters. For each user in the test set, we sample 99 negative items and pair them with the positive sample. Then each model will calculate the score for each user-item interaction in the test set.

To evaluate the recommendation performance, we employ three widely-adopted metrics: *Hit Ratio* (HR), *Normalize Discounted Cumulative Gain* (NDCG) and *Area Under Curve* (AUC) [6].

Comparison Methods. We compare our model TANN with some recent state-of-the-art (SOTA) methods, including ConvNCF [6], DeepFM [5], NRMS [21], NAML [19], DAN [23], LSTUR-ini [1]. These models are based on deep learning methods.

Implementation Details. We implement TANN based on Tensorflow. We optimize the AUC on the validation set to obtain the optimal hyper-parameter settings. They are setting as follows: we set negative sampling size $S = 3$ with random strategy; we set the sequence length $K = 10$; Margin is set as $m = 0.05$; And Adam [8] is used to optimize the parameters with learning parameter of 0.0003. The mask probability p_m is 0.7. Besides we set the $\Delta_{threshold}$ to be 1000 seconds. Regularization penalty factor λ is set to 0.005 and the dropout rate is 0.5. The batch size is 1024. We train the models at most 100 epoches.

For a fair comparison, we employ the source codes of all the SOTA methods from Github and fine-tune parameters to get the best performance for these models according to their works.

4.2   Performance Comparison

Comparisons of Different Models. First of all, we conduct experiments to compare our model with the SOTA methods on two datasets. We show the detailed results on three different metrics in Table 2. We can obtain several key observations from Table 2:

Firstly, the performance of our method is significantly improved compared to all comparative methods. Especially, NDCG@5 increases by more than 6.41% and HR@20 increases by more than 3.34%. We believe that the superior performance of the model mainly stems from its three advantages: (1) TANN considers the time interval information when learning the short-term user interest. (2) Our method considers both the long-term user interest and the short-term interest for better user representation. (3) We introduce the information of news readers and

Table 2. Comparison of different methods on two Datasets for Top-K news recommendation.

Datasets	Methods	HR@K			NDCG@K			AUC
		$K = 5$	$K = 10$	$K = 20$	$K = 5$	$K = 10$	$K = 20$	
Globo.	ConvNCF	0.7439	0.8260	0.8587	0.5548	0.5819	0.5902	0.8612
	DeepFM	0.7559	0.8223	0.8398	0.5489	0.5701	0.5745	0.8779
	NRMS	0.7536	0.8284	0.8483	0.5610	0.6172	0.6728	0.9106
	DAN	0.7543	0.8356	0.8527	0.5317	0.5802	0.6825	0.9027
	NAML	0.8301	0.8370	0.8342	0.5950	0.6276	0.6774	0.9141
	LSTUR-ini	0.8635	0.9383	0.9483	0.6172	0.6728	0.6906	0.9206
	TANN	**0.8748**	**0.9428**	**0.9631**	**0.6568**	**0.6864**	**0.7091**	**0.9601**
Adressa.	ConvNCF	0.7107	0.7895	0.8021	0.4719	0.5368	0.5583	0.8550
	DeepFM	0.7457	0.8013	0.8101	0.5334	0.5549	0.5573	0.9259
	NRMS	0.7468	0.8142	0.8233	0.5210	0.5572	0.5728	0.9316
	DAN	0.7516	0.8217	0.8438	0.5110	0.5426	0.5701	0.9231
	NAML	0.7601	0.8311	0.8561	0.5486	0.5788	0.6289	0.9308
	LSTUR-ini	0.7935	0.8257	0.8783	0.5732	0.5923	0.6406	0.9362
	TANN	**0.8088**	**0.8392**	**0.9076**	**0.5886**	**0.6049**	**0.6499**	**0.9694**

add the collaborative signals to represent news, which can improve the accuracy of news recommendations.

Secondly, the methods that consider the reading sequences of users (e.g., DAN, NRMS) is better than the collaborative-filtering based method (e.g., ConvNCF). This is because CF-based methods cannot reflect the current interest of users for news recommendation.

Thirdly, the methods which exploit to combine the long-term and short-term user representations (e.g., LSTUR-ini and TANN) outperform other methods. This may be because these methods have stronger feature extraction capabilities and can model complex and varied user preferences from reading history. For the models using a single user representation, often cannot fully reflect the users interest.

Comparisons of TANN Variants. Moreover, to show the effectiveness of the design of our method, we compare among the variants of TANN with respect several aspects: Time-aware self attention, combining of long-term and short-term user interests, and the novel representation of news. The experimental results on the Globo dataset are shown in Table 3.

As can be seen that, there is dramatic decline in performance when removing the T-SA. This proves the effectiveness of our design. In the design of T-SA, we consider the time interval information and adopt the self-attention to learn the feature from the short reading sequences. Besides, removing long-term interest representation can also lead to poor experimental results. It demonstrates that considering both long-term and short-term user interests is necessary. Moreover, we remove the novel news representation and only use news article embedding as the news representation. The decreased performance shows the rationality of our news representation method.

Table 3. Comparison of TANN Variants.

Model	HR@K		NDCG@K	
	$K = 5$	$K = 10$	$K = 5$	$K = 10$
TANN without T-SA	0.8021	0.8319	0.5248	0.5619
TANN without long-term interest	0.8428	0.8735	0.5752	0.5876
TANN without novel news representation	0.8631	0.9143	0.6252	0.6376
TANN	**0.8748**	**0.9428**	**0.6568**	**0.6864**

5 Conclusion

In this paper, we propose a novel news recommendation model, called Time-aware Attentive Neural Network with long- and short-term user representation (TANN). We propose to take the time interval of user reading into account and design the novel self-attention model to learn more accurate user short-term interest. Besides, we propose to learn the news feature from the historical readers and contents of news articles via the attention mechanism. Furthermore, we use the latent factor model to learn the short-term interests and combine the long-term and short-term interests together for better representation. From extensive experimental results, it can be proved that TANN outperforms previous advanced methods in news recommendation performance. In the future, we plan to extend TANN to other recommendation scenarios.

Acknowledgements. This work is supported by the project of National Key research and development plan (Grant No. 213) and the National Natural Science Foundation of China (Grant No. 61976160, No. 61673301), and Major project of Ministry of Public Security (Grant No. 20170004).

References

1. An, M., Wu, F., Wu, C., Zhang, K., Liu, Z., Xie, X.: Neural news recommendation with long-and short-term user representations. In: Proceedings of the 57th ACL, pp. 336–345 (2019)
2. Chen, X., Zhang, Y., Qin, Z.: Dynamic explainable recommendation based on neural attentive models. In: Proceedings of the AAAI, vol. 33, pp. 53–60 (2019)
3. Das, A.S., Datar, M., Garg, A., Rajaram, S.: Google news personalization: scalable online collaborative filtering. In: Proceedings of the 16th WWW, pp. 271–280. ACM (2007)
4. Gulla, J.A., Zhang, L., Liu, P., Özgöbek, Ö., Su, X.: The Adressa dataset for news recommendation. In: Proceedings of the International Conference on Web Intelligence, pp. 1042–1048. ACM (2017)
5. Guo, H., Tang, R., Ye, Y., Li, Z., He, X.: DeepFM: a factorization-machine based neural network for CTR prediction. arXiv preprint arXiv:1703.04247 (2017)

6. He, X., Du, X., Wang, X., Tian, F., Tang, J., Chua, T.S.: Outer product-based neural collaborative filtering. In: Proceedings of the 27th IJCAI, pp. 2227–2233. AAAI Press (2018)
7. Hu, L., Li, C., Shi, C., Yang, C., Shao, C.: Graph neural news recommendation with long-term and short-term interest modeling. Inf. Process. Manag. **57**(2), 102142 (2020)
8. Kingma, D.P., Ba, J.: Adam: a method for stochastic optimization. arXiv preprint arXiv:1412.6980 (2014)
9. Koren, Y., Bell, R., Volinsky, C.: Matrix factorization techniques for recommender systems. Computer **42**(8), 30–37 (2009)
10. Li, L., Wang, D., Li, T., Knox, D., Padmanabhan, B.: Scene: a scalable two-stage personalized news recommendation system. In: Proceedings of the 34th ACM SIGIR, pp. 125–134 (2011)
11. Li, L., Zheng, L., Yang, F., Li, T.: Modeling and broadening temporal user interest in personalized news recommendation. Expert Syst. Appl. **41**(7), 3168–3177 (2014)
12. Li, L., Chu, W., Langford, J., Schapire, R.E.: A contextual-bandit approach to personalized news article recommendation. In: Proceedings of the 19th WWW, pp. 661–670 (2010)
13. Liu, J., Dolan, P., Pedersen, E.R.: Personalized news recommendation based on click behavior. In: Proceedings of the 15th International Conference on Intelligent User Interfaces, pp. 31–40. ACM (2010)
14. Okura, S., Tagami, Y., Ono, S., Tajima, A.: Embedding-based news recommendation for millions of users. In: Proceedings of the 23rd ACM SIGKDD, pp. 1933–1942 (2017)
15. Pennington, J., Socher, R., Manning, C.D.: Glove: global vectors for word representation. In: Proceedings of the EMNLP, pp. 1532–1543 (2014)
16. Ren, H., Feng, W.: CONCERT: a concept-centric web news recommendation system. In: Wang, J., Xiong, H., Ishikawa, Y., Xu, J., Zhou, J. (eds.) WAIM 2013. LNCS, vol. 7923, pp. 796–798. Springer, Heidelberg (2013). https://doi.org/10.1007/978-3-642-38562-9_82
17. Son, J.W., Kim, A.Y., Park, S.B.: A location-based news article recommendation with explicit localized semantic analysis. In: Proceedings of the 36th ACM SIGKDD, pp. 293–302 (2013)
18. Wang, H., Zhang, F., Xie, X., Guo, M.: DKN: deep knowledge-aware network for news recommendation. In: Proceedings of WWW, pp. 1835–1844 (2018)
19. Wu, C., Wu, F., An, M., Huang, J., Huang, Y., Xie, X.: Neural news recommendation with attentive multi-view learning. arXiv preprint arXiv:1907.05576 (2019)
20. Wu, C., Wu, F., An, M., Huang, J., Huang, Y., Xie, X.: NPA: neural news recommendation with personalized attention. In: Proceedings of the 25th SIGKDD, pp. 2576–2584. ACM (2019)
21. Wu, C., Wu, F., Ge, S., Qi, T., Huang, Y., Xie, X.: Neural news recommendation with multi-head self-attention. In: Proceedings of the EMNLP-IJCNLP, pp. 6390–6395 (2019)
22. Zhang, L., Liu, P., Gulla, J.A.: A deep joint network for session-based news recommendations with contextual augmentation. In: Proceedings of the 29th on Hypertext and Social Media, pp. 201–209 (2018)
23. Zhu, Q., Zhou, X., Song, Z., Tan, J., Guo, L.: Dan: deep attention neural network for news recommendation. In: Proceedings of the AAAI, vol. 33, pp. 5973–5980 (2019)
24. Zhu, Y., et al.: What to do next: modeling user behaviors by time-LSTM. In: Proceedings of the IJCAI, pp. 3602–3608 (2017)

A Time Interval Aware Approach for Session-Based Social Recommendation

Youjie Zhang, Ting Bai, Bin Wu[✉], and Bai Wang

Beijing University of Posts and Telecommunications, Beijing, China
{molo_j,baiting,wubin,wangbai}@bupt.edu.cn

Abstract. Users on social media like Facebook and Twitter are influenced by their friends. Social recommendation exploits users' social activities for modeling their interests to enhance the activeness and retention of users. Besides, their interests change from time to time. Session-based recommendation divides users' interaction history into sessions and predict users' behaviors with the context information in each session. It's essential but challenging to model the social activities and the dynamic property in an unified model. Besides, most of existing session-based recommendation approaches model users' interaction history as ordered sequence in regardless of real timestamps of those interactions. To solve the above issues together, we design a heterogeneous graph for modeling the complex interactions among users and items and propose a Time Interval aware graph neural network-based Recommendation approach(TiRec) to model both the social activities and the dynamic property of users' interaction with items. Furthermore, to capture users' dynamic preference, we propose a time interval aware and self-attention based aggregator to model users' preference in each session. Experimental results on several real-world datasets demonstrates the effectiveness of our proposed approach over some competitive baselines.

Keywords: Session-based recommendation · Time Interval · Social recommendation

1 Introduction

People interact with items like movies, books and music on social media like Douban by purchasing, rating and leaving comments. And users often share their thoughts and communicate with others, through which users' activities tends to be influenced by their related friends. Besides, users' interests are dynamic and change from time to time. A user may be interested in sports for a period of time and then be attracted to video games. Session-based recommendation divides users' interaction records into sessions and model users' preference in each session separately. As a result, social influence among users are often context-dependent. Like when a user want to buy a camera, she tends to consult for her friends who are photography enthusiast instead of those who are interested at sports.

© Springer Nature Switzerland AG 2020
G. Li et al. (Eds.): KSEM 2020, LNAI 12275, pp. 88–95, 2020.
https://doi.org/10.1007/978-3-030-55393-7_8

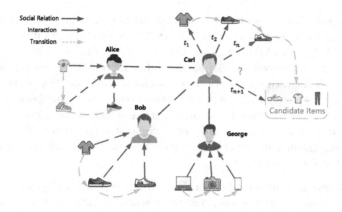

Fig. 1. The illustration of session-based social recommendation.

As the example in Fig. 1, Carl is interested in shoes in the current session while his friend Bob and Alice bought some sporting shoes recently. Then the next purchase decision of Carl may be more influenced by Bob and Alice instead of George, who is an electronic enthuasist instead. It's essential but challenging to model social activities and users' dynamic preference together. However, most of existing work focus on modeling either social activities or users' dynamic preference, instead of them together. To solve those challenges, we propose a graph neural network-based model to model the heterogeneity of users' social activities and interaction activities and the dynamic of users interests.

Our contributions are summarized as follows:

- We propose a Time Interval aware graph neural network-based Recommendation approach(TiRec) to model both the social activities and users' dynamic preference in an unified manner.
- We propose a time interval aware aggregator to model users' preference in each session which utilizes time interval information.
- Experimental results on two real-world datasets demonstrate the effectiveness of our model, showing the usefulness of modeling time interval information in social recommender systems.

The rest of this paper is organized as follows: Some related work is discussed in Sect. 2. We then present the problem definition and introduce our model in Sect. 3. Following it, Sect. 4 presents experiments results and analysis. Finally, Sect. 5 concludes this paper.

2   Related Work

Social Recommendation. Leveraging the social network information provides an effective approach to alleviate data sparsity and improve performance in social recommendation, which has received attention from researchers [2,6,12].

As interactions among users and items can be naturally modeled as a graph, Graph Neural Network(GNN) based approaches have attracted much attention recently [2,11,12]. GNN-based approach socialGCN [12] captures how users' preferences are influenced by the social diffusion process in social networks. GraphRec [2] is for rating prediction in a social recommendation setting, which aggregates information from users and items' connected neighbors. Wu et al. [11] designs a layer-wise influence propagation structure to model how users preferences evolve as the social influence propagates recursively. Our Model is also based on graph neural network, which models social influence and users' dynamic interests together.

Session-Based Recommendation. Capturing the purchase appetites of users and their evolution over time has attracted researchers' attention in recent years, which utilizes sequential information among items to capture users' dynamic interests [1,5,10]. GNN based model are also proposed for session-based recommendation [8,13]. DGRec [8] is the first work to work on session-based social recommendation. They propose a dynamic graph attention network which models the social influence and the sequential pattern of users' actions together. However, the real time interval information is neglected when predicting users' interests, as only the order information in the action sequences is utilized.

3 Model

In this section, we will introduce the problem definition and the details of our model, which is mainly composed of two modules modeling users' dynamic interests and social influence.

3.1 Problem Definition

Definition 1. *(Session-based Social Recommendation) Assume that we have a set of users and items, denoted by U and I respectively. $G_{social} = (U, E_{UU})$ is the social network, where E_{UU} is the social links between users. Given a new session $S_{T+1}^u = \{i_{T+1,t_1}^u, i_{T+1,t_2}^u, ..., i_{T+1,t_n}^u\}$ of user u, the goal of session-based social recommendation is to recommend a set of items from I that u is likely to be interested in at the next timestamp t_{n+1} by exploiting her dynamic interests (i.e., from previous sessions) and social influence received from her social links $N(u)$.*

3.2 Dynamic Interests Aggregator

Dynamic interests aggregator aims to learn user u's latent factor, denoted by $h_S(u)$, from his latest session $S_T(u)$(Our model can be easily generalized to more sessions). The challenge here is to incorporate the real time interval information and sequential information together. To address this challenge, we adopt a self-attention based aggregator, inspired by [4]. The input is user u's latest session

$S_T(u) = \{i^u_{T,t_1}, i^u_{T,t_2}, ..., i^u_{T,t_n}\}$. The sequence is truncated or padded according to a threshold as the max sequence length. As mentioned above, each user-item interaction is associated with a timestamp t_i. So each session $S_T(u)$ is also associated with a timestamp sequence $t = (t_1, t_2, ..., t_n)$. The output of dynamic interest aggregator is $H = (h_1, h_2, ..., h_n)$ in which h_i is the hidden representation of first i elements.

As two sessions may have a same item sequence but with different time intervals among items, we take not only order information between two items i^u_{T,t_j} and i^u_{T,t_k} but also the time interval $\delta_{jk} = |t_k - t_j|$ of two interactions into consideration. δ_{jk} is clipped into a proper range, since a too large time interval or too small time interval tends to be meaningless. δ_{jk} is taken as an index to get a relative temporal encoding $r(\delta_{jk}) \in R^d$. Following [4], we adopt a fixed set of sinusoid functions as basis, with a learnable linear project function $f_{Linear}(\cdot)$ to obtain δ_{jk}:

$$Base(\delta_{jk}, 2i) = sin(\delta_{jk}/10000^{2i/d}) \tag{1}$$

$$Base(\delta_{jk}, 2i+1) = cos(\delta_{jk}/10000^{2i/d}) \tag{2}$$

$$r(\delta_{jk}) = f_{Linear}(Base(\delta_{jk})) \tag{3}$$

With the relative temporal encoding, a weighted sum of the linearly transformed representations of first i elements in the sequence, the position encoding and the relative temporal encoding, denoted by $\hat{h}_i$, is computed as:

$$\hat{h}_i = \sum_{j=1}^{n} \alpha_{ij}(W^V e_i + r(\delta_{ij}) + p_j) \tag{4}$$

where e_i is the embedding of the i-th element, W^V is the linear transformation matrix, $r(\delta_{ij})$ is the relative temporal encoding and p_j is the position encoding. The position encoding is also computed with sinusoid function like the relative temporal encoding following [9].

The weighted coefficient α_{ij} is computed with a softmax function on similarity on e_{ij}, which is computed with a similarity function as:

$$e_{ij} = \frac{W^Q e_i(W^K e_j + r(\delta_{ij}) + p_j)^T}{\sqrt{d}} \tag{5}$$

where W^Q and W^K are linear transformation matrix for query and key in self-attention mechanism. And the final output is obtained performing a feed-forward network $g(\cdot)$ on $\hat{h}_i$ as $h_i = g(\hat{h}_i)$.

So far, we have obtained the representation of session $S_T(u)$ which is the short-term preference of user u. And user u's long-term preference is represented with a base embedding e_u. The final representation $h_I(u)$ of the dynamic interest aggregator is compute with a non-linear transform on the long-term interest representation e_u and short-term interest as h_n:

$$h_I(u) = ReLU(W^I[e_u; h_n]) \tag{6}$$

where $ReLU(x) = max(0, x)$ is a non-linear activation function and W_I is the transformation matrix.

3.3   Social Influence Aggregator

We observe that users receive influence from her friends when making a decision. To model such social relation, we adopt an attention-based aggregator. The output of the social influence aggregator $h_S^{(L)}(u)$ is computed as:

$$h_S^{\widetilde{(L)}}(u) = \sum_{\forall j \in N(u) \cup u} \alpha_{ij} h_S^{(L-1)}(u) \tag{7}$$

where α_{ij} is the weight coefficient, computed as follows:

$$\alpha_{ij} = \frac{exp(h_S^{(L-1)}(i) \cdot h_S^{(L-1)}(j)^T)}{\sum_{k=1}^{k=n} exp(h_S^{(L-1)}(i) \cdot h_S^{(L-1)}(k)^T)} \tag{8}$$

The final output $h_S^{(L)}(u)$ of the aggregator is computed as $h_S^{(L)}(u) = ReLU(W \cdot h_S^{\widetilde{(L)}}(u))$. Through stacking more layers of the social influence aggregator, target user u can receive influence from higher-order neighbors. $h_S^{(0)}(u) = h_I(u)$. The final output of social influence aggregator is $h_S^{(L)}(u)$.

3.4   Recommendation

The final interest representation $h(u)$ of user u is computed by combining both the short-term preference and social influence as:

$$h(u) = ReLU(W[h_I(u); h_S^{(L)}(u)] \tag{9}$$

Then the probability user u will interact with item i is computed with:

$$p(i|S_{T+1}(u), \{S_T(i), \forall k \in N(i)\}) = \frac{exp(e_i \cdot h(u)^T)}{\sum_{k=1}^{k=|I|} exp(e_k \cdot h(u)^T)} \tag{10}$$

where e_k is the item embedding for item k and $|I|$ is the number of items. Items with highest probability will be taken as candidate items.

3.5   Loss

The whole model is trained through maximizing the below loss as follows:

$$\mathcal{L} = \sum_{\forall u \in U} \sum_{t=2}^{t=T} \sum_{n=1}^{N_{u,t-1}} log(p(i_{t,n+1}^u | \{i_{T+1,t_1}^u, i_{T+1,t_2}^u, ..., i_{T+1,t_n}^u\}, \{S_T(i), \forall k \in N(i)\})) \tag{11}$$

That is, for each observed item $i_{t,n+1}^u$ in all sessions, we maximize its likelihood with previously observed items $\{i_{T+1,t_1}^u, i_{T+1,t_2}^u, ..., i_{T+1,t_n}^u\}$ in its current session and social influences $\{S_T(i), \forall k \in N(i)\}$ received from peers.

4 Experiments

In the following part, we will first introduce our experimental settings, including datasets, baselines, and evaluation metrics. Then we analyze the experimental results.

4.1 Experimental Settings

Datasets. We experiment with some representative real-world datasets in social recommendation: Delicious[1], Douban Movie[2] and Epinions[3], which contain both the users' interaction records and social relations, such as trust and following. Table 1 summarizes the statistics of the datasets.

Table 1. Statistics of the datasets.

Datasets	Delicious	Douban	Epinions
# Users	1,650	32,094	8,727
# Items	4,282	13,078	10,009
# User-Item interactions	296,705	10,077,494	41,933
# Social Links	15,328	45,107	21,406
# User-item Density	4.1994%	2.4000%	0.0480%
# User-user Density	5.6301%	0.0043%	0.0281%

Baseline Methods. In our experiment, we take Hit Ratio@k and NDCG@k as evaluation metrics. We report the average value on all testing sessions. Baseline methods are listed below:

- BPR [7]. It optimizes the MF model with a pairwise ranking loss.
- SBPR [14]: It's a social Bayesian ranking model that considers social relationships, which is improved on BPR.
- SocialReg [6]: It's a social-based model that regularizes and weighs friends' latent factors with social network information.
- GraphRec [2]: It is a graph neural network based social recommendation model. It models users in social and item domains separately.
- RNN-Session [3]: It's a recurrent neural network-based model for session-based recommendation.
- DGRec [8]: It is a graph neural network-based model which models users' dynamic interests and social influence.
- TiRec: Our proposed TiRec combines the influence from social relation, temporal information into consideration.

---

[1] https://grouplens.org/datasets/hetrec-2011/.
[2] https://www.dropbox.com/s/u2ejjezjk08lz1o/Douban.tar.gz?dl=0.
[3] https://cseweb.ucsd.edu/~jmcauley/datasets.html#social/_data.

Parameter Settings. We test different settings on validation data to obtain the best hyperparameters. For the embedding size d, we fix the embedding size of users and items as 100. We sample 10 or 15 friends for each user to accelerate computing. The batch size and learning rate are searched in $[32, 64, 128, 256, 512]$ and $[0.001, 0.005, 0.01, 0.05, 0.1]$, respectively. We user Adam for optimization due to its effectiveness. To avoid overfitting, we set the dropout rate as 0.2. We stop training if the evaluation metrics on validation set decreased for 10 successive epochs.

Table 2. Performance on the next-item recommendation task.

Datasets	Delicious		Douban		Epinions	
Models	Hits@20	NDCG@20	Hits@20	NDCG@20	Hits@20	NDCG@20
BPR	0.2765	0.2287	0.0159	0.1112	0.1174	0.0592
SBPR	0.2954	0.2393	0.0168	0.1062	0.1253	0.0623
SocialReg	0.2699	0.2283	0.0177	0.1123	0.1093	0.0543
RNN-Session	0.3453	0.2593	0.1639	0.1802	0.1289	0.0774
GraphRec	0.2745	0.2287	0.0168	0.1187	0.1183	0.0599
DGRec	0.4071	0.2935	0.1867	0.1958	0.1332	0.0898
TiRec	**0.4157**	**0.3128**	**0.1988**	**0.2123**	**0.1384**	**0.0923**

4.2 Main Results

For the session-based social recommendation task, we present the results of Hit@k and NDCG@k, (*i.e.* $K = 20$) in Table 2. We have the following observations:

(1) The models utilizes social information outperforms those don't. The improvement is more significant on the Delicious dataset than other two datasets, which has a larger density. This demonstrates that the social network information is helpful in predicting the preference of users.

(2) The models capture users' dynamic preference, including RNN-Session, DGRec and TiRec, performs better than other methods. This implies the effectiveness of capturing users' dynamic interests.

(3) Our TiRec performs better compared with all the baseline methods, which captures users' social influence and temporal information in interaction sequences with two kinds of aggregators. TiRec performs better than DGRec as TiR ec utilizes the time interval information while DGRec doesn't.

5   Conclusions

In this work, we explore combined effects of social relations and sequential information in social recommender systems and propose a graph neural network-based model. Our model models dynamic interests of users with an aggregator based on self-attention, which utilizes both sequential information between items and real time intervals. To capture social influence, we leverage an attention-based aggregator. Experimental results on several real-world datasets have demonstrated the effectiveness of our proposed model. For future work, it's also possible to incorporate rich attribute information (*i.e.* category, price) and textual description of items to provide better prediction performance.

References

1. Donkers, T., Loepp, B., Ziegler, J.: Sequential user-based recurrent neural network recommendations. In: Proceedings of the Eleventh ACM Conference on Recommender Systems, pp. 152–160. ACM (2017)
2. Fan, W., et al.: Graph neural networks for social recommendation. In: The World Wide Web Conference, pp. 417–426. ACM (2019)
3. Hidasi, B., Karatzoglou, A., Baltrunas, L., Tikk, D.: Session-based recommendations with recurrent neural networks. arXiv preprint arXiv:1511.06939 (2015)
4. Hu, Z., Dong, Y., Wang, K., Sun, Y.: Heterogeneous graph transformer. In: Proceedings of the Web Conference 2020, pp. 2704–2710 (2020)
5. Liu, Q., Zeng, Y., Mokhosi, R., Zhang, H.: Stamp: short-term attention/memory priority model for session-based recommendation. In: Proceedings of the 24th ACM SIGKDD International Conference on Knowledge Discovery & Data Mining, pp. 1831–1839. ACM (2018)
6. Ma, H., Zhou, D., Liu, C., Lyu, M.R., King, I.: Recommender systems with social regularization. In: Proceedings of the Fourth ACM International Conference on Web Search and Data Mining, pp. 287–296. ACM (2011)
7. Rendle, S., Freudenthaler, C., Gantner, Z., Schmidt-Thieme, L.: BPR: Bayesian personalized ranking from implicit feedback. In: Proceedings of the Tenty-Fifth Conference on Uncertainty in Artificial Intelligence, pp. 452–461. AUAI Press (2009)
8. Song, W., Xiao, Z., Wang, Y., Charlin, L., Zhang, M., Tang, J.: Session-based social recommendation via dynamic graph attention networks. In: Proceedings of the Twelfth ACM International Conference on Web Search and Data Mining, pp. 555–563. ACM (2019)
9. Vaswani, A., et al.: Attention is all you need. In: Advances in Neural Information Processing Systems, pp. 5998–6008 (2017)
10. Wu, C.Y., Ahmed, A., Beutel, A., Smola, A.J., Jing, H.: Recurrent recommender networks. In: Proceedings of the Tenth ACM International Conference on Web Search and Data Mining, pp. 495–503. ACM (2017)
11. Wu, L., Sun, P., Fu, Y., Hong, R., Wang, X., Wang, M.: A neural influence diffusion model for social recommendation. arXiv preprint arXiv:1904.10322 (2019)
12. Wu, L., Sun, P., Hong, R., Fu, Y., Wang, X., Wang, M.: SocialGCN: an efficient graph convolutional network based model for social recommendation. arXiv preprint arXiv:1811.02815 (2018)
13. Wu, S., Tang, Y., Zhu, Y., Wang, L., Xie, X., Tan, T.: Session-based recommendation with graph neural networks. In: Proceedings of the AAAI Conference on Artificial Intelligence, vol. 33, pp. 346–353 (2019)
14. Zhao, T., McAuley, J., King, I.: Leveraging social connections to improve personalized ranking for collaborative filtering. In: Proceedings of the 23rd ACM International Conference on Information and Knowledge Management, pp. 261–270. ACM (2014)

AutoIDL: Automated Imbalanced Data Learning via Collaborative Filtering

Jingqi Zhang, Zhongbin Sun[✉], and Yong Qi

School of Computer Science and Technology, Xi'an Jiaotong University, Xi'an, China
zhongbin725@mail.xjtu.edu.cn

Abstract. AutoML aims to select an appropriate classification algorithm and corresponding hyperparameters for an individual dataset. However, existing AutoML methods usually ignore the intrinsic imbalance nature of most real-world datasets and lead to poor performance. For handling imbalanced data, sampling methods have been widely used since their independence of the used algorithms. We propose a method named AutoIDL for selecting the sampling methods as well as classification algorithms simultaneously. Particularly, AutoIDL firstly represents datasets as graphs and extracts their meta-features with a graph embedding method. In addition, meta-targets are identified as pairs of sampling methods and classification algorithms for each imbalanced dataset. Secondly, the user-based collaborative filtering method is employed to train a ranking model based on the meta repository to select appropriate sampling methods and algorithms for a new dataset. Extensive experimental results demonstrate that AutoIDL is effective for automated imbalanced data learning and it outperforms the state-of-the-art AutoML methods.

Keywords: AutoML · Imbalanced data · Collaborative filtering

1 Introduction

With the success of machine learning, there is a growing demand to build appropriate machine learning systems for real-world datasets, which needs a large amount of human sources and efforts. To solve the problem, the development of automated machine learning (AutoML), seeking to select the best machine learning algorithms and their hyperparameters without experts intervention, has attracted increasing interest recently [5]. A number of AutoML methods have been proposed, such as Auto-sklearn [6] and OBOE [23]. However, previous AutoML methods usually assume that training data used for building models are balanced, which is not always the case in real world.

In the last two decades, a variety of methods have been proposed for tackling the class imbalance problem [7], among which sampling methods are more

---

Supported by National Natural Science Foundation of China under Grant No. 61702405 and the China Postdoctoral Science Foundation under Grant No. 2017M623176.

G. Li et al. (Eds.): KSEM 2020, LNAI 12275, pp. 96–104, 2020.
https://doi.org/10.1007/978-3-030-55393-7_9

prevalent and versatile as they are independent of the selected classification algorithms [15]. Particularly, sampling methods alter the distribution of original data either by increasing minority class cases (Oversampling) or by eliminating majority class cases (Undersampling). Up to now, various sampling methods have been proposed yet there is not a consensus about which is the best sampling method for an individual dataset. Therefore we consider that it is important and necessary to integrate sampling method selection and algorithm selection for automated imbalanced data learning.

In this study, we propose AutoIDL (Automated Imbalanced Data Learning) to select sampling method and algorithm automatically. Firstly, AutoIDL applies a k-nearest neighbor graph method to convert historical imbalanced datasets into various graphs, and the graph embedding method graph2vec [17] is then employed to extract meta-features. Furthermore, the meta-targets of historical datasets are identified and represented as pairs of sampling method and algorithm, which are integrated with corresponding meta-features into the meta-repository. Secondly, the user-based collaborative filtering method is used to train a ranking model to select appropriate sampling methods and algorithms simultaneously for new datasets. In the experiment, 70 imbalanced datasets are used and the experimental results demonstrate that AutoIDL achieves better performance compared with Auto-sklearn and OBOE.

The main contributions of present study are summarized as follows: (i) We propose AutoIDL based on collaborative filtering for automated imbalanced data learning, taking into consideration the selection of sampling methods and algorithms simultaneously. (ii) A graph-based meta-feature extraction method is proposed by transforming datasets to graphs and then representing graphs into meta-features. (iii) We empirically demonstrate the effectiveness of AutoIDL and show that AutoIDL outperforms Auto-sklearn and OBOE.

This paper is organized as follows. Section 2 introduces the related work. Section 3 presents the AutoIDL method in detail. The experimental results are provided in Sect. 4. Finally, Sect. 5 concludes present study.

2   Related Work

AutoML Methods: There has been increasing interest in AutoML in recent years. A number of AutoML methods have been proposed, most of which are based on Bayesian optimization [20], evolutionary algorithms [18] and meta-learning [6]. One of the earliest work for AutoML is Auto-sklearn [6], which based on scikit-learn library [19] in Python. To be specific, Auto-sklearn uses the meta-learning step to warmstart the Bayesian optimization procedure and stacks multiple models to derive more accurate performance. Recently, to reduce the time consuming, OBOE [23] is proposed by using a collaborative filtering method for time-constrained model selection and hyperparameters tuning.

Sampling Methods: Sampling methods modify the distribution of imbalanced datasets and they are flexible to automated model selection as the preprocessing methods. Up to now, a variety of sampling methods have been proposed, which could be divided into oversampling and undersampling methods. Particularly, oversampling methods alter the imbalance ratio of a given dataset by increasing the minority examples and the following seven oversampling methods have been employed in present study, including ROS, SMOTE [3], BSMOTE [8], ADASYN [10], SLSMOTE [2], MWMOTE [1] and KSMOTE [4]. Furthermore, undersampling methods rebalance the skewed data by eliminating majority class instances and nine undersampling methods have been used in the experiment, including RUS, CNN [9], ENN [22], Tomek [21], OSS [12], NCL [13], NearMiss2 [16], USBC [24] and CentersNN [14].

3 AutoIDL Method

3.1 Overview

Generally, the proposed AutoIDL method consists of two procedures: Meta Repository Construction and Automated Model Selection. Particularly, in the procedure of Meta Repository Construction, the meta-features and meta-targets for historical imbalanced datasets are separately obtained and uniformly maintained in the meta repository. For Automated Model Selection, the meta repository is employed to select the applicable sampling method and algorithm for a new dataset. The framework of AutoIDL is represented in Fig. 1.

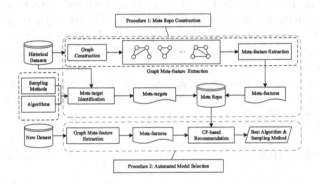

Fig. 1. General framework of the proposed AutoIDL method.

3.2 Meta Repository Construction

As shown in Fig. 1, two separate steps are included for constructing the meta repository, namely graph meta-feature extraction and meta-target identification.

Graph Meta-feature Extraction. Meta-feature plays an important role in the final performance. We argue that existing meta-features only describe the general characteristics of a dataset while ignore the interactive relationship of the instances in the dataset. Therefore, we propose a novel graph meta-feature extraction method, transforming the datasets into graphs and then using a graph embedding method to extract meta-features, which ensure that similar datasets will have similar vector representations in the embedded space.

To be specific, the graphical representations of datasets are generated by using k-nearest neighbor graph, which is recognized as a good candidate graph without sufficient domain knowledge of graph construction [25]. Furthermore, based on the principle that data points in a subregion are linear, we measure how each instance linearly relates to its closet neighbors. For each data point x_i, it is reconstructed as a linear function of its k-nearest neighbors. Specifically, it finds the weights $w_{i,j}$ such that the squared distance between x_i and $\sum_{j=1}^{N} w_{i,j}x_j$ is as small as possible, assuming $w_{i,j} = 0$ if x_j is not among the k-nearest neighbors of x_i. This step is the constrained optimization problem described as follows:

$$\hat{W} = \arg\min_{W} \sum_{i=1}^{N} ||x_i - \sum_{j=1}^{N} w_{i,j}x_j||^2$$

$$subject\ to \begin{cases} \sum_{j=1}^{N} w_{i,j} = 1, & for\ i = 1, 2, ..., N, \\ w_{i,j} = 0, & x_j \notin KNN(x_i) \end{cases} \tag{1}$$

where W is the weight matrix containing all the weights $w_{i,j}$. Therefore the weight matrix W could be regarded as the co-occurrences of data points. A graph $G = (V, E)$ is generated according to the filtered weight matrix. Nodes represent instances and edges stand for the existence of filtered weight for two instances.

With the built graph, the graph2vec [17] method is used to generate corresponding embedding representation. Particularly, graph2vec is an unsupervised method inspired by document embedding models and it finds the representation for a graph by maximizing the likelihood of graph subtrees given the graph embedding using deep learning approaches. The detailed meta-feature extraction step is presented in Algorithm 1.

Meta-target Identification. The meta-targets of these historical datasets are obtained with the popular cross-validation method. The process of identifying meta-targets is as following: Given a set of datasets D and a set of pairs $\langle S, A \rangle$ where S represents sampling methods and A represents learning algorithms, we evaluate all combinations of $d \in D$ and $\langle s, a \rangle \in \langle S, A \rangle$ and the meta-targets are denoted as $R_{d,\langle s,a \rangle}$. To integrate the meta-features and the meta-targets, a meta repository is built for maintaining the meta information of historical datasets.

3.3   Automated Model Selection

In the procedure of Automated Model Selection, the new data will be firstly preprocessed to obtain corresponding meta-features with the graph meta-feature extraction method. Then based on the constructed meta repository, a user-based collaborative filtering (CF) method [11] is employed to select the best sampling

---

Algorithm 1. Meta-Feature Extraction

Input: $D = \{D_1, ..., D_n\}$: Historical datasets;
k: The number of nearest neighbors;
θ: The threshold for filtering graphs;
δ: The dimension of latent representation;
Output: V: Vector representations
1: **for** $i = 1, 2, ..., n$ **do**
2:    $nn \leftarrow k$-nearest neighbors of data points in D_i
3:    $W \leftarrow$ weight matrix of D_i according to nn
4:    $filterd\_W \leftarrow$ filter the weak co-occurrences less than the threshold θ
5:    $G_i \leftarrow$ generate graph (V, E) for dataset D_i
6: **end for**
7: $V \leftarrow$ the δ dimension representations of G using graph2vec(G, δ)
8: **return** V

---

method and algorithm, in which Pearson correlation coefficient is used to calculate the similarity. Particularly, each user in CF is replaced by a dataset and each item is replaced by a pair of $\langle s, a \rangle$ which represents the combination of a sampling method and an algorithm (e.g. $\langle$ SMOTE, KNN-k-3 $\rangle$).

4   Evaluation

4.1   Experimental Setup

In the experiment, 16 different sampling methods have been selected, which have been introduced in Sect. 2. Four popular classification algorithms, including Naive Bayes, C4.5, KNN and Random Forest, have been employed. Furthermore, extensive experiments are conducted with the leave-one-out method on 70 imbalanced datasets, which are originated from the UCI and PROMISE repository. In addition, for evaluating the imbalanced data classification performance, the top widely used measure AUC is employed [7]. Three baselines are employed for comparison, including SMOTE with Random Forest (SMOTE_RF), Auto-sklearn [6] and OBOE [23]. The k value for generating knn graphs is set as 3 according to the corresponding study [25]. For the graph2vec method, we have selected the value $\delta = 32$ for obtaining 32 meta-features as recommended in [17].

4.2   Experimental Results

In order to compare AutoIDL with the baselines, we provide the detailed AUC values for these methods on the employed 70 imbalanced datasets in Table 1. For each dataset, the best AUC values are highlighted in boldface. From Table 1, it could be observed that AutoIDL obtains the largest number of datasets with best AUC values among the compared four methods. Among all the 70 datasets, the proposed AutoIDL method could obtain the best classification performance with 57 datasets, which outperforms the three baseline methods apparently.

Table 1. The detailed AUC values for employed methods on the 70 datasets

DataID	SMOTE_RF	Auto-sklearn	OBOE	AutoIDL	DataID	SMOTE_RF	Auto-sklearn	OBOE	AutoIDL
1	0.6985	0.7147	0.7558	**0.7878**	36	**0.7331**	0.5826	0.5826	0.7194
2	0.7592	**0.8184**	0.6694	0.6778	37	0.5906	0.6733	0.654	**0.758**
3	1	0.9783	1	1	38	0.5583	0.6378	0.6705	**0.7381**
4	1	1	1	0.9938	39	0.623	0.6202	0.5984	**0.6906**
5	0.8192	0.8118	0.8713	**0.9093**	40	0.5855	0.6678	0.6207	**0.8076**
6	0.5694	0.6311	0.6511	**0.6658**	41	0.58	0.6474	0.6468	**0.7982**
7	0.9782	**0.987**	0.9694	0.9833	42	0.5675	0.6057	**0.7515**	0.7298
8	0.5	0.4873	0.4557	**0.6765**	43	0.553	0.7152	0.6559	**0.7777**
9	0.525	0.6403	0.5667	**0.6567**	44	0.5	0.4972	0.4915	**0.658**
10	0.8411	0.8438	0.8596	**0.9224**	45	0.5	0.6158	0.7117	**0.7942**
11	0.8664	0.8653	0.8873	**0.9586**	46	0.6382	0.7115	0.641	**0.8911**
12	0.5	0.5173	0.5243	**0.701**	47	0.6972	0.6948	0.6449	**0.773**
13	**0.7639**	0.6028	0.6222	0.7593	48	0.7266	0.7368	0.7489	**0.8037**
14	0.8051	0.8295	0.7967	**0.8834**	49	0.4912	0.4737	0.5182	**0.7504**
15	0.8988	0.8807	0.8764	**0.903**	50	0.5	0.6185	0.6596	**0.6831**
16	0.652	0.6996	0.7004	**0.7385**	51	1	1	1	1
17	0.7836	0.722	0.7717	**0.88**	52	0.6151	0.6151	0.6429	**0.7804**
18	0.6604	0.7459	0.658	**0.8125**	53	0.7212	0.6303	0.6351	**0.7465**
19	0.4965	0.5313	0.5428	**0.6275**	54	0.7238	**0.7909**	0.7499	0.7844
20	0.6411	**0.6946**	0.6038	0.6148	55	0.7939	0.8336	0.8657	**0.9582**
21	0.8194	0.8351	0.8212	**0.8767**	56	0.5949	0.7406	0.6628	**0.7647**
22	0.625	0.623	0.7964	**0.8343**	57	0.6835	0.7277	0.7479	**0.8729**
23	0.5406	0.5793	0.6072	**0.7313**	58	0.4983	0.5507	0.5085	**0.7466**
24	0.8817	0.875	0.846	**0.9784**	59	0.918	0.926	0.9625	**0.9902**
25	0.5	0.475	0.5283	**0.7273**	60	**0.9015**	0.7674	0.8089	0.8858
26	0.7086	0.7286	0.7724	**0.8063**	61	0.5833	0.6667	0.6136	**0.6927**
27	0.663	0.6944	0.6944	**0.8057**	62	0.7164	0.7063	0.6423	**0.7798**
28	0.6122	0.6419	0.5837	**0.6605**	63	0.5	0.5206	0.4872	**0.6452**
29	0.5845	**0.7365**	0.5867	0.6034	64	1	0.9197	0.9765	0.9766
30	0.9286	1	1	0.9459	65	0.5435	0.5992	0.5718	**0.7616**
31	0.6757	**0.7019**	0.6438	0.6925	66	0.6865	0.7098	0.6865	**0.7248**
32	0.8392	**0.8655**	0.8363	0.7729	67	0.91	0.9307	0.9378	**0.9719**
33	0.6609	0.6461	0.6609	**0.6965**	68	0.6413	0.6731	0.6546	**0.7938**
34	0.5	0.6184	0.5732	**0.7653**	69	0.7363	0.7089	0.75	**0.7952**
35	0.5818	0.5409	0.5864	**0.7113**	70	0.7076	0.7573	0.6491	**0.7893**

Furthermore, for the purpose of evaluating the gap between AutoIDL and the best classification performance, we use the relative maximum value (RMV) measure to calculate the ratio of AUC value for AutoIDL to the best classification result. For example, if the AUC of AutoIDL is 0.8 and the best AUC is 0.9 for a specific dataset, the corresponding RMV value of AutoIDL would be 0.89. Figure 2 provides the distribution of RMV values for the four employed methods on all datasets. It can be observed that AutoIDL obtains RMV values over 90% for most datasets, which indicates that AutoIDL could usually achieve the approximate AUC values to the best results. In addition, according to the RMV values shown in Fig. 2, the three baseline methods show rather poor classification

performance on some of the datasets compared with the best results, which may
be attributed to the imbalanced data problem.

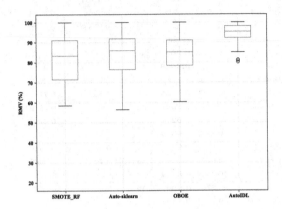

Fig. 2. Distribution of the RMV scores.

In order to evaluate the improvements of AutoIDL over existing AutoML
methods, Fig. 3 shows the improved AUC of AutoIDL against Auto-sklearn and
OBOE respectively. Note that the values are plotted in increasing order. Furthermore, in Fig. 3, the green area indicates the amount of performance improvement,
and the red area indicates the amount of performance degradation.

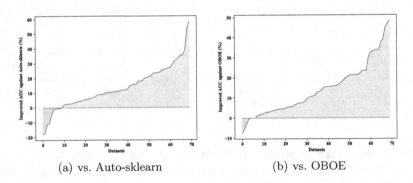

(a) vs. Auto-sklearn   (b) vs. OBOE

Fig. 3. Improved performance of AutoIDL against: (a) Auto-sklearn, (b) OBOE.

From Fig. 3, it can be observed that AutoIDL could usually outperform Auto-
sklearn and OBOE on most of the employed datasets as the two green areas
are both much larger than the corresponding red areas. What's more, Auto-
sklearn could be improved at most nearly 60% by AutoIDL while OBOE could
be improved at most nearly 50%. To be specific, based on the detailed AUC values shown in Table 1, AutoIDL outperforms Auto-sklearn for 87.14% datasets

and among these datasets, AutoIDL shows much higher performance improvement on 21 datasets, varying from 20.49% to 58.41%. Moreover, compared with OBOE, AutoIDL obtains the better classification performance in terms of AUC for 92.86% datasets. Particularly, 21 datasets have been improved more than 20%, varying from 20.43% to 48.45%.

5   Conclusion

In present study, we propose an automated imbalanced data learning method (AutoIDL) which handles the imbalanced data for AutoML. To obtain the characteristic of imbalanced datasets, they are firstly represented as graphs and the meta-features are then extracted with graph2vec. Subsequently the meta-targets of historical datasets are identified and represented as pairs of sampling methods and algorithms, which are integrated into the meta-repository as well as the corresponding meta-features. Finally based on the meta-repository, a user-based collaborative filtering method is used to build a ranking model to select appropriate models for new datasets. The extensive experiments with 70 different imbalanced datasets demonstrate that AutoIDL is effective and usually outperforms Auto-sklearn and OBOE.

References

1. Barua, S., Islam, M.M., Yao, X., Murase, K.: Mwmote-majority weighted minority oversampling technique for imbalanced data set learning. IEEE Trans. Knowl. Data Eng. **26**(2), 405–425 (2012)
2. Bunkhumpornpat, C., Sinapiromsaran, K., Lursinsap, C.: Safe-level-smote: safe-level-synthetic minority over-sampling technique for handling the class imbalanced problem. In: PAKDD, pp. 475–482 (2009)
3. Chawla, N.V., Bowyer, K.W., Hall, L.O., Kegelmeyer, W.P.: Smotesynthetic minority over-sampling technique. J. Artif. Intell. Res. **16**, 321–357 (2002)
4. Douzas, G., Bacao, F., Last, F.: Improving imbalanced learning through a heuristic oversampling method based on k-means and smote. Inf. Sci. **465**, 1–20 (2018)
5. Elshawi, R., Maher, M., Sakr, S.: Automated machine learning: state-of-the-art and open challenges. CoRR (2019). http://arxiv.org/abs/1906.02287
6. Feurer, M., Klein, A., Eggensperger, K., Springenberg, J., Blum, M., Hutter, F.: Efficient and robust automated machine learning. In: NeurIPS, pp. 2962–2970 (2015)
7. Guo, H., Li, Y., Jennifer, S., Gu, M., Huang, Y., Gong, B.: Learning from class-imbalanced data: review of methods and applications. Expert Syst. Appl. **73**, 220–239 (2017)
8. Han, H., Wang, W.Y., Mao, B.H.: Borderline-smote: a new over-sampling method in imbalanced data sets learning. In: ICIC, pp. 878–887 (2005)
9. Hart, P.: The condensed nearest neighbor rule. IEEE Trans. Inf. Theory **14**(3), 515–516 (1968)
10. He, H., Bai, Y., Garcia, E.A., Li, S.: ADASYN: adaptive synthetic sampling approach for imbalanced learning. In: IJCNN, pp. 1322–1328 (2008)

11. Herlocker, J.L., Konstan, J.A., Borchers, A., Riedl, J.: An algorithmic framework for performing collaborative filtering. In: SIGIR, pp. 227–234 (1999)
12. Kubat, M., Matwin, S., et al.: Addressing the curse of imbalanced training sets: one-sided selection. In: ICML, pp. 179–186 (1997)
13. Laurikkala, J.: Improving identification of difficult small classes by balancing class distribution. In: Quaglini, S., Barahona, P., Andreassen, S. (eds.) AIME 2001. LNCS (LNAI), vol. 2101, pp. 63–66. Springer, Heidelberg (2001). https://doi.org/10.1007/3-540-48229-6_9
14. Lin, W.C., Tsai, C.F., Hu, Y.H., Jhang, J.S.: Clustering-based undersampling in class-imbalanced data. Inf. Sci. **409**, 17–26 (2017)
15. López, V., Fernández, A., García, S., Palade, V., Herrera, F.: An insight into classification with imbalanced data: empirical results and current trends on using data intrinsic characteristics. Inf. Sci. **250**, 113–141 (2013)
16. Mani, I., Zhang, I.: KNN approach to unbalanced data distributions: a case study involving information extraction. In: Proceedings of Workshop on Learning from Imbalanced Datasets, pp. 1–7 (2003)
17. Narayanan, A., Chandramohan, M., Venkatesan, R., Chen, L., Liu, Y., Jaiswal, S.: graph2vec: learning distributed representations of graphs. CoRR (2017). http://arxiv.org/abs/1707.05005
18. Olson, R.S., Moore, J.H.: TPOT: a tree-based pipeline optimization tool for automating machine learning. In: Hutter, F., Kotthoff, L., Vanschoren, J. (eds.) Automated Machine Learning. TSSCML, pp. 151–160. Springer, Cham (2019). https://doi.org/10.1007/978-3-030-05318-5_8
19. Pedregosa, F., et al.: Scikit-learn: machine learning in Python. J. Mach. Learn. Res. **12**, 2825–2830 (2011)
20. Thornton, C., Hutter, F., Hoos, H.H., Leyton-Brown, K.: Auto-WEKA: combined selection and hyperparameter optimization of classification algorithms. In: KDD, pp. 847–855 (2013)
21. Tomek, I.: Two modifications of CNN. IEEE Trans. Syst. Man Cybern. **6**, 769–772 (1976)
22. Wilson, D.L.: Asymptotic properties of nearest neighbor rules using edited data. IEEE Trans. Syst. Man Cybern. **3**, 408–421 (1972)
23. Yang, C., Akimoto, Y., Kim, D.W., Udell, M.: OBOE: collaborative filtering for AutoML model selection. In: KDD, pp. 1173–1183 (2019)
24. Yen, S.J., Lee, Y.S.: Cluster-based under-sampling approaches for imbalanced data distributions. Expert Syst. Appl. **36**(3), 5718–5727 (2009)
25. Zhu, X.J.: Semi-supervised learning literature survey. Technical report. University of Wisconsin-Madison Department of Computer Sciences (2005)

Fusion of Domain Knowledge and Text Features for Query Expansion in Citation Recommendation

Yanli Hu$^{(\boxtimes)}$ (ID), Chunhui He (ID), Zhen Tan, Chong Zhang, and Bin Ge

Science and Technology on Information Systems Engineering Laboratory,
National University of Defense Technology,
Changsha 410073, Hunan, People's Republic of China
{huyanli,tanzhen08A,zhangchong,gebin}@nudt.edu.cn, xtuhch@163.com

Abstract. Academic citation recommendation addresses the task of recommending citations for a scientific paper. Effective citation recommendation is greatly important for literature reviewing, literature-based discovery and a wide range of applications. In this paper, we propose a query expansion framework via fusing domain-specific knowledge and text features for academic citation recommendation. Starting from an original query, domain-specific and context-aware concepts are derived to expand the query to improve the performance of citation recommendation. From the perspective of enriching knowledge structure, domain-specific concepts are extracted from domain knowledge graphs such as ACM Computing Classification System and IEEE thesaurus. From the perspective of providing query scenarios, the query is extensively considered with context-aware concepts derived from text feature extraction. Then candidate concepts are filtered via distributed representations like BERT to expand the query. Experiments of citation recommendation for papers in public data sets show that our proposed model of query expansion improves the performance of academic citation recommendation.

Keywords: Query expansion · Citation recommendation · Domain knowledge graph · Feature extraction · Distributed representation

1 Introduction

Literature reviews are in great demand arising from the rapidly-increasing publications of scientific articles as well as maintaining awareness of developments in scientific fields. Finding and evaluating relevant material to synthesize information from various sources is a crucial and challenging task.

One main method of literature searching is keyword search. However, keyword search can't meet the need of literature searching partly due to the *synonymy* and *polysemy* problems. On the other hand, literature reviews consist of correlated articles of variant words, resulting in worse performance of citation recommendation.

G. Li et al. (Eds.): KSEM 2020, LNAI 12275, pp. 105–113, 2020.
https://doi.org/10.1007/978-3-030-55393-7_10

Table 1. The number of references of surveys hit in the top 100 retrieval results.

Surveys	#references	#references hit in the top100 results
Title: A Survey of Mobility Models for Ad Hoc Network research **Keywords:** ad hoc networks, entity mobility models, group mobility models	33	0
Title: A Survey of Web Information Extraction Systems **Keywords:** Information Extraction, Web Mining, Wrapper, Wrapper Induction	47	14
Title: A Survey of Controlled Experiments in Software Engineering **Keywords:** Controlled experiments, survey, research methodology, empirical software engineering	52	13

The situation improves little despite of the development of search techniques in recently years. Take the literature searching of surveys as examples. Three surveys are randomly selected from a bibliography sharing service, CiteULike, and the retrieval performance by the search engine Google Scholar is evaluated according to the amount of references located among the searching results, as illustrated in Table 1. The amount of references of the surveys hit in the top 100 retrieval results indicates poor performance for citation recommendation.

In this paper, we propose a novel query expansion framework for academic citation recommendation. By incorporating multiple domain knowledge graphs for scientific publications, we use domain-specific concepts to expand an original query with appropriate spectrum of knowledge structure. Meanwhile, text features are extracted to capture context-aware concepts of the query. Then candidate concepts, namely domain-specific and context-aware concepts, are further filtered via distributed representations to derive the expanded query for citation recommendation.

2   Related Work

Traditional methods of query expansion choose terms from relevant/irrelevant documents. Terms are usually weighted according to their frequency in single document and in the collection, and the top terms with the highest frequency are added into the initial query. To provide structured knowledge of a topic, Ref [1] addresses the problem of citation recommendation by expanding the semantic features of the abstract using DBpedia Spotlight [14], a general purpose knowledge graph.

A content-based method for recommending citations [2] is suggested, which embeds a query and documents into a vector space, then reranks the nearest neighbors as candidates using a discriminative model. A context-aware citation recommendation model [8] is proposed with BERT and graph convolutional networks. In the context of heterogenous bibliographic networks, ClusCite [12], a cluster-based citation recommendation framework, is proposed.

Text feature extraction is fundamental to citation recommendation. *TF-IDF* [9] (Term Frequency–Inverse Document Frequency) and their variants are essential to represent documents as vectors of terms weighted according to term frequency in one document as well as in the number of documents in the corpus.

To capture the latent semantic associations of terms, Latent Semantic Analysis (LSA) [5] is proposed to analyze the relationships of documents based on common patterns of terms. To uncover semantic structures of documents, Latent Dirichlet Allocation (LDA) [3] is a generative probabilistic model to represent one document as a mixture of topics, and the words in each document imply a probabilistic weighting of a set of topics that the document embodies.

In the above representations, each term in documents is represented by a one-hot vector, in which only the corresponding component of the term is 1 and all others are zeros. To provide a better estimate for term semantics, distributed representations of terms are proposed to boost the semantics of documents. Instead of sparse vectors, embedding techniques, such as Word2Vec [10] and BERT [6], learn a dense low-dimensional representation of a term by means of its neighbors.

3   The Proposed Framework

The proposed query expansion framework for citation recommendation is composed of three major steps: (1) From the perspective of enriching knowledge structure, expanding an original query with domain-specific concepts based on multiple domain knowledge graphs. (2) From the perspective of providing query scenarios, extending the query with context-aware concepts derived from text feature extraction. (3) Filtering out the above candidate concepts via distributed representations to derive the expanded query for citation recommendation.

3.1   Model Overview

To capture diverse information needs underlying the query, we propose a query expansion framework combining domain knowledge with text features, as illustrated in Fig. 1.

Starting from an initial query q_0 in Fig. 1, domain knowledge and text features are used to expand the query, and then candidate concepts are filtered to derive the enriched query for citation recommendation.

More specifically, domain knowledge are used to provide knowledge structure in the form of domain-specific concepts, with the aid of multiple domain knowledge graphs such as ACM Computing Classification System[1] (CCS for short)

---

[1] https://dl.acm.org/ccs.

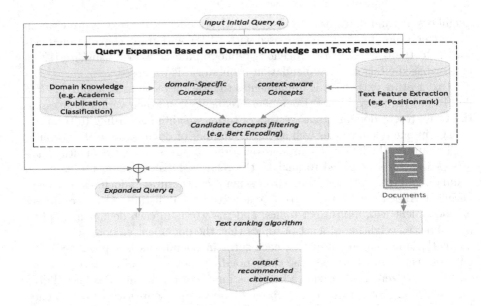

Fig. 1. The query expansion framework based on domain knowledge and text features.

and IEEE thesaurus[2]. Meanwhile, text feature extraction is utilized to derive context-aware concepts from document collections to provide query scenarios.

Among the above domain-specific and context-aware candidate concepts, filtering techniques are applied to choose a set of closely-related concepts to formulate a new query q, and citations are recommended with respect to the expanded query q.

3.2   Domain-Specific Expansion Using Multiple Knowledge Graphs

With the characteristics of citation recommendations in mind, we propose query expansion based on knowledge graphs. However, general-purpose knowledge graphs are not applicable in academic disciplines.

Take the citation recommendation for '*A Survey of Mobility Models for Ad Hoc Network research*' in Table 1 as an example (hereinafter called the survey). Using Microsoft Concept Graph[3], the related concepts for *ad hoc network* consists of *research domain, wireless network, network, system* and so on. As to the concept *mobility model*, the suggested concepts include *parameter, component, user criterion* and so on. Analyzing the references cited in the survey, the suggested concepts derived from general-purpose knowledge graphs contribute little to recommend appropriate citations. On the other hand, knowledge graphs

---

[2] https://www.ieee.org/content/dam/ieee-org/ieee/web/org/pubs/ieee-thesaurus.pdf.

[3] https://concept.research.microsoft.com/.

such as WordNet [11] have also limited contribution, as no concepts for *ad hoc network* and *mobility model.*

We utilize domain-specific knowledge graphs, such as ACM CCS and IEEE thesaurus, to expand academic concepts. Take the concept *ad hoc networks* as an example. The concept hierarchies of ACM CCS and IEEE thesaurus are illustrated in Fig. 2, which suggest structures of concepts and their relations with *ad hoc networks*. Exploring the references of the survey, highly relevant concepts appearing in the references are circled in red rectangles in Fig. 2.

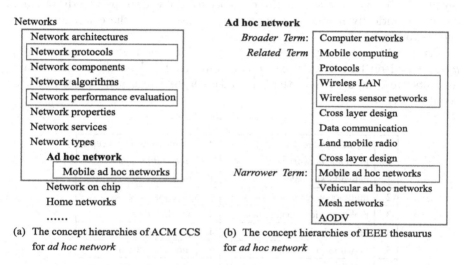

(a) The concept hierarchies of ACM CCS for *ad hoc network*

(b) The concept hierarchies of IEEE thesaurus for *ad hoc network*

Fig. 2. The diverse concept hierarchies for *ad hoc networks* in multiple domain knowledge graphs. (Color figure online)

According to IEEE thesaurus, *related terms* are also highly probably mentioned in references, such as *wireless sensor networks*. For ACM CCS, we notice that the concepts of the siblings of *network types*, a hypernym of *ad hoc networks*, are also highly related to the topics of the references, rather than the direct siblings of *ad hoc networks* like *networks on chip* or *home networks*. Here a concept's siblings refer to those concepts that share the same parents with the given concept.

Therefore, to expand domain-specific concepts of a term to the original query, we design adaptive policies to multiple domain knowledge graphs. For IEEE thesaurus, the concepts of broader terms, related terms and narrower terms are added into the expansion set of candidate concepts. For ACM CCS, a two-level policy is suggested to add the concepts of the hyponyms and the siblings of the hypernyms of the term to the original query.

Using the policy, domain-specific knowledge for *ad hoc networks* with 22 concepts are derived from the ACM CCS and IEEE thesaurus, as circled in the blue rectangles in Fig. 2. The concept *mobile ad hoc networks* is counted once.

3.3   Context-Aware Expansion Based on Text Feature Extraction

To expand a query with context awareness, the basic idea is to derive context-aware terms from document collections based on text feature extraction, as context-aware terms enriched the original query with characteristics of scenarios for better matching of relevant documents.

Key phrases are first extracted using PositionRank algorithm [7] as features of documents. Then document collections are processed to derive semantically related key phrases with terms in the original query, such as highly co-occurred key phrases. Providing the scenarios of co-occurrence with respect to the original query, semantically related key phrases are appended to the candidate set of query expansion.

Figure 3 lists ten context-aware concepts of *ad hoc networks* and *mobility model* based on text feature extraction, respectively. Here, the context-aware concepts are computed from the CiteULike corpus which will be described in detail in Sect. 4.

#	Context-aware concepts to *ad hoc networks*	#	Context-aware concepts to *mobility model*
1	networks simulator	1	mobile users
2	multihop wireless networks	2	wireless networks
3	mobile devices communicate	3	wireless links
4	bandwidth channel	4	performance differences
5	networks protocol design	5	mobility pattern
6	mobility patterns	6	random movement
7	simulation environments	7	mobile station
8	mobility models	8	cellular environment
9	wireless networks	9	traffic parameters
10	Network protocol	10	cellular systems

Fig. 3. The context-ware concepts of *ad hoc networks* and *mobility model.*

3.4   Candidate Concept Filtering via Distributed Representation

Rather than directly expanding an original query with candidate concepts in Sect. 3.2 and Sect. 3.3, the aim is to filter the set of candidate concepts to derive a subset of closely-related concepts for query expansion to reduce noise.

In order to solve this problem, we propose a candidate concept filtering method via distributed representations. The input consists of a candidate concept for expansion and the original query. Then, the vectorized representations of the inputs are concatenated via BERT distributed representations, and each input will be converted into a $1 * 1024$ dimensional vector. For detailed calculation principles, please see the BERT-as-service tool[4].

Then cosine similarity is used to calculate the distance between the vector representations of the candidate concept and the original query, and output the normalized result (between 0 and 1) as a matching score. With the matching scores between candidate concepts and query, candidate concepts are sorted, and the top-k closely-related concepts are chosen to expand the original query.

4   Experiments

4.1   Data Sets and Evaluation Metrics

In this section, we test our model for citation recommendation tasks on two public data sets. The DBLP data set contains citation information extracted by Tang et al. [13]. The CiteULike data set consists of scientific articles from CiteULike database[5]. Statistics of the two data sets are summarized in Table 2.

Table 2. Data sets overview.

Data sets	CiteULike	DBLP
#papers	24,167	149,363
#terms	6,488	30,849
#average citations per paper	8.25	5.52
#papers in training set	14501	89619
#papers in validation set	4833	29872
#papers in testing set	4833	29872

The performance is mainly evaluated in terms of precision and recall. The precision at top N results (P@N for short), and the recall at top N results (R@N for short) are reported respectively. Additionally, mean average precision (MAP) [4] is evaluated to measure the performance averaged over all queries,

$$MAP(Q) = \frac{1}{\|Q\|} \sum_{j=1}^{\|Q\|} \frac{1}{m_j} \sum_{k=1}^{m_j} Precision(R_{jk}) \qquad (1)$$

where Q is the set of queries. For the query q_j, $\{d_1, ..., d_{m_j}\}$ is the set of cited articles and R_{jk} is the set of ranked results from the top result to the article d_k.

---

[4] https://github.com/hanxiao/bert-as-service.
[5] http://static.citeulike.org/data/current.bz2.

4.2   Experimental Results

The experimental results on the CiteULike data set are presented in Table 3. We vary our model with text analysis and filtering methods. The *domain KG + TF-IDF* method expands query without filtering, namely implementing domain-specific expansion based on multiple domain knowledge graphs and context-aware expansion with TF-IDF. All candidate concepts are appended to the original query without filtering. Similarly, the *domain KG + LSA* and *domain KG + LDA* methods use domain knowledge plus LSA and LDA, respectively. In contrast to the first three methods, BERT embeddings of candidate concepts are used for filtering in the *domain KG + TF-IDF + BERT filtering* method.

Table 3. Experimental results of our model on the CiteUlike data set

CiteULike	MAP	P@10	P@20	R@20	R@50
domain KG + TF-IDF	0.211	0.448	0.364	0.277	0.313
domain KG + LSA	0.150	0.340	0.271	0.145	0.209
domain KG + LDA	0.113	0.278	0.222	0.122	0.182
domain KG + TF-IDF + BERT filtering	**0.287**	**0.694**	**0.605**	**0.329**	**0.361**

The results in Table 3 show that among the first three methods without filtering, the *domain KG + TF-IDF* method has better performance than the ones using LSA and LDA. Thus, the impacts of filtering technique on query expansion are further evaluated based on TF-IDF.

Furthermore, the method with filtering outperformed the ones without filtering. And the *domain KG + TF-IDF + BERT filtering* method contributes the best performance among all the methods on the CiteULike data set. It indicates that the expansion of domain-specific and context-ware concepts plus BERT filtering improves the performance of citation recommendation.

The experimental results on the DBLP data set also proves the effectiveness of query exapnsion using domain-specific and context-ware concepts plus BERT filtering, as shown in Table 4.

Table 4. Experimental results of our model on the DBLP data set

DBLP	MAP	P@10	P@20	R@20	R@50
domain KG + TF-IDF	0.095	0.160	0.151	0.203	0.365
domain KG + LSA	0.087	0.143	0.127	0.182	0.360
domain KG + LDA	0.073	0.137	0.122	0.172	0.282
domain KG + TF-IDF + BERT filtering	**0.168**	**0.295**	**0.228**	**0.331**	**0.439**

The above results suggest that combining domain knowledge with text features indicates remarkable advantages for citation recommendation, and filtering of candidate concepts is key to improve the performance.

5   Conclusions

In this paper, we address the problem of citation recommendation for literature review. Fusing domain knowledge and text feature to expand query is verified to improve the performance of locating citations scientific articles. Domain-specific concepts are extracted from multiple domain knowledge graphs to enrich knowledge structure for query expansion. Context-aware concepts are derived from text feature extraction to provide query scenarios. Then candidate concepts are filtered via distributed representations like BERT to expand the query with closely-related concepts. Experiments of citation recommendation on bibliographic databases show that our proposed model effectively improves the performance of citation recommendation.

Future research considers using large-scale scientific literature corpora to fine-tune the BERT pre-training vectors. In addition, the combination of named entity recognition technology to achieve the extraction of term features is also our focus.

References

1. Ayala-Gómez, F., et al.: Global citation recommendation using knowledge graphs. J. Intell. Fuzzy Syst. **34**(5), 3089–3100 (2018)
2. Bhagavatula, C., et al.: Content-based citation recommendation. arXiv preprint arXiv:1802.08301 (2018)
3. Blei, D.M., et al.: Latent Dirichlet allocation. J. Mach. Learn. Res. **3**(Jan), 993–1022 (2003)
4. Christopher, D.M., et al.: Introduction to Information Retrieval (2008)
5. Deerwester, S., et al.: Indexing by latent semantic analysis. J. Inf. Sci. **41**(6), 391–407 (1990)
6. Devlin, J., et al.: BERT: pre-training of deep bidirectional transformers for language understanding. CoRR abs/1810.04805 (2018)
7. Florescu, C., et al.: PositionRank: an unsupervised approach to keyphrase extraction from scholarly documents. In: ACL, pp. 1105–1115. ACL, July 2017
8. Jeong, C., et al.: A context-aware citation recommendation model with BERT and graph convolutional networks. arXiv preprint arXiv:1903.06464 (2019)
9. Jing, L.P., et al.: Improved feature selection approach TFIDF in text mining. In: ICMLC, vol. 2, pp. 944–946. IEEE (2002)
10. Mikolov, T., et al.: Distributed representations of words and phrases and their compositionality. In: NIPS, pp. 3111–3119 (2013)
11. Miller, G.A.: WordNet: An Electronic Lexical Database. MIT Press (1998)
12. Ren, X., et al.: Cluscite: effective citation recommendation by information network-based clustering. In: SIGKDD, pp. 821–830. ACM (2014)
13. Tang, J., et al.: Arnetminer: extraction and mining of academic social networks. In: SIGKDD, pp. 990–998. ACM (2008)
14. Xu, B., et al.: CN-DBpedia: a never-ending chinese knowledge extraction system. In: Benferhat, S., Tabia, K., Ali, M. (eds.) IEA/AIE 2017. LNCS (LNAI), vol. 10351, pp. 428–438. Springer, Cham (2017). https://doi.org/10.1007/978-3-319-60045-1_44

Robust Sequence Embedding
for Recommendation

Rongzhi Zhang[1], Shuzi Niu[2(✉)], and Yucheng Li[2]

[1] University of Chinese Academy of Sciences, Beijing, China
`zhangrongzhi17@mails.ucas.edu.cn`
[2] Institute of Software, Chinese Academy of Sciences, Beijing, China
`{shuzi,yucheng}@iscas.ac.cn`

Abstract. Sequential recommendation is a significant task that predicts the next items given user historical transaction sequences. It is often reduced to a multi-classification task with the historical sequence as the input, and the next item as the output class label. Sequence representation learning in the multi-classification task is of our main concern. The item frequency usually follows the long tail distribution in recommendation systems, which will lead to the imbalanced classification problem. This item imbalance poses a great challenge for sequence representation learning. In this paper, we propose a **Ro**bust **S**equence **E**mbedding method for the recommendation, RoSE for short. RoSE improves the recommendation performance from two perspectives. We propose a balanced k-plet sampling strategy to make each training batch balanced at the data level and propose the triplet constraint for each training sequence to make sure of balance and robust distribution in feature space at the algorithmic level. Comprehensive experiments are conducted on three benchmark datasets and RoSE shows promising results in the face of item imbalance.

Keywords: Sequence embedding · Imbalance · Recommendation systems

1 Introduction

Sequential recommendation is a fundamental problem in real world. It attempts to predict the next items based on user historical transaction sequences. At each time step, it treats the user sequences before this time step as input instances, and the current items as the output class labels. In this sense, sequential recommendation is reduced to a multi-class classification task at each time step.

Many sequential recommendation approaches [1, 11, 12] have been proposed, among which neural sequence models [1] are becoming the main stream. They roughly consist of the following two steps. First, recurrent hidden units or its variants are utilized to learn the users' sequential features from input sequence instances. Second, users' sequential features, or sequence representation, and items' features are combined to feed the final layer to determine which items will

© Springer Nature Switzerland AG 2020
G. Li et al. (Eds.): KSEM 2020, LNAI 12275, pp. 114–122, 2020.
https://doi.org/10.1007/978-3-030-55393-7_11

be bought next. In this paper, we mainly focus on the sequence representation learning in this classification task.

In the recommendation system, most items in the tail of the distribution are consumed only several times, which belong to the minority category. On the contrary, a few items are consumed many times, which belong to the majority category. Such sequential recommendation methods fail in face of the long tail distribution data.

In traditional classification task, sampling based methods [9] and cost sensitive learning methods [4] are two major kinds of techniques to deal with class imbalance problem. Therefore, both kinds of solutions are not able to be applied to sequence embedding in imbalanced classification tasks. How to learn robust sequence embedding against the item imbalance becomes our main concern.

In this paper, we propose a **Ro**bust **S**equence **E**mbedding method against the item imbalance problem in the sequential recommendation task. From the data perspective, we propose a balanced k-plet sampling strategy to generate a training batch for both tasks in an alternative manner. It aims to uniformly sample sequences with balanced target class label. From the algorithmic perspective, we introduce a triplet constraint that intra-class sequence distance is larger than the inter-class sequence distance as the auxiliary task to the basic RNN, referred to as triplet constrained RNN. The stability of the triplet constraint contributes to balance representation between the input sequences with the same and different target items. We train the sequential recommendation task and its auxiliary task in a multi-task framework.

To investigate the effectiveness of sequence representations learned from RoSE, we conduct comprehensive experiments on three benchmark recommendation data sets. Experimental results that RoSE can achieve better performance compared with other baselines. Our main contributions are summarized as follows: (1) We first investigate how the sequence representation changes when the predicted item distribution is imbalanced. (2) To solve this problem from the algorithmic perspective, we propose triplet constrained RNN to help separate the minority class of sequences. (3) To solve this problem from the data perspective, we propose balanced k-plet sampling strategy to obtain a balanced training batch.

2   Related Work

2.1   Sequential Recommendation

Existing sequential recommendation methods capture the sequential pattern based on either Markov Chains or Neural Sequence Models. Markov Chain based models [11,12] utilize the sequence as a state transform process and capture the local feature through neighbor relationship. In light of the basic Markov Chain model [12], FPMC [11] puts forward a novel concept that leverages a personalized transmission graph matrix for each user. Neural Sequence models [1,5,8] can encode the whole item sequences while decode item representations with maximum probability in the next step.

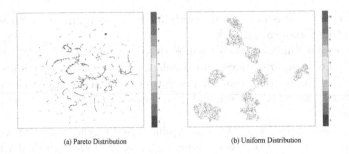

(a) Pareto Distribution   (b) Uniform Distribution

Fig. 1. UMAP of sequence representations learned from data with item frequency follows (a) Pareto Distribution and (b) Uniform Distribution respectively.

2.2 Learning from Imbalanced Data

Learning from imbalanced data is still a focus of intense research for decades [6]. Traditional studies solve this problem with data level, algorithm level and hybrid methods. **Data level methods** mainly concentrate on how to adjust the training set to balance the distribution through sampling. Oversampling and undersampling are two major kind of data-level methods. **Algorithm level methods** directly modify learning algorithms to alleviate bias towards the majority classes or correct the data distribution. The most popular branch is cost-sensitive learning. This kind of methods set a higher cost to the minority class and we boost its importance during the learning process. **Hybrid methods** combine the advantages of two previous groups. Quintuplet instance sampling and the associated triple-header hinge loss [3] are ensured to learn a robust and discriminative representation in standard deep learning framework for vision tasks.

Most existing studies on imbalanced data are for the typical classification task. In this paper, we put more emphasis on sequence representation learning from imbalanced data.

3   Sequence Embedding from Imbalanced Data

According to our survey, rare studies focus on how learned representations change in face of imbalanced data. Supervised embedding methods encode the label information in the input representation. Assuming that the item frequency follows nearly a uniform distribution, the label information does make the representation learning more discriminative. A more natural phenomenon is that the class frequency follows the long-tail distribution such as Pareto distribution in a social or scientific scenario. How the learned representations will be changed from the imbalanced data is attracting our attention.

To investigate this problem, we exploit the basic Recurrent Neural Network [1] with one layer of 10 hidden units to do comparative experiments on two synthetic datasets. Both datasets contain $11,000$ sequences with length 100, including 900 for training, 100 for validation and $10,000$ for test. Each item in

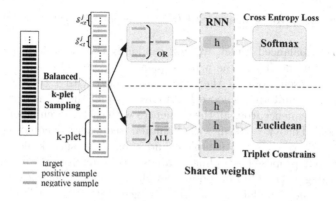

Fig. 2. Architecture of RoSE.

the sequence is sampled from $\{i\}_{i=1}^{10}$. The only difference between two datasets lies in the item frequency distribution. One is from the Pareto distribution, the other is from the Uniform distribution. We train the recurrent neural network model on both sets and obtain the sequence representations on test data shown in Fig. 1.

It is obvious that the sequence point separation is more clear in Fig. 1(b) than in Fig. 1(a). In this sense, sequence representations learned from uniform data are better than those learned from Pareto data. Another observation is that points labeled with item 1, 2, 3 almost dominate the whole dataset and other points even cannot be picked out in Fig. 1(a). These confused representations will lead to cascade errors in the following classifier layer. Thus the imbalanced data makes the sequence embedding mixed together and pose great challenge to learn robust sequence representation in this scenario.

4   Robust Sequence Embedding

To learn a robust sequence representation against item imbalance, we propose a **Ro**bust **S**equence **E**mbedding approach, referred to as RoSE. As shown in Fig. 2, the network architecture is divided into two components: (1)balanced k-plet sampling strategy; (2) Triplet Constrained RNN. First, we generate each training batch by balanced K-plet sampling strategy. Then, the training batch is fed to the triplet constrained RNN.

4.1   Balanced K-Plet Sampling Strategy

Balanced k-plet sampling strategy employs k-plet structure to make sure uniformly sample the training sequences for each item. Details are listed as follows. It generates a balanced training batch. The batch includes b balanced k-plets. Each balanced k-plet is consist of one pivot sequence, $k/2$ sequences with the same label as the pivot sequence and $k/2$ sequences with different labels from the pivot.

To sample a balanced k-plet, we first randomly choose an item label i uniformly from all items. Then, the pivot sequence $S_{\prec t}^u$ and its $k/2$ sequences $\mathcal{S}_{\prec t}^j$ are randomly chosen from all the sequences labeled with i through inverted index. Finally, we uniformly choose $k/2$ items from all the items except i, randomly choose a sequence from all the sequences with each chosen label e through inverted index. Thus we obtain $k/2$ dissimilar sequences denoted as $\bar{S}_{\prec t}^u$. So far a k-plet $(S_{\prec t}^u, \mathcal{S}_{\prec t}^u, \bar{S}_{\prec t}^u)$ is sampled, and a batch of such k-plets are sampled through our proposed sampling strategy. Different from traditional k-plet sampling strategies [7]. Our proposed sampling methods not only focuses on sequences with the same label, but also sequences with different labels in order to make training batch balanced.

4.2 Triplet Constrained RNN

Each item is represented as a one-hot vector in $\mathbb{R}^N$ is fed into the recurrent layer. For user's historical sequence $S_{\prec t}^u$, we use the Long Short Term Memory (LSTM) block as the basic unit to obtain the dynamic sequence embedding denoted as $h(S_{\prec t}^u; W)$ at time $t-1$. Parameters in this recurrent layer is denoted as W.

In order to predict which items will occur at time step t, the sequence embedding $h(S_{\prec t}^u)$ at time step $t-1$ will be passed through the softmax layer with N output units. The i-th unit is represented as the probability that item i appears at time step t shown as $P(i|S_{\prec t}^j) = \frac{\exp(f_1(h(S_{\prec t}^u; W), i; W_s))}{\sum_j \exp(f_1(h(S_{\prec t}^u; W), j; W_s))}$, where W_s is the parameter of the output layer.

Through the balanced k-plet sampling strategy, we obtain the batch $\mathcal{B}$ and organize it into the training instances for sequential recommendation task denoted as $B_s = \{(S_{\prec t}^u, s_t^u)\}$. Each training instance is a input sequence $S_{\prec t}^u$ with its target item s_t^u. Suppose these instances are independent, sequential recommendation task is to optimize the cross entropy objective function in Eq. (1).

$$\mathcal{L}_1(B_s) = \frac{1}{|B_s|} \sum_{(S_{\prec t}^j, s_t^j) \in B_s} \log P(s_t^j | S_{\prec t}^j). \tag{1}$$

To make the learned sequence embedding robust against item imbalance, we propose a triplet constrain to the sequence embedding. Triplet constrain says that the distance between sequence a and b of the same label should be smaller than the distance between sequence a and c of different labels as $d(h(a; W), h(b; W)) < d(h(a; W), h(c; W))$. $d(\cdot, \cdot)$ is the distance between two sequences. This constrain will push the sequence embedding of a and b together, while pulling the sequence embedding of a apart from c. In other words, the explicit separation constrain solves the problem that sequence representations are mixed together in face of item imbalance. In this sense, the triplet constrain will make the sequence embedding robust.

There are various ways to define the sequence distance, such as Euclidean distance and cosine distance so on. However, whether the sequence distance is suitable for this scenario remains unknown. Therefore, we learn the sequence distance measure as the auxiliary task. First, we organize the training sequence batch $\mathcal{B}$ into a triplet set B_m, and $B_m = \{(S_{\prec t}^j, C_{\prec t_c}^p, \bar{C}_{\prec t_{\bar{c}}}^n)|C_{\prec t_c}^p \in \mathcal{S}_{\prec t}^j, \bar{C}_{\prec t_{\bar{c}}}^n \in \bar{\mathcal{S}}_{\prec t}^j, (S_{\prec t}^j, \mathcal{S}_{\prec t}^j, \bar{\mathcal{S}}_{\prec t}^j) \in \mathcal{B}\}$. For each triplet $(S_{\prec t}^j, C_{\prec t_c}^p, \bar{C}_{\prec t_{\bar{c}}}^n)$, there is a pivot sequence $S_{\prec t}^j$, sequence with the same label as $S_{\prec t}^j$ is the positive sample $C_{\prec t_c}^p$, and sequence with a different label is the negative sample $\bar{C}_{\prec t_{\bar{c}}}^n$. Then we use a ratio measure [2] to classify the samples into two class. In order to satisfy the triplet constrain, mean square error is defined as the loss function in Eq. (2). The loss aims to maximize the probability of positive samples in local neighborhood.

$$\mathcal{L}_2(B_m) = \sum_{(x,x^+,x^-)\in B_m} \|(d_+, d_- - 1)\|_2^2 = const * d_+$$

$$\text{where, } d_+ = \frac{e^{\|h(x)-h(x^+)\|_2}}{e^{\|h(x)-h(x^+)\|_2} + e^{\|h(x)-h(x^-)\|_2}},$$

$$d_- = \frac{e^{\|h(x)-h(x^-)\|_2}}{e^{\|h(x)-h(x^-)\|_2} + e^{\|h(x)-h(x^+)\|_2}}$$

$$(2)$$

Finally, we learn both sequential recommendation task and the auxiliary task satisfying these triplet constrains in a multi-task framework simultaneously. Here we define the whole network structure as the Triplet Constrained RNN. We share the weight W of two tasks in the hidden layer and train two tasks by updating the weight matrix alternately. We use the training parameter α to choose the order.

5 Experiment

To investigate the proposed RoSE, we conduct comprehensive experiments on three benchmark datasets.

5.1 Experimental Setting

We conduct the experiment on three benchmark dataset, MovieLens-1M, Tmall and Amazon Movies. In our experiments, MovieLens-1M is a subset of the Movielens dataset with $6.0K$ users, $3.7K$ movies and $1.0M$ ratings. Amazon-Movies has $3.2K$ users and $24.3K$ items, and $451.8K$ ratings. Tmall is a user-purchase dataset obtained from IJCAI 2015 competition which has $182.8K$ transactions, $0.8K$ user and $4.5K$ brands. We remove users with the number of ratings less than 50 in all datasets.

We compare our model with the following baselines. **POP**: The most popular items are recommended. **UKNN**: It predicts the next items based on consumed items of the target user's k neighbors, where user neighbors are defined based on cosine distance or Euclidean distance. **BPRMF** [10]: It is a matrix factorization method and directly optimize the ranking loss. **FPMC** [11]: Factorized Personalized Markov Chains stacks state transition matrixes into a cube which satisfies the Markov property. **RNN** [1]: Recurrent Neural Network encodes transactions as item sequences into recurrent neural network and predicts the next behavior.

Table 1. Performance comparison on three datasets.

Dataset	Metric@10	POP	UKNN	BPRMF	FPMC	RNN	RoSE
MovieLens-1M	sps	0.0502	0.1237	0.0140	0.0154	<u>0.2788</u>	**0.2866**
	recall	0.0391	0.0552	0.0105	0.0468	<u>0.0756</u>	**0.0764**
	precision	0.2360	0.2305	0.0699	0.2120	<u>0.3078</u>	**0.3109**
	F1-measure	0.0671	0.0889	0.0183	0.0767	<u>0.1214</u>	**0.1227**
	NDCG	0.2450	0.2370	0.0790	0.2228	<u>0.3229</u>	**0.3263**
	$U_{c}ov$	0.7077	0.7546	0.4230	0.7657	**0.8766**	<u>0.8764</u>
Tmall	sps	0.0465	0.0465	0.0100	0.0235	0.0233	**0.0581**
	recall	0.0125	0.0230	0.0042	0.0076	0.0207	**0.0265**
	precision	**0.0860**	0.0651	0.0349	0.0209	0.0605	<u>0.0825</u>
	F1-measure	0.0218	0.0339	0.0074	0.0111	0.0308	**0.0401**
	NDCG	0.0949	0.0718	0.0388	0.0216	0.0765	**0.0983**
	$U_{c}ov$	0.5233	0.4535	0.2674	0.1860	0.4186	**0.5465**
Amazon Movies	sps	0.0065	0.0265	0.0038	0.0220	0.0228	**0.0531**
	recall	0.0078	**0.0141**	0.0031	0.0083	0.0109	<u>0.0137</u>
	precision	0.0498	0.0701	0.0199	0.0476	<u>0.0649</u>	**0.0761**
	F1-measure	0.0135	**0.0234**	0.0053	0.0141	0.0186	<u>0.0232</u>
	NDCG	0.0508	0.0713	0.0222	0.0504	0.0677	**0.0835**
	$U_{c}ov$	0.3138	**0.4431**	0.1630	0.2810	0.3399	<u>0.4285</u>

5.2   Performance on Sequential Recommendation

To obtain a whole picture of RoSE's effectiveness, we compare the performance of RoSE with baselines' in terms of short term prediction, long term prediction and generalization. Comparison results on three benchmark datasets are shown in Table 1. **Short Term Prediction**. There are two main observations. (1) Compared with the general recommender, the sps improvement of sequential recommender is over one time on both MovieLens and Amazon Movies. (2) RoSE has a significant improvement in three datasets and shows its remarkable performance in short-term recommendation. For example, RoSE outperforms the best baseline method UKNN by 24.9% on Tmall and 100% on Amazon Movies. Both observations indicate that our model obtains better sequence embedding that makes the prediction results better. **Long Term Prediction**. We observe that UKNN seems to be the best baseline in terms of recall, precision, F1-measure and ndcg. RoSE performs the best among all the baselines on three datasets. For example, the F1-measure and NDCG improvements on Tmall are 30.2% and 28.5% compared with RNN. Generally, RoSE achieves an excellent performance in long term prediction. **Generalization**. We conclude the following two points. (1) We can see that the user coverage of RoSE outperforms the best baseline on Tmall by 4.3% and similar results are obtained on MovieLens. (2) However, things are different on Amazon Movies, and UKNN achieves better performance than RoSE. The reason may lies in that we use the sample of Amazon Movies with uniform distribution and the cosine distance used for uniform data has strong capacity to represent the global interests. For most datasets, the generalization performance of RoSE achieves the best.

In general, all these observations from Table 1 show RoSE achieves highly superiority on three datasets especially for sps. Meanwhile, these results conform that the learned sequence representations are more discriminative to help separate the predicted items better, which lead to the performance increase in terms of short term and long term prediction.

6   Conclusion

In this paper, we first investigate how sequence representations change with the item imbalance. With the advent of item imbalance, sequence embeddings from minority classes will be mixed together. To solve this problem, we propose a robust representation learning framework RoSE which is composed of balanced k-plet sampling strategy and triplet constrained RNN. We generate a balanced mini-batch through balanced k-plet sampling strategy and define a triplet constrain. The triplet constrains are introduced as an auxiliary task and multi-task framework make representations with better distribution. Experimental results on three benchmark datasets show that our model outperforms baselines against the item imbalance problem.

References

1. Devooght, R., Bersini, H.: Collaborative filtering with recurrent neural networks. arXiv preprint arXiv:1608.07400 (2016)
2. Hoffer, E., Ailon, N.: Deep metric learning using triplet network. In: Feragen, A., Pelillo, M., Loog, M. (eds.) SIMBAD 2015. LNCS, vol. 9370, pp. 84–92. Springer, Cham (2015). https://doi.org/10.1007/978-3-319-24261-3_7
3. Huang, C., Li, Y., Change Loy, C., Tang, X.: Learning deep representation for imbalanced classification (2016)
4. Khan, S.H., Hayat, M., Bennamoun, M., Sohel, F.A., Togneri, R.: Cost-sensitive learning of deep feature representations from imbalanced data. IEEE Trans. Neural Netw. Learn. Syst. **29**(8), 3573–3587 (2018)
5. Ko, Y.J., Maystre, L., Grossglauser, M.: Collaborative recurrent neural networks for dynamic recommender systems. In: JMLR: Workshop and Conference Proceedings, vol. 63 (2016)
6. Krawczyk, B.: Learning from imbalanced data: open challenges and future directions. Prog. Artif. Intell. **5**(4), 221–232 (2016). https://doi.org/10.1007/s13748-016-0094-0
7. Lin, X., Niu, S., Wang, Y., Li, Y.: K-plet recurrent neural networks for sequential recommendation. In: SIGIR, ACM (2018)
8. Niu, S., Zhang, R.: Collaborative sequence prediction for sequential recommender. In: CIKM. ACM (2017)
9. Rendle, S., Freudenthaler, C.: Improving pairwise learning for item recommendation from implicit feedback. In: WSDM (2014)
10. Rendle, S., Freudenthaler, C., Gantner, Z., Schmidt-Thieme, L.: BPR: Bayesian personalized ranking from implicit feedback. In: UAI, pp. 452–461 (2009)

11. Rendle, S., Freudenthaler, C., Schmidt-Thieme, L.: Factorizing personalized Markov chains for next-basket recommendation. In: WWW (2010)
12. Zimdars, A., Chickering, D.M., Meek, C.: Using temporal data for making recommendations. In: UAI (2001)

Deep Generative Recommendation with Maximizing Reciprocal Rank

Xiaoyi Sun, Huafeng Liu, Liping Jing$^{(\boxtimes)}$, and Jian Yu

Beijing Key Lab of Traffic Data Analysis and Mining, Beijing Jiaotong University,
Beijing, China
{sunxiaoyi,huafeng,lpjing,jianyu}@bjtu.edu.cn

Abstract. Variational autoencoders (VAEs) have proven to be successful in the field of recommender systems. The advantage of this non-
linear probabilistic generative model is that it can break through the
limited modeling capabilities of linear models which dominate collaborative filtering research to a large extend. In this paper, we propose a
deep generative recommendation model by enforcing a list-wise ranking
strategy to VAE with the aid of multinomial likelihood. This model has
ability to simultaneously generate the point-wise implicit feedback data
and create the list-wise ranking list for each user. To seamlessly combine
ranking loss with VAE loss, the Reciprocal Rank (RR) is adopted here
and approximated with a smoothed function. A series of experiments on
two real-world datasets (*MovieLens-100k* and *XuetangX*) have been conducted. We show that maximizing the ranking loss will cause as many
relevant items appearing at the top of the predicted recommendation
list as possible. The experimental results demonstrated that the proposed method outperforms several state-of-the-art methods in ranking
estimation task.

Keywords: Personalized recommendation · Deep generative model ·
Learning to rank

1 Introduction

With the rapid growth of online information, recommender systems have become
increasingly indispensable in the era of information overload. The purpose of the
recommender systems is to help users find items that may be of interest to them
from a large number of items. In a typical recommender system, it is expected
to predict all the interactions between users and items based on partial observed
information. In real-world application, Top-N recommendation is much widely

This work was supported in part by the National Natural Science Foundation of
China under Grant 61822601, 61773050, and 61632004; the Beijing Natural Science
Foundation under Grant Z180006; National Key Research and Development Program
(2017YFC1703506); The Fundamental Research Funds for the Central Universities
(2019JBZ110).

G. Li et al. (Eds.): KSEM 2020, LNAI 12275, pp. 123–130, 2020.
https://doi.org/10.1007/978-3-030-55393-7_12

used for generating personalized recommendation list for each user. For example, in the field of e-commerce and online education, the system usually prefers to recommending Top-N items or courses to the user.

Collaborative filtering (CF) [1] is more widely used in recommender systems because it relies only on the past user behavior and without requiring the creation of explicit profiles. Among them, latent factor models [4,12,16] are very concerned due to its capabilities in modeling the hidden causal relationships. However, such model-based CF methods suffer from their essential linear structure. Previous works [8] have proven that recommendation performance can be significantly improved by adding non-linear features into the latent factor models. Recently, a series of methods have been proposed by applying deep neural networks to collaborative filtering [3,13,17]. Compared to the classical deep neural network, deep generative model combining deep learning with probabilistic latent variable modeling [6,11] is more flexible and has been proven to achieve better results in the recommendation field. For example, Mult-VAE [9] is a typical deep generative model with multinomial conditional likelihood, which adopts Bayesian inference to implement parameter estimation.

Even though the aforementioned methods obtain promising performance on recommendation, it is hard for them to effectively and explicitly generate a personalized ranking list for implicit feedback data. Actually, it is necessary and more important for a recommender system to provide users with only a few but valuable items in some scenarios, because an attempt to return many relevant items may reduce the chances of finding any relevant item [2].

Thus, in this paper, we propose a List-wise Ranking Variational Auto-encoder (LRVAE) by combining list-wise learning to rank mechanism with a deep generative strategy to solve the above problems. We use a deep generative method to generate a ranking coefficients vector for each user on implicit data and model the ranking coefficients by directly optimizing the Reciprocal Rank (RR) [15]. The Mean Reciprocal Rank (MRR) is the average of RR for individual users in all the recommendation lists. For domains that usually provide users with only a few but valuable recommendations, MRR is a particularly useful list-wise ranking evaluation metric since it focuses on the rule *less is more*. To efficiently handle ranking problem, we approximate RR with a smoothed function and integrate it to VAE learning framework. We drive a lower bound of the proposed mdoel combining both VAE loss and RR loss. We investigate the proposed model on two widely-used benchmark datasets in terms of several evaluation metrics. The experimental results have illustrated the advantages of our model by comparing with the state-of-the-art baselines.

2   List-Wise Ranking Variational Auto-encoder (LRVAE)

In this section, we describe the proposed List-wise Ranking Auto-encoder (LRVAE) model. This model has ability to simultaneously generate the point-wise implicit feedback data and create the list-wise ranking list for each user. The architecture of LRVAE as shown in Fig. 1. Then, we introduce a smoothed version of Reciprocal Rank (RR) and infer a lower bound of the smoothed RR.

In this work, we consider learning with implicit feedback data and binarize the click[1] matrix $X \in \mathbb{N}^{U \times I}$ as user-by-item interaction matrix. We use $u \in \{1, \ldots, U\}$ to represent users and $i \in \{1, \ldots, I\}$ to represent items. The lower case $\mathbf{x}_u = [x_{u1}, \ldots, x_{uI}]^\top \in \mathbb{N}^I$ is a binary vector indexing the click history of user u on each item.

2.1 Deep Generative Probabilistic Model

The architecture of our proposed model is consists of a variational autoencoder (VAE). Each VAE utilizes the **encoder** to compress the input to a variational distribution, and then the **decoder** accepts the latent variable sampled from the posterior to obtain a generated distribution for prediction.

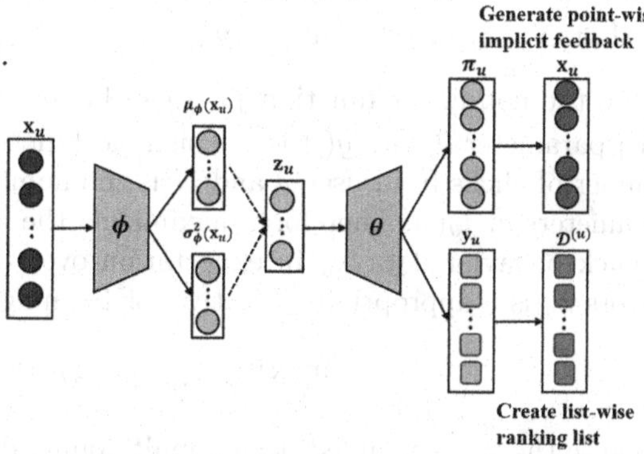

Fig. 1. The architecture of LRVAE model.

Encoder. In this part, we expect to construct a generative latent variable model for observed data. Firstly we assume that the user-by-item interaction data click history $\mathbf{x}_u$ can be generated by user latent variable $\mathbf{z}_u \in \mathbb{R}^k$ (in k-dimensional latent space), which can be sampled from a standard Gaussian prior, i.e., $\mathbf{z}_u \sim \mathcal{N}(0, \mathbf{I}_k)$.

We introduce the variational distribution $q_\phi(\mathbf{z}_u \mid \mathbf{x}_u)$, which represented by the parameterized diagonal Gaussian $\mathcal{N}(\mu_\phi, \text{diag}\{\sigma_\phi^2\})$, to approach the true posteriors $p(\mathbf{z}_u \mid \mathbf{x}_u)$. Finally, the inference process of the probabilistic encoders is as follows:

1. For user u, constructing a vector representation $\mathbf{e}_u$ of observed data: $\mathbf{e}_u = f_\phi(\mathbf{x}_u)$.
2. Parameterize the variational distribution of the user latent variables $[\mu_\phi(\mathbf{x}_u), \sigma_\phi(\mathbf{x}_u)] = l_\phi(\mathbf{e}_u) \in \mathbb{R}^{2k}$.

where the non-linear function $f_\phi(\cdot)$ is a neural network that is applicable to the observed data. $l_\phi(\cdot)$ is linear transformation used to calculate the variational distribution parameters. And ϕ is a collection of all the parameters in $f_\phi(\cdot)$ and $l_\phi(\cdot)$.

---

[1] We use the verb "click" for concreteness to indicate any type of interactions, including "watch", "purchase" or "check-in".

Decoder. In this part, we expect to generate a probability distribution of prediction through latent variables and reconstruct the observed data. The generation process of the decoder is as follows:

1. Samples $\mathbf{z}_u \in \mathbb{R}^k$ from variational posterior $q_\phi(\mathbf{z}_u \mid \mathbf{x}_u)$.
2. Generate the probabilistic distribution $\pi_u \in \mathbb{N}^I$ and ranking coefficients vector $\mathbf{y}_{u} \in \mathbb{N}^I$ of I items for each user via a multi-layer perceptron with parameter θ with softmax and sigmoid function: $\pi_u \propto \exp(f_\theta(\mathbf{z}_u)), \mathbf{y}_u = g(f_\theta(\mathbf{z}_u))$..
3. Generate reconstructed observed data $\mathbf{x}_u$ from multinomial distributions, and obtain the ranking list $\mathcal{D}_s^{(u)} \in \mathbb{N}^I$ by sorting $\mathbf{y}_u$ for each user:$\mathbf{x}_u \sim \text{Mult}(N_u, \pi_u), \mathcal{D}_s^{(u)} \sim sorted(\mathbf{y}_u)$..

where the non-linear function $f_\theta(\cdot)$ can be any type of multi-layer perceptrons with parameter θ, and $g(\cdot)$ is the sigmoid function. $N_u = \sum_{i=1}^{I} x_{ui}$ is the total number of clicks from user u and N is the number of items. θ is a collection of parameters in $f_\theta(\cdot)$. Therefore, maximizing the marginal log-likelihood function of click behavior data $\mathbf{x}_u$ in expectation over the whole distribution of latent factors $\mathbf{z}_u$ is a appropriate objective for learning latent variable distribution:

$$\max_\theta \mathbb{E}_{q_\phi(\mathbf{z}_u|\mathbf{x}_u)} [\log p_\theta(\mathbf{x}_u \mid \mathbf{z}_u)] \tag{1}$$

Because the $\mathbf{x}_u$ is sampled from a multinomial distribution with probability π_u, where the log-likelihood for user u is $\log p_\theta(\mathbf{x}_u \mid \mathbf{z}_u) = \sum_{i=1}^{I} x_{ui} \log \pi_{ui}$.

Optimization. We optimize the objective function from two aspects. On the one hand, to adjust the capacity of different users, we introduce a limitation over $q_\phi(\mathbf{z}_u \mid \mathbf{x}_u)$. If we try to match $q_\phi(\mathbf{z}_u \mid \mathbf{x}_u)$ to a uninformative prior $p(\mathbf{z}_u)$, we can control the capacity of the latent information bottleneck. The prior $p(\mathbf{z}_u)$ can be set to isotropic unit Gaussian $(p(\mathbf{z}_u) = \mathcal{N}(0, \mathbf{I}_k))$.

On the other hand, we introduce reciprocal rank (RR) as a limitation over the ranking list $\mathcal{D}_s^{(u)}$. RR is equal to the reciprocal of the rank at which the first relevant item was returned. The definition of RR can be formulated as follows:

$$RR_u = \sum_{i=1}^{I} \frac{x_{ui}}{r_{ui}} \prod_{j=1}^{I} (1 - x_{uj}\mathbb{I}(r_{uj} < r_{ui})) \tag{2}$$

where $\mathbb{I}(x)$ represents an indicator function. If x is true, $\mathbb{I}(x)$ is equal to 1, otherwise, 0. r_{ui} indicates the position of item i for user u and it can be obtained from $\mathcal{D}_s^{(u)}$. We expect the value of RR is as large as possible until equal to 1. Hence the constrained optimization problem can be written as:

$$\max_\theta \mathbb{E}_{q_\phi(\mathbf{z}_u|\mathbf{x}_u)} [\log p_\theta(\mathbf{x}_u \mid \mathbf{z}_u)],$$
$$\text{subject to } KL(q_\phi(\mathbf{z}_u \mid \mathbf{x}_u)\|p(\mathbf{z}_u)) < \varepsilon, \quad RR_u > \delta. \tag{3}$$

where ε represents the strength of the applied constraint. If the posterior distribution is equal to the uninformative prior, model will learn nothing from the data and $KL(q_\phi(\mathbf{z}_u \mid \mathbf{x}_u) \| p(\mathbf{z}_u))$ is equal to zero. δ is a constant with a range of $(0, 1)$. We assume that RR_u is a convex function, and then a convex function form equivalent to RR_u is given in Sect. 3.2. Re-writing Eq. 3 as a Lagrangian under the KKT conditions. Since $\beta, \alpha, \varepsilon, \delta > 0$ and ε, δ are constant, the optimization problem can be written as:

$$\mathcal{F}(\mathbf{x}; \theta, \phi) \simeq \mathcal{L}(\mathbf{x}; \theta, \phi) = \mathbb{E}_{q_\phi(\mathbf{z}_u | \mathbf{x}_u)} [\log p_\theta(\mathbf{x}_u \mid \mathbf{z}_u)]$$
$$- \beta KL(q_\phi(\mathbf{z}_u \mid \mathbf{x}_u) \| p(\mathbf{z}_u)) + \alpha RR_u \tag{4}$$

The multiplier β putting implicit independence pressure on the learned posterior to align with the unit Gaussian prior. And α limits the impact of learning to rank on the VAE model.

2.2 Smoothing the Reciprocal Rank

From Eq. 2 we can know that as a function of predicting correlation scores, the ranking of related items changes in a non-smooth manner, thus RR_u is a non-smooth function of model parameters. Inspired by the latest developments in the field of learning to rank [14], we have adopted smoothing the reciprocal rank approach as follow. In order to smooth the RR_u, we approximate $\mathbb{I}(r_{uj} < r_{ui})$ by $h(y_{uj} - y_{ui})$. Where $h(x)$ is *relu* function. y_{ui} denotes the ranking coefficients generated by LRVAE model. We also need to directly approximate $1/r_{ui}$ by the logistic function $g(y_{ui})$ base on a basic assumption that the lower item position value. Substituting these functions into Eq. 2, we can get a smooth version of RR_u:

$$RR_u \approx \sum_{i=1}^{I} x_{ui} g(y_{ui}) \prod_{j=1}^{I} (1 - x_{uj} h(y_{ui} - y_{uj})) \tag{5}$$

We give the equivalent variant of Eq. 5, which is of low complexity in the optimization process. We assume that the number of click items for user u in a given dataset is c_u. Notice that the model parameter that maximizes Eq. 5 is equivalent to the parameter that maximizes $\ln(\frac{1}{c_u} RR_u)$ because the monotonicity of the logarithmic function is the same as the original RR_u function. And we derive the lower bound of $\ln(\frac{1}{c_u} RR_u)$ based on Jensen's inequality and the concavity of logarithmic function as follow:

$$\ln(\frac{1}{c_u} RR_u) = \ln(\sum_{i=1}^{I} \frac{x_{ui}}{\sum_{k=1}^{I} x_{uk}} g(y_{ui}) \prod_{j=1}^{I} (1 - x_{uj} h(y_{ui} - y_{uj})))$$

$$\geq \frac{1}{c_u} \sum_{i=1}^{I} x_{ui} \ln(g(y_{ui}) \prod_{j=1}^{I} (1 - x_{uj} h(y_{ui} - y_{uj}))) \tag{6}$$

$$= \frac{1}{c_u} \sum_{i=1}^{I} x_{ui} (\ln g(y_{ui}) + \sum_{j=1}^{I} \ln(1 - x_{uj} h(y_{ui} - y_{uj})))$$

According to the definition of c_u above we can derive $c_u = \sum_{k=1}^{I} x_{uk}$. After ignoring the constant $1/c_u$ we get the convex ranking loss function:

$$RR_u \simeq L(\mathbf{y}_u) = \sum_{i=1}^{I} x_{ui}(\ln g(y_{ui}) + \sum_{j=1}^{I} \ln(1 - x_{uj}h(y_{ui} - y_{uj}))) \qquad (7)$$

We analyzed the two terms within the first summation and found that maximizing the first item helps to promote the deep generative model to more accurately predict the user click behavior, and maximizing the second term is important for obtaining a ranking list for a personalized recommendation. For example, given a clicked item, e.g., item i, the maximization of the second term in order to degrade other items relevance ratings. In summary, these two effects together promote and decentralize related projects, that is, maximizing this ranking loss will cause a part of the clicked items to be at the top of the user recommendation list.

Substituting Eq. 7 into Eq. 4, we obtain the objective function:

$$\mathcal{L}(\mathbf{x}; \theta, \phi) = \mathbb{E}_{q_\phi(\mathbf{z}_u|\mathbf{x}_u)}[\log p_\theta(\mathbf{x}_u \mid \mathbf{z}_u)] - \beta(KL(q_\phi(\mathbf{z}_u \mid \mathbf{x}_u)\|p(\mathbf{z}_u)))$$
$$+ \alpha(\sum_{i=1}^{I} x_{ui}(\ln g(y_{ui}) + \sum_{j=1}^{I} \ln(1 - x_{uj}h(y_{ui} - y_{uj})))) \qquad (8)$$

Note that the parameters need to be optimized are ϕ and θ. We can obtain an unbiased estimate of $\mathcal{L}$ by sampling $z_u \sim q_\phi$ and then use stochastic gradient ascent to optimize it. We sample $\epsilon \sim \mathcal{N}(0, \mathbf{I}_k)$ and reparameterize $\mathbf{z}_u = \mu_\phi(\mathbf{x}_u) + \epsilon \odot \sigma_\phi(\mathbf{x}_u)$ by employing *reparameterization trick* [6,11]. By using this trick, the stochasticity is avoided during the sampling process, and the gradient with respect to ϕ can be back-propagated through the sampled latent representation $\mathbf{z}_u$.

3   Experiments

In this section, we present a series of experiments to evaluate the LRVAE on two real-world datasets and compare it with the state-of-the-art methods.

3.1   Experimental Setting

Datasets: We conduct experiments using two real-world datasets from various domains: *MovieLens-100k (ML-100k)*[2] and *XuetangX*[3] [18].

Baselines: We compare results with three kinds of standard state-of-the-art collaborative filtering models, including traditional methods: (1) **Traditional models**: WMF [5] and SLIM [10]. (2) **Deep neural network recommendation models**: NCF [3] and CDAE [17]. (3) **Deep generative recommendation model**: Mult-VAE [9].

---

[2] https://grouplens.org/datasets/movielens/.
[3] https://next.xuetangx.com/.

3.2 Performance Evaluation Results

After some experiments, we acquire the best control parameter set. And then to delve into the performance of recommendation, we conducted a second experiment comparing the proposed LRVAE with five baselines. We implement the results for criteria NDCG and MRR at different numbers of recommended models and the results are recorded in Table 1. The best and second results are marked in bold and underline. As we can see from the experimental results, most of the deep methods (LRVAE, Mult-VAE, CDAE, NCF) perform better than traditional recommendation models (SLIM and WMF) in most cases, which demonstrated that non-linear features are advantageous for improving recommendation performance. But SLIM performs better than some deep methods even in all datasets because SLIM is more effective and efficient for Top-N recommendation on sparse data. Further, in most cases, the deep generative models (LRVAE, Mult-VAE) outperform other deep methods (CDAE, NCF), which demonstrated that the appropriate generative process of point-wise data matches is helpful to learn more useful latent representation. LRVAE performs better than the state-of-the-art deep methods(Mult-VAE, CDAE, NCF), which demonstrates that considering both implicit feedback generation and list-wise ranking can improve the effectiveness of recommendations.

Table 1. Comparison between various baselines and LRVAE.

Metrics	ML-100k			XuetangX		
	NDCG@1	NDCG@5	MRR	NDCG@1	NDCG@5	MRR
WMF	0.02105	0.20340	0.23435	0.06200	0.15385	0.33653
SLIM	<u>0.36761</u>	<u>0.30455</u>	<u>0.50214</u>	0.32240	0.33230	0.36562
NCF	0.14576	0.24276	0.22435	0.28485	0.33490	0.38700
CDAE	0.27147	0.29346	0.11272	0.28570	0.32622	0.35794
Mult-VAE	0.36000	0.29420	0.51971	<u>0.33150</u>	<u>0.39545</u>	<u>0.41844</u>
LRVAE	**0.37333**	**0.30729**	**0.54073**	**0.33740**	**0.40174**	**0.42472**

As we see the results are small in numerical improvement, but small improvements can lead to significant differences in recommendations in practice [7]. We conduct paired t-test (confidence 0.95) between LRVAE and two baselines (SLIM and Mult-VAE) with five-fold cross-validation results to prove the effectiveness of the proposed method. The p-values in all cases are less than 0.01, which demonstrates that the results of LRVAE showed a significant improvement compared with the other models. Therefore, based on these results, we can conclude that the LRVAE is consistently superior to the state-of-art methods and significantly improves the recommendation performance.

4 Conclusions and Future Work

In this paper, we propose a deep generative recommendation method by enforcing a list-wise ranking strategy to VAE with the aid of multinomial likelihood.

This method has the ability to simultaneously generate the point-wise implicit feedback data and create the list-wise ranking list for each user. We use a deep generative method to generate a ranking coefficients vector for each user on implicit data and model the ranking coefficients by directly optimizing the Reciprocal Rank (RR). The experiments have shown that LRVAE has the ability to generate personalized ranking lists for users and make more relevant items appear at the top of lists. In future work, we expect to further explore the trade-offs introduced by the addition of parameters β and α and improve the interpretability of the model.

References

1. Adomavicius, G., Tuzhilin, A.: Toward the next generation of recommender systems: a survey of the state-of-the-art and possible extensions. IEEE Trans. Knowl. Data Eng. **17**, 734–749 (2005)
2. Chen, H., Karger, D.R.: Less is more: probabilistic models for retrieving fewer relevant documents. In: SIGIR (2006)
3. He, X., Liao, L., Zhang, H., Nie, L., Hu, X., Chua, T.S.: Neural collaborative filtering. In: WWW (2017)
4. Hofmann, T.: Latent semantic models for collaborative filtering. ACM Trans. Inf. Syst. **22**, 89–115 (2004)
5. Hu, Y., Koren, Y., Volinsky, C.: Collaborative filtering for implicit feedback datasets. In: 2008 Eighth IEEE International Conference on Data Mining, pp. 263–272 (2008)
6. Kingma, D.P., Welling, M.: Auto-encoding variational Bayes. CoRR abs/1312.6114 (2014)
7. Koren, Y., Bell, R.M., Volinsky, C.: Matrix factorization techniques for recommender systems. Computer **42**, 30–37 (2009)
8. Liang, D., Altosaar, J., Charlin, L., Blei, D.M.: Factorization meets the item embedding: regularizing matrix factorization with item co-occurrence. In: RecSys (2016)
9. Liang, D., Krishnan, R.G., Hoffman, M.D., Jebara, T.: Variational autoencoders for collaborative filtering. In: WWW (2018)
10. Ning, X., Karypis, G.: Sparse linear methods with side information for top-n recommendations. In: RecSys (2012)
11. Rezende, D.J., Mohamed, S., Wierstra, D.: Stochastic backpropagation and approximate inference in deep generative models. In: ICML (2014)
12. Salakhutdinov, R.R., Mnih, A.: Probabilistic matrix factorization. In: NIPS (2007)
13. Sedhain, S., Menon, A.K., Sanner, S., Xie, L.: AutoRec: autoencoders meet collaborative filtering. In: WWW (2015)
14. Shi, Y., Karatzoglou, A., Baltrunas, L., Larson, M., Oliver, N., Hanjalic, A.: CLiMF: learning to maximize reciprocal rank with collaborative less-is-more filtering. In: RecSys (2012)
15. Voorhees, E.M.: The TREC-8 question answering track report. In: TREC (1999)
16. Wang, C., Blei, D.M.: Collaborative topic modeling for recommending scientific articles. In: KDD (2011)
17. Wu, Y., DuBois, C., Zheng, A.X., Ester, M.: Collaborative denoising auto-encoders for top-n recommender systems. In: WSDM (2016)
18. Zhang, J., Hao, B., Chen, B., Li, C., Chen, H., Sun, J.: Hierarchical reinforcement learning for course recommendation in MOOCs. In: AAAI (2019)

Spatio-Temporal Attentive Network for Session-Based Recommendation

Chunkai Zhang[✉] and Junli Nie

Department of Computer Science and Technology, Harbin Institute of Technology,
Shenzhen, China
ckzhang@hit.edu.cn, njl_primary@163.com

Abstract. Session-based recommendation aims to predict the user's next click behavior based on the existing anonymous session information. Existing methods either only utilize temporal information of the session to make recommendations or only capture complex item transitions from spatial perspective to recommend, they are insufficient to obtain rich item representations. Besides, user's real purpose of the session is also not emphasized. In this paper, we propose a novel session-based recommendation method, named Spatio-Temporal Attentive Session-based Recommendation, STASR for brevity. Specifically, we design a hybrid framework based on Graph Neural Network (GNN) and Gated Recurrent Unit (GRU) to obtain richer item representations from spatio-temporal perspective. During the process of constructing corresponding session graph in GNN, an individual-level skipping strategy, which considers the randomness of user's behaviors, is proposed to enrich item representations. Then we utilize attention mechanism to capture the user's real purpose involved user's initial will and main intention. Extensive experimental results on three real-world benchmark datasets show that STASR consistently outperforms state-of-the-art methods on a variety of common evaluation metrics.

Keywords: Session-based recommendation · Spatio-temporal perspective · Attention mechanism

1 Introduction

Recommendation systems help users alleviate the problem of information overload and suggest items that may be of interest to users. Traditional recommendation methods [1], such as content-based methods and collaborative filtering methods, utilize user profiles and user-item interaction records to recommend. However, in many services, such as e-commerce websites and most media sites, user profiles may be unknown, and only the on-going session is available. Under these circumstances, session-based recommendation [5] is proposed to predict user's next behavior based merely on the history click records in the current session.

© Springer Nature Switzerland AG 2020
G. Li et al. (Eds.): KSEM 2020, LNAI 12275, pp. 131–139, 2020.
https://doi.org/10.1007/978-3-030-55393-7_13

Recently, a lot of researches have begun to realize the importance of Recurrent Neural Network (RNN) for session-based recommendation. Hidasi et al. [2] first employ the Gated Recurrent Unit(GRU) to solve session-based recommendation problem. Then Tan et al. [6] improve the recommendation performance via considering data augmentation and temporal shifts of user behaviors. Recently, Li et al. [3] propose NARM based on encoder-decoder architecture to capture the sequence behavior and main purpose of the user simultaneously. In contrast to NARM, STAMP [4]aims to capture user's long-term and short-term interests to make effective recommendations. Although these RNN-based methods above have achieved significant results, they only consider single-way transitions between consecutive items and neglect the transitions among the contexts, i.e., other items in the session.

Graph Neural Network(GNN) [7] is designed to learn the representations for graph structured data. As for session-based recommendation, SRGNN [8] converts session into the form of session graph, then utilizes GNN to generate accurate item representations via capturing complex item transitions. Then Xu et al. [9] propose a graph contextualized self-attention network based on both GNN and self-attention to model local graph-structured dependencies of separated session sequences and obtain contextualized non-local representations.

Although the existing GNN-based methods achieve excellent performance, they still have some limitations. Firstly, they only emphasize complex item transitions of the current session from spatial perspective. They ignore the impact of repeated user behavior pairs on user's interests, i.e., session $[x_1, x_2, x_3, x_2, x_4]$ and session $[x_1, x_2, x_3, x_2, x_3, x_2, x_4]$ correspond to the same session graph as well as the same item representations, which confines the prediction accuracy. In other words, they neglect temporal information of the whole session. Secondly, the randomness of user's behaviors is ignored, we also call it as individual-level skip behaviors of the user in the current session, i.e., the past one behavior not only has direct impact on the next behavior but also may have direct impact on the behavior after skipping a few time steps. Thirdly, previous work does not emphasize the user's real purpose involved user's initial will and main intention of the current session.

To overcome these limitations, in this paper, we propose a novel method for session-based recommendation. The main contributions of our work can be summarized as:

– We design a hybrid framework based on GNN and GRU to obtain richer item representations from spatio-temporal perspective. During the process of constructing corresponding session graph in GNN, an individual-level skipping strategy, which considers the randomness of user's behaviors, is proposed to enrich item representations.
– We apply two attention networks to extract the user's real purpose involved user's initial will and main intention in the current session.
– We carried out extensive experiments on three real-world benchmark datasets. The results demonstrate that our proposed method performs better than state-of-art methods in terms of Recall@20 and MRR@20.

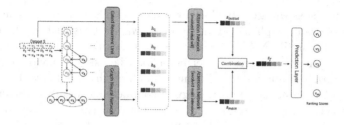

Fig. 1. The graphical model of our proposed method.

2   Related Work

Hidasi et al. [2] first design a RNN model with GRU to predict user's next behavior based on previous behaviors in the current session. Then Tan et al. [6] propose improved version of GRU4Rec to boost performance via two methods, i.e., data augmentation and accounting for shifts in the input data distribution. Recently, NARM [3] takes advantage of an encoder-decoder architecture to consider sequence behavior features and capture main purpose of the current session simultaneously. Then STAMP [4] using MLP networks and attention mechanism, is proposed to efficiently capture both the user's long-term and short-term interests of a session. SRGNN [8] is first proposed to convert the session into the form of session graph, and uses GNN to capture more complex items transitions based on the session graph. Then Xu et al. [9] propose a graph contextualized self-attention network to capture local graph-structured dependencies and contextualized non-local representations simultaneously.

3   Method

3.1   Notations

Let $V = \{v_1, v_2, ..., v_{m-1}, v_m\}$ represents the set of unique items appearing in the whole dataset. Anonymous session is denoted as $S = [x_1, x_2, ..., x_{n-1}, x_n]$, where $x_t \in V(1 \le t \le n)$ denotes the item clicked at time step t. As for any given prefix of session $[x_1, x_2, ..., x_t](1 \le t \le n)$, our proposed method aims to model the current session and predict the user's next behavior x_{t+1}. In many online services, recommendation systems provide a ranking list $y = [y_1, y_2, ..., y_{m-1}, y_m]$ over all candidate items for the user, where $y_j(1 \le j \le m)$ represents the probability of item j clicked by the user at the next time step.

3.2   Item Representation Layer

- First, we utilize GNN with an individual-level skipping strategy to obtain item representations from spatial perspective, which can capture complex item transitions and consider the randomness of user behaviors.

Construct session graph. The first part of GNN is to construct corresponding session graph. Different from the way of constructing the directed session graph in SRGNN [8], we propose an individual-level skipping strategy to consider the randomness of user behaviors (i.e., skip behaviors in the current session). To improve the robustness and keep the simplicity of our model, the individual-level skipping strategy is conducted via adding directed connections according to chronological order between two items randomly at a certain ratio in the corresponding session graph. $M^O, M^I \in \mathbb{R}^{n \times n}$ represent weighted connections of outgoing and incoming edges in the session graph. For example, given a session $S = [v_1, v_3, v_2, v_4, v_5, v_6, v_2]$, to consider the randomness of user behaviors, we apply skipping strategy to obtain corresponding session graph and matrices M^O, M^I, which are shown in Fig. 2. Following previous work [8], since some items maybe appear repeatedly in a session, the normalization process is often used when constructing the outgoing matrix and incoming matrix, which is calculated as the occurrence of the edge divided by the outdegree of that edge's start node.

Node representation learning. Here, we describe how to update node representations based on the constructed session graph and matrices M^O, M^I, as for each node in the session graph, its update functions are given as follows:

$$a_v^{(t)} = A_{v:}^T [h_1^{(t-1)T} ... h_{|V|}^{(t-1)T}]^T + b \tag{1}$$

$$z_v^t = \sigma(W^z a_v^{(t)} + U^z h_v^{(t-1)}) \tag{2}$$

$$r_v^t = \sigma(W^r a_v^{(t)} + U^r h_v^{(t-1)}) \tag{3}$$

$$\tilde{h}_v^{(t)} = tanh(W a_v^{(t)} + U(r_v^t \odot h_v^{(t-1)})) \tag{4}$$

$$h_v^{(t)} = (1 - z_v^t) \odot h_v^{(t-1)} + z_v^t \odot \tilde{h}_v^{(t)} \tag{5}$$

$A_{v:}$ is defined as the combination of the two columns corresponding to node v from the outgoing and incoming matrices. $a_v^{(t)}$ extracts the contextual information of adjacent nodes for node v. z_v^t and r_v^t are update gate and reset gate respectively. $\tilde{h}_v^{(t)}$ represents the newly generated information, and $h_v^{(t)}$ is the final updated node state. In addition, $\sigma(\cdot)$ denotes the logistic sigmoid function and $\odot$ denotes element-wise multiplication.

- Then, we utilize GRU to consider temporal information including repeated user behavior pairs of the current session into account. GRU is more simplified than the standard RNN and its update formulas are as follows:

$$z_t = \sigma(W_z \cdot [h_{t-1}, v_t]) \tag{6}$$

$$r_t = \sigma(W_r \cdot [h_{t-1}, v_t]) \tag{7}$$

$$\tilde{h}_t = tanh(W \cdot [r_t * h_{t-1}, v_t]) \tag{8}$$

$$h_t = (1 - z_t) * h_{t-1} + z_t * \tilde{h}_t \tag{9}$$

z_t, r_t represent update gate and reset gate respectively. And h_t denotes the activation of GRU which is a linear interpolation between the previous activation h_{t-1} and the candidate activation $\tilde{h}_t$. Here, we essentially use the

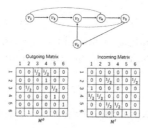

Fig. 2. An example of session graph structure and connection matrices M^O, M^I after applying individual-level skipping strategy.

corresponding hidden states $[h_1, h_2, ..., h_{t-1}, h_t]$ of the input as the item representations of the current session sequence from temporal perspective.

Finally, we could obtain richer item representations from spatio-temporal perspective, which are combined as the unified item representations later.

3.3   Attention Layer

Here, we apply two item-level attention networks to dynamically choose more important items and linearly combine each part of the input for the user's initial will and main intention respectively. The formulas are defined as follows:

$$s_{initial} = \sum_{j=1}^{t} \alpha_{tj} h_j \qquad (10)$$

$$s_{main} = \sum_{j=1}^{t} \beta_{tj} h_j \qquad (11)$$

where

$$\alpha_{tj} = v^T \sigma(W_1 h_1 + W_2 h_j + c_1) \qquad (12)$$

$$\beta_{tj} = q^T \sigma(W_3 h_t + W_4 h_j + c_2) \qquad (13)$$

α_{tj} and β_{tj} determine the importance of each item in the session when we consider user's initial will and main intention respectively. In detail, α_{tj} is used to compute the similarity between h_1 and the representation of previous item h_j. And β_{tj} computes the similarity between the final item representation h_t and the representation of previous item h_j. $\sigma(\cdot)$ is an activate function and matrices W_1, W_2, W_3, W_4 control the weights. Finally, we obtain the session representation s_f by taking linear transformation over the concatenation of $s_{initial}$ and s_{main}.

$$s_f = W_5[s_{initial}; s_{main}] \qquad (14)$$

Matrix W_5 is used to compress these two combined embedding vectors into the latent space.

3.4   Prediction Layer

Here we calculate corresponding probability of each candidate item v_i being clicked at the next time step. The computing formula can be defined as:

$$\hat{z}_i = s_f{}^T v_i \qquad (15)$$

Then we apply a softmax function to get the output vector of the model:

$$\hat{y} = softmax(\hat{z}) \qquad (16)$$

For each session, its loss function can be defined as cross-entropy of the prediction and the ground truth:

$$L(\hat{y}) = -\sum_{i=1}^{m} y_i log(\hat{y}_i) + (1 - y_i) log(1 - \hat{y}_i) \qquad (17)$$

y_i denotes the one-hot encoding vector of the ground truth item.

Finally, our proposed model is trained by Back-Propagation Through Time (BPTT) algorithm in the learning process.

Table 1. Statistics of the datasets used in our experiments

Datasets	#clicks	#train	#test	#items	avg.length
Diginetica	982961	719470	60858	43097	5.12
Yoochoose1/64	557248	369859	55898	16766	6.16
Retailrocket	710856	433648	15132	36968	5.43

4   Experiments and Analysis

4.1   Settings

Datasets. Yoochoose is a public dataset released on RecSys Challenge 2015. Diginetica is obtained from CIKM Cup 2016 competition, in our experiment, we only use the click records dataset. Retailrocket comes from an e-commerce company, we select the user's browsing history records dataset in the experiment. We also filter out sessions with length of 1 and items appearing less than 5 times in all datasets. For Yoochoose dataset, we select the sessions from the last day as the test set and the other as the training set, for the Diginetica and Retailrocket datasets, we select the sessions from the last week as the test set and the others as the training set. The statistics of the datasets is shown in Table 1.

Evaluation Metrics. Recall is an evaluation of unranked retrieval results, which represents the proportion of correctly recommended items among Top-N items. MRR(Mean Reciprocal Rank) is an evaluation metric of ranked list, which indicates the correct recommendations in the Top-N ranking list.

Baselines. To show the effectiveness of our proposed method, we compare it with six methods: POP, S-POP, GRU4Rec, NARM, STAMP and SRGNN. POP and S-POP are traditional recommendation methods. GRU4Rec, NARM and STAMP are RNN-based methods. SRGNN is a session-based recommendation method with GNN.

Parameter Setup. We initially set the dimensionality of latent vectors as 160 on Yoochoose dataset and 100 on Diginetica and Retailrocket datasets. All parameters are initialized using a Gaussian distribution with a mean of 0 and a standard deviation of 0.1. The initial learning rate is set to 0.001 and will decay by 0.1 after every 3 epochs. The number of epochs is set to 30 and 10% of the training data is used as validation set.

4.2 Comparison Results

The results of all methods over three real-world datasets in terms of Recall@20 and MRR@20 are shown in Table 2.

Table 2. Comparison of our proposed method with baseline methods over three real-world datasets

Methods	Diginetica		Yoochoose1/64		Retailrocket	
Measures	Recall@20	MRR@20	Recall@20	MRR@20	Recall@20	MRR@20
POP	0.89	0.2	6.71	1.65	1.24	0.32
S-POP	21.06	13.68	30.44	18.35	40.48	32.04
GRU4Rec	29.45	8.33	60.64	22.89	55.59	32.27
NARM	49.7	16.17	68.32	28.63	61.79	34.07
STAMP	45.64	14.32	68.74	29.67	61.08	33.1
SRGNN	50.73	17.59	70.57	30.94	62.79	34.49
Ours	**52.58**	**18.28**	**71.32**	**31.01**	**64.15**	**35.23**

We have the observation from the results that compared to all the baseline methods, our proposed method achieves better performance among all the methods on three real-world datasets in terms of Recall@20 and MRR@20.

4.3 Model Analysis and Discussion

To verify the performance of different components in our model, we conduct a series of experiments. The results are shown in Table 3, we can observe that the hybrid framework plays an important role in recommending results. A possible reason is that we make recommendations based on item representations, so it is important to obtain rich item representations considering various user's behaviors from different perspectives. From Table 3, we can also observe that only considering a single feature, i.e., user's initial will or main intention, does not perform better than considering both features.

Table 3. The performance of our proposed method with and without different components in terms of Recall@20 and MRR@20.

Methods	Diginetica		Yoochoose1/64		Retailrocket	
Measures	Recall@20	MRR@20	Recall@20	MRR@20	Recall@20	MRR@20
w/ hybrid framework	52.58	18.28	71.32	31.01	64.15	35.23
w/o hybrid framework	51.59	17.93	70.68	30.77	63.09	34.59
only w/ main intention	52.36	18.04	70.79	30.86	63.31	34.94
only w/ initial will	52.31	17.96	70.80	30.84	63.27	34.65

5   Conclusion

In this paper, we propose a novel method named STASR for session-based recommendation. Specifically, we design a hybrid framework based on GNN with individual-level skipping strategy and GRU to obtain richer item representations from spatio-temporal perspective. Besides, user's real purpose involved user's initial will and main intention is considered for accurate recommendation. On three datasets, our proposed method can consistently outperform other state-of-art methods.

Acknowledgments. This work was supported by Natural Science Foundation of Guangdong Province, China (Grant NO.2020A1515010970) and Shenzhen Research Council (Grant NO.GJHZ20180928155209705).

References

1. Adomavicius, G., Tuzhilin, A.: Toward the next generation of recommender systems: a survey of the state-of-the-art and possible extensions. IEEE Trans. Knowl. Data Eng. **6**, 734–749 (2005)
2. Hidasi, B., Karatzoglou, A., Baltrunas, L., Tikk, D.: Session-based recommendations with recurrent neural networks. arXiv preprint arXiv:1511.06939 (2015)
3. Li, J., Ren, P., Chen, Z., Ren, Z., Lian, T., Ma, J.: Neural attentive session-based recommendation. In: Proceedings of the 2017 ACM on Conference on Information and Knowledge Management, pp. 1419–1428. ACM (2017)
4. Liu, Q., Zeng, Y., Mokhosi, R., Zhang, H.: Stamp: short-term attention/memory priority model for session-based recommendation. In: Proceedings of the 24th ACM SIGKDD International Conference on Knowledge Discovery & Data Mining, pp. 1831–1839. ACM (2018)
5. Schafer, J.B., Konstan, J., Riedl, J.: Recommender systems in e-commerce. In: Proceedings of the 1st ACM Conference on Electronic Commerce, pp. 158–166. ACM (1999)
6. Tan, Y.K., Xu, X., Liu, Y.: Improved recurrent neural networks for session-based recommendations. In: Proceedings of the 1st Workshop on Deep Learning for Recommender Systems, pp. 17–22. ACM (2016)
7. Wang, X., He, X., Wang, M., Feng, F., Chua, T.S.: Neural graph collaborative filtering. arXiv preprint arXiv:1905.08108 (2019)

8. Wu, S., Tang, Y., Zhu, Y., Wang, L., Xie, X., Tan, T.: Session-based recommendation with graph neural networks. In: Proceedings of the AAAI Conference on Artificial Intelligence, vol. 33, pp. 346–353 (2019)
9. Xu, C., Zhao, P., Liu, Y., Sheng, V.S., Xu, J., Zhuang, F., Fang, J., Zhou, X.: Graph contextualized self-attention network for session-based recommendation. In: Proceedings of 28th International Joint Conference on Artificial Intelligence (IJCAI), pp. 3940–3946 (2019)

Social Knowledge Analysis
and Management

Category-Level Adversarial Network for Cross-Domain Sentiment Classification

Shaokang Zhang[1,2], Huailiang Peng[1,2(✉)], Yanan Cao[1,2], Lei Jiang[1,2], Qiong Dai[1,2], and Jianlong Tan[1,2]

[1] Institute of Information Engineering, Chinese Academy of Sciences, Beijing, China
{zhangshaokang,penghuailiang,caoyanan,jianglei,daiqiong,
tanjianlong}@iie.ac.cn
[2] School of Cyber Security, University of Chinese Academy of Sciences, Beijing, China

Abstract. Cross-domain sentiment classification utilizes useful information in the source domain to improve the sentiment classification accuracy in the target domain which has few or no labeled data. Most existing methods based on single domain classifier only consider the global alignment without taking category-level alignment into consideration, which can lead to the mismatch of category-level features and reduce classification accuracy. To slove the above problem, we propose the Category-level Adversarial Network (CAN). On the basis of single domain classifier, CAN adds K category-wise domain classifiers which can achieve fine-grained alignment of different data distributions by combining the label information and document representations. Specifically, we obtain document representations by introducing transferable attention network which mirrors the hierarchical structure of documents and transfers attentions across domains. Experiments results demonstrate that CAN model outperforms state-of-the-art methods on the Amazon and Airline datasets.

Keywords: Category-level Adversarial Network · Cross-domain sentiment classification · Domain adaptation · Hierarchical Attention

1 Introduction

Sentiment classification, which identifies the sentiment polarity of the review or a sentence, has attracted more and more research attention over the past decades. Traditional sentiment classification methods generally have good performance for a specific domain with abundant labeled data [1–3]. However, because labeling data is expensive and time-consuming, many domains lack of sufficient labeled data, which make traditional methods don't work well.

To address the problem, cross-domain sentiment classification has been proposed. It uses the knowledge from source domain with sufficient labeled data to enhance the prediction accuracy of target domain with few or no labeled

© Springer Nature Switzerland AG 2020
G. Li et al. (Eds.): KSEM 2020, LNAI 12275, pp. 143–154, 2020.
https://doi.org/10.1007/978-3-030-55393-7_14

data. Researchers have proposed many methods to solve the cross-domain sentiment classification problem. Learning domain-shared and domain-specific features have been presented, which utilize the words with high co-occurrence in different domains and domain-independent words [4,5]. While these methods require to manually extract domain-independent words. Recently, some methods can learn better sample features by deep neural network [6–9]. Domain-Adversarial training of Neural Networks (DANN) [8] which adds adversarial mechanism into the training of deep neural network. It introduces a domain classifier which can minimize the discrepancy between the source and target domain by gradient reversal. Most of the previous efforts only focus on aligning the global marginal distribution, while ignoring the category-level alignment. As shown in Fig. 1 (left), the positive/negative data aligns the negative/positive data from different domains. This mismatch promotes negative transfer and reduces classification accuracy.

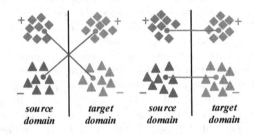

Fig. 1. "+" and "−" denote positive and negative samples respectively. **Left**: domain adaptation without category-level alignment. **Right**: domain adaptation with category-level alignment

To overcome the sample mismatch issue, we propose the Category-level Adversarial Network (CAN) for cross-domain sentiment classification. CAN achieves the category-level alignment by introducing the category-wise domain classifiers, as shown in Fig. 1 (right). CAN constructs category-level adversarial network by combining the label information and document representations. This method can decide how much each document should be sent to the category-wise domain classifiers by utilizing the probability distribution over the label space. Besides, the word with sentiment polarity usually has higher contribution for document representation. CAN utilizes the hierarchical attention transfer mechanism, which automatically transfers word-level and sentence-level attentions across domains. In summary, the main contributions of our work are summarized as follows:

- We introduce the category-level information to achieve fine-grained alignment of different data distributions.
- We propose a CAN method which achieves category-level alignment. It adds the category-wise domain classifiers which joint the label information and

document representations. Besides, the hierarchical attention transfer mechanism can transfer attentions by assigning different weights to words and sentences.

- The experimental results clearly demonstrate that our method outperforms other state-of-the-art methods.

2   Related Work

Domain Adaptation: Domain adaptation has a large number of works in natural language processing over the past decades. Among them, Blitzer et al. [4] proposed the Structural Correspondence Learning (SCL) which produces correspondences among the features from different domains. Pan et al. [5] proposed the Spectral Feature Alignment (SFA) which solves mismatch of data distribution by aligning domain-specific words. Unfortunately, these methods mentioned above highly rely on manually selecting domain-shared features.

Recently, deep learning methods have obtained better feature representations for cross-domain sentiment classification. Glorot et al. [6] proposed the Stacked Denoising Auto-encoders (SDA) which successfully learns feature representations of a document from different domains [6]. [7] Chen et al. proposed Marginalized Stacked Denoising Autoencoder (mSDA) which reduces computing cost and improves the scalability to high-dimensional features. Kim, Yu et al. [10, 11] used two auxiliary tasks to produce the sentence embedding based on convolutional neural network. DANN leverages the adversarial training method to produce feature representations [8]. Li et al. [12] proposed the Adversarial Memory Network (AMN) which automatically obtains the pivots by using attention mechanism and adversarial training. Li et al. [9] proposed Hierarchical Attention Transfer Network (HATN) which transfers word-level and sentence-level attentions. Zhang et al. [13] proposed Interactive Attention Transfer Network (ITAN) which provides an interactive attention transfer mechanism by combining the information of sentence and aspect. Peng et al. [14] proposed the CoCMD which simultaneously extracts domain specific and invariant representations. Sharma et al. [15] proposed a method which can identify transferable information by searching significant consistent polarity (SCP) words. But these methods only align the global marginal distribution by fooling domain classifiers, which bring the category-level mismatch. To solve this problem, we align the category-level distribution by adding the label information.

Attention Mechanism: The contribution of each word in a document is different. To address this problem, attention mechanism is also used in many other tasks, such as machine translation [16], sentiment analysis [3], document classification [17], question prediction [18]. Besides, the hierarchical structure has superior performance than word-level attention which captures better feature representations since it expresses the hierarchical structure of the document.

3 Category-Level Adversarial Network

In this section, we first introduce the problem definition of cross-domain sentiment classification, followed by an summary of the model. Finally we present
the details of CAN model.

3.1 Problem Definition

We assume that there are two domains D_s and D_t which denote a source domain
and a target domain respectively. We further suppose that we give a set of labeled
training data $\mathbf{X_s^l} = \{x_s^i, y_s^i\}_{i=1}^{N_l}$ and unlabeled training data $\mathbf{X_s^u} = \{x_s^j\}_{j=N_s^l+1}^{N_s}$
from the source domain, where N_s^l is the number of labeled data and N_s is
the all data of source domain. Besides, we give a set of unlabeled training data
$\mathbf{X_t} = \{x_t^j\}_{j=1}^{N_t}$ in the target domain, where N_t is the number of unlabeled data.
The goal of cross-domain sentiment classification is training a robust model on
labeled data and adapts it to predict the sentiment label on unlabeled data.

3.2 An Overview of CAN

We introduce the overview of the CAN as shown in Fig. 2. Firstly, we obtain
document representations by Transferable Attention Network (TAN). Then we
utilize the Category-level Adversarial (CA) classifiers which combine the label
information and document representations on adversarial process. Finally, we
use the data in source domain and unlabeled data in target domain to train
CA classifiers. Meanwhile, we train sentiment classifier using the labeled data.
Finally, TAN and sentiment classifier predict the sentiment label.

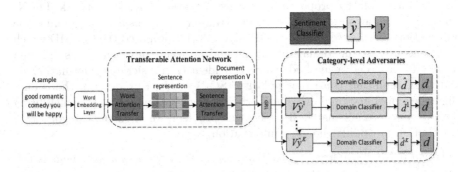

Fig. 2. The architecture of the CAN model, where $\hat{y}$ is the predicted sentiment label
and $\hat{d}$ is the predicted domain label; y and d are the ground truth.

3.3 Components of CAN

CAN mainly includes three parts, i.e., the TAN transfers the word-level and sentence-level attentions across domains, sentiment classifier predicts the sentiment label and CA classifiers align the category from source to target domain. TAN includes the word attention transfer and sentence attention transfer. The document representations and labels are obtained by TAN and sentiment classifier respectively. On the basis of single domain classifier, CA classifiers add K category-wise domain classifiers, which combine the document representations and pseudo labels to avoid the mismatch. We describe the components of CAN successively.

Transferable Attention Network: To transfer the important words across domains, we use the hierarchical attention network that assign different weights to words [17]. Assuming that a document has L sentences and each sentence s_p contains Q words, where w_{pq} is the q-th word in the p-th sentence, $q \in [1, Q]$, We map words into dense vectors for representing sentences through an embedding matrix M, $x_{pq} = M w_{pq}$. The sentence vector s_p summarizes all words' representation by word attention transfer and the document vector v which summarizes the all sentences' information by sentence attention transfer.

Category-Level Adversarial: In domain adaptation problems, the data distribution is usually very complicated and it is difficult to achieve complete alignment. Incorrect alignment may be prone to under transfer or negative transfer. To enhance the positive transfer and combat negative transfer, we urgently need a technology which can align the data distribution for improving the transfer effect.

We propose the CA classifiers. On the basis of single domain classifier G_d, we add K category-wise domain classifiers $G_d^k (k = 1, 2...K)$, where K is the number of category. Since target domain data is unlabeled, we use the output of sentiment classifier $\hat{y} = G_c(v)$ as the probability of target domain. Similarly, it is also utilized on source domain. It is similar to attention mechanism which indicates the probability of the data from source or target doamin. In other words, this method take category-level alignment into account during the adversarial process. The document vector v is modeled by the weight $\hat{y}$ as the input of domain classifier G_d^k. We use the data X_s^l, X_s^u and X_t to train domain classifiers G_d and G_d^k, which predict domain labels. The goal of domain classifiers is to distinguish the two domains as accurately as possible. However, we want to learn the common features which can not be distinguished by domain classifiers. To address this problem, we introduce the Gradient Reversal Layer (GRL) [8] to reverse the gradient in the training process. The feedward and backpropagation process are as follows:

$$G(x) = x, \frac{\partial G(x)}{\partial x} = -\lambda I \tag{1}$$

The v and $\widehat{vy^k}$ are the input of domain classifiers G_d and G_d^k respectively. Through the GRL as $G(v) = \tilde{v}_d$ and $G(\widehat{vy^k}) = \tilde{v}_d^k$. Then we feed it to the corresponding domain classifiers.

$$\widetilde{y_d} = softmax(fc(fc(\widetilde{v_d}))) \tag{2}$$

$$\widetilde{y_d^k} = softmax(fc(fc(\widetilde{v_d^k}))) \tag{3}$$

where fc is the fully connected layer.

Sentiment Classifiers: The sentiment classifier utilizes the unlabeled data from source domain. According to the document representation v, we calculate the output of sentiment classifier as follows:

$$\widetilde{y_s} = softmax(fc(fc(fc(v)))) \tag{4}$$

Training Strategy: The cross-domain sentiment classifier needs to use the domain-shared features to predict sentiment labels. In order to achieve this goal, CAN has two tasks i.e., domain classification and sentiment classification. We introduce cross-entropy loss functions for training CA classifiers and sentiment classifier respectively:

$$L_{dom_{con}} = -\frac{1}{N_s + N_t} \sum_{i=1}^{N_s+N_t} \widetilde{y_d} ln y_d + (1 - \widetilde{y_d}) ln(1 - y_d) \tag{5}$$

$$L_{dom_{class}} = -\frac{1}{N_s + N_t} \sum_{k=1}^{K} \sum_{i=1}^{N_s+N_t} \widetilde{y_d^k} ln y_d + (1 - \widetilde{y_d^k}) ln(1 - y_d) \tag{6}$$

$$L_{CA} = L_{dom_{con}} + L_{dom_{class}} \tag{7}$$

$$L_{sen} = -\frac{1}{N_s^l} \sum_{i=1}^{N_s^l} \widetilde{y_s} ln y_s + (1 - \widetilde{y_s}) ln(1 - y_s) \tag{8}$$

where y_d and y_s are the ground truth. Besides, we add the squared l_2 regularization for sentiment classifier and CA classifiers. Finally, the objective function is as follows:

$$L = L_{CA} + L_{sen} + \rho L_{reg} \tag{9}$$

where L_{reg} is the regularization and prevents the overfitting, ρ is the regularization parameter. CAN model aims to minimize L expect the GRL part which is maximized. Besides, we optimize the parameters by SGD.

4   Experiments

4.1   Dataset Preparation

In this section, we utilize two datasets to evaluate the ability of CAN method, one is the Amazon reviews dataset, the other is Airline reviews dataset. Table 1 summarizes the all dataset. We select the data from four domains: Books (B), DVD (D), Electronics (E) and Kitchen (K). Each domain contains 6000 labeled reviews with 3000 positive samples (higher than 3 stars) and 3000 negative samples (lower than 3 stars). Additionally, the dataset also contains lots of unlabeled data. Here we randomly extract 8,000 unlabeled reviews as training data. These unlabeled data are only used to train the domain classifier which distinguishes the data comes from different domains. We choose 1000 reviews from the target domain as the testing data. We conduct the cross-domain experiments between every two domains and get 12 cross-domain sentiment classification tasks: B → D, B → E, B → K, D → B, D → E, D → K, E → B, E → D, E → K, K → B, K → D, K → E. For example, the B → D is the task which transfers from the source domain B to the target domain D.

Airline reviews dataset is scraped from Skytraxs Web portal which has one of the most popular review sites within the air travel industry. It is the labeled data that includes 41396 reviews for Airline(AL), 17721 Reviews for Airport(AP), 1258 reviews for Seat(S), 2264 reviews for Lounge(L). We select the 3000 positive reviews (recommended value is 1) and 3000 negative reviews (recommended value is 0) from the Airline and Airport to be consistent with the Amazon reviews dataset. Besides, we randomly extract 8000 labeled reviews to train the domain classifier. We construct 8 cross-domain sentiment classification tasks: B → AL, D → AL, K → AL, E → AL, B → AP, D → AP, K → AP, E → AP.

Table 1. Statistics of Amazon and Airline datasets

Domain	B	D	K	E	AL	AP
#Train	5000	5000	5000	5000	5000	5000
#Test	1000	1000	1000	1000	1000	1000
#Unlabel	8000	8000	8000	8000	8000	8000

4.2 Implementation Details

We adopt the 300-dimensional word2vec vectors with the skip-gram model to initialize the embedding matrix M [19]. The maximum sentence length L and maximum word length Q are 20 and 28 respectively. All weight matrices are randomly initialized by a uniform distribution $U[-0.01, 0.01]$. The dimensional of GRU hidden states is set to 70. The regularization weight ρ and dropout rate are set to 0.005 and 0.6 respectively. We utilize the stochastic gradient descent with momentum rate 0.9 to optimize model during the training process. Because the different training sizes for different classifiers we set batch size $b_s = 50$ for the sentiment classifier and batch size $b_d = 260$ for the domain classifier. The adaptation rate is $\lambda = \frac{2}{1+exp(-10p)-1}$, where $p = \frac{n}{N}$. The n and N are current epoch and the maximum epoch respectively. The N is set to 100. The learning rate is $\eta = max(0.003 * 0.1^{\lfloor \frac{n}{10} \rfloor}, 0.0005)$.

4.3   Benchmark Methods

Naive (Hochreiter et al. 1997): it only uses source domain data based on LSTM.

SCL (Blitzer et al. 2006): it aims to identify the correlation between pivot and non-pivot by using multiple pivot prediction tasks.

SFA (Pan et al. 2010): it aims to solve mismatch of data distribution by aligning domain-specific words.

DANN (Ganin et al. 2015): it uses the domain adaptation with the 5000-dimension feature representations by adversarial training.

AMN (Li et al. 2017): it uses memory networks and adversarial training to get domain-shared representations.

HATN & HATN^n (Li et al. 2018): it extracts pivots and non-pivots by the hierarchical attention network across domains. HATN does not contain the hierarchical positional encoding and HATN^h does.

ITAN & ITAN^n (Zhang et al. 2019): it uses interactive attention which combines aspects and sentences information. ITAN^n does not contain the aspects information and ITAN does.

CAN^s & CAN^c & CAN: it is our methods that include the single domain classifier, category-wise classifiers and the CA classifiers respectively.

Table 2. Classification results on the Amazon dataset.

S T	Naive	SCL	SFA	DANN	AMN	HATN	HATN^h	IATN	IATN^n	CAN^s	CAN^c	CAN
B D	0.786	0.807	0.813	0.832	0.855	0.858	0.861	0.868	0.854	0.869	0.871	**0.874**
B E	0.752	0.763	0.776	0.764	0.824	0.853	0.857	0.865	0.849	0.861	0.859	**0.867**
B K	0.737	0.771	0.785	0.790	0.811	0.849	0.852	0.859	0.838	0.866	0.865	**0.871**
D B	0.756	0.782	0.788	0.805	0.846	0.858	0.863	0.870	0.848	0.871	0.871	**0.876**
D E	0.734	0.754	0.758	0.796	0.812	0.849	0.856	**0.869**	0.855	0.858	0.863	0.867
D K	0.767	0.779	0.786	0.814	0.827	0.853	0.862	0.858	0.839	0.858	0.861	**0.865**
E B	0.696	0.716	0.724	0.735	0.766	0.808	0.810	0.818	0.768	0.832	0.835	**0.841**
E D	0.722	0.745	0.754	0.786	0.827	0.838	0.840	0.841	0.825	0.844	0.845	**0.850**
E K	0.787	0.817	0.825	0.841	0.857	0.868	0.879	0.887	0.859	0.888	0.893	**0.894**
K B	0.686	0.713	0.724	0.752	0.805	0.824	0.833	0.847	0.828	0.841	0.843	**0.851**
K D	0.723	0.752	0.758	0.776	0.812	0.841	0.845	0.844	0.835	0.838	0.841	**0.848**
K E	0.807	0.818	0.825	0.843	0.867	0.868	0.870	0.876	0.864	0.878	0.881	**0.882**
Avg	0.746	0.768	0.776	0.794	0.825	0.847	0.851	0.859	0.837	0.858	0.861	**0.866**

The experimental results are shown in Table 2. Comparing with others methods, CAN model has achieved the best performances on most tasks. The Naive model performance is badly at every task because it only uses the source domain data. SFA model achieves 77.6% on average since the features of the review are manually extracted and the linear classifier is not sufficient to express the model. HATN^h model achieves 85.1% on average because it automatically learns the domain-shared and domain-specific features. IATN model achieves 85.9%

on average because it combines the sentences and aspects information. However, HATN[h] and IATN model does not take the category-level alignment into account. Compared with CAN[s] model, CAN[c] achieves 86.1% on average, which is 0.3% higher than CAN[s] because the categories from different domains are aligned. Comparing with CAN[c] model, CAN model has better performance. The reason is that the pseudo-labels may lead to the incorrect alignment of data distributions. For the hard transfer task E → B, E → D, the performance of CAN[c] is very close to the CAN[s]. But for the simple transfer task E → K, K → E, the promotion is very obvious since the classification accuracy of target domain is higher. Finally, CAN model achieves 86.6% on average since we combine the single domain and category-wise classifiers.

The cross-domain sentiment classification tasks in Table 2 come from different domains with the same origin, such as E → K and B → D task. They are highly relevant. To show the performance of CAN on hard transfer tasks, we construct 8 new tasks which come from different origins. The experimental results are reported in Table 3. The Naive method without target domain samples performs the worst and the classification accuracy is 8.1% lower than CAN. The CAN and CAN[c] improves the classification accuracy by 0.9% and 0.3% than CAN[s], respectively. The results show that our method is effective on difficult transfer tasks.

Table 3. Classification results on the Airline dataset.

S T	B AL	D AL	E AL	K AL	B AP	D AP	E AP	K AP	Avg
Naive	0.702	0.694	0.718	0.704	0.584	0.582	0.603	0.579	0.646
CAN[s]	0.768	0.787	0.792	0.794	0.651	0.659	0.644	0.649	0.718
CAN[c]	0.773	0.792	0.789	0.801	0.646	0.661	0.646	0.658	0.721
CAN	**0.776**	**0.795**	**0.794**	**0.811**	**0.658**	**0.669**	**0.652**	**0.659**	**0.727**

4.4   Visualization of Features

To better illustrate the effectiveness of CAN, we visualize the feature of penultimate layer as shown in Fig. 3. We choose the task E → K and B → D. The visualization results show that CAN model has better distinguishable features. The features of CAN[s] model are not well discriminated clearly for lacking the category-wise alignment.

CAN model considers complex structures of the data distributions. The different domains are more indistinguishable and different categories are more discriminated under the CAN model. The experimental result is superior for cross-domain sentiment classification because we add the category-wise domain classifiers on training process.

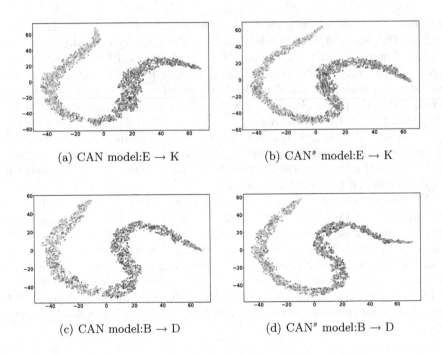

(a) CAN model:E → K

(b) CAN^s model:E → K

(c) CAN model:B → D

(d) CAN^s model:B → D

Fig. 3. The t-SNE visualization of features extracted by CAN and CAN^s model for E → K and B → D task. The red, blue, yellow and green points denote the source positive, source negative, target positive and target negative examples correspondingly (Color figure online)

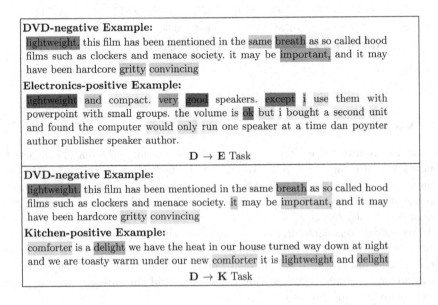

Fig. 4. Visualization of the CAN on D → E and D → K tasks

4.5 Visualization of Attention

We also visualize the word attention transfer in Fig. 4. We choose the D → E and D → K and highlight the words according to the attentions. Deeper red words mean that it has the heavier weight than others. Figure 4 shows that the words with sentiment polarity have higher attentions. Generally, the word "lightweight" intuitively indicates a negative sentiment in DVD domain. In contrast, the word "lightweight" usually indicates a positive sentiment in the electrics or kitchen domain. Since there are some words (good, delight) which have opposite sentiment polarity in the document, CAN model still correctly predicts labels on target domain. The document representation takes these words into account by hierarchical attention mechanism.

5 Conclusion

In this paper, we propose the CAN model. CA classifiers further align the category-level data distribution by combining the label information and document representations on adversarial process. Besides, we transfer word-level and sentence-levels attentions under TAN. The experimental results show that the CAN model effectively improves classification accuracy on Amazon reviews dataset and Airline reviews dataset.

Acknowledgments. This paper is Supported by National Key Research and Development Program of China under Grant No.2017YFB0803003 and National Science Foundation for Young Scientists of China (Grant No. 61702507).

References

1. Wang, G., Sun, J., Ma, J., Xu, K., Gu, J.: Sentiment classification: the contribution of ensemble learning. Decis. Support Syst. **57**, 77–93 (2014)
2. Wang, S., Manning, C.D.: Baselines and bigrams: simple, good sentiment and topic classification. In: Proceedings of the 50th Annual Meeting of the Association for Computational Linguistics: Short Papers, vol. 2, pp. 90–94. Association for Computational Linguistics (2012)
3. Tang, D., Qin, B., Feng, X., Liu, T.: Target-dependent sentiment classification with long short term memory. arXiv preprint arXiv:1512.01100 (2015)
4. Blitzer, J., McDonald, R., Pereira, F.: Domain adaptation with structural correspondence learning. In: Proceedings of the 2006 Conference on Empirical Methods in Natural Language Processing, pp. 120–128. Association for Computational Linguistics (2006)
5. Pan, S.J., Ni, X., Sun, J.T., Yang, Q., Chen, Z.: Cross-domain sentiment classification via spectral feature alignment. In: Proceedings of the 19th International Conference on World Wide Web, pp. 751–760. ACM (2010)
6. Glorot, X., Bordes, A., Bengio, Y.: Domain adaptation for large-scale sentiment classification: a deep learning approach. In: Proceedings of the 28th International Conference on Machine Learning (ICML 2011), pp. 513–520 (2011)

7. Chen, M., Xu, Z., Weinberger, K., Sha, F.: Marginalized denoising autoencoders for domain adaptation. arXiv preprint arXiv:1206.4683 (2012)
8. Ganin, Y., Lempitsky, V.: Unsupervised domain adaptation by backpropagation. arXiv preprint arXiv:1409.7495 (2014)
9. Li, Z., Wei, Y., Zhang, Y., Yang, Q.: Hierarchical attention transfer network for cross-domain sentiment classification. In: Thirty-Second AAAI Conference on Artificial Intelligence (2018)
10. Yu, J., Jiang, J.: Learning sentence embeddings with auxiliary tasks for cross-domain sentiment classification. In: Proceedings of the 2016 Conference on Empirical Methods in Natural Language Processing, pp. 236–246 (2016)
11. Kim, Y.: Convolutional neural networks for sentence classification. arXiv preprint arXiv:1408.5882 (2014)
12. Li, Z., Zhang, Y., Wei, Y., Wu, Y., Yang, Q.: End-to-end adversarial memory network for cross-domain sentiment classification. In: IJCAI, pp. 2237–2243 (2017)
13. Zhang, K., Zhang, H., Liu, Q., Zhao, H., Zhu, H., Chen, E.: Interactive attention transfer network for cross-domain sentiment classification. In: Thirty-Third AAAI Conference on Artificial Intelligence (2019)
14. Peng, M., Zhang, Q., Jiang, Y.G., Huang, X.J.: Cross-domain sentiment classification with target domain specific information. In: Proceedings of the 56th Annual Meeting of the Association for Computational Linguistics (Volume 1: Long Papers), pp. 2505–2513 (2018)
15. Sharma, R., Bhattacharyya, P., Dandapat, S., Bhatt, H.S.: Identifying transferable information across domains for cross-domain sentiment classification. In: Proceedings of the 56th Annual Meeting of the Association for Computational Linguistics (Volume 1: Long Papers), pp. 968–978 (2018)
16. Bahdanau, D., Cho, K., Bengio, Y.: Neural machine translation by jointly learning to align and translate. arXiv preprint arXiv:1409.0473 (2014)
17. Yang, Z., Yang, D., Dyer, C., He, X., Smola, A., Hovy, E.: Hierarchical attention networks for document classification. In: Proceedings of the 2016 Conference of the North American Chapter of the Association for Computational Linguistics: Human Language Technologies, pp. 1480–1489 (2016)
18. Huang, Z., et al.: Question difficulty prediction for reading problems in standard tests. In: Thirty-First AAAI Conference on Artificial Intelligence (2017)
19. Mikolov, T., Sutskever, I., Chen, K., Corrado, G.S., Dean, J.: Distributed representations of words and phrases and their compositionality. In: Advances in Neural Information Processing Systems, pp. 3111–3119 (2013)

Seeds Selection for Influence Maximization Based on Device-to-Device Social Knowledge by Reinforcement Learning

Xu Tong[1], Hao Fan[1], Xiaofei Wang[1(✉)], Jianxin Li[2], and Xin Wang[1]

[1] School of Computer Science and Technology,
College of Intelligence and Computing, Tianjin University, Tianjin, China
{xutong,haofan,xiaofeiwang,wangx}@tju.edu.cn
[2] School of Information Technology, Deakin University, Geelong, Australia
jianxin.li@deakin.edu.au

Abstract. Recently, how to use Device-to-Device (D2D) social knowledge to reduce the network traffic on mobile networks has become a hot topic. We aim to leverage D2D social knowledge to select influential users (seed users or seeds) for influence maximization to minimize network traffic. Lots of work has been done for seeds selection in a single community. However, few studies are about seeds selection in multiple communities. In this paper, we build a Multi-Community Coverage Maximization (MCCM) model to maximize the D2D social coverage so that the cellular network traffic can be minimized. We transform it into a resource allocation problem and use a Reinforcement Learning (RL) approach to tackle it. Specifically, we present a novel seeds allocation algorithm based on Value Iteration method. To reduce the time delay, we design an edge-cloud computing framework for our method by moving part of the computing tasks from the remote cloud to adjacent base stations (BSs). The experiment results on a realistic D2D data set show our method improves D2D coverage by 17.65% than heuristic average allocation. The cellular network traffic is reduced by 26.35% and the time delay is reduced by 63.53%.

Keywords: Seeds selection · Social knowledge · Device-to-Device · Traffic offloading · Edge computing

1 Introduction

In recent years, with mobile communication technology rapidly developing, it is getting more popular for people to use mobile devices such as smart phones and tablets to satisfy their entertainment and social demands in life and work [1]. This trend has caused the explosive growth of traffic load on mobile networks. There are severe problems with repeated downloads of popular content, which seriously waste the resources of communication networks [2,3].

© Springer Nature Switzerland AG 2020
G. Li et al. (Eds.): KSEM 2020, LNAI 12275, pp. 155–167, 2020.
https://doi.org/10.1007/978-3-030-55393-7_15

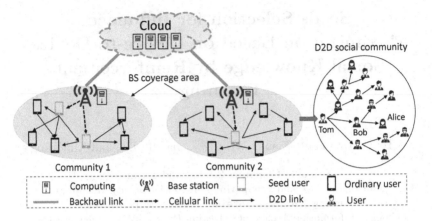

Fig. 1. Seeds selection in D2D communities under edge computing environment.

An effective solution to reduce the repeated downloads within cellular networks is to use D2D opportunistic sharing mechanism to encourage users to obtain the content they need from mobile devices in proximity [4]. Users who use D2D technology to communicate with each other can form a D2D community, as depicted in Fig. 1. To better reduce cellular network traffic, we propose to mine influential users (seed users or seeds) in D2D communities. As depicted in Fig. 1, seed users download content via cellular links and share them with other users. Given an integer k, it is important to find k most influential users to maximize the D2D coverage (or D2D influence), so as to minimize the cellular network traffic. There are many studies on seeds selection in one community [5–8]. Though, little work has been done for multiple communities. As depicted in Fig. 1, one of our goal is to figure out how to allocate k seeds to maximize the D2D coverage. Fortunately, reinforcement learning (RL) is a strong tool to solve resource allocation problems like seeds allocation [9,10]. Therefore, we use this technology to tackle the seeds allocation problem.

To provider better Quality of Service (QoS) and user experience, we devise a special edge-cloud computing framework for our solution. Edge-cloud computing is a new paradigm to offload part of computing tasks from remote cloud to edge servers. The cloud servers process tasks that require global data as input or need more computing resources. The edge servers process local data so as to reduce unnecessary data transmission on backhaul networks and time delay [11–13].

We summarize the contributions of this paper as follows.

1) We use a reinforcement learning method and propose a seeds allocation algorithm based on Value Iteration to maximize D2D social coverage.
2) We innovatively devise an edge-cloud computing infrastructure to offload part of computing tasks from the remote cloud to adjacent BSs to decrease the time delay.
3) The experiment results on a realworld D2D dataset show our method improves D2D coverage by 17.65% than heuristic average allocation. The cellular network traffic is reduced by 26.35% and the time delay is reduced by 63.53%.

The remainder of this paper is organized as follows. After reviewing related work in Sect. 2, we build the optimization model in Sect. 3. Then we propose an edge computing assisted approach in Sect. 4. We show the comparative experiments in Sect. 5. Finally, we conclude this paper in Sect. 6.

2   Related Work

It has been pointed out that a large amount of traffic load is caused by repeated downloads of popular content [2]. How to effectively offload cellular network traffic by applying D2D sharing mechanism is gaining much attention [3]. With D2D communications, users in close proximity can directly communicate with each other via D2D social links, instead of accessing BSs. It is also a feasible solution to cache popular content at the edge of networks to reduce the traffic and energy consumption of backhaul networks [14].

Seeds selection for Influence Maximization (IM) is to find influential communicators in social networks (or social communities) who are more likely to spread the information widely through word of mouth effect [15]. There are lots of effective methods on seeds selection in a single social community, such as degree centrality [5], PageRank [6], the Greedy algorithm [7] with performance guarantee ratio of $(1 - 1/e)$, and the Weighted LeaderRank with Neighbors (WLRN) algorithm [8]. In [16] and [17], some new models and algorithms are proposed to ensure the fairness the seeds allocation problem in competitive influence maximization. However, their theoretical diffusion models are not practical for real-world D2D social communities. For seeds selection in multiple communities, the key is to find the optimal seeds allocation scheme. Fortunately, reinforcement learning (RL) is a strong tool to solve resource allocation problems like the seeds allocation problem [9,10].

Edge-cloud computing is a new paradigm to offload part of computing tasks from the remote cloud to edge servers [11]. The cloud servers process tasks that require global data as input or need more computing resources. The edge servers process local data so as to reduce redundant data transmission on backhaul networks and time delay [12]. The computing delay is reduced due to the distributed nature of edge computing. And the transmission delay is decreased because part of the data is not transmitted to the cloud but processed at edge servers [13].

3   Optimization Model to Maximize D2D Social Coverage

To reduce cellular network traffic by encouraging users to use D2D communication, we aim to mine seed users. Equivalently, we present to maximize the D2D social coverage in D2D communities. Before modeling the coverage maximization problem, we define the D2D social coverage function.

3.1 D2D Social Coverage

In a D2D community, when user (device) u shares content to user (device) v, we say, u has directly *covered* v via D2D social links. We define D2D coverage of user u as the number of users covered by u. It includes directly covered users and indirectly covered users. For instance, user u uses D2D technology to transmit a video to v and v re-shares it to v', then v' is the directly covered user of v and the indirectly covered user of u in the meantime. The directly covered users of u can be defined as follows.

$$O(u) = \{u\} \cup \{v | e_{uv}^t \in E\}, \tag{1}$$

where E contains all the D2D social links and e_{uv}^t means u shares content to v at time t_{uv}. We define the D2D coverage of u as $C(u)$ as follows recursively.

$$C(u) = O(u) \cup (\cup_{v \in O(u)} C(v)), \tag{2}$$

satisfying $t_{uv} \le t_{vv'}$ for all $v \in O(u)$ and $v' \in O(v)$ with t_{uv} denoting the time of u sharing content to v and $t_{vv'}$ the time of v sharing content to v'. The coverage function through D2D propagation of users set S is defined as follows.

$$f(S) = |C(S)| = |\cup_{u \in S} C(u)|. \tag{3}$$

3.2 Multi-community Coverage Maximization

It has been well studied to select seed users in one community. Given a community $G = (V, E)$ where V is users set and E is edges set, and an integer k, the seeds selection problem is to find out k users from G to maximize their coverage. We can denote S as a function of G and k, namely $S = H(V, k)$, where H maps (V, k) into a subset of V with size k. That is $H : V \times \mathbb{N} \to 2^V$. Then $f(S) = f(H, V, k)$ and H determines how to select seeds in one community.

There is little work for multiple communities. Given that there are n communities, $G_1, G_2, ..., G_n$, where $G_i = (V_i, E_i)$, the Multi-Community Coverage Maximization (MCCM) problem is to find out $k_1, k_2, ..., k_n$ users respectively in $G_1, G_2, ..., G_n$ so that they can cover as many users as possible through D2D communication. We denote $K = (k_1, k_2, ..., k_n)$ and it satisfies $\sum_{i=1}^n k_i = k$. Denoting S_i as the set of k_i seed users selected from G_i. Let $S = \cup_{i=1}^n S_i$. Due to the geographic restriction in D2D communication, users from different D2D communities have a much lower probability to communicate with each other than those in the same community. Consequently, it is rational to assume that $C(S_1) \cap C(S_2) = \varnothing$, for any two seeds sets S_1 and S_2. This assumption indicates the coverage of seed users from different communities don't overlap. Therefore, the objective function can be written as follows.

$$f(S) = \sum_{i=1}^n f(S_i) = \sum_{i=1}^n f(H, V_i, k_i) \tag{4}$$

To find a seeds selection method H has been well explored in many researches [5–8]. Therefore, in this study, we concentrate on finding the optimal allocation scheme $K^* = (k_1^*, k_2^*, ..., k_n^*)$. Once H is fixed, the Multi-Community Coverage Maximization (MCCM) problem can be summarized as follows.

$$\max_{K} \sum_{i=1}^{n} f(V_i, k_i; H).$$
$$s.t. \sum_{i=1}^{n} k_i \leq k; \ k_i \in \mathbb{N}, \ i = 1, 2, ..., n. \tag{5}$$

The optimization problem Eq. (5) is NP-hard.

Proof. We reduce an NP-hard problem, the knapsack problem, to problem Eq. (5). There is a knapsack with capacity w and n sorts of items. Each item has weight $w_i \in \mathbb{N}^+$ and value $v_i \geq 0$. The goal of the knapsack problem is to put items into this knapsack to maximize the total weight, formulated as follows.

$$\max_{\Theta} \sum_{i=1}^{n} \theta_i v_i.$$
$$s.t. \sum_{i=1}^{n} \theta_i w_i \leq w; \theta_i \in \mathbb{N}, \ i = 1, 2, ..., n. \tag{6}$$

Here $\Theta_i = \{\theta_i\}_{i=1}^{n}$ refers to the number of each selected item. Let

$$k = w; k_i = \theta_i w_i, i = 1, 2, ..., n;$$
$$f(V_i, k_i; H) = \sum_{i=1}^{n} \theta_i v_i = \sum_{i=1}^{n} \frac{k_i}{w_i} v_i. \tag{7}$$

Then the knapsack problem Eq. (6) is reduced to the MCCM problem Eq. (5). Therefore, problem Eq. (5) is NP-hard. ∎

4   Edge Computing Assisted Approach to MCCM

4.1   Seeds Allocation Algorithm with Reinforcement Learning

To tackle MCCM is to find the optimal seeds allocation scheme. Reinforcement learning is a strong tool to solve resource problems like the seeds allocation problem [9,10]. Therefore, we use this technology to tackle the MCCM problem. We model the seeds allocation process as a Markov Decision Process as follows.

State. The seeds allocation state is written as $s = (i, q)$, $i, q \in \mathbb{N}$, $1 \leq i \leq n$ and $0 \leq q \leq k$. State $s = (i, q)$ represents $k - q$ seeds have been allocated to communities $\{G_1, G_2, ..., G_{i-1}\}$ and there are q seeds we are to allocate to communities $\{G_i, G_{i+1}, ..., G_n\}$.

Action. The seeds allocation action is $a = j$, $j \in \mathbb{N}$. Action $a = j$ under state $s = (i, q)$ $(j \leq q)$ means to select j users by function H from G_i to form seeds set $S_i = H(V_i, k_i)$.

Reward. The reward of action $a = j$ under state $s = (i, q)$ is the increased coverage by selecting j seeds in community G_i. The reward function $r(s, a)$ is as follows. It determines which action to take under a certain state so that the reward can reach the highest.

$$r(s, a) = f(S_i) = f(V_i, j; H). \tag{8}$$

Policy. The policy in MDP is represented as a function, $\pi : S \to A$, which gives the next move $a = \pi(s)$ under state s. We denote the *value* of state s under policy π as $V_\pi(s)$, which means the expected return when starting from state s and following π thereafter till the decision process ends. We define $V_\pi(s)$ as:

$$V_\pi(s) = E_\pi[\sum_{i=0}^{\infty} \gamma^i r_i | s_0 = s], \tag{9}$$

where $E_\pi[\cdot]$ is the expected value of a random variable, r_i denotes the reward by taking the next i^{th} action following π. $0 \leq \gamma \leq 1$ is the discount factor which determines how important the future reward is. We call V_π *the state value function for policy* π. In our allocation problem, state $s = (i, q)$ means the q seeds will be allocated to $\{G_i, G_{i+1}, ..., G_n\}$. We view $V_\pi(s)$ as the expected D2D coverage by allocating q seed users to $\{G_i, G_{i+1}, ..., G_n\}$ following policy π. It is shown in Eq. (10) as follows,

$$V_\pi(s = (i, q)) = \sum_{j=i}^{n} f(S_j) = \sum_{j=i}^{n} f(V_j, \pi(s_j); H), \tag{10}$$

where $s_j = (j, q_j)$ is the seeds allocation state for G_j.

A key property of the state value function V is the particular recursive relationship [9]. Given a state s and a policy π, the expected reward by taking the action $a = \pi(s)$ following policy π starting from s is the sum of the current reward and the future reward. The recursive relationship suggests the state value function satisfies the Bellman Equation:

$$V_\pi(s) = E_\pi[r(s'|s, a) + \gamma V_\pi(s') | s_0 = s] = \sum_{s' \in S} p(s'|s, a)[r(s, a) + \gamma V_\pi(s')]. \tag{11}$$

The agent starts from the first community and continue one by one to the last community. One round of decision process is shown as follows, where s_i is the state and a_i is the action. The agent starts from community G_1 with the budget k, and select k_1 seeds from G_1. The agent stops till there is no seed budget left.

$$s_1 = (1, k) \xrightarrow{a_1 = k_1} s_2 = (2, k - k_1) \xrightarrow{a_2 = k_2} ... \xrightarrow{a_{n-1} = k_{n-1}} s_n = (n, k - \sum_{i=1}^{n-1} k_i) \tag{12}$$

Algorithm 1. Seeds allocation method based on Value Iteration

---

Require: n, k, H, γ, T, $G = \{G_1, G_2, ..., G_n\}$
1: Initialize $S = \{(i,q), 1 \leq i \leq n, 0 \leq q \leq k, i,q \in \mathbb{N}\}$ and V arbitrarily.
2: **for** $s \in S$ **do**
3:     $A_s \leftarrow \{0, 1, ..., s[1]\}$; // Get possible actions under state s.
4:     **for** $a \in A_s$ **do**
5:         $S_{s[0]} \leftarrow H(V_i, a)$; $r(s,a) \leftarrow |C(S_{s[0]})|$; // Calculate reward function.
6:     **end for**
7: **end for**
8: **for** $iter = 1$ to T **do** // Train the model for T rounds to get optimal value function.
9:     **for** $s \in S$ **do**
10:         $V(s) \leftarrow \max_a [r(s,a) + \gamma V(s')]$;
11:     **end for**
12: **end for**
13: **for** $s \in S$ **do**
14:     $\pi^*(s) \leftarrow \arg\max_a r(s,a) + \gamma V(s')$; // Get the optimal policy π^*.
15: **end for**
16: $s \leftarrow (1, k)$; $allocation \leftarrow [\ ]$;
17: **while** $s[0] \leq n$ **do** // Form allocation scheme and seeds sets.
18:     $k_i^* \leftarrow \pi^*(s)$; $s \leftarrow (i+1, k - k_i^*)$;
19:     $allocation.\text{add}(k_i^*)$;
20:     $Seed_{s[0]} \leftarrow H(V_{s[0]}, k_i^*)$;
21: **end while**
22: $Seeds \leftarrow \{Seed_1, Seed_2, ..., Seed_n\}$ // $Seed_i$ is the seeds set of the $i-th$ community.
23: **return** $Seeds, allocation$.

Let $s_i = (i, q_i)$ and $s_j = (j, q_j)$ be two states, the state transition probability $p(s_j|s_i, a)$ is 1 if $j = i + 1$ and 0 otherwise. Thus, Eq. (11) can be simplified as:

$$V_\pi(s) = r(s, a) + \gamma V_\pi(s'). \tag{13}$$

There are many model-free methods and model-based methods to solve an MDP. Monte Carlo methods, Q-learning and Sarsa are model-free methods that don't need to know the environment. Model-based methods like Policy Iteration and Value Iteration, can be used when the environment is determined. Since the state transition probability in the defined MDP is known in our problem, model-based methods can be applied to solve this MDP. Policy Iteration and **Value Iteration** are two model-free methods based on dynamic programming [9,10]. The shortcoming of Policy Iteration method is its policy evaluation in each iteration may cause the extra computation to scan the state space S. In Value Iteration method, the policy evaluation step of Policy Iteration can be removed without the loss of convergence. Value Iteration truncates policy evaluation according to the following Bellman Optimal Equation.

$$V^*(s) = \max_{a \in A}[r(s, a) + \gamma V^*(s')]. \tag{14}$$

And the corresponding optimal policy π^* is:

$$\pi^*(s) = \arg\max_{a \in A}[r(s,a) + \gamma V^*(s')]. \tag{15}$$

Therefore, we leverage Value Iteration to solve the seeds allocation problem. Based on Value Iteration, we summarize the proposed seeds allocation method in **Algorithm 1**. According to Eq. (14), an agent starts from an initial state s and finds the best action a that maximizes $r(s,a) + \gamma V_\pi(s')$. After numerous rounds of iterations, the value function V converges to V^*. Then we use Eq. (15) to find out the optimal policy π [9]. Note that Algorithm 1 is based on the seeds selection method H. The time complexity is $O(|States| \cdot |Actions|) = O(nk^2)$.

4.2   The Proposed Edge-Cloud Computing Framework

Edge-cloud computing is a new paradigm to offload part of computing tasks from remote cloud to edge servers. To decrease the time delay and provide better user experience, we devise an edge-cloud computing framework to realize our method. Our framework is shown in Fig. 2. There are six steps in our edge-cloud computing infrastructure, including three computing processes and three data transmission processes. There are three main computing tasks listed as follows.

- C_1: Each BS executes the method H within each community to rank users' social influence in its coverage.
- C_2: Each BS calculates the reward function with collected D2D logs according to Eq. (3) and Eq. (8).
- C_3: The remote cloud server performs the proposed Algorithm 1 in a centralized manner.

The three data transmission processes are as follows.

- T_1: The BS collects D2D communication traces from users' devices through cellular links in format $< MD5, sender, receiver, timestamp, content\ size >$.
- T_2: The results of C_1 and C_2 are transmitted to the remote cloud center.
- T_3: The remote cloud returns the final allocation scheme to each BS.

The time delay in our edge-cloud computing framework is composed of data transmission delay and computing delay. The data transmission delay can be modeled as follows [18].

$$d_t = \frac{s}{r}, \tag{16}$$

where s is the data size and r is the upstream rates of cellular networks or backhaul networks. The computing delay can be modeled as follows [19].

$$d_c = \frac{c_k}{f_k}, \tag{17}$$

where c_k is the number of CPU cycles the server needs to finish computing task C_k and f_k is the CPU cycles assigned to task C_k per second.

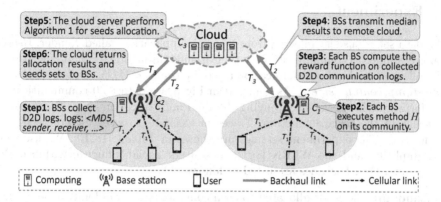

Fig. 2. The workflow of edge-cloud computing framkework.

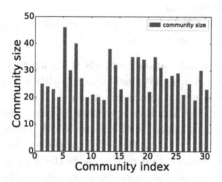

Fig. 3. Statistical analysis on D2D social communities.

Table 1. Experiment settings.

Symbol	Value	Description
n	30	Number of D2D social communities
N_B	30	Number of base stations
$|V|$	822	Total number of users
$|E|$	1049	Total number of D2D social links
k	[5, 60]	Total number of seed users
H	PageRank, WLRN	The method to rank users
γ	1	The discount factor
T	100	Iteration times
r_c	40 Mbps	Cellular network capacity
r_b	80 Mbps	Backhaul network capacity
f_k^B	2 G cycles/s	CPU of edge servers
f_k^C	4 G cycles/s	CPU of cloud servers

5   Experiments

The data set we use is collected from a mobile application called Xender. Users can use Xender to communicate with their friends through D2D communication. We collect 13 weeks of communication traces in format $<MD5, sender, receiver, timestamp, content\ size>$. As depicted in Fig. 3, we detect 30 communities in the former eight weeks by applying *Union Find*. Traces in week 9 are used to generate the reward function. Traces in week 10 to week 14 are used for verification. The experiment settings are shown in Table 1 [18,19]. We use the PageRank [6], and the WLRN [8] to rank users' social influence within each community. Four baselines are listed as follows [20].

1) **Random:** The seeds allocated to each community are randomly generated.
2) **Average:** The seeds are equally allocated to the communities.
3) **Proportion:** The number of seeds allocated to each community is proportional to the community size.
4) **Without allocation (WOA):** Regard n communities as one social community G and select top k users by H.

The coverage of different allocation methods is shown in Fig. 4. Our proposed allocation method outperforms baselines. Our algorithm improves the D2D coverage by 17.65% than Average allocation method. Average allocation achieves better performance than the Random algorithm because it gives each community the equal chance for propagation. Furthermore, WOA method does not consider the structural property of communities. The proposed algorithm considers the actual propagation process, which directly affects the final propagation coverage.

It is shown in Fig. 5 that the more seeds we invest, the more data we can offload from cellular networks. When we select 60 seeds out of 822 users in 30 communities as relays, under the WLRN setting, the cellular network traffic can be reduced by 26.35%. In Fig. 4 and Fig. 5, we show that the more users we encourage to use D2D communication, the more traffic we can offload from cellular networks to D2D networks.

Figure 6 depicts the time delay under two computing frameworks. We only show the case under $H =$ WLRN since the running time of WLRN and PageRank are nearly the same. The time delay is reduced by 63.53% than merely cloud computing. In edge-cloud computing framework, task C_1 and C_2 are computed in a distributed manner at each BS respectively. Then the computing delay is decreased. And massive D2D traces are not transmitted to cloud but processed at adjacent BSs. Therefore, the transmission delay is reduced.

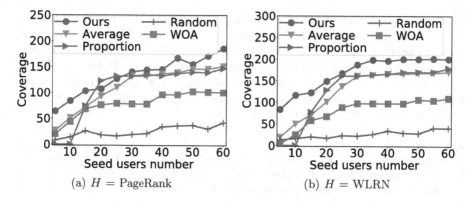

Fig. 4. D2D social coverage

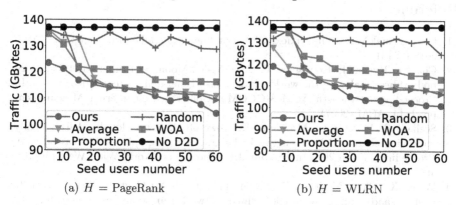

Fig. 5. Cellular network traffic

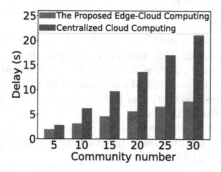

Fig. 6. Time delay.

6   Conclusions

In this paper, to leverage D2D social knowledge to reduce cellular network traffic, we build an optimization model for the MCCM problem. We put our focus on selecting seeds as content relays in multiple communities. We use the reinforcement learning method to solve the seeds allocation problem. To realize lower time delay, we propose an edge-computing framework for cloud computing offloading. The results show our proposed method is valid to maximize D2D social coverage and reduce both network traffic and time delay. Training the reinforcement learning model could be time consuming. Therefore, more efficient ways to solve our MDP model will be explored in future work.

Acknowledgement. This work was supported by the National Key Research and Development Program of China under Grant 2019YFB2101901 and 2018YFC0809803, the National Natural Science Foundation of China under Grant 61702364 and 61972275, and Australia Research Council Linkage Grant LP180100750.

References

1. Cisco Visual Networking Index. Global mobile data traffic forecast update, 2013–2018 (2014)
2. Cha, M., Kwak, H., Rodriguez, P., Ahn, Y.-Y., Moon, S.: I tube, you tube, everybody tubes: analyzing the world's largest user generated content video system. In: Proceedings of the 7th ACM SIGCOMM Conference on Internet Measurement, pp. 1–14. ACM (2007)
3. Scellato, S., Mascolo, C., Musolesi, M., Crowcroft, J.: Track globally, deliver locally: improving content delivery networks by tracking geographic social cascades. In: Proceedings of the 20th International Conference on World Wide Web, pp. 457–466. ACM (2011)
4. Wang, X., Chen, M., Han, Z., Kwon, T.T., Choi, Y.: Content dissemination by pushing and sharing in mobile cellular networks: an analytical study. In: 2012 IEEE 9th International Conference on Mobile Ad-Hoc and Sensor Systems (MASS 2012), pp. 353–361. IEEE (2012)
5. Freeman, L.C.: Centrality in social networks conceptual clarification. Soc. Netw. **1**(3), 215–239 (1978)
6. Page, L., Brin, S., Motwani, R., Winograd, T.: The PageRank citation ranking: bringing order to the web. Technical report, Stanford InfoLab (1999)
7. Kempe, D., Kleinberg, J., Tardos, J.: Maximizing the spread of influence through a social network. In: Proceedings of the Ninth ACM SIGKDD International Conference on Knowledge Discovery and Data Mining, pp. 137–146. ACM (2003)
8. Fan, H., Tong, X., Zhang, Q., Zhang, T., Wang, C., Wang, X.: Identifying influential users in mobile device-to-device social networks to promote offline multimedia content propagation. In: 2019 IEEE International Conference on Multimedia and Expo (ICME), pp. 1720–1725. IEEE (2019)
9. Sutton, R.S., Barto, A.G., et al.: Introduction to Reinforcement Learning, vol. 2. MIT Press, Cambridge (1998)
10. Reyes, M.G.: Reinforcement learning in *A Marketing Game*. In: Arai, K., Bhatia, R., Kapoor, S. (eds.) CompCom 2019. AISC, vol. 997, pp. 705–724. Springer, Cham (2019). https://doi.org/10.1007/978-3-030-22871-2_49
11. Chen, X., Jiao, L., Li, W., Xiaoming, F.: Efficient multi-user computation offloading for mobile-edge cloud computing. IEEE/ACM Trans. Netw. **24**(5), 2795–2808 (2015)

12. Wang, X., Yang, L.T., Xie, X., Jin, J., Deen, M.J.: A cloud-edge computing framework for cyber-physical-social services. IEEE Commun. Mag. **55**(11), 80–85 (2017)
13. Wang, P., Yang, L.T., Li, J.: An edge cloud-assisted CPSS framework for smart city. IEEE Cloud Comput. **5**(5), 37–46 (2018)
14. Gregori, M., Gómez-Vilardebó, J., Matamoros, J., Gündüz, D.: Wireless content caching for small cell and D2D networks. IEEE J. Sel. Areas Commun. **34**(5), 1222–1234 (2016)
15. Lü, L., Chen, D., Ren, X.-L., Zhang, Q.-M., Zhang, Y.-C., Zhou, T.: Vital nodes identification in complex networks. Phys. Rep. **650**, 1–63 (2016)
16. Lu, W., Bonchi, F., Goyal, A., Lakshmanan, L.V.: The bang for the buck: fair competitive viral marketing from the host perspective. In: Proceedings of the 19th ACM SIGKDD International Conference on Knowledge Discovery and Data Mining, pp. 928–936. ACM (2013)
17. Yu, Y., Jia, J., Li, D., Zhu, Y.: Fair multi-influence maximization in competitive social networks. In: Ma, L., Khreishah, A., Zhang, Y., Yan, M. (eds.) WASA 2017. LNCS, vol. 10251, pp. 253–265. Springer, Cham (2017). https://doi.org/10.1007/978-3-319-60033-8_23
18. Tao, X., Ota, K., Dong, M., Qi, H., Li, K.: Performance guaranteed computation offloading for mobile-edge cloud computing. IEEE Wirel. Commun. Lett. **6**(6), 774–777 (2017)
19. Liu, D., Yang, C.: Energy efficiency of downlink networks with caching at base stations. IEEE J. Sel. Areas Commun. **34**(4), 907–922 (2016)
20. Cao, T., Wu, X., Wang, S., Hu, X.: Maximizing influence spread in modular social networks by optimal resource allocation. Expert Syst. Appl. **38**(10), 13128–13135 (2011)

CIFEF: Combining Implicit and Explicit Features for Friendship Inference in Location-Based Social Networks

Cheng He[1], Chao Peng[1(✉)], Na Li[2], Xiang Chen[3], Zhengfeng Yang[1], and Zhenhao Hu[1]

[1] Shanghai Key Laboratory of Trustworthy Computing,
Software Engineering Institute, East China Normal University,
Shanghai, China
chenghe28@qq.com,
{cpeng,zfyang}@sei.ecnu.edu.cn,
hzh@stu.ecnu.edu.cn
[2] School of Data Science and Engineering, East China Normal University,
Shanghai, China
nali0606@foxmail.com
[3] School of Information Science and Technology, Nantong University,
Nantong, China
xchencs@ntu.edu.cn

Abstract. With the increasing popularity of location-based social networks (LBSNs), users can share their check-in location information more easily. One of the most active problems in LBSNs is friendship inference based on their rich check-in data. Previous studies are mainly based on co-occurrences of two users, however, a large number of user pairs have no co-occurrence, which weakens the performance of previous proposed methods. In this paper, we propose a method CIFEF that *C*ombines the *I*mplicit *F*eatures and a *E*xplicit *F*eature for friendship inference. Specifically, based on whether a user has different trajectory patterns on weekdays and weekends, we take the embedding technique to learn implicit weekdays' trajectory features and weekends' trajectory features from their check-in trajectory sequences, respectively, which can work effectively even for user pairs with no co-occurrence. Moreover, we propose a new explicit feature to capture the explicit information of user pairs who have common locations. Extensive experiments on two real-world LBSNs datasets show that our proposed method CIFEF can outperform six state-of-the-art methods.

Keywords: Location-based social networks · Implicit features · Explicit features · Friendship inference

This research is supported by the Scientific and Technological Innovation 2030 Major Projects under Grant 2018AAA0100902.

G. Li et al. (Eds.): KSEM 2020, LNAI 12275, pp. 168–180, 2020.
https://doi.org/10.1007/978-3-030-55393-7_16

1   Introduction

In the past decade, with the rising popularity of smart phones, applications on location-based social networks (LBSNs) [5] have attracted tens of millions of users. In LBSNs, users can share their location information (i.e., check-in) when they find a new place or take part in social activities. Huge volume check-in data of users are collected from location-based social networking services, which provides an opportunity for researchers to study various social behaviors.

One interesting question is whether we can infer the social relationship between two users based on their check-in data in LBSNs. According to social homophily principle [12], friends tend to visit a certain number of same locations to participate in some social activities, such as attending the wedding of their common friends or having dinner together at a restaurant. Inferring the relationship between two users have been largely adopted in friend recommendation [7], social influence analysis [18], and targeted marketing [11]. Therefore, given the check-in data of two users, inferring whether they are friends or strangers attracts a lot of researches [4, 10, 15, 16, 19–22].

However, previous studies have three major shortcomings. Firstly, they mined some co-occurrence (refer to Definition 2) features of user pairs to address the friendship prediction problem. In this situation, if friend pairs share co-occurrences rarely, it will affect the performance of these methods. In Table 1, we show the ratio of the friend pairs who have at least one co-occurrence on Gowalla dataset and Brightkite dataset [13], respectively. We can observe that even if we set the time threshold to 120 min and distance threshold to 200 m of co-occurrence, only 30.25% and 47.17% of friend pairs have at least one co-occurrence. Therefore, the performance of the friendship inference method based on co-occurrence is not satisfactory.

Secondly, in previous studies, He et al. [10] considered the distances between the two locations where users checked in most frequently on weekdays and weekends. However, as shown in Fig. 1, the users generate significantly different trajectory patterns on weekdays and on weekends. Therefore, modelling the users trajectories on weekday and weekend separately is more reasonable.

Thirdly, previous work used location entropy to measure the popularity of common place between two users. However, the time interval between two users' visit should be considered. If two users visit a same place, but there is a long time interval between their visits, they may not have any relationship.

To address the aforementioned issues, we propose a method that combining implicit and explicit features (CIFEF) for inferring friendship in LBSNs. Specifically, we first exploit the embedding learning technique to capture each user's contextual trajectory information of weekdays and weekends, respectively. In this way, we can get the implicit vector representation of each user, which does not rely on the co-occurrences. Besides, we further propose a new explicit feature by introducing check-in time factor into location entropy, which can mine the explicit information of user pairs who have visited a common place. In summary, to our best knowledge, the major contributions of this paper are as follows:

1. We exploit the embedding technique to learn latent vector representation of user's trajectory, which can work effectively even for user pairs with no co-occurrence. Moreover, because a user has different trajectory patterns on weekdays and weekends, we learn implicit representation of users weekday's trajectory and weekend's trajectory, respectively.
2. We further propose a new feature named *twcle* to measure the importance of each common place of user pairs by introducing the time interval of check-ins into location entropy.
3. We propose an effective method CIFEF to infer friendship in LBSNs, which combines implicit features and explicit feature for inferring friendship.
4. We conduct extensive experiments on two real-world datasets to evaluate the performance of our proposed method and the experiment results show that our method is superior to six state-of-art methods.

Table 1. The ratio of friend pairs who have at least one co-occurrence under different time threshold and distance threshold

(Time threshold, Distance threshold)	Gowalla	Brightkite
(10 min, 0 m)	14.21%	19.94%
(10 min, 100 m)	16.21%	22.62%
(10 min, 200 m)	17.60%	24.11%
(30 min, 0 m)	19.04%	27.72%
(30 min, 100 m)	21.35%	31.44%
(30 min, 200 m)	23.03%	33.60%
(60 min, 0 m)	22.57%	33.13%
(60 min, 100 m)	24.92%	37.52%
(60 min, 200 m)	26.85%	40.20%
(120 min, 0 m)	25.73%	38.88%
(120 min, 100 m)	28.13%	44.05%
(120 min, 200 m)	30.25%	47.17%

The rest of the paper is organized as follows. In Sect. 2, we give a brief review on related work. In Sect. 3, we introduce the preliminaries and the detail of our method. In Sect. 4, we report our experimental study. In Sect. 5, we discuss the contribution of the implicit features and explicit feature to our method. We will also analysis the sensitivity of our method to the parameter *embedding size*. In Sect. 6, we conclude our work and discuss the future work.

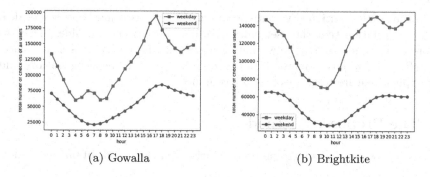

(a) Gowalla   (b) Brightkite

Fig. 1. Trajectory patterns at different hours on weekday and weekend

2   Related Work

Inferring social relationships from location-based check-in data has been a hot research topic in the past few years. The existing studies can be roughly classified into two categories based on the features they consider: co-occurrence based approaches and others. The co-occurrence based methods rely on the explicit co-occurrence [4,10,16,19,21], these methods mainly mine some co-occurrence based features of user pairs. However, these approaches need to set time thresholds and distance thresholds for co-occurrence. What's worse, a large number of user pairs do not have co-occurrences in real life as shown in Table 1. Some another approaches measure the similarity of user pairs by well-designed location features, such as the distance of home[20], the Jaccard similarity of check-in sequences and the number of common locations, etc. Pham et al. [19] proposed an entropy-based model, which designed two features: *location diversity* and *weight frequency*. The *location diversity* measures how diverse the co-occurrences of a user pair are. The *weight frequency* uses location entropy to measure how important the user pair's co-occurrences. Wang et al. [21] argued that not all co-occurrences are equally important in predicting user pair's friendship. They considered three features, including *personal factor*, *global factor* and *temporal factor*. The *personal factor* aims to determine the significance of a co-occurrence between two users, they think that if two users meet at a place where they frequently visit, then this meeting event is likely to happen by chance. The *global factor* is to reward the co-occurrence at private locations and penalize those at public locations. The *temporal factor* aims to penalize a co-occurrence if it is temporally close to other events. Njoo et al. [16] proposed two features *stability* and *duration* in the temporal domain of the co-occurrences, which can reflect the consistency and the total duration of the co-occurrences between two users. Cheng et al. [4] proposed a feature called *weighted number of co-occurrences*, which were aimed to strengthen the co-occurrence happened at private locations and while weaken the co-occurrence happened at popular locations. He et al. [10] designed 12 spatiotemporal features from four aspects to infer friendship.

In the above mentioned works, the co-occurrence based methods cannot deal with the situation that the user pairs have no co-occurrence. Besides, They all ignored the time interval between two users who have visited a same place. Therefore, we propose a method that combines implicit embedding features and one explicit feature to address the above issues.

3   The Proposed Method

In this section, we introduce the preliminaries and the details of our method.

3.1   Preliminaries

Definition 1 (Check-in Triplet). *When user u checks in location l at time t, the information can be called a check-in triplet $c_u =< u, l, t >$. Given user u, all his check-ins form a trajectory sequence $S_u = \{< u, l_1, t_1 >, \cdots, < u, l_n, t_n >\}$.*

Definition 2 (Co-occurrence). *A user u_1 and a user u_2 have a co-occurrence if their check-in distance is less than a distance threshold and the time interval is less than a time threshold.*

3.2   Implicit Features

As shown in Table 1, most friend pairs have no co-occurrences. Besides, hand-designed features cannot capture some implicit information of user's check-in. Therefore, it prompts us to learn the latent information from user's check-in data. *Word2vec* [14] is a very effective method to learn embedding representations in word sequences, which achieves a great success in recommendation systems [2]. Therefore, we also adopt *word2vec* to learn the trajectory embedding of each user. Concretely, we view each check-in location as a "word" and each "sentence" represents a user trajectory, then using skip-gram [14] to learn a location latent vector. In skip-gram model, given a user's check-in trajectory sequence S_u and the window size k, we need to maximize the following objective function:

$$J(S_u) = \frac{1}{|S_u|} \sum_{l_i \in S_u} \sum_{-k \leq j \leq k} (\log(P(l_{i+j}|l_i)) \tag{1}$$

where l_i represents the target location and l_{i+j} denotes the context location. Then we adopt a softmax function to formulate the probability $P(l_{i+j}|l_i)$:

$$P(l_{i+j}|l_i) = \frac{exp(I'_{i+j} \cdot I_i)}{\sum_{l_i \in L} exp(I'_{l_i} \cdot I_i)} \tag{2}$$

where I' and I denote output and input vector of location, respectively. $|L|$ is the number of all locations in the dataset. From Eq. 2, we can know its time complexity is very high, we apply the negative sampling method [14] to speed up learning process and the objective function can be redefined as follows:

$$J(S_u) = \frac{1}{|S_u|} \sum_{l_i \in S_u} \sum_{-k \leq j \leq k} (\log(\sigma(I'_{i+j} \cdot I_i)) + \sum_{h=1}^{H} \log(\sigma(-I'_h \cdot I_i))) \tag{3}$$

where $\{I'_h | h = 1...H\}$ are sampled from L, and $\sigma(\cdot)$ is the sigmoid function.

We train the above model using stochastic gradient descent and finally obtain an M-dimensional embedding vector for each location. Therefore, for the trajectory sequence S_u, we can get a $N \times M$ matrix $W_u = (w_{l_1}, w_{l_2}, \cdots, w_{l_N})^T$ and take the maximum value for each column of the W_u. Finally, we can get the vector representation of user u:

$$V_u = [v_u^1, v_u^2, \cdots, v_u^n]^T \tag{4}$$

3.3 Explicit Feature

Location entropy [6] was used to measure the popularity of a location, which can be defined as follows:

$$loc\_entropy(l) = -\sum_{u \in \Phi_l} \frac{C_u^l}{C^l} log \frac{C_u^l}{C^l} \tag{5}$$

where Φ_l represents all users who have checked in location l, C_u^l represents the check-in frequency of the user u at l and C^l is the total number of check-ins that all users have at place l. If two users meet at a popular location, they may be strangers because their meeting is more likely to happen by chance.

However, the location entropy ignored the check-in time interval of two users. For example, there are three users $< u_1, u_2, u_3 >$, u_1 checked in at location l_1 at 3 p.m. on December 5, 2019, u_2 checked in at location l_1 at 10 a.m. on December 1, 2018 and u_3 checked in at location l_1 at 3:30 p.m. on December 5, 2019. Although $< u_1, u_2 >$, $< u_1, u_3 >$ both have one common place, respectively, it is intuitive that $< u_1, u_3 >$ is more likely to be friends than the $< u_1, u_2 >$ because the check in time of $< u_1, u_3 >$ is closer.

Based on the above consideration, by introducing the time interval of check-ins into location entropy, we propose a new feature named *Time-weight-common-location-entropy (twcle)*. it can be defined as follows:

$$twcle(u, v, l) = loc\_entropy(l) \times e^{\rho \times |t_u^l - t_v^l|} \tag{6}$$

where t_u^l is the check-in time of the user u at location l. The parameter ρ is a constant coefficient to prevent the result of exponential function from overflowing. In our experiment, we set it to 0.001. Note that the time unit is a day. From the above formula, we can know that if two users visit a same private location at a closer time, their *twcle* value is lower. For two users u and v, the set of their common places is $L_{u,v}$, their *twcle* value can be computed as follows:

$$twcle(u, v, L_{u,v}) = \min_{l \in L_{u,v}} (loc\_entropy(l) \times e^{\rho \times |t_u^l - t_v^l|}) \qquad (7)$$

To verify the effectiveness of *twcle* in differentiate the friend pairs and stranger pairs, we plot the cumulative distribution function (CDF) of *twcle* between friends and strangers. As shown in Fig. 2, the friend pairs and stranger pairs are easily separable on two dataset, respectively. Specifically, on Gowalla dataset, more than 80% of friend pairs' *twcle* values are less than 3, while only less than 40% of stranger pairs' *twcle* values are below 3. On Brightkite dataset, the maximum gap is 45%. These indicate the *twcle* feature is an effective measure to differentiate the friend pairs and stranger pairs.

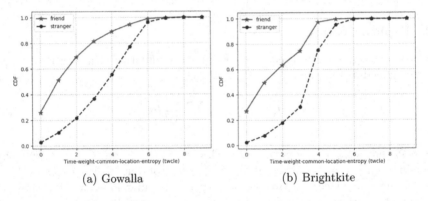

(a) Gowalla   (b) Brightkite

Fig. 2. The CDF of *twcle* between friend pairs and stranger pairs

3.4 The Detail of Our Method CIFEF

In this subsection, we introduce the details of our friendship inference method CIFEF. Figure 3 shows the overview of CIFEF. Firstly, we divide each user's check-in trajectory into weekday trajectory and weekend trajectory, and utilize *word2vec* to learn the embedding vector, respectively. Then, we apply element-wise max operations to the list of weekday location embedding vectors to get weekday trajectory embedding vector, weekend location does the same operation. Thirdly, for each user pair, their weekday vector and weekend vector are sent to the interaction layer for interacting (i.e., element-wise multiplication and element-wise subtraction). Next, each user pair gets four vectors: *weekday_hm*, *weekday_sub*, *weekend_hm* and *weekend_sub*. Together with their weekday vector and weekend vector, we finally get eight feature vectors. Moreover, we extract *twcle* feature according to Eq. 7. Finally, based on the above eight feature vectors and the *twcle* feature, we train the friendship inference model.

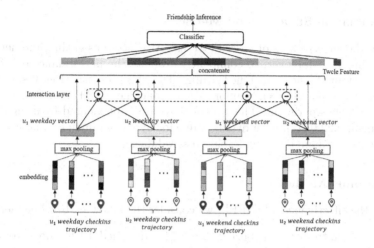

Fig. 3. The overview of our proposed method CIFEF

4 Experiment Study

In this section, we show experimental study in detail, including datasets, evaluation strategy and metric, baseline methods and experiment result.

4.1 Dataset

We conduct our experiment on two public real-world datasets: Gowalla dataset and Brightkite dataset [13], which have been widely used in previous studies. Table 2 shows the detailed statistics of the two datasets. We first select the active users who have more than 100 check-ins from two original dataset, getting 54,713 friend pairs and 54,221 friend pairs from two datasets, respectively. Then we double the dataset size by randomly sample the same number of stranger pairs in the two processed dataset, respectively. Finally, we conduct our experiment on processed Gowalla dataset with 109,426 samples and processed Brightkite dataset with 108,442 samples.

Table 2. Statistics of Datasets

Dataset	#Users	#Check-ins	#Friend pairs
Gowalla	107,092	6,442,890	950,327
Brightkite	58,228	4,491,143	214,078

4.2   Evaluation Strategy and Metric

To verify the effectiveness of CIFEF, we use AUC score as evaluation metric in our experiments. Moreover, to confidently evaluate the performance of CIFEF, we consider four classical machine learning algorithms: Logistic Regression (LR), K-Nearest Neighbor (KNN), Support Vector Machine (SVM) and Random Forest (RF) in our experiment. These algorithms are implemented by using Scikit-learn version 0.22.0 [17], with default value for hyper parameters. Lastly, all the experiments are done by using 5-fold cross validation.

4.3   Baseline Methods

To show the effectiveness of CIFEF, we consider the following baseline methods.

1. **STIF:** As introduced in Sect. 2. STIF [10] designed 12 features from four aspects for inferring friendship. This method include some co-occurrence based features and some spatiotemporal features.
2. **PGT:** PGT [21] used 6 features by considering the combinations of three features described in Sect. 2. It relies on the co-occurrences of user pairs.
3. **SCI:** SCI [16] extracted three features based on the co-occurrences as introduced in Sect. 2. it also relies on the co-occurrences of user pairs.
4. **SCI+:** SCI+ [15] is an improved version of SCI, which generalizes the temporal model by accommodating all meeting events between users instead of just considering the last meeting event. Besides, they also use two previous features: co-occurrences frequency and location popularity.
5. **OSFLC:** Bayrak et al. [1] aimed to reduce the time cost of friendship inference by feature selection. The 19 location-based features were collected from previous friendship inference papers in LBSNs. We use 15 features of them (there are 4 features we cant use because they rely on location category, which is not available in the dataset we use.) as a baseline. This method does not rely the co-occurrences of user pairs.
6. **CIFEF_T:** This is a baseline we designed, which uses a user's complete trajectory sequence to train the embedding vector instead of dividing the trajectory into workdays trajectory and weekends trajectory. The other parts are consistent with CIFEF method.

4.4   Comparison with Baseline Methods

In this Subsection, we report the experiment results of baseline methods and our method CIEFE. Table 3 shows the AUC scores of CIEFE and the above baseline methods. We firstly analyze the performance of baseline methods that based on co-occurrence, i.e., SCI, SCI+ and PGT. From Table 3, we can see that these three methods have a relatively poor performance, mainly because there are very few user pairs who have co-occurrences. Therefore, features extracted by these methods contain lots of missed value, which weakens their performance.

Table 3. AUC for different supervised classifiers on the two datasets

	Gowalla					Brightkite				
	LR	KNN	SVM	RF	Average	LR	KNN	SVM	RF	Average
CIFEF	**0.861**	**0.889**	**0.859**	**0.939**	**0.887**	**0.834**	**0.871**	**0.832**	0.903	**0.860**
CIFEF_T	0.847	0.887	0.849	0.911	0.874	0.823	0.856	0.820	0.876	0.844
OSFLC	0.804	0.865	0.773	0.914	0.839	0.752	0.863	0.746	**0.911**	0.818
STIF	0.733	0.826	0.771	0.873	0.801	0.715	0.799	0.740	0.876	0.783
PGT	0.599	0.599	0.600	0.599	0.599	0.564	0.581	0.557	0.581	0.571
SCI	0.551	0.551	0.551	0.551	0.551	0.556	0.537	0.541	0.556	0.548
SCI+	0.597	0.596	0.597	0.596	0.597	0.576	0.575	0.576	0.575	0.576

Secondly, we discuss the baseline STIF. Table 3 shows that on Gowalla dataset, the average AUC value of STIF has a 33.7%, 45.4%, and 34.2% performance improvement over those of PGT, SCI and SCI+, respectively. On Brightkite dataset, compared to PGT, SCI and SCI+, STIF improves AUC by 37.1%, 42.9% and 35.9%, respectively. The reason why the performance of STIF can be so significantly improved is that it not only contains some co-occurrence based features, but also some other spatiotemporal features, which makes it possible to work well even without co-occurrence.

Thirdly, we discuss the baseline OSFLC, which does not depend on co-occurrence. As shown in Table 3, although the performance of OSFLC is worse than our method CIFEF, it achieves much better performance than the above four baseline methods. It is worth mentioning that on the Brightkite dataset, for the random forest classifier, OSFLC method has achieved better performance than all other methods. However, OSFLC method needs a large number of well-designed features, which requires strong expertise knowledge.

Nextly, from Table 3, we can observe that CIFEF_T achieves better performance compared to SCI, SCI+, PGT, STIF and OSFLC. However, it performs worse than CIFEF by 1.5% and 1.9% on Gowalla dataset and Brightkite dataset, respectively. This shows that dividing user's trajectory into weekdays' trajectory and weekends' trajectory is good for friendship inference in LBSNs.

Finally, except the case of RF classifier on Brightkite, our method CIFEF outperforms all baseline methods in all cases. Compared to OSFLC method, the average AUC score of CIFEF has a 5.7% and 5.1% improvement on Gowalla dataset and Brightkite dataset, respectively. The major reason is that our method is not only able to capture the explicit information of user pairs who share common place, but also learns latent multi-grain trajectory information, which don't rely on user pair's co-occurrence. However, on RF classifier, CIFEF performs slightly worse than OSFLC method by 0.8%. We observe that compared with Gowalla dataset, the performance of all methods on the Brightkite dataset decreased at different degrees. According to our statistics, on Gowalla dataset, 12% of users check in more than 50 times with the interval less than

5 min, while the number is 25% on Brightkite dataset. Therefore, it means that there is more noise on Brightkite dataset.

5   Discussion

In this section, we first discuss the contributions of the implicit features and explicit feature to CIFEF. Then, we analyze the sensitivity of our method to the parameter *embedding size*.

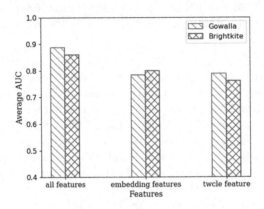

Fig. 4. The contributes of implicit features and explicit feature to CIFEF

5.1   The Contributions of Implicit Features and Explicit Feature

In this Subsection, we study the contributions of implicit features and explicit feature (i.e., the embedding features and the *twcle* feature) to CIFEF. Figure 4 shows the average AUC score of the aforementioned four classifiers. We can see that both features are useful to improve the performance of CIFEF. Concretely, on Gowalla dataset, the contribution of *twcle* feature is slightly greater than the embedding features, while the embedding features' contribution is greater than *twcle* feature on Brightkite dataset. Moreover, the average AUC score of embedding features on two datasets are 0.784 and 0.799, which are all worse than the average AUC score of OSFLC. However, the AUC score is greatly improved by adding the *twcle* feature, which is significantly better than OSFLC.

5.2   Paramter Sensitivity Analysis

We try several values of *embedding size*: {4,8,10,12,14,16,32,64,128}. From the Fig. 5, we can observe that our method achieves the best performance when embedding size is set to 8 on Gowalla dataset and 14 on Brightkite dataset. It seems to be a good balance for two datasets when setting the embedding size to 32.

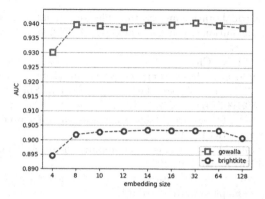

Fig. 5. The AUC of different embedding sizes on RF classifier

6   Conclusion and Future Work

In this paper, we study the problem of friendship inference based on users' check-in data in LBSNs. We adopt the embedding method to learn implicit features of user's weekday trajectory and weekend trajectory, respectively, which works effectively even if user pairs have no co-occurrence. Meanwhile, we propose a new feature *twcle*, which measures the importance of user pair's common place based on the time interval of check-in and location entropy. We have conducted extensive experiments on two public real-world datasets, the experiment results demonstrate the superiority of our method over state-of-the-art baseline methods. For the future work, we plan to design a more effective end-to-end deep learning model[3,8,9] for friendship inference in LBSNs.

References

1. Bayrak, A.E., Polat, F.: Mining individual features to enhance link prediction efficiency in location based social networks. In: ASONAM, pp. 920–925 (2018)
2. Caselles-Dupré, H., Lesaint, F., Royo-Letelier, J.: Word2vec applied to recommendation: hyperparameters matter. In: RecSys, pp. 352–356. ACM (2018)
3. Chen, S., Peng, C., Cai, L., Guo, L.: A deep neural network model for target-based sentiment analysis. In: IJCNN2018, pp. 1–7. IEEE (2018)
4. Cheng, R., Pang, J., Zhang, Y.: Inferring friendship from check-in data of location-based social networks. In: ASONAM, pp. 1284–1291. ACM (2015)
5. Cho, E., Myers, S.A., Leskovec, J.: Friendship and mobility: user movement in location-based social networks. In: SIGKDD, pp. 1082–1090. ACM (2011)
6. Cranshaw, J., Toch, E., Hong, J., Kittur, A., Sadeh, N.: Bridging the gap between physical location and online social networks. In: UbiComp2010, pp. 119–128 (2010)
7. Ding, D., Zhang, M., Li, S.Y., Tang, J., Chen, X., Zhou, Z.H.: Baydnn: friend recommendation with bayesian personalized ranking deep neural network. In: CIKM, pp. 1479–1488. ACM (2017)
8. Ding, J., Zhang, X., Chen, M., Xue, K., Zhang, C., Pan, M.: Differentially private robust ADMM for distributed machine learning. In: BigData, pp. 1302–1311 (2019)

9. Gu, H., et al.: A collaborative and sustainable edge-cloud architecture for object tracking with convolutional siamese networks. IEEE Trans. Sustain. Comput. (2019)
10. He, C., Peng, C., Li, N., Chen, X., Guo, L.: Exploiting spatiotemporal features to infer friendship in location-based social networks. In: Geng, X., Kang, B.-H. (eds.) PRICAI 2018. LNCS (LNAI), vol. 11013, pp. 395–403. Springer, Cham (2018). https://doi.org/10.1007/978-3-319-97310-4_45
11. Hu, Z., Peng, C., He, C.: Io-aware factorization machine for user response prediction. In: IJCNN2020. IEEE (2020)
12. Kossinets, G., Watts, D.J.: Origins of homophily in an evolving social network. Am. J. Sociol. **115**(2), 405–450 (2009)
13. Leskovec, J., Krevl, A.: SNAP Datasets: Stanford large network dataset collection, June 2014. http://snap.stanford.edu/data
14. Mikolov, T., Sutskever, I., Chen, K., Corrado, G.S., Dean, J.: Distributed representations of words and phrases and their compositionality. In: NeurIPS, pp. 3111–3119 (2013)
15. Njoo, G.S., Hsu, K.W., Peng, W.C.: Distinguishing friends from strangers in location-based social networks using co-location. PMC **50**, 114–123 (2018)
16. Njoo, G.S., Kao, M.-C., Hsu, K.-W., Peng, W.-C.: Exploring check-in data to infer social ties in location based social networks. In: Kim, J., Shim, K., Cao, L., Lee, J.-G., Lin, X., Moon, Y.-S. (eds.) PAKDD 2017. LNCS (LNAI), vol. 10234, pp. 460–471. Springer, Cham (2017). https://doi.org/10.1007/978-3-319-57454-7_36
17. Pedregosa, F., et al.: Scikit-learn: machine learning in python. JMLR **12**, 2825–2830 (2011)
18. Peng, S., Yang, A., Cao, L., Yu, S.: Social influence modeling using information theory in mobile social networks. Inf. Sci. **379**, 146–159 (2017)
19. Pham, H., Shahabi, C., Liu, Y.: EBM: an entropy-based model to infer social strength from spatiotemporal data. In: SIGMOD, pp. 265–276. ACM (2013)
20. Scellato, S., Noulas, A., Mascolo, C.: Exploiting place features in link prediction on location-based social networks. In: SIGKDD, pp. 1046–1054. ACM (2011)
21. Wang, H., Li, Z., Lee, W.C.: PGT: Measuring mobility relationship using personal, global and temporal factors. In: ICDM, pp. 570–579. IEEE (2014)
22. Zhang, Y., Pang, J.: Distance and friendship: a distance-based model for link prediction in social networks. In: Cheng, R., Cui, B., Zhang, Z., Cai, R., Xu, J. (eds.) APWeb 2015. LNCS, vol. 9313, pp. 55–66. Springer, Cham (2015). https://doi.org/10.1007/978-3-319-25255-1_5

A Knowledge Enhanced Ensemble Learning Model for Mental Disorder Detection on Social Media

Guozheng Rao[1,3,5] , Chengxia Peng[1] , Li Zhang[2] , Xin Wang[1(✉)] ,
and Zhiyong Feng[1,4,5]

[1] College of Intelligence and Computing, Tianjin University, Tianjin 300350, China
{rgz,wangx,zyfeng}@tju.edu.cn, pengchengxia@163.com
[2] School of Economics and Management, Tianjin University of Science
and Technology, Tianjin 300222, China
zhangli2006@tust.edu.cn
[3] School of New Media and Communication, Tianjin University, Tianjin 300072, China
[4] Shenzhen Research Institute of Tianjin University, Shenzhen 518000, China
[5] Tianjin Key Laboratory of Cognitive Computing and Applications, Tianjin 300350, China

Abstract. With the rapid development of social media in recent decades, a large amount of data posted by people with mental disorders has become available for researchers to use deep learning methods to detect potential mental disorders automatically. However, the current methods usually neglect to consider commonsense knowledge for detection tasks. Meanwhile, uneven data distribution is a great challenge for the current models. To address these problems, we propose a knowledge enhanced ensemble learning model for mental disorder detection on social media. First, we inject the knowledge triples into DailyPosts (Posts in one day posted online by a user) and then pass the reconstructed DailyPosts through a hierarchical model. The hierarchical model employs the BERT (Bidirectional Encoder Representations from Transformers) as the word embedder to integrate the BiGRU (Bidirectional Gated Recurrent Unit) network with an attention mechanism. Second, we combine the knowledge enhanced hierarchical model with the AdaBoost ensemble learning algorithm for the data imbalance. The proposed model was evaluated on two mental disorders (depression and anorexia) detection tasks. The model achieved competitive performance for depression and anorexia detection. The experimental results indicate that our model can obtain a robust performance in realistic scenarios.

Keywords: Knowledge enhanced · Ensemble learning · Mental disorder · Depression detection · Anorexia detection

1 Introduction

The widespread use of social media provides ways for people with mental disorders to share their lives and express their moods. Thus, social media data becomes an important resource for mental disorder detection tasks. Recently, there are lots of studies analyzing

© Springer Nature Switzerland AG 2020
G. Li et al. (Eds.): KSEM 2020, LNAI 12275, pp. 181–192, 2020.
https://doi.org/10.1007/978-3-030-55393-7_17

mental disorders through social media data. For instance, Huang et al. combined topic modeling, distributional semantics, and specialized lexicon to identify suicidal users on social media [1]. Yates et al. employed CNN (Convolutional Neural Network) to create user representation [2]. Then, they used the representation to analyze whether the user is depressed or not.

Although these works based on machine learning or deep learning methods have achieved promising results [3], commonsense knowledge is neglected to incorporate for language understanding. For example, in the sentence "I've been smoking salvia for a year," salvia is a plant. Without knowing the external knowledge that salvia is a hallucinogenic drug with transient psychoactive properties, it is difficult to identify the user suffering from mental disorders for the model. In order to learn more about prior knowledge, we propose a knowledge fusion component in which knowledge triples are injected into the DailyPosts(Posts in one day posted online by a user). In this way, the model can learn the semantics of the DailyPosts comprehensively.

In addition to incorporating prior knowledge, reliable language representation also contributes to discovering mental disorders in an early stage. To detect if a user suffers from mental disorders, we need to study all posts published within a certain period of time. In this case, all posts posted by the user in a period of time are called UserPosts. Posts in one day are called DailyPosts. Recently, a state-of-the-art model of pre-training language representations named BERT has obtained groundbreaking results on a series of NLP (natural language processing) tasks [4]. The BERT can yield dynamic representation based on the context, which motivates us to generate dynamic DailyPosts representations through this model. After obtaining the contextualized DailyPosts representations, we feed them into the BiGRU network with an attention mechanism to form a UserPosts representation. Finally, we use the UserPosts representation to identify if the user suffers from mental disorders.

Moreover, the disorder-related domain usually has the problem of data imbalance. The number of samples with mental disorders is usually much smaller than that of samples without disorders. In this case, the classifier tends to identify all samples as the non-disorder category to obtain a high overall performance. Therefore, the classifier performs badly on the disorder category. However, the performance of the classifier on the disorder category is exactly what we are concerned about. In this paper, we apply the AdaBoost ensemble learning algorithm [5] to reduce the impact of data imbalance. During each iteration, the algorithm will increase the weight of samples classified incorrectly and decrease the weight of samples classified correctly. In this way, the model will pay more attention to the misclassified samples and then improve the performance of the classifier on the misclassified category. Since the samples with mental disorders are more likely to be misclassified, the AdaBoost algorithm can improve the performance of the classifier on the disorder category.

In this work, we build the knowledge enhanced ensemble learning model to identify users with mental disorders on social media. First, we build the knowledge enhanced model as the base classifier. In the DailyPosts-level, we inject knowledge triples into DailyPosts and then pass reconstructed DailyPosts through the BERT embedder to obtain the DailyPosts embedding matrix. In the UserPosts-level, the matrix is encoded into UserPosts embedding by the BiGRU network with an attention mechanism. Then, we apply the AdaBoost ensemble learning algorithm to leverage the base model. We evaluate the proposed model with two mental disorders (the depression and anorexia) detection tasks and compare its performance with that of other methods.

The main contributions of our work are summarized as follows:

- We incorporate the external commonsense knowledge into the classifier by injecting knowledge triples into the DailyPosts, which promotes the model to identify more individuals with mental disorders in an early stage.
- We use a hierarchical structure to construct the base classifier. The BERT model is used to model the dynamic DailyPosts embedding. The BiGRU network with an attention mechanism is utilized to model the UserPosts embedding.
- We apply the AdaBoost ensemble learning algorithm to leverage the base classifier, which reduces the impact of data imbalance on the classifier.

2 Related Work

With the rapid development of social media, it has become an essential channel for users to express their opinions and emotions. More and more studies detect mental disorders through social media data. Recently, CLEF eRisk 2018 (Conference and Labs of the Evaluation Forum Early Risk Prediction on the Internet) introduced two different shared tasks (the depression and anorexia detection). In these two tasks, participants explored various NLP methods for identifying depressed users or anorexic users. For example, FHDO constructed four base models that employed user-level linguistic metadata, Bag-of-words (BoW), neural word embeddings, and convolutional neural networks. They integrated an ensemble model that combined four base predictions [15]. IRIT combined linguistic features and vectorization representations in different ways. Then, they trained a number of machine learning models to identify depressed or anorexic users [15]. Besides, Rao et al. [6] proposed hierarchical SGL-CNN (Single-Gated LeakyReLU CNNs) and MGL-CNN (Multi-Gated LeakyReLU CNNs) to model the user representation for depression detection.

However, these machine learning or general deep learning methods usually neglect to consider prior knowledge. To tackle this problem, our model merges the external prior knowledge with the classifier by incorporating knowledge triples extracted from the Wikidata. There are many other works based on combining KGs (Knowledge Graphs) with deep learning methods. For instance, Zhang et al. [7] identified named entity tokens in text and then aligned these tokens to the corresponding entities in KGs. Then, they used TransE [8] to encode the KGs and incorporated the informative entity embeddings to train an ERNIE (Enhanced Language Representation Model). K M et al. [9] proposed a convolution-based model that could capture relevant prior facts from KGs by using an attention mechanism. The proposed KG-augmented model combined the facts with the features learned from the training data. The model showed significant improvement in text classification and natural language inference tasks. Peters et al. [10] inserted multiple knowledge bases into a pre-trained model with KAR (Knowledge Attention and Recontextualization). Then, they enhanced language representations with structured knowledge.

Simultaneously, we exploit a hierarchical structure to build the base model. The BERT model is utilized as the word embedder to generate high-quality DailyPosts representation. There has been much research that employs the BERT model to encode semantic relations in context. Lee et al. [11] utilized the BERT as the word embedding method. Then, they integrated the BiLSTM with an attention mechanism for medical text inferences. Their best score ranked the top-third among all submissions in the natural language inference task at MEDIQA 2019. Han et al. [12] used the BERT model as the word embedder and stacked BiLSTM on top of the embedder. Then, the hidden states from the BiLSTM layer went through the attention layer and CRF layer to detect opinion expression.

Moreover, we apply the AdaBoost ensemble learning algorithm to reduce the impact of data imbalance. There are also other works employing ensemble learning to strengthen the model. Zhang et al. [13] combined the BiLSTM with an attention mechanism to build a base model for sentiment analysis. Then, the AdaBoost algorithm was implemented on the base model for the task with unevenly distributed data. Wu et al. [14] proposed a hybrid ensemble learning approach. This approach combined the AdaBoost algorithm and LSTM network for financial time series forecasting.

3   Methodology

In this section, we describe our knowledge enhanced ensemble learning model in detail. The model combines the KFB-BiGRU-Att (Knowledge Fusion BERT BiGRU-Att) base model and the AdaBoost ensemble learning algorithm, called KFB-BiGRU-Att-AdaBoost. The overall architecture of the proposed model is illustrated in Fig. 1. α_i represents the weight of the base model based on its classification performance.

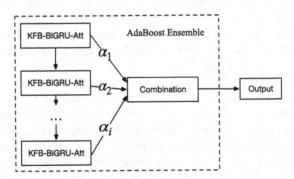

Fig. 1. The architecture of our knowledge enhanced ensemble learning model.

3.1   KFB-BiGRU-Att Base Model

The KFB-BiGRU-Att base model consists of a knowledge fusion layer, a BERT embedding layer, a BiGRU layer, an attention layer, and an output layer. The architecture of the base model is shown in Fig. 2.

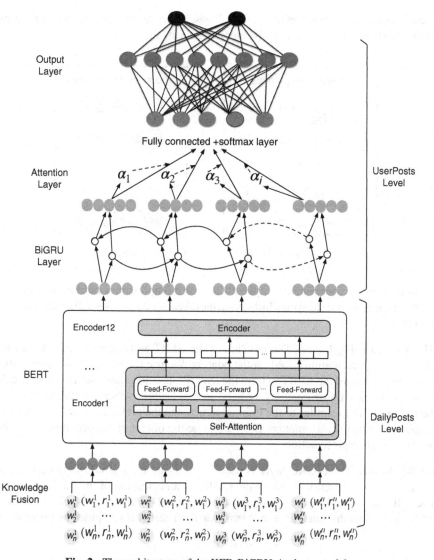

Fig. 2. The architecture of the KFB-BiGRU-Att base model.

Knowledge Fusion Layer. The knowledge fusion layer is used to inject knowledge triples into original DailyPosts. Then, we use the new UserPosts composed of a sequence of recombinant DailyPosts as input.

Specifically, we assume that the original UserPosts has m DailyPosts. Each DailyPosts has n tokens. Given a DailyPosts $S = \left(w_1^i, w_2^i, w_3^i, \ldots, w_n^i\right)$ and the corresponding knowledge triples $T = \left(\left(w_1^i, r_1^i, w_1^i\right) \ldots \left(w_n^i, r_n^i, w_n^i\right)\right)$, the knowledge fusion layer generates the recombinant DailyPosts $R = \left(w_1^i\left(r_1^i, w_1^i\right), w_2^i\left(r_2^i, w_2^i\right), w_3^i\left(r_3^i, w_3^i\right), \ldots, w_n^i\left(r_n^i, w_n^i\right)\right)$.

The w_n^i means the n_{th} token in the i_{th} DailyPosts. The r_n^i means the relation between two entities in a triple. The process of this layer is shown in Fig. 3.

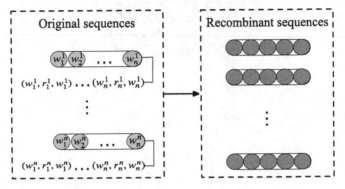

Fig. 3. The process of knowledge fusion.

In this paper, a recombinant DailyPosts can have multiple triples, but the number of each token's triple is no more than one. In this way, the token in a DailyPosts will not derive triples iteratively.

BERT Embedding Layer. The function of the BERT embedding layer is to produce DailyPosts representation based on the context. We use the output of the penultimate transformer layer of the BERT to form the DailyPosts embedding matrix. Theoretically, the output of each transformer layer can be used as DailyPosts representation. However, the last layer is too similar to the original target, and the previous layers may not fully extract the semantic information. So, we choose the output of the penultimate layer as our DailyPosts representation.

BiGRU Layer. The BiGRU layer is used to aggregate the DailyPosts representations containing the bidirectional information. GRU is an extension of the traditional RNN. The GRU cell uses reset gate r_t and update gate z_t to solve the disadvantage of traditional RNN on gradient disappearance problem. The reset gate r_t is computed by Formula (1).

$$r_t = \sigma\left(W^r x_t + U^r h_{t-1}\right), \tag{1}$$

where σ is the logistic sigmoid function. x_t and h_{t-1} are separately the input and the previous hidden state. W^r and U^r are weight matrices.

Similarly, the update gate z_t is computed by Formula (2).

$$z_t = \sigma\left(W^z x_t + U^z h_{t-1}\right). \tag{2}$$

The last hidden state h_t is computed by Formula (3).

$$h_t = (1 - z_t) * h_{t-1} + z_t * \tilde{h}_t, \tag{3}$$

Where

$$\tilde{h}_t = \tanh(W x_t + r_t U W h_{t-1}). \tag{4}$$

For DailyPosts x_t, a forward GRU extracts the DailyPosts feature as $\overrightarrow{h_t}$, and a backward GRU extracts the feature as $\overleftarrow{h_t}$. Then, we obtain the final sequence of DailyPosts features $H = (h_1, h_2, h_t, \ldots, h_n)$ where h_t is calculated by $h_t = [\overrightarrow{h_t}, \overleftarrow{h_t}]$.

Attention Layer. The attention layer is exploited to capture the distinguishing effect of the DailyPosts and form the UserPosts representation. The final UserPosts representation s_t is the weighted sum of the probability distribution α_t and the DailyPosts feature h_t.

$$s_t = \sum_n \alpha_t h_t, \alpha_t = \frac{exp(u_t u_w)}{\sum_n exp(u_t u_w)}, u_t = \tanh(W_w h_t + b_w) \tag{5}$$

u_w is the context vector. W_w and b_w are the weight and bias.

Output Layer. We use a fully connected softmax layer to output the probability distribution over the final classes. The fully connected softmax layer maps a n-dimension UserPosts vector into a K-dimension vector composed of probabilities between 0 and 1. Meantime, the sum of probabilities equals 1.

3.2 Knowledge Enhanced Ensemble Learning Model

We use the AdaBoost algorithm to strengthen the KFB-BiGRU-Att base model. Given a training dataset $T = (x_i, y_i), i = 1, 2, \ldots, N$, the steps of the ensemble learning model KFB-BiGRU-Att-AdaBoost can be illustrated as follows:

(i) Initialize the weight distribution of the training data. Each training sample is initially given the same weight $1/N$. The first weight distribution D_1 is calculated as Formula (6).

$$D_1 = (w_{11}, w_{12} \ldots w_{li} \ldots, w_{1N}), w_{li} = \frac{1}{N}, (i = 1, 2, \ldots, N) \tag{6}$$

where N is the number of training samples.

(ii) Update the weight distribution and do multiple rounds of iteration. $m = 1, 2, \ldots, M$ represents the round of iteration. $G_m(x)$ represents the base model trained by the weight distribution D_m. The error rate of $G_m(x)$ on the training data is calculated by $e_m = P(G_m(x_i) \neq y_i) = \sum_{i=1}^{N} w_{mi} I(G_m(x_i) \neq y_i)$. e_m is the sum of the weights of the samples misclassified by $G_m(x)$. The importance of $G_m(x)$ in the final classifier is calculated by $\alpha_m = \frac{1}{2} \ln \frac{1-e_m}{e_m}$. When $e_m \leq 1/2$, $\alpha_m \geq 0$. α_m increases with the decrease of e_m. The weight distribution is updated by Formula (7).

$$D_{m+1} = (w_{m+1,1}, w_{m+1,2} \ldots w_{m+1,i} \ldots, w_{m+1,N}),$$

$$w_{m+1,i} = \frac{w_{mi}}{Z_m} \exp(-\alpha_m y_i G_m(x_i)), (i = 1, 2, \ldots, N), \tag{7}$$

where $Z_m = \sum_{i=1}^{N} w_{mi} \exp(-\alpha_m y_i G_m(x_i))$ is a normalization factor.

(iii) Combine all base models to get the ensemble learning model. The ensemble learning model is calculated by $G(x) = sign(f(x)) = sign\left(\sum_{m=1}^{M} \alpha_m G_m(x)\right)$.

4 Experiments

4.1 Dataset

We used two datasets, the depression [2] and the anorexia [15], to evaluate the performance of our model. The depression dataset includes 9,210 depressed users and 107,274 non-depressed users. The training set, validation set and testing set are over 6G json files. All posts released on mental health-related subreddits or including depression keywords were deleted. The detailed statistics are shown in Table 1.

Table 1. The depression dataset.

Item	Depressed users	Non-depressed users
Training	3,070	35,753
Validation	3,070	35,746
Testing	3,070	35,775
Total	9,210	107,274

Table 2. The anorexia dataset.

Categories	Training		Testing	
	Anorexia	Non-anorexia	Anorexia	Non-anorexia
Num. users	20	132	41	279
Num. posts	7,452	77,514	17,422	151,364
Avg num. posts per user	372.6	587.2	424.9	542.5
Avg num. days from first to last post	803.3	641.5	798.9	670.6
Avg num. words per post	41.2	20.9	35.7	20.9

The anorexia dataset consists of 61 anorexic users and 411 non-anorexic users. The training set contains 25.6M XML files. The testing set contains 49.6M XML files. Although the number of users is not large, each user has a long history of submissions. The average number of days from the first submission to the last is more than 600. Table 2 shows the specific statistics of the anorexia dataset.

4.2 Experimental Settings

We train the model using the binary cross-entropy loss function. We use the Adam optimizer and the ReLu activation function. The dropout ratio is 0.5 at the BERT embedding layer, the last hidden state of the BiGRU layer, and the fully connected layer. In the depression dataset, the maximum number of DailyPosts processed by BiGRU layer is set to 600. Similarly, the maximum number of DailyPosts in the anorexia dataset is set to 300. Since the BERT embedding layer accepts variable-length tokens, we need not set the maximum number of tokens in a DailyPosts. We use the precision, recall, and F_1-score to evaluate the performance of our model.

$$precision = \frac{TP}{TP + FP}, \quad recall = \frac{TP}{TP + FN}, \quad F_1 = \frac{2 * precision * recall}{precision + recall} \quad (1)$$

4.3 Baselines

Baselines of the RSDD Dataset. We compare the performance of our proposed model with the following baselines. BoW-MNB and BoW-SVM [2] represented MNB or SVM classifier combined with sparse Bag-of-words features. Feature-rich-MNB and Feature-rich-SVM [2] represented MNB or SVM classifier combined with a set of features including Bag-of-words features, psychology-based lexicon features caught by LIWC, and emotion lexicon features. User model-CNN [2] employed a convolutional network and several dense layers to process user's posts. SGL-CNN and MGL-CNN [6] respectively used hierarchical Single-Gated LeakyReLU CNNs and Multi-Gated LeakyReLU CNNs to capture both post-level and user-level features. The previous best performance is obtained by the MGL-CNN model.

Baselines of the Anorexia Dataset. We compare the performance of our proposed model with that of other studies. FHDO-BCSGE [15] combined different base models that employed Bag-of-words, convolutional neural networks, linguistic metadata, and neural word embeddings. LIIRB [15] used different machine learning methods and combined two textual representations (linguistic features and vectorization) in different ways. UNSLD [15] estimated the association of each user and the corresponding category based on a sequential incremental classification approach. UPFC [15] consisted of different types of features including domain-specific vocabularies, linguistic features, and psycholinguistic lexicon features. It also produced a dynamic and incremental representation of the user's submissions.

4.4 Results

Results of the Depression Dataset. The results of our models and baselines are shown in Table 3. We can observe that the F_1-score of the base model KFB-BiGRU-Att is 0.54 which is equal to that of the state-of-the-art MGL-CNN model. Compared with the B-BiGRU-Att model which does not combine the prior knowledge, the F_1-score of the KFB-BiGRU-Att model is increased by 1%. Although the precision for identifying depressed samples is reduced, the overall F_1-score performance is still increased. The results show that the model recalls more depressed users after considering the prior knowledge in KGs. The best-performing F_1-score is 0.56 obtained by our KFB-BiGRU-Att-AdaBoost model. It is 2% higher than the F_1-score of the KFB-BiGRU-Att base model. In terms of precision and recall, the results of the KFB-BiGRU-Att-AdaBoost model are respectively 1% and 3% higher than that of the KFB-BiGRU-Att model. These improvements can obviously be attributed to the addition of the AdaBoost ensemble learning algorithm. The great performance shows the ability of the ensemble learning algorithm in reducing the impact of data imbalance.

Table 3. The results of the depression dataset.

Model	Precision	Recall	F_1
BoW-MNB [2]	0.44	0.31	0.36
BoW-SVM [2]	**0.72**	0.29	0.42
Feature-rich-MNB [2]	0.69	0.32	0.44
Feature-rich-SVM [2]	0.71	0.31	0.44
User model-CNN [2]	0.59	0.45	0.51
SGL-CNN [6]	0.51	**0.56**	0.53
MGL-CNN [6]	0.63	0.48	0.54
B-BiGRU-Att	0.61	0.47	0.53
KFB-BiGRU-Att	0.57	0.51	0.54
KFB-BiGRU-Att-AdaBoost	0.58	0.54	**0.56**

Compared with the previous best-performing model MGL-CNN, we can see that the recall and F_1-score of the KFB-BiGRU-Att-AdaBoost model are 6% and 2% higher than that of the MGL-CNN model, respectively. Nevertheless, the precision is 5% lower than that of the MGL-CNN model. In terms of depression detection task, a higher recall means that more individuals with depression disorder can be detected. After that, a comprehensive assessment from clinicians can further diagnose depressed individuals accurately. Thus, the aim of our model is to improve the recall as high as possible under the premise of ensuring precision.

Results of the Anorexia Dataset. The results of our models and the current best-performing methods are shown in Table 4. We can see that the maximum recall and F_1-score are respectively 0.87 and 0.86 both achieved by the KFB-BiGRU-Att-AdaBoost model. Compared with the performance of KFB-BiGRU-Att, the precision of KFB-BiGRU-Att-AdaBoost is increased by 3%. The recall and F_1-score are both increased by 2%. It can be seen that the performance has been significantly improved after applying the AdaBoost algorithm. The improved performance proves that the application of the AdaBoost algorithm reduces the impact of the data imbalance between anorexic samples and non-anorexic samples. Compared with the B-BiGRU-Att model without the knowledge fusion layer, the precision of the KFB-BiGRU-Att model is decreased by 2%. But the recall and F_1-score of the KFB-BiGRU-Att model are increased by 5% and 2%, respectively. Obviously, the ability of the model in recalling more anorexic individuals has been significantly improved after considering the prior knowledge. Although the precision of the model is slightly decreased due to the increase of recalled samples, the F_1-score is still increased.

Table 4. The results of the anorexia dataset.

Model	Precision	Recall	F_1
FHDO-BCSGE [15]	0.87	0.83	0.85
LIIRB [15]	0.79	0.73	0.76
UNSLD [15]	**0.91**	0.71	0.79
UPFC [15]	0.76	0.71	0.73
B-BiGRU-Att	0.85	0.80	0.82
KFB-BiGRU-Att	0.83	0.85	0.84
KFB-BiGRU-Att-AdaBoost	0.86	**0.87**	**0.86**

Compared with the state-of-the-art FHDO-BCSGE model, the precision and F_1-score of the KFB-BiGRU-Att model are respectively 4% and 1% lower than the FHDO-BCSGE, but the recall is 2% higher than the FHDO-BCSGE model. The precision of the KGB-BiGRU-Att-AdaBoost model is reduced by 1%, but its recall and F_1-score are increased by 4% and 1%, respectively. Although the model has a slight decrease in precision, the overall F_1-score is increased. Moreover, the significant increase in recall allows the model to identify more anorexic users. Then, the suicide caused by anorexia may be reduced after further comprehensive treatment.

5   Conclusions

We present a knowledge enhanced ensemble learning model for detecting mental disorders on social media. Firstly, we obtain the reconstructed DailyPosts containing the prior knowledge by injecting triples into the original DailyPosts. Secondly, we pass the reconstructed DailyPosts through the BERT embedder and the BiGRU network with an attention mechanism to build the base model. Finally, we apply the AdaBoost ensemble learning algorithm to leverage the base model. We evaluate the performance of our model with two mental disorders (the depression and anorexia) detection tasks. The proposed model accomplishes the highest F_1-score for both tasks. The performance indicates that our proposed model is expected to be competitive in other disorder detection tasks.

Acknowledgement. This work was supported by the National Natural Science Foundation of China (NSFC) under Grant 61373165, 61972275. The work described in this paper is partially supported by Shenzhen Science and Technology Foundation (JCYJ20170816093943197).

References

1. Huang, X., Li, X., Liu, T., Chiu, D., Zhu, T., Zhang, L.: Topic model for identifying suicidal ideation in Chinese microblog. In: Proceedings of the 29th Pacific Asia Conference on Language, Information and Computation, pp. 553–562 (2015)
2. Yates, A., Cohan, A., Goharian, N.: Depression and self-harm risk assessment in online forums. In: Proceedings of the 2017 Conference on Empirical Methods in Natural Language Processing, pp. 2968–2978 (2017)
3. Rao, G., Huang, W., Feng, Z., Cong, Q.: LSTM with sentence representations for document-level sentiment classification. Neurocomputing. **308**, 49–57 (2018)
4. Devlin, J., Chang, M.W., Lee, K., Toutanova, K.: BERT: pre-training of deep bidirectional transformers for language understanding. In: Proceedings of the 2019 Conference of the North American Chapter of the Association for Computational Linguistics: Human Language Technologies, Volume 1 (Long and Short Papers), pp. 4171–4186 (2019)
5. Freund, Y., Schapire, R.E.: A decision-theoretic generalization of on-line learning and an application to boosting. J. Comput. Syst. Sci. **55**, 119–139 (1997)
6. Rao, G., Zhang, Y., Zhang, L., Cong, Q., Feng, Z.: MGL-CNN: a hierarchical posts representations model for identifying depressed individuals in online forums. IEEE Access **8**, 32395–32403 (2020)
7. Zhang, Z., Han, X., Liu, Z., Jiang, X., Sun, M., Liu, Q.: ERNIE: enhanced language representation with informative entities. In: Proceedings of the 57th Annual Meeting of the Association for Computational Linguistics, pp. 1441–1451 (2019)
8. Bordes, A., Usunier, N., Garcia-Duran, A., Weston, J., Yakhnenko, O.: Translating embeddings for modeling multi-relational data. In: Advances in Neural Information Processing Systems, pp. 2787–2795 (2013)
9. Annervaz, K.M., Chowdhury, S.B.R., Dukkipati, A.: Learning beyond datasets: knowledge graph augmented neural networks for natural language processing. In: Proceedings of the 2018 Conference of the North American Chapter of the Association for Computational Linguistics: Human Language Technologies, Volume 1 (Long Papers), pp. 313–322 (2018)
10. Peters, M.E., et al.: Knowledge enhanced contextual word representations. In: Proceedings of the 2019 Conference on Empirical Methods in Natural Language Processing and the 9th International Joint Conference on Natural Language Processing (EMNLP-IJCNLP), pp. 43–54 (2019)
11. Lee, L.H., Lu, Y., Chen, P.H., Lee, P.L., Shyu, K.K.: NCUEE at MEDIQA 2019: medical text inference using ensemble BERT-BiLSTM-Attention model. In: Proceedings of the 18th BioNLP Workshop and Shared Task, pp. 528–532 (2019)
12. Han, W.B., Kando, N.: Opinion mining with deep contextualized embeddings. In: Proceedings of the 2019 Conference of the North American Chapter of the Association for Computational Linguistics: Student Research Workshop, pp. 35–42 (2019)
13. Zhang, Y., Wang, J., Zhang, X.: YNU-HPCC at SemEval-2018 Task 1: BiLSTM with attention based sentiment analysis for affect in tweets. In: Proceedings of The 12th International Workshop on Semantic Evaluation, pp. 273–278 (2018)
14. Wu, Y., Gao, J.: AdaBoost-based long short-term memory ensemble learning approach for financial time series forecasting. Curr. Sci. **115**, 159–165 (2018)
15. Losada, D.E., Crestani, F., Parapar, J.: Overview of eRisk 2018: early risk prediction on the internet (extended lab overview). In: Proceedings of the 9th International Conference of the CLEF Association, CLEF, pp. 1–20 (2018)

Constrained Viral Marketing
in Social Networks

Lei Yu, Guohui Li, and Ling Yuan(✉)

Huazhong University of Science and Technology, Wuhan, China
{LYU91,guohuili,cherryyuanling}@hust.edu.cn

Abstract. Influence maximization is to find a small number of influencers to maximize product adoptions. In reality, companies are usually more concerned about marketing cost that pay the influencers to reward their contributions on product promotion and specified marketing deadline. However, the former studies have not fully considered such an issue. In this paper, we focus on a more practical constrained viral marketing problem, a generalization of influence maximization. We propose a greedy algorithm with approximate guarantee. Due to the poor computational efficiency, we further propose an efficient heuristic algorithm and several effective optimization strategies to speed up the seed selection. Extensive experiments over real-world social networks of different sizes demonstrate the effectiveness and efficiency of the proposed methods.

Keywords: Social networks · Viral marketing · Heuristic algorithm

1 Introduction

Social networks (e.g., Facebook, Twitter, etc.) provide a new opportunity for information dissemination by the powerful word-of-mouth effect. Motivated by the viral marketing, influence maximization problem aims to find a seed set to maximize the number of activated nodes in a network. Kempe et al. [2] propose the widely used Independent Cascade (IC) model and Linear Threshold (LT) model, and show that the problem is NP-hard under both models.

Although a line of models and computational methods are actively studied, they still have not completely solved the influence maximization problem under the strict constraints of marketing cost and deadline at the same time. Firstly, it exists an impractical assumption in most of the previous literature that there is unit cost for each user. However, this assumption seldom holds in practice due to it is directly related to the final revenue. In the viral marketing, companies are usually forced to pay few seeders (i.e., the initial adopters) a certain amount of fees to reward their contributions on product promotion, in the hope that they can quickly spread the information to more users. Generally, users tend to have different abilities in information dissemination. As a result, the reward paid to each user is different. Therefore, the vary cost for different users should be significantly taken into account in influence diffusion. In addition, another impractical

© Springer Nature Switzerland AG 2020
G. Li et al. (Eds.): KSEM 2020, LNAI 12275, pp. 193–201, 2020.
https://doi.org/10.1007/978-3-030-55393-7_18

assumption is that the process of information propagation is considered infinite. Whereas it is just an ideal situation. In many actual marketing campaigns, the specified deadline is also a crucial aspect for successful and effective marketing.

We take a real scenario to illustrate the above situations. A football match will be held on July 2nd 2021. To attract more people to buy tickets and watch the live, the organizers engage some advertisers to propagate the match and enlarge its influence among people. Due to the limited budget, they are concerned about how to choose few key advertisers without exceeding the predetermined budget to influence as many users as possible before the opening date. Inspired by such realistic demands, it is very essential to further explore influence maximization that considers the marketing cost and specified deadline simultaneously.

In this paper, we are interested in a new Constrained Viral Marketing (CVM) problem in social networks, which asks for identifying a small set of influential users to influence the maximal number of users under the constraints of marketing cost and deadline simultaneously. We show that the problem is NP-hard and the influence spread function is monotone and submodular. We propose a greedy algorithm with approximate guarantee to solve the problem effectively. Due to its poor computational efficiency, we propose an efficient heuristic method to approximate the influence spread calculation, and devise several effective pruning techniques to accelerate the seed selection without compromising its accuracy.

2   Related Work

Several work has recognized the importance of time information in the influence diffusion. Chen et al. [5] consider some time delay, where a user needs to wait for a meeting event to happen to influence neighbors. Liu et al. [4] assign a probability distribution of influencing delay to each edge, which indicates how likely one node influences another in each time unit. [8,9] explore influence maximization in a continuous time dynamic diffusion network. Meanwhile, few work on budgeted influence maximization is studied. Nguyen et al. [6] consider the problem of selecting few seeds to maximize the influence spread at a total cost no more than the budget. Han et al. [7] develop three simple strategies to tackle the problem. However, none of the above work has taken into account both marketing cost and deadline information together to find the optimal set of users in influence maximization. In fact, the CVM problem is considered an important complement to these work, and more closely mirrors the realistic marketing campaigns.

3   Problem Definition

Due to both marketing cost and deadline are included in the CVM problem, the IC model is inapplicable. We define a Constrained Independent Cascade (CIC) model that extends the IC model. For a social network modeled by a directed graph $\mathcal{G} = (\mathcal{V}, \mathcal{E})$, a seed set $\mathcal{S}$ is activated initially. Let $\mathcal{S}_t$ be the set of activated nodes at the time step $t(t \geq 0)$, and $\mathcal{S}_0 = \mathcal{S}$. Each node in $\mathcal{S}_t$ attempts to independently activate each of its non-active neighbor v with a probability at the time step $t+1$. If the attempt succeeds within a bounded time, v is activated and continues to activate its neighbors similar to the above process in the next time step. The propagation terminates at a time step t' when the total cost of seeds in $\mathcal{S}_{t'}$ exceeds a predetermined marketing budget, or there is no more nodes to be activated within the constrained marketing budget and bounded time.

Given a marketing budget ν and a bounded time τ where $0 < \tau, \nu < l$ (l is an integer), the cost of a node u selected as a seed is defined as $\mathcal{C}_u$. The influence probability on edge $(u,v) \in \mathcal{E}$ is $\omega(u,v) = 1/d_{in}(v)$, where $d_{in}(v)$ is the indegree of v. If $(u,v) \notin \mathcal{E}$, $\omega(u,v) = 0$. Let $\mathcal{R}_\tau^\nu(\cdot) : 2^V \to \mathbb{R}$ be a set function such that $\mathcal{R}_\tau^\nu(\mathcal{S})$ is the expected number of activated nodes by a seed set $\mathcal{S}$ within ν and τ. The CVM problem is to find an optimal seed set $\mathcal{S}^*$ in $\mathcal{G}$ to maximize the influence spread $\mathcal{R}_\tau^\nu(\mathcal{S})$ under the CIC model, i.e., $\mathcal{S}^* = argmax\{\mathcal{R}_\tau^\nu(\mathcal{S})|\mathcal{S} \subseteq V, \sum_{u \in \mathcal{S}} \mathcal{C}_u \leq \nu, 0 < \tau, \nu < l\}$. The CVM problem is NP-hard since the traditional influence maximization problem can be regarded as its special case when node cost is unit cost and the bounded time is infinite.

4 Approximate Algorithms

4.1 Greedy Algorithm

For any two sets $\mathcal{S}_1$ and $\mathcal{S}_2$ where $\mathcal{S}_1 \subseteq \mathcal{S}_2 \subseteq V$, a set function $\mathcal{F} : 2^V \to \mathbb{R}$ is monotone if $\mathcal{F}(\mathcal{S}_1) \leq \mathcal{F}(\mathcal{S}_2)$. Meanwhile, for any node $w \in V \setminus \mathcal{S}_2$, the set function $\mathcal{F}$ is submodular if $\mathcal{F}(\mathcal{S}_1 \cup \{w\}) - \mathcal{F}(\mathcal{S}_1) \geq \mathcal{F}(\mathcal{S}_2 \cup \{w\}) - \mathcal{F}(\mathcal{S}_2)$.

Theorem 1. *Influence spread function $\mathcal{R}_\tau^\nu(\mathcal{S})$ is monotone and submodular.*

Proof. We briefly remind the "live-edge" model in [2]. For each edge (u,v) in $\mathcal{E}$, it flips a coin once with bias $\omega(u,v)$. The edge is "living" with probability $\omega(u,v)$ and "blocking" with probability $1-\omega(u,v)$. It can generate a random graph that includes all nodes in V and the selected "living" edges from $\mathcal{E}$. Let $\mathcal{X}$ be the set of all possible random graphs generated from $\mathcal{G}$. For any z in $\mathcal{X}$, the influence spread $\mathcal{R}_\tau^\nu(\mathcal{S})$ is $\sum_{z \in \mathcal{X}} \mathcal{P}(z)\mathcal{R}_\tau^\nu(\mathcal{S}; z)$, where $\mathcal{P}(z)$ is the probability of generating z and $\mathcal{R}_\tau^\nu(\mathcal{S}; z)$ is the influence spread of the seed set $\mathcal{S}$ within ν and τ in z.

For any $v \in V \setminus \mathcal{S}$, it can see that $\mathcal{R}_\tau^\nu(\mathcal{S}; z) \leq \mathcal{R}_\tau^\nu(\mathcal{S} \cup \{v\}; z)$. Due to $\mathcal{P}(z) \in (0,1]$, $\mathcal{R}_\tau^\nu(\mathcal{S})$ is monotone. Additionally, for the sets $\mathcal{S}_1$ and $\mathcal{S}_2$ and node w, it considers that v is reachable from $\mathcal{S}_2 \cup \{w\}$ in z, which means that there exists at least one "living" path from $\mathcal{S}_2 \cup \{w\}$ to v within τ and ν, but not from $\mathcal{S}_2$. Since $\mathcal{S}_1 \subseteq \mathcal{S}_2$, v must not be reachable from $\mathcal{S}_1$, but can be reachable from $\mathcal{S}_1 \cup \{w\}$ within τ and ν in z. It has $\mathcal{R}_\tau^\nu(\mathcal{S}_1 \cup \{w\}; z) - \mathcal{R}_\tau^\nu(\mathcal{S}_1; z) \geq \mathcal{R}_\tau^\nu(\mathcal{S}_2 \cup \{w\}; z) - \mathcal{R}_\tau^\nu(\mathcal{S}_2; z)$. Therefore, $\mathcal{R}_\tau^\nu(\mathcal{S}; z)$ is submodular. Because $\mathcal{R}_\tau^\nu(\mathcal{S})$ is a non-negative linear combination of the submodular functions, $\mathcal{R}_\tau^\nu(\mathcal{S})$ is submodular.

In practice, it is natural to consider that a user with larger influence is always assigned higher cost. Accordingly, we define Profit Cost Ratio (PCR) of a node u with a seed set $\mathcal{S}$ as $PCR(u|\mathcal{S}) = \Delta\mathcal{R}_\tau^\nu(u|\mathcal{S})/\mathcal{C}_u$, where $\Delta\mathcal{R}_\tau^\nu(u|\mathcal{S})$ is the incremental influence spread that equals $\mathcal{R}_\tau^\nu(\mathcal{S} \cup \{u\}) - \mathcal{R}_\tau^\nu(\mathcal{S})$. The PCR has

---

ALGORITHM 1: Greedy Algorithm

Input: $\mathcal{G} = (\mathcal{V}, \mathcal{E}), \tau, \nu$.
Output: $\mathcal{S}^*$.
Initialize: $\mathcal{S} \leftarrow \emptyset$;
while *true* **do**
    for *each node* $u \in \mathcal{V}$ **do**
        $PCR(u|\mathcal{S}) \leftarrow \Delta\mathcal{R}_\tau^\nu(u|\mathcal{S})/\mathcal{C}_u$;
    end
    $v \leftarrow argmax\{PCR(u|\mathcal{S})|u \in \mathcal{V}\}$;
    $\nu \leftarrow \nu - \mathcal{C}_v$;
    if $\nu < 0$ **then**
        break;
    end
    $\mathcal{S} \leftarrow \mathcal{S} \cup \{v\}$;
    $\mathcal{V} \leftarrow \mathcal{V} \setminus v$;
end
$w \leftarrow argmax\{\mathcal{R}_\tau^\nu(\{u\})|u \in \mathcal{V}\}$;
$\mathcal{S}^* \leftarrow argmax\{\mathcal{R}_\tau^\nu(\mathcal{T})|\mathcal{T} \in \{\mathcal{S}, \{w\}\}\}$;
return $\mathcal{S}^*$.

---

relative trade-off between profit and cost in the seed selection. Whereas it may cause bad solution based solely on PCRs of nodes in the CVM problem. We give a counter example. Consider a directed graph with nodes u_1, u_2, u_3, u_4, u_5 and edges $(u_1, u_2), (u_2, u_3)$, it assumes that $\omega(u_i, u_{i+1})(i = 1, 2)$ is 1, τ is 2, ν is 4, $\mathcal{C}_{u_j}(j = 1, 2, 3, 4)$ is 4 and $\mathcal{C}_{u_5}$ is 1. When selecting one seed from the graph, it has $PCR(\{u_5\}) = 1$ and $PCR(\{u_1\}) = 3/4$, which means that u_5 is selected as the seed. Whereas u_1 is actually the seed due to it achieves the larger influence spread within ν and τ. To overcome this issue, we propose an effective greedy algorithm presented in Algorithm 1. Moreover, the algorithm can provide a solution with $(1 - e^{-1/2} - \varepsilon)$ approximation ratio [6] for the CVM problem.

4.2 Calculation of the Influence Spread

We devise an efficient heuristic method that utilizes the arborescence structure to calculate the influence spread $\mathcal{R}_\tau^\nu(\mathcal{S})$ in the greedy algorithm. For a pair of nodes u and v, an influence path from u to v is expressed as $P_{u,v} = (u = q_1, q_2, \ldots, q_m = v)$, which is a non-cyclic sequence of nodes. Its influence propagation probability is $ipp(P_{u,v}) = \prod_{i=1}^{m-1} \omega(q_i, q_{i+1})$ and length is $l(P_{u,v}) = \sum_{i=1}^{m-1} l(q_i, q_{i+1})$. Let $MIP_{u,v}$ be the path with maximum influence propagation probability from u to v. Since it may be multiple paths between them, we only choose $MIP_{u,v}$ where it provides the greatest opportunity for u to influence v. If the influence probability $\omega(u, w)$ on each edge (u, w) is translated into a distance weight $-log(\omega(u, w))$, computing $MIP_{u,v}$ is equivalent to identifying the shortest path between nodes u and v. Moreover, within the tolerance of computational error, we also employ a pre-defined threshold ξ to remove the maximum influence path that has too small propagation probability, i.e., if $ipp(MIP_{u,v}) < \xi$, it has $ipp(MIP_{u,v}) = 0$.

In the CIC model, we can calculate $\mathcal{R}_\tau^\nu(\mathcal{S})$ as $\sum_{u \in \mathcal{V}} \mathcal{P}_\tau^\nu(u|\mathcal{S})$, where $\mathcal{P}_\tau^\nu(u|\mathcal{S})$ is the probability that u can be activated by $\mathcal{S}$ within ν and τ. Let $\mathcal{P}(u|\mathcal{S};\nu,t)$ be the probability of $\mathcal{S}$ activate u at the time step t. To calculate $\mathcal{P}_\tau^\nu(u|\mathcal{S})$ effectively, it needs a summation of all possible time steps from 0 to τ at which u is activated by $\mathcal{S}$, i.e., $\mathcal{P}_\tau^\nu(u|\mathcal{S}) = \sum_{t=0}^\tau \mathcal{P}(u|\mathcal{S};\nu,t)$. Therefore, $\mathcal{R}_\tau^\nu(\mathcal{S})$ is $\sum_{u \in \mathcal{V}} \sum_{t=0}^\tau \mathcal{P}(u|\mathcal{S};\nu,t)$ finally. The remaining issue is to calculate $\mathcal{P}(u|\mathcal{S};\nu,t)$ under the CIC model. To tackle this issue, we use the maximum influence in(out) arborescence structure [1], which assembles the maximum influence paths to(from) a given node over all other nodes and their length does not exceed τ. In this situation, $\mathcal{P}(u|\mathcal{S};\nu,t)$ can be efficiently calculated in the arborescences by utilizing a dynamic programming-based algorithm [5]. More specifically, when the time step t is 0, it has $\mathcal{P}(u|\mathcal{S};\nu,t) = 1$ for $u \in \mathcal{S}$ and $\mathcal{P}(u|\mathcal{S};\nu,t) = 0$ for $u \notin \mathcal{S}$. Meanwhile, when $t \in (0,\tau]$ and $u \in \mathcal{S}$, $\mathcal{P}(u|\mathcal{S};\nu,t)$ is 0. Otherwise $\mathcal{P}(u|\mathcal{S};\nu,t)$ can be recursively calculated as $\prod_{v \in \mathcal{N}_i(u)} (1 - \sum_{i=0}^{t-2} \mathcal{P}(v|\mathcal{S};\nu,i)\omega(v,u)) - \prod_{v \in \mathcal{N}_i(u)} (1 - \sum_{i=0}^{t-1} \mathcal{P}(v|\mathcal{S};\nu,i)\omega(v,u))$, where $\mathcal{N}_i(u)$ is the set of in-neighbour nodes of u.

4.3   Optimization Strategy

However, it exists a severe limitation that the greedy algorithm must traverse each node to compute their $PCRs$ in each iteration. In fact, there may be many unpromising nodes that do not need to be computed exactly. Therefore, we devise two pruning techniques to further improve the efficiency in the seed selection.

In the first iteration, it can calculate the PCR of each node in the arborescences. For each subsequent iteration, the greedy algorithm checks each node in descending order of their $PCRs$. We define $r_i(u_m)$ as a dynamic influence threshold, which represents the largest PCR with the current seed set $\mathcal{S}_{i-1}$ for node u_m in the $i(i > 1)$ iteration and is constantly updated. Specially, $r_i(u_m)$ is 0 initially. When checking a node v_k, if its $PCR(\{v_k\})$ is less than $r_i(u_m)$, it no longer needs to evaluate this node and all remaining nodes exactly. The process of seeking a new seed in this iteration terminates, and it returns u_m as the seed.

Lemma 1. *In the i iteration, if $PCR(\{v\})$ of a node v is less than $r_i(u_m)$, it can avoid computing the $PCRs$ of node v and all remaining nodes with the seed set $\mathcal{S}_{i-1}$ exactly, and this iteration terminates.*

Additionally, by fully using the upper bound of the $PCRs$ of nodes, it can further prune some unpromising nodes. Let $PCR_{up}(v|\mathcal{S}_{i-1})$ be the corresponding upper bound of $PCR(v|\mathcal{S}_{i-1})$ of node v, i.e., $PCR(v|\mathcal{S}_{i-1}) \le PCR_{up}(v|\mathcal{S}_{i-1})$. If $PCR_{up}(v|\mathcal{S}_{i-1})$ is less than $r_i(u_m)$, node v must not become a seed in the i iteration. Therefore, we also do not need to compute $PCR(v|\mathcal{S}_{i-1})$ exactly.

Lemma 2. *In the i iteration, it can avoid computing $PCR(v|\mathcal{S}_{i-1})$ of node v with the seed set $\mathcal{S}_{i-1}$ exactly when its upper bound $PCR_{up}(v|\mathcal{S}_{i-1})$ is less than the current $r_i(u_m)$.*

Table 1. Statistics of four social networks

Networks	Wiki-vote	NetPHY	Amazon	LiveJournal
No. of nodes	7115	37K	262K	1.3M
No. of edges	104K	181K	1.23M	4.47M
Direction	Directed	Undirected	Directed	Directed

The key is to estimate $PCR_{up}(v|S)$ at little computational cost. When the seed set S and node v have no co-influence to other nodes in influence diffusion, it satisfies $\Delta \mathcal{R}_\tau^\nu(v|S) = \mathcal{R}_\tau^\nu(\{v\})$. Otherwise $\Delta \mathcal{R}_\tau^\nu(v|S) < \mathcal{R}_\tau^\nu(\{v\})(1 - \mathcal{P}_\tau^\nu(v|S))$. Therefore, we have $\Delta \mathcal{R}_\tau^\nu(v|S) \leq \mathcal{R}_\tau^\nu(\{v\})(1 - \mathcal{P}_\tau^\nu(v|S))$. Because the cost $\mathcal{C}_v$ is greater than 0, it has $PCR(v|S) \leq \frac{1}{\mathcal{C}_v}\mathcal{R}_\tau^\nu(\{v\})(1 - \mathcal{P}_\tau^\nu(v|S))$. Therefore, we can get the upper bound $PCR_{up}(v|S)$, i.e., $PCR_{up}(v|S) = \frac{1}{\mathcal{C}_v}\mathcal{R}_\tau^\nu(\{v\})(1 - \mathcal{P}_\tau^\nu(v|S))$, where $\mathcal{R}_\tau^\nu(\{v\})$ has been calculated in the first iteration and can be used directly.

5   Experiments

5.1   Experimental Setup

Four social network datasets [3] shown in Table 1 are used. The following methods are evaluated. Largest Degree (LD). It selects nodes with the largest degrees. Random. It randomly selects nodes. MIA. One of the state-of-the-art heuristics for traditional influence maximization [1]. For the purpose of comparisons, it does not include both node cost and bounded time. Monte Carlo (MC). It applies MC simulations to estimate the influence spread. Maximum Influence Propagation Path based method with Optimization Strategy (MIPP-OS). It uses the maximum propagation path and optimization strategy to select seeds. Additionally, due to node cost has a positive correlation with its influence in practice, the cost of a node u is linear cost $\mathcal{C}_u = \frac{|\mathcal{V}|}{3*|\mathcal{E}|}d_u + 1$ or logarithmic cost $\mathcal{C}_u = log(d_u + e)$, where d_u is the degree of u. The threshold ξ is empirically set to $1/160$.

5.2   Experimental Results and Analysis

Quality of Seed Set. The quality of seed set is evaluated based on the influence spread. We run 10,000 MC simulations for each method, and take the average value. Figures 1 and 2 show the influence spread for different ν and τ when node cost is logarithmic cost. In Figure 1, MIPP-OS method achieves the similar influence spread with MC method, which is larger than other methods. It verifies the effectiveness of MIPP-OS method for solving the CVM problem. In most cases, MIA, LD and Random methods achieve the lower influence spread. Furthermore, we can observe the similar results in Figure 2 when τ varies.

Running Time. Figure 3 shows the running time with varying ν when node cost is logarithmic cost and τ is 5. We do not include Random and LD methods

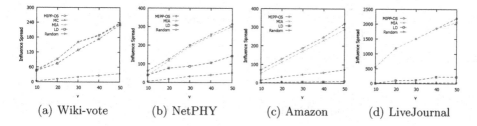

Fig. 1. The results of the influence spread with $\tau = 5$ over all social networks

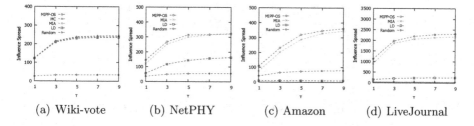

Fig. 2. The results of the influence spread with $\nu = 50$ over all social networks

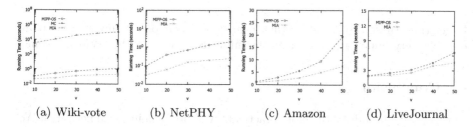

Fig. 3. The results of running time over all social networks

because their time consumption is too trivial. MC method needs the largest amount of time, while MIA method takes relatively less time. MIPP-OS method is far faster than MC method, and is about several orders of magnitude more efficient than it. Therefore, it verifies the efficiency of MIPP-OS method.

Affect of Optimization Strategy. Figure 4 shows the influence spread for MIPP-OS and MIPP (i.e., it does not include optimization strategies) methods with varying ν when node cost is logarithmic cost and τ is 5. As ν increases, MIPP-OS method achieves almost the same influence spread as MIPP method. It demonstrates that the optimization strategies still maintain the solution quality.

Affect of Different Node Costs. Figure 5 shows the influence spread for different node costs over the Wiki-vote social network when τ is 5. In this figure, the influence spread of each evaluated method grows with ν increases under both node costs. Moreover, the influence spread under the logarithmic cost is much

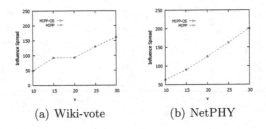

(a) Wiki-vote   (b) NetPHY

Fig. 4. The results of the influence spread for MIPP-OS and MIPP methods

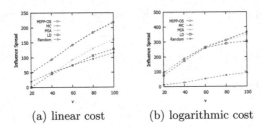

(a) linear cost   (b) logarithmic cost

Fig. 5. The results of the influence spread for different node costs

larger than those results under the linear cost. Therefore, it implies that node cost has very important impact on the influence spread for the CVM problem.

6   Conclusion

In this work, we address the more practical CVM problem in social networks. We develop an effective greedy algorithm. Moreover, we propose an efficient heuristic method and several effective pruning techniques to greatly accelerate selecting the seeds. The extensive experiments demonstrate that our proposed methods outperform intuitive baselines in the effectiveness and efficiency, and can scale to large networks. In future, we further explore how to assign the cost for different users more accurately based on real-world available traces and profiles of users.

References

1. Chen, W., Wang, C., Wang, Y.: Scalable influence maximization for prevalent viral marketing in large-scale social networks. In: 16th ACM SIGKDD International Conference on Knowledge Discovery and Data Mining, pp. 1029–1038. ACM, Washington (2010)
2. Kempe, D., Kleinberg, J., Tardos, E.: Maximizing the spread of influence through a social network. In: 9th ACM SIGKDD International Conference on Knowledge Discovery and Data Mining, pp. 137–146. ACM, Washington (2003)
3. SNAP Datasets. http://snap.stanford.edu/data

4. Liu, B., Cong, G., Xu, D., Zeng, Y.: Time constrained influence maximization in social networks. In: 12th IEEE International Conference on Data Mining, pp. 439–448. IEEE, Brussels (2012)
5. Chen, W., Lu, W., Zhang, N.: Time-critical influence maximization in social networks with time-delayed diffusion process. In: 26th AAAI Conference on Artificial Intelligence, pp. 592–598. AAAI, Toronto (2012)
6. Nguyen, H., Zheng, R.: On budgeted influence maximization in social networks. IEEE J. Sel. Areas Commun. **31**(6), 1084–1094 (2013)
7. Han, S., Zhuang, F., He, Q., Shi, Z.: Balanced seed selection for budgeted influence maximization in social networks. In: Tseng, V.S., Ho, T.B., Zhou, Z.-H., Chen, A.L.P., Kao, H.-Y. (eds.) PAKDD 2014. LNCS (LNAI), vol. 8443, pp. 65–77. Springer, Cham (2014). https://doi.org/10.1007/978-3-319-06608-0_6
8. Du, N., Song, L., Gomez-Rodriguez, M., Zha, H.: Scalable influence estimation in continuous-time diffusion networks. In: 27th Annual Conference on Neural Information Processing Systems, pp. 3147–3155. ACM, Nevada (2013)
9. Xie, M., Yang, Q., et al.: DynaDiffuse: a dynamic diffusion model for continuous time constrained influence maximization. In: 29th AAAI Conference on Artificial Intelligence, pp. 346–352. AAAI, Texas (2015)

A Multi-source Self-adaptive Transfer Learning Model for Mining Social Links

Kai Zhang[1], Longtao He[1,2(✉)], Chenglong Li[1(✉)], Shupeng Wang[2], and Xiao-yu Zhang[2]

[1] National Computer Network Emergency Response Technical Team/Coordination Center of China, Beijing, China
{zhangkai,hlt,lichenglong}@cert.org.cn
[2] Institute of Information Engineering, Chinese Academy of Sciences, Beijing, China
{wangshupeng,zhangxiaoyu}@iie.ac.cn

Abstract. How to mine social links with sparse data is a traditional problem. Researchers have ever studied the issue with transfer learning and achieved good results on telephone communication network or other acquaintances networks. However, the method would not work well on social networks sometime, as the social relationships in many social networks are not necessarily acquaintances. In this paper, we propose a new model, **MS-TrBPadaboost**, based on transfer learning and multiple social networks to solve the issue. The model could transfer information from multiple sources reasonably and self-adaptively so as to select more suitable knowledge and samples. The experimental results shows that our model yields better performance than state-of-the-art baselines, demonstrating the effectiveness of the model.

Keywords: Social network · Link mining · Transfer learning

1 Introduction

Rapid development of social network makes communication more convenient, and people would prefer online interaction with WeChat, Weibo or other apps nowadays. As a result, a mass of data about social links are generated, which facilitates the study of social-link mining. However, as most of users in many social networks are inactive, the quantity of individual information is sparse typically. In fact, short of known links is a traditional problem when mining social links. Researchers have tried to solve the problem by transferring information from other networks in past years, but there exists several problems when dealing with social network. Traditionally, most researchers studied the issue of social-link mining with telephone communication networks, cooperation networks of scientists, criminal partnership networks, etc. Although these networks are encoded in different forms, all of them are constituted by acquaintance links. Therefore, transferring information from other networks is easy as these networks have similar structures.

© Springer Nature Switzerland AG 2020
G. Li et al. (Eds.): KSEM 2020, LNAI 12275, pp. 202–209, 2020.
https://doi.org/10.1007/978-3-030-55393-7_19

However, social links in social network don't always mean acquaintances. Besides, acquaintance social network and stranger social network couldn't be distinguished strictly in many cases, making it more hard to use information of other social networks when mining social links. Therefore, current methods or models don't suit the situation of social networks necessarily. How to evaluate the contribution of information from multiple sources is an urgent issue.

In this paper, we design a new model namely **MS-TrBPadaboost**, which is based on transfer learning and multiple social networks. The main idea is to make full use of samples and knowledge of sources to enhance the performance of model for the target. The main contributions of this paper are as follows:

- We propose to transfer knowledge from multiple sources based on their performance self-adaptively so as to select more suitable knowledge;
- We propose to transfer knowledge probabilistically to ensure the diversity of weak classifiers and the stability of the model;
- We propose to transfer samples with weights from multiple sources, and especially pay attention to the transfer of positive samples to reduce the influence of data skew.

The rest of this paper is organized as follows. Section 2 introduces related work for mining social links. Section 3 describes the design process of MS-TrBPadaboost. Experiments are described in Section 4. We make conclusions in Section 5. Acknowledgments are given below Sect. 5.

2   Related Work

The importance of social relations has drawn worldwide attentions for many years. Wu et al.[1] proposed a probabilistic approach to fuse social theories for jointly modeling user temporal behaviors in social network service to mine social links. Liu et al. [2] proposed a new model for mining social connections based on deep belief networks. Methods mentioned above are able to mine social links and have achieved satisfying results. However, they couldn't take advantage of useful information provided by a mass of other social networks. With the rapid development of machine learning, transfer learning methods have overcomed this issue [3]. Tang et al. [4] designed a model TranFG with transfer learning to mine social links across heterogeneous networks and ideal results are achieved. However, only single source is considered to transfer the information. Yao et al. [5] brought out two models namely MultiSourceTrAdaBoost and TaskTrAdaBoost, while Fang et al. [6] proposed SSL-MSTL with multi-source transferring learning. However, they didn't pay enough attention to the different contributions of each source. Besides, as adaboost is designed with ensemble learning, the diversity of weak classifiers is the fundamental. But current studies about multi-source transfer learning seldom consider it and often transfer parameters from the source which provides the most valuable information arbitrarily. In this way, the diversity of weak classifiers is hard to ensure, which results in unstable performance of model. Therefore, a multi-source self-adaptive transfer learning model for mining social links MS-TrBPadaboost is proposed in this paper to solve the issues.

3   Design of MS-TrBPadaboost

3.1   Preliminaries

In this paper, $\mathcal{D} = \{\mathcal{X}, P(X)\}$ is exploited to denote domain, where $\mathcal{X}$ is space feature and $P(X)$ is the marginal probability distribution. $\mathcal{T} = \{\mathcal{Y}, f(\cdot)\}$ is used to denote task, where $\mathcal{Y}$ is label space and $f(\cdot)$ is the objective predictive function. Therefore, the sample dataset could be denoted as $D = \{\mathcal{X}, \mathcal{Y}\}$. Detailed notations are listed in Table 1.

Table 1. Summary of the basic notations used in this section.

Notation	Description
$\mathcal{D}_{S_u}$	Domain of source u
$\mathcal{D}_T$	Domain of target
$\boldsymbol{D}_{S_u}$	Sample dataset of source u
$\boldsymbol{D}_T$	Sample dataset of target
$\boldsymbol{A}$	Knowledge matrix for transferring
$\mathcal{L}$	Transferred knowledge matrix
$\mathcal{F}(\cdot)$	function for choosing knowledge
$\boldsymbol{B}$	Probability matrix for transferring knowledge
$\boldsymbol{Q}$	Weight distribution matrix of samples
α	Weight for representation of weak classifier
$h(x)$	Weak classifier
$H(x)$	Strong classifier

3.2   Theoretical Analysis

The core idea of MS-Trbpadaboost is to realize the transfer of samples and knowledge from multiple sources reasonably and effectively. Therefore, the model consists of two parts: Firstly, extract samples and knowledge for transferring; Secondly, make full use of them. We design the model based on BPadaboost. BPadaboost is a kind of ensemble learning by taking Back-Propagation neural network (BP) as weak classifiers. It can be seen as the combination of BP and adaboost, where BP is used to extract transfer knowledge, and the idea of adaboost is exploited to transfer information effectively.

As mentioned above, the process of MS-Trbpadaboost could be viewed as the solution for $\boldsymbol{A}, \boldsymbol{B}$ and $\boldsymbol{Q}$. Therefore, we would introduce methods for computing the three matrices in this part.

Computation of Knowledge Matrix for Transferring (A)

As Back-Propagation neural network is the basic classifier, its parameters, ν, ω, γ and θ are the knowledge which needs to be transferred. The meaning of these parameters are as follows:

ν_{ih} - connection weight of neuron i in input layer and neuron h in hidden layer;

ω_{hj} - connection weight of neuron h in hidden layer and neuron j in output layer;

γ_h- trigger threshold of neuron h in hidden layer;

θ_j - trigger threshold of neuron j in output layer.

Therefore, we need to train BP models with samples in each source domain,and then the knowledge matrix for transferring could be expressed as $A = [\nu, \omega, \gamma, \theta]$.

Computation of Probability Matrix for Transferring Knowledge (B)

Assume there exists U datasets in source domain $\mathcal{D}_S$, and could be expressed as $\boldsymbol{D}_S = \{\boldsymbol{D}_{S_1}, \boldsymbol{D}_{S_2}, \cdots, \boldsymbol{D}_{S_u}, \cdots, \boldsymbol{D}_{S_U}\}$. Here $\boldsymbol{D}_{S_u} = \{(x_1^{S_u}, y_1), (x_2^{S_u}, y_2),$ $\cdots, (x_{n_u}^{S_u}, y_{n_u})\}$ is denoted as dataset u in source domain, and the sum of n_u is n. Similarly, assume the dataset in target domain $\mathcal{D}_T$ is $\boldsymbol{D}_T = \{(x_1^T, y_1), \cdots, (x_m^T, y_m)\}$. Then we could compute the error rates of $h_l(x)$ on target domain $\boldsymbol{D}_T, \epsilon_l^T$, and on each source domain of $\boldsymbol{D}_s, \epsilon_l^{S_u}$, respectively.

$$\epsilon_l^T = P_{\boldsymbol{x} \sim D_T}(h_l(\boldsymbol{x}) \neq f(\boldsymbol{x})) \quad = \sum_{i=1}^m q_{li}^T \mathbb{I}(h_l(\boldsymbol{x}) \neq f(\boldsymbol{x})) \tag{1}$$

$$\epsilon_l^{S_u} = P_{\boldsymbol{x} \sim D_{S_u}}(h_l(\boldsymbol{x}) \neq f(\boldsymbol{x})) \quad = \sum_{i=1}^m q_{li}^{S_u} \mathbb{I}(h_l(\boldsymbol{x}) \neq f(\boldsymbol{x})) \tag{2}$$

Here $h_l(\boldsymbol{x})$ is denoted as the weak classifier in iteration l, while $\mathbb{I}(\cdot)$ is indicator function. When x is true, $\mathbb{I}(x) = 1$; otherwise $\mathbb{I}(x) = 0$. The sum of error rates on $\boldsymbol{D}_S, \epsilon_l^S$, is

$$\epsilon_l^S = \sum_{i=1}^U \epsilon_l^{S_u} \quad = \sum_{i=1}^U \sum_{i=1}^{n_u} q_{li} \mathbb{I}(h_l(\boldsymbol{x}) \neq f(\boldsymbol{x})) \tag{3}$$

We could obtain the voting weight of $h_l(\boldsymbol{x})$, α_l^T, on target domain $\boldsymbol{D}_T$ and $\alpha_l^{S_u}$, on each source domain of $\boldsymbol{D}_S$ as

$$\alpha_l^T = \frac{1}{2} ln \frac{1 - \epsilon_l^T}{\epsilon_l^T} \tag{4}$$

$$\alpha_l^{S_u} = \frac{1}{2} ln \frac{1 - \epsilon_l^{S_u}}{\epsilon_l^{S_u}} \tag{5}$$

Then the probability distribution matrix for transferring B_{l+1} in iteration $l + 1$ could be computed as

$$B_{l+1} = \{\alpha_l^{S_1}/Z_l^S, \alpha_l^{S_2}/Z_l^S, \cdots, \alpha_l^{S_U}/Z_l^S\} \tag{6}$$

Here Z_l is the normalization factor and can be calculated as follows:

$$Z_l^S = \sum_{i=1}^{U} \sum_{i=1}^{n_u} q_{li}^{S_u} \mathbb{I}(BP_l(\boldsymbol{x}) \neq f(\boldsymbol{x})) \tag{7}$$

At the first iteration, B needs to be initialized for startup. In this paper, we set $\boldsymbol{B_1} = \{\frac{1}{U}, \frac{1}{U}, \cdots, \frac{1}{U}\}$, which means that the probabilities for transferring knowledge from each source domain are the same.

Computation of Weight Distribution Matrix of Samples (Q)
Similarly, weight distribution matrix of samples, $\boldsymbol{Q}$, also needs to be updated in each iteration. $\boldsymbol{Q_{l+1}} = (q_{l+1,1}, \cdots, q_{l+1,i}, \cdots, q_{l+1,m+n})$ is the matrix for iteration $l + 1$, and the method for computing $q_{l+1,i}$ is

$$q_{l+1,i} = \begin{cases} \frac{q_{li}^T}{Z_l} exp(-\alpha_l^T f(\boldsymbol{x}) \cdot BP_l(\boldsymbol{x})), \\ \qquad if \ i = 1, 2, \cdots, m \\ \frac{q_{li}^{S_u}}{Z_l} exp(-\alpha_l^{S_u} f(\boldsymbol{x}) \cdot BP_l(\boldsymbol{x})), \\ \qquad if \ i = m+1, \cdots, m+n \end{cases} \tag{8}$$

Here Z_l is the normalization factor to make $\boldsymbol{Q_{l+1}}$ be a probability distribution. Z_l is calculated as

$$Z_l = \sum_{i=1}^{m} q_{li}^T \mathbb{I}(BP_l(\boldsymbol{x}) \neq f(\boldsymbol{x})) \ + \ \sum_{i=1}^{U} \sum_{i=1}^{n_u} q_{li}^{S_u} \mathbb{I}(BP_l(\boldsymbol{x}) \neq f(\boldsymbol{x})) \tag{9}$$

Similar with $\boldsymbol{B}$, $\boldsymbol{Q}$ also needs to be initialized for startup at the first iteration. In this paper, we set $\boldsymbol{Q_1} = \{q_{11}, q_{12}, \cdots, q_{1,m+n}\}$ as

$$q_{1i} = \begin{cases} \frac{1}{m}, \ if \ i = 1, 2, \cdots, m \\ \frac{1}{n}, \ if \ i = m+1, \cdots, m+n \end{cases}$$

It means the weights of samples in target domain are equal, and those on each source domain are also equal.

The pseudocode of MS-Trbpadaboost is shown in Algorithm 1.

4   Experiments

Selection of Dataset
Three datasets collected from social network website are used in experiments, namely Foursquare[7], Gowalla and Brightkite [8]. Dataset Brightkite is taken as target domain for testing, while the datasets Gowalla and Fourquare are taken as source domains.

Approaches for Comparison
As **BPadaboost** is used to design model, so it is chosen as the basic comparison model. Besides, single-source transfer learning model, TrBPadaboost, is

---

Algorithm 1: MS-TrBPadaboost

Input:
- D_S:sample datasets in multi-source domain;
- D_U:sample dataset in target domain ;
- $\mathcal{F}(\cdot)$:The function for choosing transfer knowledge;

Output:
- $H(x)$:strong classifier

Process:
1　Train multiple BP models with D_S and obtain A;
2　Obtain dataset D for training target model by combining D_T and D_S;
3　Initialize the weight distribution of samples in dataset D, Q_1;
4　Initialize the probability matrix for transferring, B_1, and obtain $\mathcal{L}$;
5　**Repeat**
6　　Compute the error rates of $h_l(x)$ on $\mathcal{D}_T$, ϵ_l^T, and on $\mathcal{D}_S, \epsilon_l^{S_u}$;
7　　Compute the voting weight of $h_l(x)$ on $\mathcal{D}_T, \alpha_l^T$, and on $\mathcal{D}_S, \alpha_l^{S_u}$;
8　　Update the weight distribution matrix of samples Q_{l+1};
9　　Update the probability matrix for transferring knowledge B_{l+1};
10　**Until** End of iteration

---

exploited for comparison. There are two source domains applied in this paper, so it is necessary to make a distinction for single-source transfer learning model. **TrBPadaboost-F** and **TrBPadaboost-G** are used to denote models trained with source domain dataset Fourquare and Gowalla respectively. In addition, a multi-source transfer learning model without self-adaptation is used for comparison referred to as **M-TrBPadaboost**. M-TrBPadaboost works by combining multiple source domains and training model like TrBPadaboost later. CN (Common friends), JC (Jaccard Coefficient), AA (Adamic- Adar Coefficient) are taken as features to train models in this paper.

Setting of Parameters

The transfer learning parameters need to be set. Connection weight ν is chosen as the transferred knowledge. For M-TrBPadaboost, we set the number of transfer learning samples as the same number of training samples in target domain; for MS-TrBPadaboost, we set the number of transfer learning samples as half (@0.5), the same (@1) and double (@2) number of training samples in target domain. Both of their ratios of positive samples and negative samples are set to be 1:1.

Results

Four performance metrics, precision (P), recall (R), F-measure $(F1)$ and accuracy (A) are calculated when estimating our model. Experiments are performed in two views, and we use $F1$ to evaluate MS-TrBPadaboost. The results and corresponding analysis are as follows.

　　In view 1, we take the number of training samples as dependent variable, set 5% of total links as known ones, and train model to mine the other 95% ones. The proportion of training samples to the total is set as 1%, 2%, 5% and 10% respectively. As shown in Fig. 1, the performance of models designed with transfer learning method outperforms the basic BPadaboost model. Besides, the advantage of our model is more obvious when training samples are less sufficient. When the proportion is 1%, the $F1$ of MS-TrBPadaboost is 17.65% greater than BPadaboost, and our model also acquires an improvement of 2.11% against M-TrBPadaboost.

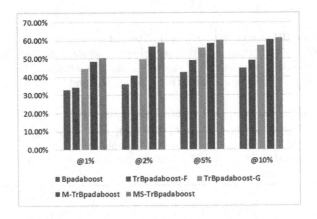

Fig. 1. $F1$ of models when the proportion of train samples in the total varies.

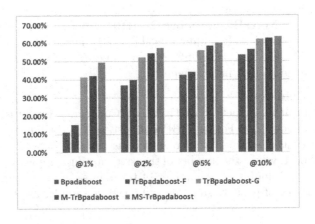

Fig. 2. $F1$ of models when the proportion of known links in the total varies.

In view 2, we take the number of known links as dependent variable, set 5% of total pairs in target domain as training samples, and train model to confirm whether the other 95% pairs are social links or not. The proportion of known social links to the total is set as 1%, 2%, 5% and 10% respectively. Similarly, from Fig. 2 we can see that the advantage of our model is more obvious when known links are sparser. When the proportion is 1%, the $F1$ of MS-TrBPadaboost is 38.20% greater than BPadaboost, and our model also obtains an improvement of 7.23% against M-TrBPadaboost.

5 Conclusion

In this paper, we introduce a new model namely MS-TrBPadaboost for mining social links in social networks. The model could transfer knowledge from multiple sources adaptively, which can enhance the performance for mining uncovered social links. We believe our method for resolving the issue is not limited to the study of social link. This work opens up new opportunities for transfer learning with multiple sources and could be spread to other fields.

Acknowledgments. This study was supported by Key-Area Research and Development Program of Guangdong Province (No. 2019B010137003).

References

1. Wu, L., Ge, Y., Liu, Q., Chen, E., Long, B., Huang, Z.: Modeling users' preferences and social links in social networking services: a joint-evolving perspective. In: Thirtieth AAAI Conference on Artificial Intelligence, pp. 279–286 (2016)
2. Liu, F., Liu, B., Sun, C., Liu, M., Wang, X.: Deep learning approaches for link prediction in social network services. In: Lee, M., Hirose, A., Hou, Z.-G., Kil, R.M. (eds.) ICONIP 2013. LNCS, vol. 8227, pp. 425–432. Springer, Heidelberg (2013). https://doi.org/10.1007/978-3-642-42042-9_53
3. Yang, P., Gao, W.: Multi-view discriminant transfer learning. In: International Joint Conference on Artificial Intelligence, pp. 1848–1854 (2013)
4. Tang, J., Lou, T., Kleinberg, J.: Inferring social ties across heterogenous networks, pp. 743–752 (2012)
5. Yao, Y., Doretto, G.: Boosting for transfer learning with multiple sources. In: Computer Vision and Pattern Recognition, pp. 1855–1862 (2010)
6. Fang, M., Zhang, X., Zhang, X., Li, X.: Multi-source transfer learning based on label shared subspace. Pattern Recogn. Lett. **51**(C), 101–106 (2015)
7. Cho, E., Myers, S.A., Leskovec, J.: Friendship and mobility: user movement in location-based social networks. In: ACM SIGKDD International Conference on Knowledge Discovery and Data Mining, San Diego, CA, USA, August, pp. 1082–1090 (2011)
8. Yuan, Q., Cong, G., Ma, Z., Sun, A., Thalmann, N.M.: Time-aware point-of-interest recommendation. In: International ACM SIGIR Conference on Research and Development in Information Retrieval, pp. 363–372 (2013)

Text Mining and Document Analysis

Multi-hop Syntactic Graph Convolutional Networks for Aspect-Based Sentiment Classification

Chang Yin[ID], Qing Zhou[(✉)][ID], Liang Ge, and Jiaojiao Ou

College of Computer Science, Chongqing University, 400030 Chongqing, China
{changyin,tzhou,geliang,jiaojiaoou}@cqu.edu.cn

Abstract. Sentiment analysis is widely applied to online and offline applications such as marketing, customer service and social media. Aspect-based sentiment classification is a fine-grained sentiment analysis that identifies the sentiment polarity of a specific aspect in a given sentence. In order to model syntactical constraints and word dependencies in a sentence, graph convolutional network (GCN) has been introduced for aspect-based sentiment classification. Though achieved promising results, GCN becomes less effective when the aspect term is far from the key context words on the dependency tree. To tackle this problem, we propose a Multi-hop Syntactic Graph Convolutional Networks model, in which a syntactic graph convolutional network is constructed according to transmission way of information in the sentence structure. Then a multi-range attention mechanism is applied to deepen the number of layers of the model to aggregate further information on the dependency tree. Experiments on benchmarking collections show that our proposed model outperforms the state-of-the-art methods.

Keywords: Aspect-based sentiment classification · Graph convolutional networks · Multi-hop · Syntactic structure

1 Introduction

Aspect-based sentiment classification is a fundamental task in sentiment analysis. The purpose of aspect-based sentiment classification is to analyze the sentiment polarity of a specific target in a given sentence. For example, in the sentence *"The food in this restaurant is delicious, but the service attitude is not good."*, sentiment polarities about food and service are positive and negative respectively.

Early traditional methods usually trained classifiers based on artificially designed features[6,14] for aspect-based sentiment classification. To avoid the use of artificial feature engineering, deep learning has gradually become popular. Most methods based on recurrent neural networks (RNN) [15,17] usually utilize LSTM to encode word vectors, and then employ attention mechanism to find the association between context and aspect term. These vectors can obtain

© Springer Nature Switzerland AG 2020
G. Li et al. (Eds.): KSEM 2020, LNAI 12275, pp. 213–224, 2020.
https://doi.org/10.1007/978-3-030-55393-7_20

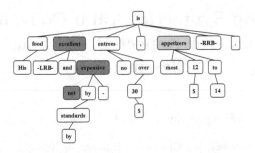

Fig. 1. Example of the key context words are far away from aspect term on dependency tree. Aspect term is *"appetizers"*. The key context words are *"excellent"* and *"not expensive"*.

semantic information between words. Compared with RNN-based models, a few methods based on convolutional neural networks (CNN) [4,16] can avoid the need of complex parameters and longer training time. Memory network-based methods [8,13] can keep context information into memory, clearly capture the importance of each word in context when inferring the sentiment popularity.

However, all the above methods obtain semantic information by considering the order of words, and don't consider the syntactic structure of sentences. Studies have shown that syntactic information obtained from dependency parsing can effectively resolve long-distance word dependencies [18]. Dependency tree can be regraded as a special form of graph. Graph convolutional networks (GCN) can aggregate syntactic information according to the dependency tree and capture the structural information of the sentences. How far GCN can capture information depends on how many convolutional layers it has. But most GCN-based models show that GCN with two layers can achieve the best results. As the number of GCN layers increases, GCN will not bring additional information due to over-smoothing problem. However, in some cases of aspect-based sentiment classification, the key context words are more than two hops away from aspect term on the dependency tree. For example, the sentence is *"His food is excellent -LRB- and not expensive by NYC standards - no entrees over $ 30, most appetizers $ 12 to 14 -RRB- ."*, and *"appetizers"* is aspect term (see Fig.1). On the dependency tree, the key context words *"not expensive"* are more 2-hops away from aspect term, which can cause critical information not to be effectively delivered to the destination.

To solve this problem and capture the syntactic structure of the sentences, we propose a Multi-hop Syntactic Graph Convolutional Network (MHSGCN) model. Syntactic GCN is designed according to the different ways of information delivery in the sentence structure, and meanwhile the multi-range attention mechanism calculates the attention coefficient of the output of each layer in the model, and finally linearly combines the outputs of all layers, which can effectively deepen the number of layers and avoid over-smoothing problem. The main contributions of this paper can be summarized as follows:

- To capture syntactic structure, we take the dependency tree as a special graph form and employ syntactic GCN to aggregate syntactic information.
- We propose MHSGCN to overcome the limitation that key context words are far away from aspect term on the dependency tree.
- We conduct experiments on public datasets, and experiments show that our method can achieve new state-of-the-art results.

2   Related Work

In recent years, neural network-based methods are introduced in aspect-based sentiment classification. Tang et al. [12] applied two target related LSTM to model the left and right contexts, and concatenated the hidden state of the two parts for aspect-based sentiment classification. In addition, attention mechanism have proven effective in NLP. Wang et al. [15] combined the LSTM and attention mechanism to propose an ATAE-LSTM model. Ma et al. [9] considered that in the aspect-based sentiment classification task, aspect term and context should be interactive. Huang et al. [5] proposed AOA that drew on the concept of *"attention-over-attention"* in machine translation, and model aspect and sentences in a joint way. Li et al. [7] believed that position information can help improve results, and then proposed a HAPN model, which can introduce position embedding to learn the position-aware representation of sentences. In addition to RNN-based methods, some other neural network-based methods are also used to solve aspect-based sentiment classification. MemNet proposed by Tang et al. [13] is based on memory attention network, which clearly captures the importance of each word through multiple layers. Besides, researchers have also tried to solve this task with CNN [4,16]. For a sentence, CNN can be used to control the flow of information generated by aspect term [5].

3   Method

The task of aspect-based sentiment classification is defined as follows: given a n-words sentence $s = \{w_1^s, w_2^s, ..., w_{v+1}^s, ..., w_{v+m}^s, ..., w_n^s\}$ to identify the sentiment polarity of the specific aspect term. The m-words aspect term is a substring of the sentence, and starts at $(v+1)$-th word in the sentence. The overall framework of Multi-hop Syntactic Graph Convolutional Networks (MHSGCN) can be shown in Fig.2. It contains input embedding layer, bidirectional LSTM layer, syntactic GCN, position layer, multi-range attention, aspect-context attention and output layer. Next, we will introduce each module of the model separately.

3.1   Input Embedding Layer

The input embedding layer can map each word to a high-dimensional vector. Pre-trained embedding matrix GloVe [10] or pre-trained BERT [3] model are used to convert the i-th word into the embedding vector $e_i \in \mathbb{R}^{d_{emb} \times 1}$, where d_{emb} is the dimension of embedding vector. So after passing the input embedding layer, the sentence representation $E_s \in \mathbb{R}^{d_{emb} \times n}$ is obtained.

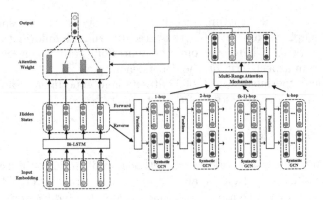

Fig. 2. The framework of MHSGCN.

3.2 Bidirectional LSTM Layer

The Bidirectional LSTM (Bi-LSTM) layer is located above the input embedding layer, and the context information is obtained by considering the sequence of words. After passing the Bi-LSTM layer, the forward hidden state $\overrightarrow{h_i^s} \in \mathbb{R}^{d_h \times 1}$ and the backward hidden state $\overleftarrow{h_i^s} \in \mathbb{R}^{d_h \times 1}$ are obtained, where d_h is the dimension of hidden state in the Bi-LSTM layer. Connect forward and backward hidden state to get the final representation $h_i^s = [\overrightarrow{h_i^s}, \overleftarrow{h_i^s}] \in \mathbb{R}^{2d_h \times 1}$.

3.3 Syntactic GCN

Sentence structure can be transformed into graph form according to dependency tree[1]. Build a bidirectional connection edge between parent and child nodes on the dependency tree to obtain adjacency matrix $A \in \mathbb{R}^{n \times n}$, and next apply GCN to capture the syntactic information of sentence. However on the dependency tree, information can be passed not only forward along syntactic dependent arcs, but also reverse along syntactic dependent arcs. Therefore, we can utilize two different weight matrices to capture different delivery modes. So the representation of i-th word is updated by:

$$f_i^l = ReLU((\sum_{u \in N(i)} W_f^l f_u^{l-1})/D_i + b_f^l) \tag{1}$$

$$r_i^l = ReLU((\sum_{u \in N(i)} W_r^l r_u^{l-1})/D_i + b_r^l) \tag{2}$$

$$x_i^l = [f_i^l, r_i^l] \tag{3}$$

where f_i^l, r_i^l represent the forward and reverse representation of the i-th word in the l-th layer respectively, and finally connect f_i^l and r_i^l to form the syntactic representation $x_i^l \in \mathbb{R}^{2d_n \times 1}$. $D_i - 1 = \sum_{j=1}^n A_{ij}$ represents the degree of i-th

---

[1] spaCy toolkit: https://spacy.io/.

node, and $N(i)$ stands for the neighbor node set of node i. W_f^l, $W_r^l \in \mathbb{R}^{d_n \times d_n}$ are trainable weight matrices, and b_f^l, $b_r^l \in \mathbb{R}^{d_n \times 1}$ are trainable bias terms (d_n is the dimension of hidden state). Note that $f_i^0 = r_i^0 = h_i^s$ ($i = 1, ..., n$), $W_f^0 \in \mathbb{R}^{d_n \times 2d_h}$ and $W_r^0 \in \mathbb{R}^{d_n \times 2d_h}$.

Receiving information from all neighbor nodes uniformly will cause lots of useless information to be aggregated. To solve the problem, we introduce gate control mechanism:

$$gf_u^l = \sigma(\hat{W}_f^l f_u^l + \hat{b}_f^l) \tag{4}$$

$$gr_u^l = \sigma(\hat{W}_r^l r_u^l + \hat{b}_r^l) \tag{5}$$

where $\widehat{W}_f^l$, $\widehat{W}_r^l \in \mathbb{R}^{1 \times d_n}$ are shared weight matrix, $\hat{b}_f^l$, $\hat{b}_r^l \in \mathbb{R}$ are bias terms, σ is the logistic sigmoid function. So Eq.1 and Eq.2 can be rewritten as follows:

$$f_i^l = ReLU((\sum_{u \in N(i)} gf_u^l(W_f^l f_u^{l-1}))/D_i + b_f^l) \tag{6}$$

$$r_i^l = ReLU((\sum_{u \in N(i)} gr_u^l(W_r^l r_u^{l-1}))/D_i + b_r^l) \tag{7}$$

3.4   Position Layer

According to intuition, the closer words in the sentence are to the aspect term, the greater the impact on aspect term. To avoid noise interference caused by dependency tree, we need design a function to weight each word in the sentence before each time through the syntactic GCN. Specifically, the function $\Gamma(\cdot)$ is designed as follows:

$$a_i = \begin{cases} 0 & dis = 0 \\ 1 - \dfrac{dis}{n} & 1 \leq dis \leq n - m \end{cases} \tag{8}$$

$$\Gamma(v_i^l) = a_i v_i^l \tag{9}$$

where v_i^l is equivalent to f_i^l, r_i^l, and dis is the distance from the context words to aspect term. Especially when $dis = 0$, the context words is equal to aspect term. Finally the output $X^l = [x_1^l, x_2^l, ..., x_{v+1}^l, ..., x_{v+m}^l, ..., x_n^l] \in \mathbb{R}^{2d_n \times n}$ of l-th layer syntactic GCN is obtained.

3.5   Multi-range Attention

In order to solve the problem that the key context words are too far away from the aspect term on the dependency tree, so that the key information cann't reach the target word. We introduce the multi-range attention mechanism [2]. We keep the output of each layer of syntactic GCN $\{X^1, X^2, ..., X^k\}$, which k is the number of layer. First a shared weight matrix $W_a \in \mathbb{R}^{2d_n \times 2d_n}$ is applied to the output of each layer of syntactic GCN, then calculate attention coefficient through u and $W_a x_i^l$, and finally use softmax to normalize the attention coefficient. Note that $u \in \mathbb{R}^{1 \times 2d_n}$ represents the context embedding in the range of k hops, which needs to be learned. The calculation process can be described as:

$$e_i^l = uW_a x_i^l \tag{10}$$

$$q_i^l = \frac{\exp(e_i^l)}{\sum_{l=1}^{k} \exp(e_i^l)} \tag{11}$$

Each layer of each node is linearly combined as:

$$o_i = \sum_{l=1}^{k} q_i^l x_i^l \tag{12}$$

After capturing multi-hop information, we can obtain the representation of the context $O = [o_1, o_2, ..., o_{v+1}, ..., o_{v+m}, ..., o_n] \in \mathbb{R}^{2d_n \times n}$.

3.6 Aspect-Context Attention

The representation of the aspect term can be extracted from O, which can contain syntactic information. Then attention mechanism is applied to capture the association between aspect term and context. This process can be described as follows:

$$\beta_j = \sum_{i=v+1}^{v+m} h_j^{s\top} o_i \tag{13}$$

$$\alpha_j = \frac{\exp(\beta_j)}{\sum_{j=1}^{n} \exp(\beta_j)} \tag{14}$$

$$z = \sum_{j=1}^{n} \alpha_j h_j^s \tag{15}$$

where we employ the dot product to calculate the similarity of context $h_j^{s\top}$ and aspect term o_i. So d_n and d_h be set to equal. z is the weighted representation of all words in the context. By now, z can be utilized for aspect-based sentiment classification.

3.7 Output Layer

We feed z into the fully connected layer, and then apply the softmax normalization layer to generate the probability distribution $P \in \mathbb{R}^{d_p}$ of the sentiment polarity space:

$$P = softmax(W_p z + b_p) \tag{16}$$

where $W_p \in \mathbb{R}^{d_p \times 2d_h}$, $d_p \in \mathbb{R}^{d_p \times 1}$ are the learned weight and bias respectively. Note that d_p is the same as the dimension of sentiment polarities.

3.8 Training

Our model is trained by the cross-entropy loss and L_2-regularization:

$$loss = \sum_{j=1}^{C} P_j \log(P_j) + \lambda \parallel \Theta \parallel \tag{17}$$

where C denotes the number of sentiment polarities, P_j for the j-th polarity, Θ represents all trainable parameters, λ is the coefficient of L_2-regularization.

Table 1. Dataset statistics.

Label	Lap14		Rest14		Rest15		Rest16		Twitter	
	Train	Test	Train	Test	Train	Test	Train	Test	Train	Test
Positive	994	341	2164	728	912	326	1240	469	1560	173
Neutral	464	169	637	196	36	34	69	30	3126	346
Negative	870	128	807	196	256	182	439	117	1560	173

4 Experiments

4.1 Datasets and Experimental Settings

We conduct experiments on five public datasets: Twitter dataset comes from comments on Twitter, and Lap14, Rest14, Rest15, and Rest16 datasets come from SemEval 2014 task 4, SemEval 2015 task 12, and SemEval 2016 task 5 respectively. Aspect term in the reviews are marked with three polarities: positive, neutral, and negative (we remove the *"conflicting"* label). The statistical result of the datasets are shown in Table 1. Pre-trained matrix GloVe [10] and pre-trained BERT [3] model are utilized to initialize word embeddings. For GloVe, the dimension of each word vector is 300. For BERT, the dimension of each word vector is 768. All weight matrices and biases in our model are initialized using a xavier uniform distribution. The dimension of hidden state in Bi-LSTM and syntactic GCN are both set to 300. In Lap14, Rest15, Rest16 datasets, the number of layers in the syntactic GCN is set to 4, and the number of layers is set to 2 in the remaining datasets. We use adam with learning rate of 0.0001 as the optimizer. Batch size is set to 32 and the coefficient of L2-regularization is set to 10^{-5}. Accuracy and Macro-F1 are employed to evaluate the performance of our model.

4.2 Benchmark Models for Comparison

- **SVM** [6] utilizes artificial features to train SVM for aspect-based sentiment classification.
- **TD-LSTM** [12] can utilize two target related LSTM to model the left and right contexts.
- **ATAE-LSTM** [15] combines the LSTM and attention mechanism to solve aspect-based sentiment classification task.
- **IAN** [9] utilizes two attention mechanism to achieve the interaction between context and aspect term.
- **MemNet** [13] can solve long-distance word dependencies through a multi-layer memory network structure.
- **RAM** [1] combines multiple attention mechanisms and RNN to extract more complex features.
- **AOA** [5] draws on the concept of *"attention-over-attention"* in machine translation.

Table 2. Performance comparison of MHSGCN and benchmark methods (%). Acc and Macro-F1 are used to evaluate the effectiveness of the models and the best results are in bold.

Embedding	Model	Lap14		Rest14		Rest15		Rest16		Twitter	
		Acc	Macro-F1	Acc	Macro-F1	Acc	Macro-F1	Acc	Macro-F1	Acc	Macro-F1
	SVM	70.49	N/A	80.16	N/A	N/A	N/A	N/A	N/A	63.40	63.30
GloVe	TD-LSTM	69.91	64.96	77.76	68.11	74.17	49.73	86.04	54.72	69.82	67.87
	ATAE-LSTM	70.69	65.20	79.20	68.35	79.15	53.37	85.55	55.24	69.73	68.54
	MemNet	71.94	66.73	79.23	68.79	77.49	52.07	85.66	56.23	71.10	69.58
	IAN	70.64	65.43	78.81	69.36	76.75	58.06	84.63	63.72	70.91	69.09
	RAM	70.85	65.44	78.48	68.64	75.09	50.83	86.53	55.43	71.39	69.45
	AOA	72.72	67.79	79.70	69.87	79.33	58.10	87.50	65.77	**72.30**	**70.20**
	MHSGCN	**75.24**	**71.40**	**81.25**	**72.95**	**81.18**	**61.52**	**89.39**	**71.34**	72.25	70.17
BERT	BERT-SPC	78.58	74.83	83.04	75.50	79.89	**67.91**	88.47	**74.63**	**73.99**	**72.13**
	AEN-BERT	78.79	74.79	81.79	72.06	**81.92**	66.82	88.80	73.16	73.30	71.86
	MHSGCN-BERT	**80.10**	**75.84**	**83.30**	**75.67**	81.37	65.54	**89.61**	72.26	72.69	70.67

- **AEN-BERT** [11] applies attention-based encoder modeling aspect term and context.
- **BERT-SPC** [3] constructs the sentence *"[CLS] + context+ [SEP] + aspect + [SEP]"*, then enter sentence as initial data into the BERT [3] model.

4.3   Result

Table 2 shows the experimental results of the models. In order to avoid interference caused by different word embeddings, we compared the GloVe-based models and the BERT-based models respectively. The feature-based SVM shows good results, illustrating the importance of feature engineering for aspect-based sentiment classification. Among all GloVe-based methods, although TD-LSTM considers the role of aspect term, it can't make full use of the correlation information between aspect term and context, so the results are the worst in all models. Compared with TD-LSTM, ATAE-LSTM, IAN and AOA have improved the experimental results to some extent due to the attention mechanism. MemNet and RAM can capture long-distance words information because of their multi-layer structure. MHSGCN introduces syntactic information through dependency tree, and overcome the limitation that key context words are far away from aspect term. Therefore, we can achieve new state-of-the-art results in the Lap14, Rest14, Rest15, Rest16. As for the poor performance in the Twitter dataset, we suspect that Twitter dataset is simple and restricts the performance of MHSGCN. Due to the strong prediction power of BERT, MHSGCN-BERT has achieved a certain improvement.

4.4   Ablation Study

Effect of the Syntactic GCN. In order to prove the validity of the syntactic GCN, we compare SGCN (MHSGCN without multi-range attention) with

Table 3. The comparison of BiLSTM-Attn, SGCN-GCN and SGCN (%). The number behind the model represents the number of layers. All experimental results are the average of three runs with random initialization.

Model	Lap14		Rest14		Rest15		Rest16		Twitter	
	Acc	Macro-F1	Acc	Macro-F1	Acc	Macro-F1	Acc	Macro-F1	Acc	Macro-F1
BiLSTM-Attn	72.83	67.82	79.85	70.03	78.97	58.18	87.28	**68.18**	71.24	69.55
SGCN-GCN (4)	73.46	68.98	80.36	70.42	79.40	61.58	**87.93**	67.12	72.01	70.16
SGCN (4)	**74.09**	**69.65**	**80.60**	**71.22**	**80.01**	**62.62**	87.82	66.76	**72.30**	**70.68**

Table 4. The comparison of GCN, CGCN and MHGCN (%). The number behind the model represents the number of layers. All experimental results are the average of three runs with random initialization.

Model	Lap14		Rest14		Rest15		Rest16		Twitter	
	Acc	Macro-F1	Acc	Macro-F1	Acc	Macro-F1	Acc	Macro-F1	Acc	Macro-F1
GCN (4)	73.46	68.98	80.36	70.42	**79.40**	61.58	87.93	67.12	**72.01**	70.16
CGCN (4)	73.82	69.56	80.68	72.03	76.94	57.40	**88.96**	68.28	71.34	69.69
MHGCN (4)	**74.45**	**69.66**	**80.92**	**72.30**	79.34	**63.95**	**88.96**	**69.72**	71.82	**70.30**

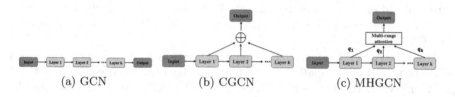

(a) GCN   (b) CGCN   (c) MHGCN

Fig. 3. The illustration of GCN, CGCN and MHGCN.

two methods: (1) BiLSTM-Attn is considered as the benchmark method, which employs two LSTM for context and aspect term to encoding. (2) SGCN-GCN which applies GCN instead of syntactic GCN is regarded as a comparative experiment of SGCN. Table 3 shows the comparison results. BiLSTM-Attn performs the worst because it doesn't take syntactic information into account. Compared with SGCN-GCN, SGCN can select the information of neighboring nodes through the gating mechanism, thereby eliminating the noise caused by the dependency tree and improve experimental results to a certain extent.

Effect of the Multi-range Attention. To prove the effectiveness of the multi-range attention mechanism, we compare MHGCN (MHSGCN without syntactic GCN) with different variants which connect multiple layers of GCN in different ways (see Fig. 3). (1) GCN takes the output of $(l-1)$-th layer as the input of l-th layer. (2) CGCN contacts the output of each layer. Table 4 shows the comparison results. Since GCN doesn't take multi-range information into account, the experimental performance is the worst. Compared to CGCN, MHGCN can capture multiple ranges information more effectively.

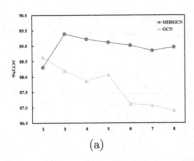
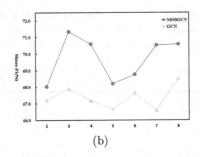

(a)   (b)

Fig. 4. Impact of layers on GCN and MHSGCN.

4.5 Impact of Layer Number

The number of syntactic GCN layers is an important experimental parameter that determines the performance of MHSGCN. We conduct experiments with different numbers of MHSGCN and GCN layers from 2 to 8 using GloVe as initial word embedding on the Rest16 dataset. The experimental results are shown in Fig. 4. It can be seen from Fig. 4(a) and Fig. 4(b) that MHSGCN performs better than GCN.

From the Acc (%) index, as the number of layers increases, the performance of GCN starts to decline rapidly because too many layers will cause over-smoothing problem. When the number of layers is 3, MHSGCN performs best. As the number of layers is greater than 3, the MHSGCN experiment results decline more slowly than GCN. This is because the existence of a multi-range attention mechanism can reduce noise to some extent.

4.6 Case Study

To understand how MHSGCN works, we present a case study from the Rest14 dataset. Moreover, we visualize the attention weights of the words by MemNet, IAN and MHSGCN(see Fig. 5). The testing example is *"His food is excellent -LRB- and not expensive by NYC standards - no entrees over $ 30, most appetizers $ 12 to 14 -RRB-."*, in which aspect term word is *"appetizers"* and sentiment polarity is *"positive"*. Both MemNet and IAN models predict *"neutral"*.

MemNet	His	food	is	excellent	-LRB-	and	not	expensive	by	NYC	standards	-	no	entrees
	over	$	30	,	most	appetizers	$	12	to	14	-RRB-	.		

IAN	His	food	is	excellent	-LRB-	and	not	expensive	by	NYC	standards	-	no	entrees
	over	$	30	,	most	appetizers	$	12	to	14	-RRB-	.		

MHSGCN	His	food	is	excellent	-LRB-	and	not	expensive	by	NYC	standards	-	no	entrees
	over	$	30	,	most	appetizers	$	12	to	14	-RRB-	.		

Fig. 5. The illustration of attention weights obtained by MemNet, IAN and MHSGCN respectively.

In the MemNet, it shows that attention score has little difference. While IAN is better at capturing different aspect terms and qualifiers in sentences, such as *"not"*, *"NYC"* and *"entrees"*. MHSGCN can accurately predict result, which can capture long-distance information *"expensive"* away from *"appetizers"* and the negative effect of *"not"* in syntax.

5   Conclusion

In this paper, we propose a novel model MHSGCN for aspect-based sentiment classification. We use the syntactic GCN to capture the structure information of sentences and MHSGCN solves the problem that the key words are too far away from the aspect term on the dependency tree. Experiments based on public datasets have demonstrated the effectiveness of MHSGCN. Visualization cases show that our proposed model can capture the relationship between long-distance words as well as local syntactic information. In future work, we will explore how to better capture the structure information of sentence. One potential solution is to constructing graphs with fine-grained semantic information.

References

1. Chen, P., Sun, Z., Bing, L., Yang, W.: Recurrent attention network on memory for aspect sentiment analysis. In: Proceedings of the 2017 Conference on Empirical Methods in Natural Language Processing, pp. 452–461 (2017)
2. Chen, W., Chen, L., Xie, Y., Cao, W., Gao, Y., Feng, X.: Multi-range attentive bicomponent graph convolutional network for traffic forecasting. arXiv preprint arXiv:1911.12093 (2019)
3. Devlin, J., Chang, M.W., Lee, K., Toutanova, K.: BERT: pre-training of deep bidirectional transformers for language understanding. arXiv preprint arXiv:1810.04805 (2018)
4. Huang, B., Carley, K.M.: Parameterized convolutional neural networks for aspect level sentiment classification. arXiv preprint arXiv:1909.06276 (2019)
5. Huang, B., Ou, Y., Carley, K.M.: Aspect level sentiment classification with attention-over-attention neural networks. In: Thomson, R., Dancy, C., Hyder, A., Bisgin, H. (eds.) SBP-BRiMS 2018. LNCS, vol. 10899, pp. 197–206. Springer, Cham (2018). https://doi.org/10.1007/978-3-319-93372-6_22
6. Kiritchenko, S., Zhu, X., Cherry, C., Mohammad, S.: NRC-Canada-2014: detecting aspects and sentiment in customer reviews. In: Proceedings of the 8th International Workshop on Semantic Evaluation (SemEval 2014), pp. 437–442 (2014)
7. Li, L., Liu, Y., Zhou, A.: Hierarchical attention based position-aware network for aspect-level sentiment analysis. In: Proceedings of the 22nd Conference on Computational Natural Language Learning, pp. 181–189 (2018)
8. Liu, Q., Zhang, H., Zeng, Y., Huang, Z., Wu, Z.: Content attention model for aspect based sentiment analysis. In: Proceedings of the 2018 World Wide Web Conference, pp. 1023–1032. International World Wide Web Conferences Steering Committee (2018)
9. Ma, D., Li, S., Zhang, X., Wang, H.: Interactive attention networks for aspect-level sentiment classification. arXiv preprint arXiv:1709.00893 (2017)

10. Pennington, J., Socher, R., Manning, C.D.: Glove: global vectors for word representation. In: Proceedings of the 2014 Conference on Empirical Methods in Natural Language Processing (EMNLP), pp. 1532–1543 (2014)
11. Song, Y., Wang, J., Jiang, T., Liu, Z., Rao, Y.: Attentional encoder network for targeted sentiment classification. arXiv preprint arXiv:1902.09314 (2019)
12. Tang, D., Qin, B., Feng, X., Liu, T.: Effective LSTMs for target-dependent sentiment classification. arXiv preprint arXiv:1512.01100 (2015)
13. Tang, D., Qin, B., Liu, T.: Aspect level sentiment classification with deep memory network. arXiv preprint arXiv:1605.08900 (2016)
14. Wagner, J., et al.: DCU: Aspect-based polarity classification for SemEval task 4 (2014)
15. Wang, Y., Huang, M., Zhao, L., et al.: Attention-based LSTM for aspect-level sentiment classification. In: Proceedings of the 2016 Conference on Empirical Methods in Natural Language Processing, pp. 606–615 (2016)
16. Xue, W., Li, T.: Aspect based sentiment analysis with gated convolutional networks. arXiv preprint arXiv:1805.07043 (2018)
17. Zhang, M., Zhang, Y., Vo, D.T.: Gated neural networks for targeted sentiment analysis. In: Thirtieth AAAI Conference on Artificial Intelligence (2016)
18. Zhang, Y., Qi, P., Manning, C.D.: Graph convolution over pruned dependency trees improves relation extraction. arXiv preprint arXiv:1809.10185 (2018)

A Matching-Integration-Verification Model for Multiple-Choice Reading Comprehension

Luxi Xing[1,2], Yue Hu[1,2(✉)], Yuqiang Xie[1,2], Chunhui Wang[3], and Yong Hu[3]

[1] Institute of Information Engineering, Chinese Academy of Sciences, Beijing, China
{xingluxi,huyue,xieyuqiang}@iie.ac.cn
[2] School of Cyber Security, University of Chinese Academy of Sciences, Beijing, China
[3] Effyic Intelligent Technology (Beijing) Co., Ltd., Beijing, China
{wangch,huyong}@effyic.com

Abstract. Multiple-choice reading comprehension is a challenging task requiring a machine to select the correct answer from a candidate answers set. In this paper, we propose a model following a matching-integration-verification-prediction framework, which explicitly employs a verification module inspired by the human being and generates judgment of each option simultaneously according to the evidence information and the verified information. The verification module, which is responsible for recheck information from matching, can selectively combine matched information from the passage and option instead of transmitting them equally to prediction. Experimental results demonstrate that our proposed model achieves significant improvement on several multiple-choice reading comprehension benchmark datasets.

Keywords: Machine reading comprehension · Multiple-choice · Attention

1 Introduction

Machine Reading Comprehension (MRC) is a rising frontier research field in Question Answering (QA) and is a crucial yet challenging step towards realizing Artificial Intelligence (AI). MRC aims at enabling the machine to read a text and answer any question about the text. In this paper, we mainly concern the Multiple-Choice Reading Comprehension (MCRC) task.

In the MCRC task, given one passage, there is a set of candidate answers attached to each question and the machine is asked for selecting the correct one answer from the candidate set, as shown in Table 1. The questions in the MCRC task are more challenging in that they can not be answered directly from the given text surface information and the candidate answers are mostly unseen in the given passage. Besides, the hardest hamper of the MCRC task

© Springer Nature Switzerland AG 2020
G. Li et al. (Eds.): KSEM 2020, LNAI 12275, pp. 225–237, 2020.
https://doi.org/10.1007/978-3-030-55393-7_21

lies in modeling the interaction process among the passage, question, candidate answers triples without loss of information.

Previous works usually regard the passage as one sentence and encode the whole passage in the word-level, as well as the question and the candidate answers. Then various matching mechanisms are employed to generate correlation information between the selected two elements in the passage, question, and candidates by one pass [13,16,17]. It is usually to directly use the final representation of the passage or the candidates when calculating a score of each candidate. In addition, there are some works [10,18] fully utilize relationship comparing each option with the others. However, the error options are not always useful to the question or completely mismatched with the passage, and partial information in it may cause the inconsistency with the purpose of the question. Moreover, treating the matched information of passage and option equally may distract the model from focusing on the content of the question. Previous works underutilize the question during matching and ignore the question review or verification mechanism which the human will adopt for guaranteeing the integrity of the collected supporting information of each candidate.

Table 1. Examples in RACE and MCScript datasets. The correct answers are in bold.

RACE	MCScript
Passage: ... A few years ago, I was dining out with a person who kept going on and on about himself, completely unaware of the fact that I was sitting there in misery. It wasn't his company that disappointed me. I was recovering from a bad grade in my study. I could have collapsed in tears right there at the table	**Passage:** ... We put a folded napkin to the left side of each plate. On top of each napkin we put a fork. Except for Tom, he gets a small fork. On the right side of the plate we put a knife and spoon. We need to be careful with the knives. Closer to the center of the table we put a glass by each plate. Tom gets a plastic glass. ...
Q: Why did the author say she was in misery?	**Q:** When was the silverware put on the table?
Candidate Answers: A: Because she got into trouble with her friend B: Because she couldn't enjoy her meal quietly C: Because her friend complained so much D: **Because she was doing poorly in study**	**Candidate Answers:** A: **Before the glasses were put on.** B: It was put on the table first

When human solving the MCRC task, people usually read through the passage with the question to capture the connection and locate probable evidence information. Then read the candidate answers to judge which one could correspond with the evidence and generate a preliminary result. For ensuring the preciseness, people will look back to the question and verify the initial supporting information. In the last, they integrate evidence and updated information from verification to get the best answer. Moreover, rethinking the essence of the question in the MCRC task, we hypothesize that there is an information gap between the question and the given passage, and the task goal is to orient the correct answer to bridge the gap based on the evidence in the passage. The question is the pivot for connecting the passage and the options.

Enlightened by reading strategies of human and above analysis, we propose a model following a matching-integration-verification-prediction framework, which can gather supporting evidence and clarify the question-guided discriminative information. The proposed model contains evidence **M**atching, multi-side **I**ntegration, and **V**erification (**MIV**) three main modules and a common encoding layer and prediction layer. Specifically, in the encoding layer, we employ the pre-trained language model, such as BERT [2], to calculate the original contextual representation for the passage, question, and options. The evidence matching module is designed to extract relevant information in the passage according to the question and option. Then, the multi-side integration module establishes the connection between each option and the question and passage respectively, and generates an initial judgment state for each option. It aims at collecting supporting information based on the question side and passage side for each option and gauging the consistency of these two sides focused on the content in the option. Next, the verification module reviews the question and utilizes the question to recheck the matching information of passage and the state of each option in case that the option will overly depends on the passage because of the high similarity based on text-form. As the core of the whole model, the verification module is also responsible for calculating the weighted combination of evidence information and the option state which is taken as the final state of each option. Eventually, in the prediction layer, we take both the question-answer pair information and verified option state into account, and make a final score for each option.

The contributions of this paper can be summarized as follows: (a) We introduce a Matching-Integration-Verification (**MIV**) model to recheck and integrate various evidence information for MCRC task; (b) We propose a verification module to selectively combine the matched information of passage and supporting information of option based on the review of the question; (c) We evaluate our approach on three MCRC datasets, RACE [3], MCScript [5] and COIN-MCS [7]. Experimental results illustrate our proposed model achieves a strong performance on these benchmarks. We additionally perform an ablation study to show the effectiveness of our MIV model.

2   Related Work

The massive progress of machine reading comprehension largely benefited from the availability of large-scale benchmark datasets, which makes training the sizable neural network in the end-to-end fashion possible. One representative dataset is SQuAD [9], which provides challenging questions required models to extract an answer span from the original passage. To refrain from solving the question solely depends on the word-level matching, the RACE [3] was released in the form of multiple-choices style reading comprehension task. The RACE dataset requires deeper reasoning about passage content, which covers various topics, and the candidate answers do not appear in the original passage.

In the previous works on MCRC task, researchers usually focus on the improvement of matching process. [14] propose the Dynamic Fusion Network that uses reinforcement learning to determine the selection of multiple attention for reasoning. [1] propose to calculate multiple attention among the passage, question and candidates and dynamically modify the values of attention through the trainable weights. Both [18] and [10] utilize the comparison and correlation information between the candidate answers to select the correct answer.

In the recent studies, the pre-trained deep language models, such as GPT [8], BERT [2], XLNet [15] and RoBERTa [4], have shown their astonishing improvement on the MRC tasks. This kind of deep language models leverages large-scale corpora to train their network, which is consist of multiple transformer [12] layers, in the unsupervised fashion. After the unsupervised pre-training procedure, the GPT or BERT is equipped with an additional task-specific classifier and fine-tunes on the target task dataset. Utilizing the pre-trained language models, [11] brings three reading strategies into the fine-tuning process of GPT and leads a significant improvement on multiple MCRC tasks. [10,16,17] apply the pre-trained BERT as an encoder and stack the attention-based interaction layer on the above of BERT encoder, and then the whole networks are trained together with relatively small learning rate and achieve the state-of-the-art score on the RACE benchmark dataset. However, these works take information after attention as the input of the prediction layer without selection.

3   Model

In this section, we first introduce the formalized definition of the MCRC task and the core elements in it. Then, we will give a detailed description of the proposed model and elaborate the information flow between each component. The architecture of the proposed model is depicted in Fig. 1.

3.1   Task Description

In the MCRC task, given a passage $\mathbf{P}$, a question $\mathbf{Q}$ and a set of candidate answers $\mathcal{A}$, the machine $\mathcal{M}$ is asked to select the correct answer $\mathbf{a}^*$ from the candidate set, i.e.,

$$\mathbf{a}^* = \arg\max_{A_i \in \mathcal{A}} \mathcal{M}(A_i | \mathbf{P}, \mathbf{Q}) \tag{1}$$

The triplets $\langle \mathbf{P}, \mathbf{Q}, \mathcal{A} \rangle$ are the core elements in the MCRC task and the input of the model. We denote the passage $\mathbf{P}$ with n_p tokens as $\mathbf{P} = \{w_1^p, w_2^p, ..., w_{n_p}^p\}$, the question $\mathbf{Q}$ with n_q tokens as $\mathbf{Q} = \{w_1^q, w_2^q, ..., w_{n_q}^q\}$, the candidate answer set $\mathcal{A}$ with K options as $\{A_1, ..., A_K\}$ and each candidate $A_k = \{w_1^a, w_2^a, ..., w_{n_k}^a\}$, where n_k is the length of candidate A_k.

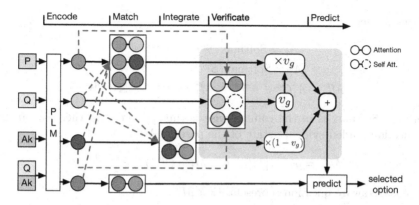

Fig. 1. The architecture of the proposed **Matching-Integration-Verification** (**MIV**) model. The verification module is highlighted with the pale yellow area.

3.2 Contextual Encoding

Firstly, we obtain the contextual encoding of each tokens in passage, question and candidate answers. We consider employing the pre-trained language models as the contextual encoder and take the last layer output of the encoder as the contextual representation of each input as follows:

$$C^p = \text{PLM}(\mathbf{P}) \in \mathbb{R}^{n_p \times d_h}, C^q = \text{PLM}(\mathbf{Q}) \in \mathbb{R}^{n_q \times d_h} \tag{2}$$

$$C^{a_k} = \text{PLM}(A_k) \in \mathbb{R}^{n_k \times d_h} \tag{3}$$

where d_h is the dimension of the contextual representation.

Besides, we also compute a contextual representation for the question-answer pair through the same encoder: $C^{qa_k} = \text{PLM}([\mathbf{P}; \mathbf{A_k}]) \in \mathbb{R}^{(n_q+n_k) \times d_h}$, where $[;]$ denotes concatenation operation.

3.3 Evidence Matching

After the contextual encoding, it is crucial to orient the most relevant evidence information in the passage with respect to the question and option. This is the purpose of clarifying what the question focuses on the passage and finding supportive information about the option.

Aiming at aligning two sequences, we employ the bidirectional attention to build connection. Taken the matching processing between passage and question as an elaborated example, we first calculate the alignment matrix $S^{pq} \in \mathbb{R}^{n_p \times n_q}$:

$$S^{pq}_{ij} = \text{F}_{Att}(C^p_i, C^q_j) = C^p_i W_1 (C^q_i)^T \tag{4}$$

where $\text{F}_{Att}(\cdot)$ is the semantic similarity score function and $W_1 \in \mathbb{R}^{d_h \times d_h}$ is a trainable weight matrix. Then the attentive passage and question representations are built up as:

$$A_1^q = \text{Softmax}(S^{pq}) \in \mathbb{R}^{n_p \times n_q} \tag{5}$$

$$A_1^p = \text{Softmax}(S^{pqT}) \in \mathbb{R}^{n_q \times n_p} \tag{6}$$

$$M_1^p = A_1^q C^q \in \mathbb{R}^{n_p \times d_h}, M_1^q = A_1^p C^p \in \mathbb{R}^{n_q \times d_h} \tag{7}$$

where the Softmax is in the column-wise normalization. And then we can get the question-guided evidence state of the passage $M_1^{p'}$:

$$M_1^{p'} = g(F^p([M_1^p - C^p; M_1^p \circ C^p])) \in \mathbb{R}^{n_p \times d_h} \tag{8}$$

and passage-aware question representation $M_1^{q'}$:

$$M_1^{q'} = g(F^q([M_1^q - E^q; M_1^q \circ E^q])) \in \mathbb{R}^{n_q \times d_h} \tag{9}$$

where $g(\cdot)$ is ReLU activation function, and $F^*(\cdot)$ is one-layer fully connected neural network. The range from Eq. 4 to Eq. 9 constitutes a complete bidirectional matching process.

Follow the above, we conduct the same matching operation, which owns the same computing but with different parameters, between passage and option, as well as passage and question-answer pair. Specifically, we can obtain the option-guided passage evidence state $M_2^{p'} \in \mathbb{R}^{n_p \times d_h}$, passage-aware option representation $M_2^{a_k'} \in \mathbb{R}^{n_k \times d_h}$ through the matching between C^p and C^{a_k}, and question-answer pair guided passage evidence state $M_3^{p'} \in \mathbb{R}^{n_p \times d_h}$ and passage-aware question-answer pair representation $M_3^{qa_k'} \in \mathbb{R}^{(n_q+n_k) \times d_h}$ through the bidirectional matching between C^p and C^{qa_k}.

To gather the multiple perspective evidence information, we combine the three passage states and conduct a column-wise pooling to produce the final passage evidence vector as follow:

$$M^p = \text{MaxPool}([M_1^{p'} : M_2^{p'} : M_3^{p'}]) \in \mathbb{R}^{d_h} \tag{10}$$

where [:] stands for concatenation on the row-wise.

3.4   Multi-side Integration

In this module, we collect the supportive information about each candidate option according to the matched evidence information in the passage side and question side which is vital to discrimination of the correct option.

We regard the passage-aware option representation $M_2^{a_k'}$ as the supportive information of option from the passage side. Simultaneously, we also need to collect supportive information for option from the question side. Firstly, we construct the alignment matrix $S^{qa_k} \in \mathbb{R}^{n_q \times n_k}$ between the question and the option:

$$S_{mn}^{qa_k} = F_{Att}(C_m^q, C_n^{a_k}) = C_m^q W_2 (C_n^{a_k})^T \tag{11}$$

where $W_2 \in \mathbb{R}^{d_h \times d_h}$ is a trainable weight matrix. The question attentive option representation is calculated as:

$$A_2^q = \text{Softmax}(S^{qa_k \, T}) \in \mathbb{R}^{n_k \times n_q} \tag{12}$$

$$M_1^{a_k} = A_2^q C^q \in \mathbb{R}^{n_k \times d_h} \tag{13}$$

Then we can get the question-aware option representation $M_3^{a_k \prime}$:

$$M_3^{a_k \prime} = g(F^a([M_3^{a_k} - C^{a_k}; M_3^{a_k} \circ C^{a_k}])) \in \mathbb{R}^{n_k \times d_h} \tag{14}$$

At last, to gather the supportive information of option from both the passage side and the question side, we summarize the passage-aware option representation and question-aware option representation into one supportive state vector for the k-th option as follow:

$$M^{a_k} = \text{MaxPool}([M_2^{a_k \prime} : M_3^{a_k \prime}]) \in \mathbb{R}^{d_h} \tag{15}$$

3.5   Verification

For improving the reliability of the final output, we introduce a verification module for rechecking the evidence information from the passage and the supportive information of each candidate option. The verification module is the core of our model as shown in the pale yellow area in Fig. 1. It selectively outputs useful information according to the question to the next prediction layer instead of treating different kinds of information equally.

We first review the question and calculate the self-attentive vector of question as follows:

$$A^{qq} = \text{Softmax}(C^q v_q) \in \mathbb{R}^{n_q}, V_0^q = A_{qq} C^q \in \mathbb{R}^{d_h} \tag{16}$$

where $v_q \in \mathbb{R}^{d_h}$ is a trainable vector.

We have already calculated the passage attentive question representation $M_1^{q \prime}$ in Sect. 3.3, thus we can get the passage attentive question state vector: $V_1^q = \text{MaxPool}(M_1^{q \prime}) \in \mathbb{R}^{d_h}$ Besides, we reuse the S^{qa_k} to calculate the option attentive question representation as follows:

$$A_2^{a_k} = \text{Softmax}(S^{qa_k}) \in \mathbb{R}^{n_q \times n_k} \tag{17}$$

$$M_2^q = A_2^{a_k} C^{a_k} \in \mathbb{R}^{n_q \times d_h} \tag{18}$$

$$M_2^{q \prime} = g(F^q([M_2^q - C^q; M_2^q \circ C^q])) \in \mathbb{R}^{n_q \times d_h} \tag{19}$$

and get a option attentive question state vector: $V_2^q = \text{MaxPool}(M_2^{q \prime}) \in \mathbb{R}^{d_h}$. Then, we utilize the three question state vectors to set up the recheck query vector: $V^q = [V_0^q; V_1^q; V_2^q] \in \mathbb{R}^{3d_h}$.

Under the guidance of the content of the question, which is the recheck query vector, we need to determine the proportion of evidence information and supportive information for the final prediction. Thus, we conduct a dynamic gate mechanism to calculate the weighting factor and merge all information to get the informed state representation $M_g^k \in \mathbb{R}^{d_h}$ for each option:

$$v_g = \sigma(V^q W_g + b_g) \in \mathbb{R} \tag{20}$$

$$M_g^k = v_g \cdot M^p + (1 - v_g) \cdot M^{a_k} \tag{21}$$

3.6   Prediction

Combining the passage-aware question-answer pair representation $M_3^{qa_k'}$ and the rechecked information state M_g^k, we get the final representation R_k for each candidate answer:

$$M^{qa_k} = \text{MaxPool}(M_3^{qa_k'}) \in \mathbb{R}^{d_h} \tag{22}$$

$$R_k = [M^{qa_k}; M_g^k] \in \mathbb{R}^{2d_h} \tag{23}$$

The final probability for the candidate answer A_k to be selected as correct is:

$$P(A_k|\mathbf{P}, \mathbf{Q}) = \frac{\exp(W_r^T R_k)}{\sum_{j=1}^{|\mathcal{A}|} \exp(W_r^T R_j)} \tag{24}$$

where $W_r \in \mathbb{R}^{2d_h}$ is a trainable vector. We use the categorical cross entropy to calculate loss between model prediction and the ground truth answer label.

4   Experiments

In this section, we first introduce the datasets used for evaluating our approach. Then we report the details of implementation and training. Lastly, we present the experiment results of our approach.

4.1   Datasets

We conduct experiments on three typical datasets: a) **RACE**: The RACE [3] is a large-scale MCRC dataset collected from English examinations for Chinese students which is designed by domain experts. Each question has four candidate choices. The portion of questions in RACE (including both RACE-Middle and RACE-High) for training/development/test set is $87,866/4,887/4,934$. b) **MCScript**: The MCScript [5], which is also known as the benchmark used in the SemEval 2018 Task 11 [6], is collected from narrative about everyday activities. It contains $2,119$ texts and a total of $13,939$ questions, each question has two candidate choices. The MCScript dataset is split into training (9731 questions), development (1411 questions), and test set (2797 questions). c) **COIN-MCS**: The COIN-MCS [7] is also collected from the narrative about daily but is harder than MCScript. This dataset contains a total of $19,821$ questions where each question has two candidate answers. The dataset is split into training/development/test set with $14,191/3,610/2,020$ questions respectively.

4.2   Implementation Details

We implement our approach with Pytorch framework[1]. We use two kinds of pre-trained language models, the BERT [2] and RoBERTa [4], as the contextual encoder respectively. The BERT have base and large version, and we only

---

[1] We will release our code upon publication.

employ the large version for a fair comparison. The large version consists of 24 transformer layers and the intermediate hidden size is 1024. The component of RoBERTa is the same with the large version of BERT. The d_h is the same with intermediate hidden size of PLM. We use the BertAdam optimizer with initial learning rate setting to 1e−5 to optimize the model and execute a linear learning rate decay schedule. For the RACE, the batch size is 16 and the model is trained for 5 epochs. For the MCScript and COIN-MCS datasets, the batch size is 32 and the model is trained for 6 epochs. Due to the limitations of our GPU devices, the maximum sequence length is set to 400.

Table 2. Experimental results on RACE test set. * indicates that the results are from the official RACE leaderboard.

Model	RACE-M	RACE-H	RACE
HCM	55.8	48.2	50.4
CSA	52.2	50.3	50.9
GPT*	62.9	57.4	59.0
BERT$^*_{large}$	76.6	70.1	72.0
RoBERTa*	86.5	81.8	83.2
RSM*	69.2	61.5	63.8
OCN*	76.7	69.6	71.7
DCMN*	77.6	70.1	72.3
BERT+DCMN+*	79.3	74.4	75.8
BERT-ours	75.7	69.1	71.0
RoBERTa-ours	87.0	81.1	82.8
MIV+BERT	80.3	72.8	**75.0**
MIV+RoBERTa	**88.0**	**82.7**	**84.2**

http://www.qizhexie.com/data/RACE_leaderboard

4.3 Main Results

We use accuracy as the metric to measure the performance of the models. On the RACE dataset, the results of our MIV model are presented in Table 2. The results in the first group in Table 2 are the attention-based models without using PLMs. Both the HCM and CSA models employ the recurrent neural network to encode contextual information. And the results of the second group are the fine-tune based PLMs with the max sequence length setting to 512. Fine-tune based PLMs directly utilize the output of the first token to perform prediction through a simple one layer feed-forward linear layer. The results in the third group are the attention-based models taking PLMs as its encoder. These models adopt the task-specific attention modules to make the interaction between passage, question, and option. Though the BERT+DCMN+ model rank high compared to our MIV+BERT model, the BERT+DCMN+ only takes several related sentences of the passage which make it easier to fit the correct answer. However, the DCMN shares the same architecture with BERT+DCMN+ and uses the standard passage as its input. It is fairer that making a comparison between our MIV model with DCMN when we utilize BERT as the contextual encoder. Due to the limitations of our GPU devices, we select a short length of 400. Thus, we fine-tuned the same PLMs with this shorter length and the results are exhibited in the first two rows of the last group. It is reasonable that our results of fine-tuning are lower than the original results in the second group. In the last group, it is obvious that our MIV model obtains a significant improvement compared with previous PLM-based baselines. With different PLMs as our encoder, our approach can always promote the performance. The MIV model achieves 4.0 increase when using BERT and achieves 1.4 increase when using RoBERTa. Compared with OCN and DCMN which utilize BERT as their encoder, our MIV+BERT model obtains higher scores on both RACE middle and high sets which can also prove the effectiveness of our model.

Table 3. Experimental results on MCScript and COIN-MCS.

Model	MCScript dev	MCScript test	COIN-MCS dev	COIN-MCS test
BERT+DCMN+	-	91.8	-	88.8
BERT-ours	88.2	89.9	86.8	84.9
RoBERTa-ours	91.2	92.9	92.3	91.6
MIV+BERT	**89.0**	**90.6**	**88.8**	**87.2**
MIV+RoBERTa	**95.1**	**94.8**	**94.6**	**93.1**

In order to evaluate ability of the proposed model, we also conduct experiments on the MCScript and COIN-MCS datasets, and the results of our approach are presented in the Table 3. The proposed MIV model obtains the best performance on both datasets which also proves that our proposed model is capable of dealing with the general MCRC task and is not confined to a specific domain.

4.4   Performance on Different Types of Questions

To deeply analysis the performance of the MIV model, we make an evaluation of the model's performance on various question types on development sets. Based on the wh-words in the question, the questions are roughly categorized into seven types which include *what, when, where, who, why, how* and the *rest*. The question type of *rest* includes the remaining questions which can not be classified into the other types. As illustrated in Fig. 2, it demonstrates the detailed performance of the proposed MIV model on various types of questions, as well as comparison with the RoBERTa model. On the RACE benchmark, the MIV model obtains improvement on the five out of seven types of questions, especially the who and how questions. On the MCScript and COIN-MCS benchmarks, the MIV model surpasses the RoBERTa model on all question types. It is remarkable that the MIV model makes progress about the *why* and *how* questions which are harder to answer and the MIV model obtains a large increase on the why questions of the COIN-MCS dataset.

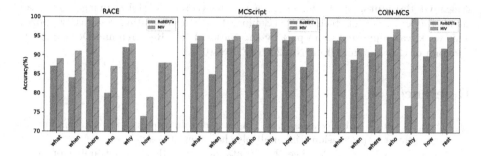

Fig. 2. The performance on different types of questions.

Table 4. Ablation results on the development set of RACE, MCScript and COIN-MCS.

Model	RACE	MCScript	COIN-MCS
MIV+RoBERTa	**84.5**	**95.1**	**94.6**
w/o verification	83.8 ($\downarrow$0.7)	94.2 ($\downarrow$0.9)	93.2 ($\downarrow$1.4)

4.5 Ablation Study

To inspect the effect of the proposed model, we conduct ablation study to illustrate the impact of the verification module. We remove the verification module of our MIV model and directly input the M^p, M^{a_k} and M^{qa_k} to the prediction module. As shown in Table 4, it is clear that once we remove the verification module, the performance of the model drops nearly 1.0 on the MCScript and COIN-MCS datasets. Therefore, it can be proofed that our proposed model with the verification module can promote the performance of the MCRC task through this ablation study.

5 Conclusion and Future Work

In this paper, we propose a matching-integration-verification model inspired by strategy drawn from human being to solve multiple-choice reading comprehension task. Our proposed model achieve a significant performance on three benchmark datasets, and the analysis and ablation study demonstrate the effectiveness of the proposed model. Although the machine can obtain high scores on these MCRC datasets, there is still a large gap between machine and human performance. Besides, the results generated by the machine do not possess interpretation. In the next step, we should put more effort into the more explanative and robust machine reading comprehension.

Acknowledgments. We thank the reviewers for their insightful comments. We also thank Effyic Intelligent Technology (Beijing) for their computing resource support. This work was supported by in part by the National Key Research and Development Program of China under Grant No. 2016YFB0801003.

References

1. Chen, Z., Cui, Y., Ma, W., Wang, S., Hu, G.: Convolutional spatial attention model for reading comprehension with multiple-choice questions (2018)
2. Devlin, J., Chang, M.W., Lee, K., Toutanova, K.: BERT: pre-training of deep bidirectional transformers for language understanding. In: NAACL, pp. 4171–4186 (2019)
3. Lai, G., Xie, Q., Liu, H., Yang, Y., Hovy, E.: Race: large-scale reading comprehension dataset from examinations. In: Proceedings of the 2017 Conference on Empirical Methods in Natural Language Processing, pp. 785–794 (2017)
4. Liu, Y., et al.: Roberta: a robustly optimized BERT pretraining approach. CoRR (2019)
5. Ostermann, S., Modi, A., Roth, M., Thater, S., Pinkal, M.: Mcscript: a novel dataset for assessing machine comprehension using script knowledge. In: Proceedings of the Eleventh International Conference on Language Resources and Evaluation (2018)
6. Ostermann, S., Roth, M., Modi, A., Thater, S., Pinkal, M.: Semeval-2018 task 11: machine comprehension using commonsense knowledge. In: Proceedings of The 12th International Workshop on Semantic Evaluation, pp. 747–757 (2018)
7. Ostermann, S., Roth, M., Pinkal, M.: Mcscript2.0: a machine comprehension corpus focused on script events and participants. In: Proceedings of the Eighth Joint Conference on Lexical and Computational Semantics, *SEM@NAACL-HLT 2019, Minneapolis, MN, USA, 6–7 June 2019, pp. 103–117 (2019)
8. Radford, A., Narasimhan, K., Salimans, T., Sutskever, I.: Improving language understanding by generative pre-training (2018)
9. Rajpurkar, P., Zhang, J., Lopyrev, K., Liang, P.: Squad: 100,000+ questions for machine comprehension of text. In: Proceedings of the 2016 Conference on Empirical Methods in Natural Language Processing, pp. 2383–2392 (2016)
10. Ran, Q., Li, P., Hu, W., Zhou, J.: Option comparison network for multiple-choice reading comprehension. arXiv preprint arXiv:1903.03033 (2019)
11. Sun, K., Yu, D., Yu, D., Cardie, C.: Improving machine reading comprehension with general reading strategies. arXiv Computation and Language (2018)
12. Vaswani, A., et al.: Attention is all you need. In: Advances in Neural Information Processing Systems, pp. 5998–6008 (2017)
13. Wang, S., Yu, M., Jiang, J., Chang, S.: A co-matching model for multi-choice reading comprehension. In: Gurevych, I., Miyao, Y. (eds.) Proceedings of the 56th Annual Meeting of the Association for Computational Linguistics, ACL 2018, Melbourne, Australia, 15–20 July 2018, vol. 2: Short Papers, pp. 746–751 (2018)

14. Xu, Y., Liu, J., Gao, J., Shen, Y., Liu, X.: Dynamic fusion networks for machine reading comprehension. arXiv preprint arXiv:1711.04964 (2017)
15. Yang, Z., Dai, Z., Yang, Y., Carbonell, J.G., Salakhutdinov, R., Le, Q.V.: Xlnet: generalized autoregressive pretraining for language understanding. In: Advances in Neural Information Processing Systems 32: Annual Conference on Neural Information Processing Systems 2019, NeurIPS 2019, Vancouver, BC, Canada, 8–14 December 2019, pp. 5754–5764 (2019)
16. Zhang, S., Zhao, H., Wu, Y., Zhang, Z., Zhou, X., Zhou, X.: Dual co-matching network for multi-choice reading comprehension. arXiv preprint arXiv:1901.09381 (2019)
17. Zhang, S., Zhao, H., Wu, Y., Zhang, Z., Zhou, X., Zhou, X.: Dcmn+: dual co-matching network for multi-choice reading comprehension. In: The Thirty-Fourth AAAI Conference on Artificial Intelligence (AAAI) (2020)
18. Zhu, H., Wei, F., Qin, B., Liu, T.: Hierarchical attention flow for multiple-choice reading comprehension. In: Thirty-Second AAAI Conference on Artificial Intelligence (2018)

How to Interact and Change? Abstractive Dialogue Summarization with Dialogue Act Weight and Topic Change Info

Jiasheng Di[iD], Xiao Wei[(✉)], and Zhenyu Zhang

School of Computer Engineering and Science, Shanghai University, Shanghai, China
{dijia123,xwei,zhenyuzhang}@shu.edu.cn

Abstract. Conventional sequence-to-sequence frameworks in neural abstractive summarization treat every document as a single topic text without interaction, so the results are often unsatisfactory when given dialogues. To tackle this problem, we propose a novel method to make the model generate a better summary with multiple topics and interactions among people. Compared with traditional abstractive summarization models, our approach makes two contributions: (1) To solve the interactive problem, we use *dialogue act* as an interactive pattern to represent important external information and propose Dialogue Act Weight to elaborately calculate the weight that dialogue acts can provide; (2) To make the model handle multiple topics better, we introduce Topic Change Info as a signal to indicate which utterance in dialogue is important and whether the topic has been changed. Compared to baselines, experiments show that our methods significantly improve the abstractive summarization performance on the modified AMI meeting corpus (The source code is available at https://github.com/d1jiasheng/DialogueSum).

Keywords: Dialogue summarization · Interactive pattern · Dialogue act · Multiple topics · AMI corpus

1 Introduction

Text summarization can be divided into two parts: *extractive summarization* and *abstractive summarization*. The extractive summarization mainly focuses on selecting existing sentences and choosing the most important sentences or words to form a new summarization [19,30]. While in abstractive summarization, the final summary is always produced through the deep neural network [1,8]. It means that words in a summary are not only from the original text but also from the additional vocabulary, which makes the abstractive summarization more flexible.

However, the dialogue is quite different from the text. A text is more like a single speaker, which has only one topic and do not have any interactions among speakers. On the contrary, the dialogue always has more than two speakers, so we

© Springer Nature Switzerland AG 2020
G. Li et al. (Eds.): KSEM 2020, LNAI 12275, pp. 238–249, 2020.
https://doi.org/10.1007/978-3-030-55393-7_22

cannot simply apply text summarization methods to dialogue summarization. In the past years, many methods have been put forward for dialogue summarization [13, 22]. Seq2seq and its transformation forms have achieved promising results. However, as mentioned above, almost all past works focus on the contents rather than interactions and multi-topics.

Therefore, we should pay more attention to the interactions in dialogue summarization. Some methods are proposed to use other information, such as prosodic and semantic features [6], spoken information [7], etc. and these methods seem reasonable. However, the external information they use still focus on contents instead of interactions. We find an interaction information called *dialogue act* [3] which is the effect of an utterance on the context. It includes some conversational interactions and provides important information to understand dialogue. In [11], researchers use the dialogue act for abstractive dialogue summarization. They propose a model with a sentence-gated mechanism which uses the final summarization to predict the dialogue act and significantly improve the performance. Meanwhile, they point out that summarization in dialogue has not yet been fully explored due to the lack of suitable benchmark data. So the AMI corpus [4], which contains many dialogues and annotated topics, has been used as their dataset.

Although interactive information has been used in the model, some problems still exist. Essentially, the model only uses dialogue act for classification and prediction but neglects the interaction with the dialogue content. Furthermore, each dialogue in AMI corpus mostly has only one topic, but we may talk more than one topic in a conversation or meeting in our daily life. So we need to make some changes in AMI corpus and add more information to guide the model to generate multiple topics.

In this paper, we apply the model with sentence-gated mechanism in [11], which has been mentioned above, as our basic model. Besides, we define **Dialogue Act Weight** as the weight of each sentence calculated by dialogue act and use this approach to make the model pay more attention to important sentences. Based on our daily situations, we also modify the AMI corpus to make each data have multiple topics. Finally, because the corpus has been changed, each dialogue may have more than one topic or summarization, so it is also a challenging task for us to find which sentence has changed the topic of the dialogue. We then derive a method to calculate the possibility that each sentence may change the topic, which is defined as **Topic Change Info**, to make the model focus on the change point in the summary of decoder part. In general, our model uses a seq2seq architecture with attention mechanism [1] as a basic summary generator. At the encoder part, Dialogue Act Weight is added for better interaction. At the decoder part, we use the Topic Change Info to judge whether the utterance is a turning point.

In summary, our works focus on how to use the interactive information to get better dialogue summarization. Meanwhile, we make some modifications in basic corpus and introduce Topic Change Info for more flexible conditions. To sum up, our contributions can be summarized as three-fold:

- We use dialogue act as external information and calculate **Dialogue Act Weight** to find important sentences.
- We modify the AMI corpus and propose the **Topic Change Info** to deal with the multiple topics situations.
- To the best of our knowledge, our method achieves state-of-the-art performance in dialogue summarization on the new AMI corpus.

Table 1. An example of a dialogue with dialogue act

Dialogue	Dialogue act
A: okay yeah, that's right	Assess
B: okay	Assess
C: okay, uh	Stall
C: now my personal preferences	Offer
B: uh using the standards, basic	Suggest
C: um i think that we should stand out uh unique, being unique with the design	Suggest
C: so when you are in the shop and you see our t_v_ controller hanging, that it stands out	Inform
A: not just in the row when you see all the same uh remote controls	Inform
Summary: look and usability	

2 Related Work

AMI Meeting Corpus [4]. The AMI meeting corpus is a multi-model dataset with different annotations, which contains 100 h of conference recordings. The recordings use a series of signals to synchronize the common timeline. These recordings are collected from short-range and far-field microphones, individuals and room video cameras, output of slide projectors and electronic whiteboards. During the meeting, participants may also use the unsynchronized pen to record the contents. The meeting uses three rooms with different acoustic properties to record English dialogues. Most of the presenters are non-native speakers. Around two-thirds of the data is drawn from scenarios where participants play different roles in the design teams. The rest of it includes natural meetings in various fields. The AMI corpus contains a wide range of annotations, including dialogue acts, topic descriptions, named entities, hand gestures, and gaze direction. Table 1 is an example of a dialogue with dialogue act from the AMI corpus.

In [11], researchers use sliding windows of 50 words in size to split a meeting into several dialogue samples. Because there is no summary annotation in the AMI data, they treat annotated topic descriptions as summaries of the dialogues.

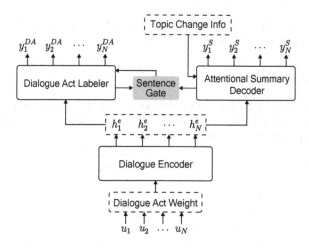

Fig. 1. The main framework of our model

They extract 7824 samples from 36 meeting recordings and split them into three groups: 7024 samples for training, 400 samples for validating and 400 samples for testing. In addition, there are 15 dialogue act labels in the dataset. Our corpus is modified on the basis of this.

3   Method

3.1   Modifications on AMI Corpus

We make some simple modifications to AMI corpus for a better fitting to our works. We randomly select two or three dialogues in corpus without repetition and reassemble them into a new piece of data. We repeat the above steps until all the data has been selected. At the same time, the dialogue act and topic description should be connected in the same order. The new dataset we generate has more than one topic in each dialogue and will be used in our experiments.

3.2   General Framework

Our general framework is depicted in Fig. 1. Given a dialogue document, it can be viewed as a sequence of utterances $\mathbf{u} = (u_1, ..., u_N)$, where N is the length of dialogue. Each utterance can be divided as a sequence of words and the embedding of each utterance can be acquired by averaging the word embedding. The utterance embedding is inputted into Dialogue Act Weight, which combines the utterance with information from dialogue act. Then, the dialogue encoder encodes the new utterance embedding and finally gets the hidden state.

The model has two decoder parts. The first one, Dialogue Act Labeler, uses the hidden state to predict the dialogue act of each utterance. The second part not only uses the hidden state but also uses Topic Change Info to add information about whether the topic changes. Between these two parts, a mechanism named sentence-gated is added for a better performance.

3.3 Details in Basic Model

In the encoder part, given a sequence of utterances $\mathbf{u} = (u_1, ..., u_N)$, the bidirectional long short-term memory (BiLSTM) [23] is employed to encode and obtain basic interactive information from dialogue by

$$\overrightarrow{h_i^e} = LSTM_r(u_1, ..., u_N) \tag{1}$$

$$\overleftarrow{h_i^e} = LSTM_l(u_N, ..., u_1) \tag{2}$$

$$h_i^e = [\overrightarrow{h_i^e}; \overleftarrow{h_i^e}] \tag{3}$$

where $\overrightarrow{h_i^e}$ denotes the forward hidden state from the input, while $\overleftarrow{h_i^e}$ denotes the backward one. The final hidden state h_i^e at step time i is the concatenation of $\overrightarrow{h_i^e}$ and $\overleftarrow{h_i^e}$, which is the final encoding with context information.

In the first decoder part, the hidden-state is inputted into an attention mechanism, from which the decoder hidden state h_i^{DA} is obtained by

$$e_{i,j} = sigmoid(W_{he}^{DA} \cdot h_j^e) \tag{4}$$

$$\alpha_{i,j}^{DA} = exp(e_{i,j})/\sum\nolimits_{n=1}^{N} exp(e_{i,k}) \tag{5}$$

$$h_i^{DA} = \sum_{j=1}^{N} \alpha_{i,j}^{DA} \cdot h_j^e \tag{6}$$

where W_{he}^{DA} is the weight matrix and $\alpha_{i,j}^{DA}$ is the attention weight. Finally, the hidden state from encoder and decoder are combined together to generate the dialogue act, specifically:

$$y_i^{DA} = softmax(W \cdot (h_i^e + h_i^{DA})) \tag{7}$$

Note that y_i^{DA} is the i-th sentence's dialogue act label and W is a weight matrix.

The second part, summary decoder, is quite similar to the first part. Given the hidden state after encoding, unidirectional LSTM with attention mechanism is used for decoding and the generated summary also combines the hidden state of the encoder and decoder:

$$y_i^S = softmax(W \cdot (h_i^e + h_i^S)) \tag{8}$$

The calculation of the h_i^S is similar with h_i^{DA}.

The sentence-gated mechanism, which uses the summary to predict the dialogue act, is then applied to the model. It derives a context vector from the dialogue act and summary to guide the generation of the summary. In [11], the author proposed two mechanisms, one is full-attention-sentence-gated and the

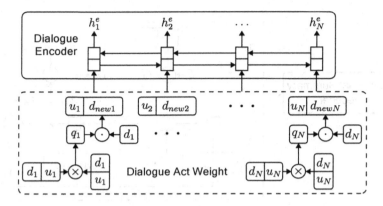

Fig. 2. The specific structure of the "**Dialogue Act Weight**" we propose. It will be added to the encoder part to make utterance has interactive with dialogue act.

other is summary-attention-sentence-gated. We choose the better one, summary attention. First, the average of the hidden state h^S is calculated from summary decoder:

$$h^S = \frac{1}{N} \sum_{n=1}^{N} h_i^S \tag{9}$$

Then, the h_i^e (hidden state from encoder) is combined with h^S as the input of the gate:

$$g = \sum v \cdot tanh(h_i^e + W \cdot h^S) \tag{10}$$

where W and v are trainable variables. g can be considered as the weighted feature of the dialogue act and summary. Note that g is used to predict the dialogue act and (7) can be replaced by:

$$y_i^{DA} = softmax(W \cdot (h_i^e + h_i^e \cdot g)) \tag{11}$$

The model parameters are adjusted by the y_i^{DA} in training, which is ultimately used for better guiding the generation of summary.

3.4 Dialogue Act Weight

In order to make the generated hidden state to include more information in the encoder part, we notice that the dialogue act can be added to the corresponding utterance as additional information. For different utterances in a dialogue, they may have the same dialogue act, but these same dialogue act don't necessarily provide the same information when considering utterances' structure and semanteme.

Inspired by the self-attention [28], we design a new method. The structure is shown in Fig. 2. Given a dialogue with N utterances, we have a sequence of dialogue acts $\mathbf{d} = (d_1, ..., d_N)$, and we calculate the Dialogue Act Weight by:

$$u'_i = [u_i; d_i] \tag{12}$$

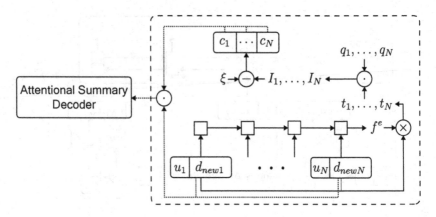

Fig. 3. The architecture of the "**Topic Change Info**" we propose. It will be added to guide the decoder when the model generates the final summary.

$$q_{i\_weight} = softmax((W_1u'_i + H_1) \times (W_2u'^{T}_i + H_2)) \tag{13}$$

where u' is the concatenation of the input utterances and corresponding dialogue act, T stands for transposition, W_1, W_2 are weight matrices and H_1, H_2 are bias vectors. The model learns the internal structure of text and dialogue act through (13). After passing through the activation function, we get q_{i_weight}, which represents the percentage of the dialogue act of each utterance in an entire dialogue. Then the input dialogue act is multiplied by q_{i_weight} to get a new input dialogue act and concatenate it with utterance:

$$d_{i\_new} = q_{i\_weight} \cdot d_i \tag{14}$$

$$u_{i\_new} = [u_i; d_{i\_new}] \tag{15}$$

As the input of the encoder, u_{i_new} combines the information of the utterance and the dialogue act.

3.5   Topic Change Info

In the scenario of our daily life, the topic of a conversation is not static all the time. So we design Topic Change Info as additional information to determine whether the topic of the current sentence has been changed. The specific structure is shown in Fig. 3. First, the utterances, $u_{new} = (u_{1\_new}, ..., u_{N\_new})$, which have been changed in the previous section is inputted into an unidirectional LSTM and get the final output f^e:

$$f^e = LSTM(u_{1\_new}, ..., u_{N\_new}) \tag{16}$$

where f^e roughly contains the entire dialogue information. Then, to find the relationship between each utterance and the entire dialogue, we have

$$t_{i\_weight} = softmax(u_{new\_i} \times f^e) \tag{17}$$

where t_{i_weight} is the interaction of the i-th utterance and the overall dialogue. At the same time, because of using the softmax activation function, it can also represent the weight ratio of each utterance in the entire dialogue. In Sect. 3.4, for the i-th utterance, we get the weight that the corresponding dialogue act can provide for this utterance, q_{i_weight}. Now, we have the weight of the information that each utterance can provide in the dialogue, t_{i_weight}. Then we have

$$I_{i\_weight} = t_{i\_weight} \otimes q_{i\_weight} \tag{18}$$

where "$\otimes$" represents adding two pieces of information(we use and compare addition and multiplication in experiments), and I_{i_weight} shows the importance of the utterance in the dialogue after taking the dialogue act and the content information into account. But when the topic changes, the model is not concerned with the utterance which has a bigger weight of I. On the contrary, the smaller the utterance of I, the more likely it is the turning point of the topic. For example, utterances like "um, ok", "yeah, so that we should talk about another question". These two sentences don't contain important information, and the weight of the dialogue act is not high, but they are more likely to change the topic of the discussion, that is, the role of the link. So a new method, Topic Change Info, is designed to get another extra information, and make the model pay more attention to such utterance:

$$c_{i\_weight} = softmax(\xi - I_{i\_weight}) \tag{19}$$

where ξ is a hyperparameter that can be chosen to reverse the weight (18). A large c_{i_weight} indicates that the corresponding utterance is more likely to be the key sentence to change the topic. Finally, this information is added to the summary decoder to give some guidance when generating the summary. The (8) is replaced as below:

$$y_i^S = softmax(W \cdot (h_i^e + h_i^S + c_{i\_weight} \cdot u_{i\_new})) \tag{20}$$

4 Experiments

4.1 Setup

In all experiments, we set the maximum epoch to 100 and the batch-size to 16. The size of hidden layer is set to 256 and all the compared models have a similar parameter size. We use AdamOptimizer as our optimizer. All experiments are performed and compared on our new dataset.

4.2 Evaluation Metrics

We use ROUGE score [9], which is widely applied for summarization evaluation [5], to evaluate our experimental results. The ROUGE metrics use ROUGE-1, ROUGE-2, ROUGE-3 and ROUGE-L to compare generated summary with reference summary.

Table 2. Experimental results with different parameters and superposition methods in our proposed method.

ξ	$\otimes$	ROUGE-1	ROUGE-2	ROUGE-3	ROUGE-L
1.0	$\times$	55.88	36.16	28.23	45.46
0.9	$\times$	55.49	33.80	25.21	43.33
0.8	$\times$	55.80	34.91	26.30	44.64
1.0	$+$	**55.90**	35.35	26.86	44.97
0.9	$+$	54.82	34.38	26.54	45.35
0.8	$+$	55.35	**37.24**	**29.38**	**46.50**
0.7	$+$	54.26	32.92	25.26	44.20

4.3 Baselines

To evaluate the performance of our proposed model, we compare it with the following baselines: (1) **S2S+Attn** [25]: Sequence-to-sequence model with attention mechanism, which is the most common method in abstractive summary. (2) **BiLSTM+S2S+Attn** [23]: a sequence-to-sequence model with BiLSTM can extract context information when encoding. (3) **Pointer-Generator Network** [26]: the method proposed for combining output sequences with input elements. (4) **S2S+Sentence-Gated** [11]: The model with sentence-gated mechanism. It is a basic model we refer to.

4.4 Results

We first conduct the experiment with the proposed model under different parameters and superposition methods. The experimental results are shown in Table 2.

Table 2 shows the results of the experiment with different parameters and superposition methods. When ξ is set to 1.0 and $\otimes$ is "+", we can get the best results on ROUGE-1. And when we set ξ to 0.8 and $\otimes$ to "+", we get the best results on the other three evaluation metrics.

To show the superiority of our methods, we compare it with the other four baseline models. The ξ is set to 0.8 and $\otimes$ is "+" in this experiment. The experimental results are shown in Table 3.

Table 3 shows that the method we proposed significantly outperforms all baselines, and improves results on all evaluation metrics. We can see that after adding the Dialogue Act Weight into the inputs, the experimental result achieves a 2.3%, 4.6%, 5.5%, 4.1% increment over the model use sentence-gated mechanism in terms of ROUGE-1, ROUGE-2, ROUGE-3 and ROUGE-L. It demonstrates that the dialogue act provides useful information for the final summary. When the Topic Change Info is added on the model, it performs slightly better than the model only add Dialogue Act Weight. Compared with the sentence-gated mechanism, the model with Dialogue Act Weight and Topic Change Info

Table 3. Experimental results of different baseline and the method we proposed. We set ξ to 0.8, $\otimes$ to "+".

Method	ROUGE-1	ROUGE-2	ROUGE-3	ROUGE-L
Attentional Seq2Seq [25]	33.93	30.56	21.59	40.91
Attentional Seq2Seq+BiLSTM [23]	36.25	32.15	24.75	41.98
Pointer-Generator [26]	30.01	24.33	23.59	30.90
Sentence-Gated [11]	54.50	34.28	26.60	44.22
Our proposed (add dialogue act weight only)	**55.75**	35.85	28.07	46.03
Our proposed (add dialogue act and topic change)	55.35	**37.24**	**29.38**	**46.50**

finally achieves a 1.5%, 8.6%, 10.4%, 5.1% increment in terms of the four evaluation metrics, which means Topic Change Info can guide the model to generate the summary more accurately.

4.5 Analysis and Discussion

Comparing with the sentence-gated mechanism, the method we proposed seems to have rarely improvement on ROUGE-1. In our opinion, ROUGE-1 is an evaluation metric that only considers words rather than phrases, which means it cares little about the relationship between the generated words. While ROUGE-2, ROUGE-2 and ROUGE-L are evaluation metrics that consider the structure of phrases and sentences. In our experiments, we just focus on whether the final generated summary contains multiple topic information rather than some words that are irrelevant to the topic. Therefore, we believe that the improvements on ROUGE-2, ROUGE-3, ROUGE-L has greater significance for evaluating the effectiveness of our method.

Table 4. Several examples of outputs compare with references

Comparing outputs with references
pre: drawing exercise . components designer presentation and energy sources .
cor: drawing animals on the whiteboard . components , materials and energy sources
pre: components , materials and energy sources . opening issues
cor: look and usability components , materials and energy sources . opening
pre: components materials , and energy sources . look usability
cor: components , materials and energy sources . look and usability
pre: evaluation of process and materials components
cor: evaluation of project process . components

Table 4 shows the results of several experimental outputs we selected and the comparisons with the references. Note that sentences after "pre:" are generated by the model and sentences after "cor:" means the correct summary. The word with red font indicates that the topics generated by the model are the same as references. We can see that the model can accurately generate a dialogue summary of multiple topics. And there is an interesting thing in the fourth example, where "and" is used in the connection of multiple topics instead of ".", making us believe that the model learns to use "and" when connecting two different topics, not just ".". This also illustrates that the Topic Change Info we develop has played a guiding role.

5   Conclusion

In this paper, we propose a dialogue summary method by using dialogue act as external information to join the interactive information of the dialogue. We made some changes to the AMI corpus so that each piece of data can contain multiple topics. Then, we propose Topic Change Info to guide the model to observe the topic changes. The final experiment achieves promising results and confirms the validity of our ideas.

Acknowledgement. This work was supported by the National Key Research and Development Program of China (No. 2018YFB0704400). We also would like to be grateful to the partners in the laboratory who have given our generous support and helpful advice for this research.

References

1. Rush, A.M., Chopra, S., Weston, J.: A neural attention model for abstractive sentence summarization. In: Proceedings of the 2015 Conference on Empirical Methods in Natural Language Processing (2015)
2. Ang, J., Yang, L., Shriberg, E.: Automatic dialog act segmentation and classification in multiparty meetings (2015)
3. Bunt, H.: Context and dialogue control. Think Q. **3**(1), 1–39 (1970)
4. Carletta, J., et al.: The AMI meeting corpus: a pre-announcement. In: Renals, S., Bengio, S. (eds.) MLMI 2005. LNCS, vol. 3869, pp. 28–39. Springer, Heidelberg (2006). https://doi.org/10.1007/11677482_3
5. Chen, X., Gao, S., Tao, C., Song, Y., Yan, R.: Iterative document representation learning towards summarization with polishing (2018)
6. Chen, Y.N., Huang, Y., Kong, S.Y., Lee, L.S.: Automatic key term extraction from spoken course lectures using branching entropy and prosodic/semantic features. In: Spoken Language Technology Workshop (2011)
7. Chen, Y.N., Metze, F.: Intra-speaker topic modeling for improved multi-party meeting summarization with integrated random walk. In: Conference of the North American Chapter of the Association for Computational Linguistics: Human Language Technologies (2012)

8. Chopra, S., Auli, M., Rush, A.M.: Abstractive sentence summarization with atten-tive recurrent neural networks. In: Conference of the North American Chapter of the Association for Computational Linguistics: Human Language Technologies (2016)
9. Flick, C.: Rouge: a package for automatic evaluation of summaries. In: Workshop on Text Summarization Branches Out (2004)
10. Gao, S., Chen, X., Li, P., Ren, Z., Yan, R.: Abstractive text summarization by incorporating reader comments (2018)
11. Goo, C.W., Chen, Y.N.: Abstractive dialogue summarization with sentence-gated modeling optimized by dialogue acts (2018)
12. Gu, J., Lu, Z., Hang, L., Li, V.O.K.: Incorporating copying mechanism in sequence-to-sequence learning (2016)
13. Harwath, D., Hazen, T.J.: Topic identification based extrinsic evaluation of sum-marization techniques applied to conversational speech. In: IEEE International Conference on Acoustics (2012)
14. Hermann, K.M., et al.: Teaching machines to read and comprehend (2015)
15. Hsu, W.T., Lin, C.K., Lee, M.Y., Min, K., Sun, M.: A unified model for extractive and abstractive summarization using inconsistency loss (2018)
16. Hu, B., Chen, Q., Zhu, F.: LCSTS: a large scale Chinese short text summarization dataset. Computer Science, pp. 2667–2671 (2015)
17. Samuel, K., Carberry, S., Vijay-Shanker, K.: Dialogue act tagging with transformation-based learning. In: Proceedings of COLING (1998)
18. Klüwer, T., Uszkoreit, H., Xu, F.: Using syntactic and semantic based relations for dialogue act recognition. In: International Conference on Computational Linguis-tics: Posters (2010)
19. Kågebäck, M., Mogren, O., Tahmasebi, N., Dubhashi, D.: Extractive summariza-tion using continuous vector space models. In: CVSC at EACL (2014)
20. Li, W., Yao, J., Tao, Y., Li, Z., Qiang, D.: A reinforced topic-aware convolu-tional sequence-to-sequence model for abstractive text summarization. In: Twenty-Seventh International Joint Conference on Artificial Intelligence IJCAI 2018 (2018)
21. Lin, J., Xu, S., Ma, S., Qi, S.: Global encoding for abstractive summarization (2018)
22. Maskey, S., Hirschberg, J.: Comparing lexical, acoustic/prosodic, structural and discourse features for speech summarization. In: INTERSPEECH, pp. 621–624 (2005)
23. Mesnil, G., Dauphin, Y., Yao, K., Bengio, Y., Zweig, G.: Using recurrent neural networks for slot filling in spoken language understanding. IEEE/ACM Trans. Audio Speech Lang. Process. **23**(3), 530–539 (2015)
24. Miao, Y., Blunsom, P.: Language as a latent variable: discrete generative models for sentence compression (2016)
25. Nallapati, R., Xiang, B., Zhou, B.: Sequence-to-sequence RNNs for text summa-rization (2016)
26. See, A., Liu, P.J., Manning, C.D.: Get to the point: summarization with pointer-generator networks (2017)
27. Sutskever, I., Vinyals, O., Le, Q.V.: Sequence to sequence learning with neural networks (2014)
28. Vaswani, A., et al.: Attention is all you need (2017)
29. Vinyals, O., Fortunato, M., Jaitly, N.: Pointer networks. In: International Confer-ence on Neural Information Processing Systems (2015)
30. Yin, W., Pei, Y.: Optimizing sentence modeling and selection for document sum-marization. In: International Conference on Artificial Intelligence (2015)

Chinese Text Classification
via Bidirectional Lattice LSTM

Ning Pang[1], Weidong Xiao[1,2], and Xiang Zhao[1,2(✉)]

[1] Science and Technology on Information Systems Engineering Laboratory,
National University of Defense Technology, Changsha, China
{pangning14,wdxiao,xiangzhao}@nudt.edu.cn
[2] Collaborative Innovation Center of Geospatial Technology, Wuhan, China

Abstract. In this paper, we investigate a bidirectional lattice LSTM (Bi-Lattice) network for Chinese text classification. The new network is different from the standard LSTM in adding shortcut paths which link the start and end characters of words, to control the information flow. Character-level features can flow into word-level by an extra gate, and word-level features are integrated into character-level via a weighted manner by another gate. Previous models take as input embeddings pre-trained by Skip-Gram model, we utilize word sememes in HowNet to further improve the word representation learning in our proposal. Our experiments show that Bi-Lattice gives better results compared with the state-of-the-art methods on two Chinese text classification benchmarks. Detailed analyses are conducted to show the success of our model in feature fusion, and the contribution of each component.

Keywords: Chinese text classification · Lattice LSTM · Word representation learning

1 Introduction

Text classification is a crucial task in natural language processing (NLP) to construct conceptual networks and knowledge trees [1, 15]. It is defined as tagging a given text with a pre-defined tag set. The recent popularity of neural networks has led to the interest of neural text classification models [16, 19, 25].

For languages without nature delimiters, e.g., Chinese, mainstream methods based on English texts cannot be adopted directly. Existing work on Chinese text classification is mainly word-based [20, 28, 29], which needs to first use NLP tools to perform word segmentation. However, this type of models suffers from segmentation errors caused by word segmentation tools. For example, in Fig. 1, the correct word sequence expresses a 'Location' type, but the wrong one tends

Supported by NSFC under grants Nos. 61872446, 61902417, 61701454, and 71971212, NSF of Hunan Province under grant No. 2019JJ20024, Postgraduate Scientific Research Innovation Project of Hunan Province (CX20190036), and basic foundation with no. 2019JCJQJJ231.

© Springer Nature Switzerland AG 2020
G. Li et al. (Eds.): KSEM 2020, LNAI 12275, pp. 250–262, 2020.
https://doi.org/10.1007/978-3-030-55393-7_23

to describe the type of 'Study'. To avoid this error propagation, some work [5] explores to perform character-level text classification. For character-based methods, the input sentence is regarded as a character sequence. One drawback of character-based models is that the powerful representation of words is not fully exploited.

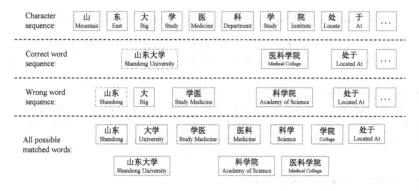

Fig. 1. Segmented word sequence.

Recent work [19,29] tries to integrate these two-granularity features via a concatenation or addition manner, but these strategies are shallow, since they fail to exploit the inner dependency relationship between words and characters. In fact, word sequence and character sequence are interdependent. For example, word '医科学院 (Medical College)' starts at character '医 (Medicine)', and thus they share the same historical context information. Besides, since word '医科学院 (Medical College)' ends at character '院 (Institute)', word-level and character-level information should be integrated at this position.

To this end, we are motivated to solve two challenges:

- **Obtaining word sequence**: Different from previous word-based models [20,28,29] applying word segmentation tools to cut sentences, we extract all matched words in the sentences with the help of an external lexicon as shown in Fig. 1. In this way, all word-level information is maintained without introducing segmentation error. For each matched word, a shortcut tunnel is added to link its start and end characters in the character sequence.
- **Obtaining fused feature**: To exploit the dependency relationship between words and characters, we apply a lattice Long Short Term Memory (Lattice LSTM) network to encode the raw texts. In this network, a standard LSTM is used to extract character-level information, an additional gate controls character-level features to flow into word-level, and another gate dynamically integrates word-level and character-level features, as shown in Fig. 2.

Based on above solutions, we propose the bidirectional lattice LSTM (Bi-Lattice) for Chinese text classification. In most neural models, words or characters are first mapped into a low-dimensional vector space by Skip-Gram model [13]. Recent work [14] finds that word semeses can further improve the word and character representation learning. Therefore, we use the word and character embeddings pre-trained by SAT model [14] with the help of HowNet [3].

2   Related Work

2.1   Traditional and Embedding-Based Text Classification

Traditional text classification methods apply feature engineering to represent texts, and then, SVM [4] is harnessed as the classifier. Among the feature designs, bag-of-words and TFIDF features [2] are the most commonly used. Different from feature engineering, inspired by word embedding [13], some work [7,10] first maps sentences into vectors, then feeds these sentence embeddings into a classifier. More recent literature jointly trains word, sentence, and sentence label embeddings to improve the quality of sentence embeddings [21].

2.2   Neural Text Classification

Recent development in deep learning leads to the trend of neural text classification. Kim et al. [8] first use CNN as the encoder for sentence classification. Afterwards, character-level CNN [26] and deep pyramid word-level CNN [6] are designed to mine different levels of information within the text.

Another group of studies [12,18] use LSTM, an extension form of RNN, to learn text representation. Among these methods, Luo et al. [12] harness a standard LSTM-based neural model for classifying clinical notes. Considering some trivial information in texts for the target, attention mechanism [24] is introduced to generate text representations in a more flexible way.

Other work adopts more advanced neural networks, such as GCN [25] and CapsuleNet [17], to encode texts for classification. Besides, some language models are also fine-tuned to classify texts, including XLNet [23], and ALBERT [9]. However, these models based on English texts cannot be adapted to Chinese text classification directly.

2.3   Chinese Text Classification

Existing Chinese text classification models are almost word-based or character-based [5,11,20]. Due to the error caused by the word segmentation tools and the limited knowledge representation of character, Zhou et al. [29] use two separate bidirectional LSTMs to extract word-level and character-level features, then integrate them via a concatenation manner. Observing that Chinese is a kind of hieroglyphics, on the basis of these two levels of information, radicals of words and characters are explored to form four-granularity features via an attentional manner for Chinese text classification [19].

To mitigate the segmentation error problem, Zhang et al. [27] investigate a lattice LSTM for Chinese named entity recognition. Lattice LSTM explicitly leverages word and character information by multiple gates, which controls to select most relevant words and characters from a sentence. This sequence labeling model is also adopted to the task of Chinese word segmentation [22], which achieves advanced results on several benchmarks. Our work is connected to these works, the major difference is that we formulate the lattice LSTM model into a new task.

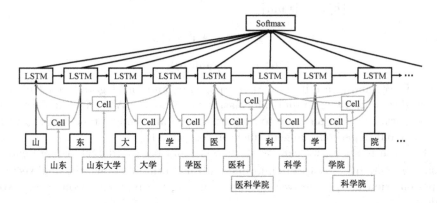

Fig. 2. The model architecture. Only forward lattice LSTM is illustrated.

3   Methodology

3.1   Input Representation

Before fed into neural networks, discrete characters and words should be mapped into low-dimensional vectors [13,14], denoted as:

$$\mathbf{x}_i = \mathbf{E}^c(c_i), \tag{1}$$

$$\mathbf{w}_i = \mathbf{E}^w(w_i). \tag{2}$$

$\mathbf{E}^c \in \mathbb{R}^{|V^c| \times d_c}$ is the character embedding table, and $|V^c|$ is the vocabulary size for characters. $\mathbf{E}^w \in \mathbb{R}^{|V^w| \times d_w}$ is the word embedding table with a vocabulary size of $|V^w|$. d_c and d_w denote the dimensionalities of characters and words.

3.2   Baselines

We take the character-based (Char-Baseline) and word-based (Word-Baseline) bidirectional LSTM framework as our baselines. For an input sentence with m characters $s = \{c_1, c_2, ..., c_m\}$ or n words $s = \{w_1, w_2, ..., w_n\}$, where c_i (resp. w_i) denotes ith character (resp. ith word) in the sentence, a bidirectional LSTM layer is harnessed as the encoder. By Eq. 1 and 2, a sentence can be transformed into $\{\mathbf{x}_1, \mathbf{x}_2, ..., \mathbf{x}_m\}$ for Char-Baseline, or $\{\mathbf{w}_1, \mathbf{w}_2, ..., \mathbf{w}_n\}$ for Word-Baseline.

Since both Char-Baseline and Word-Baseline use the same encoder, we take Char-Baseline for example to illustrate the mechanism of bidirectional LSTM. The forward LSTM representation of character c_i can be calculated by:

$$
\begin{bmatrix} \mathbf{i}_i^c \\ \mathbf{o}_i^c \\ \mathbf{f}_i^c \\ \widetilde{\mathbf{c}}_i^c \end{bmatrix} = \begin{bmatrix} \sigma \\ \sigma \\ \sigma \\ \tanh \end{bmatrix} ([\mathbf{W}^{c\top}, \mathbf{U}^{c\top}] \begin{bmatrix} \mathbf{x}_i \\ \overrightarrow{\mathbf{h}}_{i-1}^c \end{bmatrix} + \mathbf{b}^c),
\tag{3}
$$

$$
\mathbf{c}_i^c = \mathbf{f}_i^c \odot \mathbf{c}_{i-1}^c + \mathbf{i}_i^c \odot \widetilde{\mathbf{c}}_i^c,
\tag{4}
$$

$$
\overrightarrow{\mathbf{h}}_i^c = \mathbf{o}_i^c \odot \tanh(\mathbf{c}_i^c),
\tag{5}
$$

where $\mathbf{i}^c$, $\mathbf{o}^c$, and $\mathbf{f}^c$ are input gate, output gate, and forget gate to control the reserved, forgotten, and input information respectively. $\mathbf{W}^{c\top}$, $\mathbf{U}^{c\top}$ and $\mathbf{b}^c$ are model parameters for all gates to be learned. $\sigma(\cdot)$ and $\tanh(\cdot)$ represent sigmoid function and hyperbolic tangent function respectively.

For each input $\mathbf{c}_i$, we can get its forward and backward LSTM representations, $\overrightarrow{\mathbf{h}}_i^c$ and $\overleftarrow{\mathbf{h}}_i^c$. To capture information from both directions, the final vector representation of c_i is derived by concatenating the representations from two sides:

$$
\mathbf{h}_i^c = [\overrightarrow{\mathbf{h}}_i^c : \overleftarrow{\mathbf{h}}_i^c].
\tag{6}
$$

The character sequence $\{c_1, c_2, ..., c_m\}$ can be encoded as $\{\mathbf{h}_1^c, \mathbf{h}_2^c, ..., \mathbf{h}_m^c\}$. Similarly, for Word-Baseline, the encoded sequence is represented as $\{\mathbf{h}_1^w, \mathbf{h}_2^w, ..., \mathbf{h}_n^w\}$.

3.3 Our Lattice LSTM Approach

Our lattice LSTM encoder is an extension of the standard LSTM, aiming at fully capturing character-level and word-level features. Different from the baselines, the lattice LSTM model takes as input the character sequence $\{c_1, c_2, ..., c_m\}$ and all subsequences $\{w_{b,e} = \{c_b, ..., c_e\} \mid w_{b,e} \in \mathbb{D}\}$ which are matched words in a lexicon $\mathbb{D}$. For character-level input, each character c_i can be transformed into a vector $\mathbf{x}_i$ by Eq. 1. And each input word $w_{b,e}$ can be denoted as $\mathbf{w}_{b,e}$ by Eq. 2.

In this framework, shortcut path is added for each matched word from its beginning character to its end one, as shown in Fig. 2. For character sequence, hidden vectors can be obtained using Eq. 3, 4, and 5. Besides, an extra gate controls the character-level information to flow into word-level, which take as input the hidden vector $\overrightarrow{\mathbf{h}}_{b-1}^c$ of character c_{b-1} and word embedding $\mathbf{w}_{b,e}$. The calculation is denoted as:

$$
\begin{bmatrix} \mathbf{i}_{b,e}^w \\ \mathbf{f}_{b,e}^w \\ \widetilde{\mathbf{c}}_{b,e}^w \end{bmatrix} = \begin{bmatrix} \sigma \\ \sigma \\ \tanh \end{bmatrix} ([\mathbf{W}^{l\top}, \mathbf{U}^{l\top}] \begin{bmatrix} \mathbf{w}_{b,e} \\ \overrightarrow{\mathbf{h}}_{b-1}^c \end{bmatrix} + \mathbf{b}^l),
\tag{7}
$$

$$
\mathbf{c}_{b,e}^w = \mathbf{f}_{b,e}^w \odot \mathbf{c}^{c_{b-1}} + \mathbf{i}_{b,e}^w \odot \widetilde{\mathbf{c}}_{b,e}^w,
\tag{8}
$$

where $\mathbf{c}_{b,e}^w$ is the cell state of word $w_{b,e}$. $\mathbf{W}^{lT}$, $\mathbf{U}^{lT}$, and $\mathbf{b}^l$ are parameters used in this LSTM cell.

To incorporate the character-level and word-level information, the cell state of character c_e will consider all matched words ending at c_e. Another gate is utilized to control the contribution of each word:

$$\mathbf{e}_{b,e}^w = \sigma([\mathbf{W}^\top, \mathbf{U}^\top] \begin{bmatrix} \mathbf{x}_e \\ \mathbf{c}_{b,e}^w \end{bmatrix} + \mathbf{b}). \tag{9}$$

Therefore, we can calculate the cell state of c_e by:

$$\mathbf{c}_e^c = \sum_{b \in \{b' | w_{b',e} \in \mathbb{D}\}} (\alpha_{b,e}^w \odot \mathbf{c}_{b,e}^w) + \alpha_e^c \odot \widetilde{\mathbf{c}}_e^c, \tag{10}$$

where $\alpha_{b,e}^w$ and α_e^c are defined as follows:

$$\begin{aligned} \alpha_{b,e}^w &= \frac{\exp(\mathbf{e}_{b,e}^w)}{\exp(\mathbf{i}_e^c) + \sum_{b' \in \{b'' | w_{b'',e} \in \mathbb{D}\}} \exp(\mathbf{e}_{b',e}^w)}, \\ \alpha_e^c &= \frac{\exp(\mathbf{i}_e^c)}{\exp(\mathbf{i}_e^c) + \sum_{b' \in \{b'' | w_{b'',e} \in \mathbb{D}\}} \exp(\mathbf{e}_{b',e}^w)}. \end{aligned} \tag{11}$$

$\mathbf{i}_i^c$ can be obtained by Eq. 3, which is the input gate for character c_e.

Finally, the i-th forward hidden vector $\overrightarrow{\mathbf{h}}_i$, which contains both character-level and word-level information, can be calculated by Eq. 5. The backward hidden vector $\overleftarrow{\mathbf{h}}_i$ could be learned with the similar mechanism. By Eq. 6, we acquire the bidirectional hidden vector sequence $\{\mathbf{h}_1, \mathbf{h}_2, ..., \mathbf{h}_m\}$.

3.4 Classifier and Optimization

After the hidden vector sequence is obtained, we feed it into a character-level attention module to produce the sentence-level feature vector, denoted as $\mathbf{g} \in \mathbb{R}^{d_h}$, where d_h is the dimension of hidden vectors. The final representation $\mathbf{g}$ is the weighted sum of the hidden vector sequence:

$$\begin{aligned} \mathbf{g} &= \sum_{i=1}^m u_i \cdot \mathbf{h}_i, \\ u_i &= \frac{\mathbf{v}^\top \mathbf{h}_i}{\sum_{k=1}^m \mathbf{v}^\top \mathbf{h}_k}, \end{aligned} \tag{12}$$

where $\mathbf{v} \in \mathbb{R}^{d_h}$ is a trainable parameter. We select this attention form since it achieves the best performance in several alternatives.

Then, $\mathbf{g}$ is transmitted into a fully-connected layer to compute the confidence score of each type:

$$\mathbf{o} = \mathbf{W}_o \mathbf{g} + \mathbf{b}_o, \tag{13}$$

where $\mathbf{W}_o \in \mathbb{R}^{K \times d_h}$ is a learnable transformation matrix, $\mathbf{b}_o \in \mathbb{R}^K$ is a bias vector, and K is the number of text types. The probability of text s belonging to type y is computed as:

$$p(y \mid s) = \frac{\exp(o_y)}{\sum_{i=1}^{K} \exp(o_i)}. \tag{14}$$

Finally, we take the cross-entropy loss as our objective function. Given the training set $T = \{(s^i, y^i)\}$, the loss is defined as:

$$J(\Theta) = -\sum_{i=1}^{|T|} \log p(y^i \mid s^i; \Theta), \tag{15}$$

where Θ denotes all parameters used in the model. In the implementation, we harness stochastic gradient descent (SGD) optimizer to minimize the loss function by optimizing Θ iteratively until convergence.

4 Experiments

4.1 Experiment Settings

Evaluation Metrics and Datasets. To give an overall evaluation of models, we apply accuracy (Acc), macro-precision (Pre), macro-recall (Rec), and macro-F1 (F1) as our evaluation metrics in the experiments. To verify the effectiveness of our proposal, experiments are conducted on two Chinese datasets, which have gold classification labels. The first dataset[1] is collected from Chinese news titles with 32 classes, where $47,952$ titles are labeled for training, and $15,986$ for testing [29] (i.e., # Dataset 1). The second dataset is a benchmark on Chinese CLUE[2] (i.e., # Dataset 2), which contains $53,360$ instances for training, $10,000$ instances for validation, and $10,000$ instances for testing.

Implementation Details. We tuned the hyper-parameters of our model by grid searching using three-validation on the training set, and the optimal hyper-parameters are shown in bold. We selected the learning rate for SGD $\lambda \in \{0.01, 0.012, 0.014, \mathbf{0.016}, 0.018, 0.02\}$, and the LSTM hidden size $d_h \in \{100, 150, \mathbf{200}, 250\}$. Since other hyper-parameters have little effect on the overall results, we set them empirically. The size of character embedding and word embedding are set to 100 and 200 respectively. The dropout rate is set to 0.5. For word-based competitors in our experiments, we used *jieba* [3] segmentation tool to cut Chinese texts into word sequences.

---

[1] http://pan.baidu.com/s/1mgBTFOO.
[2] https://github.com/CLUEbenchmark/CLUE.
[3] https://github.com/fxsjy/jieba.

Table 1. Effect of lattice encoder.

Methods	# Dataset 1				# Dataset 2			
	Pre	Rec	F1	Acc	Pre	Rec	F1	Acc
Char-Baseline	71.64	71.43	71.38	71.37	53.14	51.45	51.92	55.50
+bichar	72.71	72.39	72.42	72.38	53.83	52.31	52.54	56.13
+softword	73.86	73.81	73.88	73.75	54.76	53.29	53.44	57.68
Word-Baseline	76.30	76.19	75.99	76.13	54.48	52.70	53.17	54.68
+char CNN	76.72	76.58	76.58	76.55	54.88	52.73	53.48	55.29
+char LSTM	76.77	76.83	76.66	76.79	55.08	53.40	53.97	55.51
Bi-Lattice	**82.31**	**82.21**	**82.13**	**82.17**	**62.64**	**61.00**	**61.65**	**63.62**

4.2 Effect of Lattice LSTM Encoder

Experiments in this section aim to show the success of our model in fusing the character-level and word-level features. Two types of strong baselines are implemented by replacing the bidirectional lattice LSTM with the standard bidirectional LSTM to compete with our proposed Bi-Lattice. For character-based baselines, Char-Baseline is mentioned in Sect. 3.2. To enhance the character-level representation, we add extra word-level features, i.e., bichar and softword. These two variants are implemented by concatenating unigram character embedding with bigram character embedding and softword (word where the current character is located) embedding respectively. Among word-based baselines, Word-Baseline takes as input the word sequence of a sentence and employs the standard bidirectional LSTM to encode the sequence as introduced in Sect. 3.2. Besides, the character-level feature is learned by CNN or bidirectional LSTM, and concatenated with the corresponding word embedding as the additional semantic feature. As thus, both types of strong baselines use both character and word information to classify texts.

Table 1 shows the experimental results of our proposal and all baselines. From it, we can see that: (1) For character-based baselines, word-level embedding, including bichar and softword, brings improvements to all metrics on two datasets. Word-based baselines also gain when incorporated with character-level features extracted by CNN or LSTM. (2) Although integrating information at different levels is useful for improving the performance of two baselines, the effect of concatenation method is relatively limited, especially for Word-Baseline. (3) Our proposed Bi-Lattice outperforms all baselines over 6% on # Dataset 1 and # Dataset 2, indicating that our Bi-Lattice is a more effective strategy to fuse character-level and word-level features by dynamically controlling the semantic feature flow.

4.3   Overall Results

In this part, we compare our proposal with multiple state-of-the-art methods, which are listed as follows:

- **TFIDF+SVM**: TFIDF+SVM is a bag-of-word model with term frequency inverse document frequency weighting. SVM is employed as the classifier.
- **FastText**: FastText [7] is a strong text classification baseline, which averages n-grams embeddings as sentence embeddings, then feeds sentence embeddings into a linear classifier.
- **LEAM**: LEAM [21] embeds words and labels in a unified space for text classification. In our comparison, this model is tested at word-level and character-level.
- **GCN**: GCN [25] builds the heterogeneous graph containing sentence nodes and word (or character) nodes, and utilizes the GCN to encode nodes in the graph. We tested word-based GCN and character-based GCN.
- **CharCNN**: CharCNN [26] explores empirically the use of character-level convolutional neural networks on the task of text classification.
- **TextCNN**: TextCNN [8] applies a CNN layer on the top of pre-trained word vectors for sentence-level classification, which is a word-level competitor in our experiments.
- **DPCNN**: DPCNN [6] is a deep pyramid CNN structure, which is built at word-level, due to the powerful representation of words.
- **BLSTM-C**: BLSTM-C [11] is a word-level neural model for Chinese text classification. It uses a hybrid neural architecture, composed of a bidirectional LSTM layer and a CNN layer, for the feature extraction.
- **HAN**: In HAN [24], a hierarchical attention mechanism is applied in a GRU-based sequence encoder for word-level document classification.
- **RAFG**[4]: RAFG [19] integrates four-granularity features, including characters, words, character radicals, and word radicals, for Chinese text classification.
- **Albert-tiny**: Albert-tiny [9] is the tiny version of a pre-trained lite BERT language model, and is fine-tuned for Chinese text classification.

Among these competitors, feature-based methods include TFIDF+SVM, and FastText. Embedding-based models include LEAM, and GCN. Neural models include CharCNN, TextCNN, DPCNN, BLSTM-C, HAN, and RAFG. And Albert-tiny is a fine-tuned language model. Table 2 shows the results of each model on two datasets. Form the comparison, we can see that: (1) Some state-of-the-art methods based on English texts show unpromising performance in Chinese text classification. (2) The pre-trained language model, Albert-tiny is the strongest competitor, our Bi-Lattice outperforms it about 1% on all metrics (Table 3).

---

[4] Since the data of radical feature is not available, we copy the test results from the paper directly.

Table 2. Overall results of different models.

Methods	# Dataset 1				# Dataset 2			
	Pre	Rec	F1	Acc	Pre	Rec	F1	Acc
TFIDF+SVM	63.81	63.78	63.75	63.29	46.79	43.74	44.73	46.86
FastText	78.70	78.29	78.34	78.27	56.96	53.90	54.81	56.99
LEAM (char)	75.49	75.64	75.83	75.33	58.44	56.85	57.62	59.67
LEAM (word)	76.29	76.20	76.21	76.51	62.45	60.87	61.71	63.87
GCN (char)	74.77	74.56	74.55	74.54	55.95	53.81	54.57	58.28
GCN (word)	77.38	77.54	77.39	77.56	59.25	57.74	58.27	63.12
CharCNN	74.28	74.35	74.21	74.30	58.11	57.09	57.37	61.25
TextCNN	69.21	68.26	68.38	68.20	57.13	54.67	55.18	57.76
DPCNN	74.90	72.67	73.07	72.61	52.94	52.22	52.82	55.84
TextRNN	74.48	74.57	74.33	74.57	54.60	52.38	52.96	56.37
BLSTM-C	77.20	76.72	76.50	76.66	55.74	53.57	53.78	57.08
HAN	77.28	77.74	77.67	77.86	58.29	55.54	56.58	58.26
Albert-tiny	81.28	81.24	81.23	81.22	61.49	59.46	60.12	62.79
RAFG*	81.81	81.87	81.81	–	–	–	–	–
Bi-Lattice	**82.31**	**82.21**	**82.13**	**82.17**	**62.64**	**61.00**	**61.65**	**63.62**

4.4 Ablation Study

To show the effect of each component, we conducted an ablation study by removing one component at a time to understand its impacts on the performance. The first variant is constructed by removing backward lattice LSTM (i.e., -Backward). The second one considers character embedding and word embedding pre-trained by Skip-Gram model [13] without word sememes (i.e., -Sememes). The last tested variant is established by replacing the character-level attention module with mean-pooling operation (i.e., -Attention).

Table 3. Ablation study.

Methods	# Dataset 1				# Dataset 2			
	Pre	Rec	F1	Acc	Pre	Rec	F1	Acc
Bi-Lattice	**82.31**	**82.21**	**82.13**	**82.17**	62.64	**61.00**	**61.65**	**63.62**
-Backward	82.14	81.99	81.93	81.96	**63.25**	60.48	61.03	63.12
-Semeses	81.68	81.59	81.55	81.58	62.12	58.73	59.71	61.68
-Attention	81.28	81.29	81.15	81.25	61.91	59.31	60.25	62.57

From the results, we can observe that: (1) After the removal of backward lattice LSTM, all metrics on two datasets suffer a reduction, demonstrating that contexts in both directions contribute to the understanding of texts. (2) When word semeses are not considered, the performance decreases more obviously than the other two variants, demonstrating that the quality of word embedding influences the downstream task. (3) All metrics decrease when character-level attention module is removed, because it is unreasonable to regard all characters equally. Some common characters, which appear in all types of texts, should be deemphasized. On the other hand, the character-level attention module provides a dynamic way to fuse all hidden vectors of characters in a weighted sum manner, which is encouraged to focus more on informative characters.

5   Conclusion

In this study, we investigate the deep integration of character-level and word-level features by the bidirectional lattice LSTM for Chinese text classification. Different from previous research on this task, we use the word semeses retrieved from HowNet to improve the quality of pre-trained embeddings of characters and words. Results on two datasets show that our proposed Bi-Lattice achieves better performance than various advanced text classification models. In the comparison with character-based and word-based baselines, we can see the success of our proposal in fusing the two-granularity features. The ablation study proves that each component in Bi-Lattice contributes to the overall performance. In the future work, we will try to solve the sense ambiguity of ploysemy.

References

1. Aggarwal, C.C., Zhai, C.: A survey of text classification algorithms. In: Aggarwal, C., Zhai, C. (eds.) Mining text data, pp. 163–222. Springer, Boston (2012). https://doi.org/10.1007/978-1-4614-3223-4_6
2. Blei, D.M., Ng, A.Y., Jordan, M.I.: Latent dirichlet allocation. J. Mach. Learn. Res. **3**(Jan), 993–1022 (2003)
3. Dong, Z., Dong, Q.: Hownet-a hybrid language and knowledge resource. In: International Conference on Natural Language Processing and Knowledge Engineering, 2003. Proceedings, pp. 820–824. IEEE (2003)
4. Drucker, H., Wu, D., Vapnik, V.N.: Support vector machines for spam categorization. IEEE Trans. Neural Netw. **10**(5), 1048–1054 (1999)
5. Huang, W., Wang, J.: Character-level convolutional network for text classification applied to chinese corpus. CoRR (2016)
6. Johnson, R., Zhang, T.: Deep pyramid convolutional neural networks for text categorization. In: Proceedings of the 55th Annual Meeting of the Association for Computational Linguistics (ACL 2017) Volume 1: Long Papers, Vancouver, Canada, 30 July–4 August (2017)
7. Joulin, A., Grave, E., Bojanowski, P., Mikolov, T.: Bag of tricks for efficient text classification. arXiv preprint arXiv:1607.01759 (2016)
8. Kim, Y.: Convolutional neural networks for sentence classification. arXiv preprint arXiv:1408.5882 (2014)

9. Lan, Z., Chen, M., Goodman, S., Gimpel, K., Sharma, P., Soricut, R.: ALBERT: a lite BERT for self-supervised learning of language representations. CoRR (2019)
10. Le, Q., Mikolov, T.: Distributed representations of sentences and documents. In: International Conference on Machine Learning, pp. 1188–1196 (2014)
11. Li, Y., Wang, X., Xu, P.: Chinese text classification model based on deep learning. Future Internet **10**(11), 113 (2018)
12. Luo, Y.: Recurrent neural networks for classifying relations in clinical notes. J. Biomed. Inform. **72**, 85–95 (2017)
13. Mikolov, T., Sutskever, I., Chen, K., Corrado, G.S., Dean, J.: Distributed representations of words and phrases and their compositionality. In: Advances in Neural Information Processing Systems, pp. 3111–3119 (2013)
14. Niu, Y., Xie, R., Liu, Z., Sun, M.: Improved word representation learning with sememes. In: Proceedings of the 55th Annual Meeting of the Association for Computational Linguistics (Volume 1: Long Papers), pp. 2049–2058 (2017)
15. Pang, B., Lee, L., et al.: Opinion mining and sentiment analysis. Found. Trends® Inf. Retrieval **2**(1–2), 1–135 (2008)
16. Ren, F., Deng, J.: Background knowledge based multi-stream neural network for text classification. Appl. Sci. **8**(12), 2472 (2018)
17. Sabour, S., Frosst, N., Hinton, G.E.: Dynamic routing between capsules. In: Advances in Neural Information Processing Systems, pp. 3856–3866 (2017)
18. Tai, K.S., Socher, R., Manning, C.D.: Improved semantic representations from tree-structured long short-term memory networks. In: Proceedings of the 53rd Annual Meeting of the Association for Computational Linguistics (ACL 2015) Volume 1: Long Papers, 26–31 July 2015, Beijing, China (2015)
19. Tao, H., Tong, S., Zhao, H., Xu, T., Jin, B., Liu, Q.: A radical-aware attention-based model for chinese text classification. In: The Thirty-Third AAAI Conference on Artificial Intelligence, (AAAI 2019), USA, 27 January–1 February 2019
20. Tian, J., Zhu, D., Long, H.: Chinese short text multi-classification based on word and part-of-speech tagging embedding. In: Proceedings of the 2018 International Conference on Algorithms, Computing and Artificial Intelligence, pp. 1–6 (2018)
21. Wang, G., et al.: Joint embedding of words and labels for text classification. arXiv preprint arXiv:1805.04174 (2018)
22. Yang, J., Zhang, Y., Liang, S.: Subword encoding in lattice lstm for chinese word segmentation. arXiv preprint arXiv:1810.12594 (2018)
23. Yang, Z., Dai, Z., Yang, Y., Carbonell, J.G., Salakhutdinov, R., Le, Q.V.: XLNet: Generalized autoregressive pretraining for language understanding. In: Advances in Neural Information Processing Systems 32: Annual Conference on Neural Information Processing Systems 2019, NeurIPS 2019, 8–14 December 2019, Vancouver, BC, Canada (2019)
24. Yang, Z., Yang, D., Dyer, C., He, X., Smola, A., Hovy, E.: Hierarchical attention networks for document classification. In: Proceedings of the 2016 Conference of the North American Chapter of the Association for Computational Linguistics: Human Language Technologies, pp. 1480–1489 (2016)
25. Yao, L., Mao, C., Luo, Y.: Graph convolutional networks for text classification. In: Proceedings of the AAAI Conference on Artificial Intelligence, vol. 33, pp. 7370–7377 (2019)
26. Zhang, X., Zhao, J., LeCun, Y.: Character-level convolutional networks for text classification. In: Advances in Neural Information Processing Systems, pp. 649–657 (2015)

27. Zhang, Y., Yang, J.: Chinese NER using lattice LSTM. In: Proceedings of the 56th Annual Meeting of the Association for Computational Linguistics (ACL 2018) Volume 1: Long Papers, Melbourne, Australia, 15–20 July 2018, pp. 1554–1564 (2018)
28. Zhou, J., Lu, Y., Dai, H.N., Wang, H., Xiao, H.: Sentiment analysis of chinese microblog based on stacked bidirectional LSTM. IEEE Access **7**, 38856–38866 (2019)
29. Zhou, Y., Xu, B., Xu, J., Yang, L., Li, C.: Compositional recurrent neural networks for chinese short text classification. In: 2016 IEEE/WIC/ACM International Conference on Web Intelligence (WI), pp. 137–144. IEEE (2016)

MG-BERT: A Multi-glosses BERT Model for Word Sense Disambiguation

Ping Guo[1,2], Yue Hu[1,2(✉)], and Yunpeng Li[1,2]

[1] Institute of Information Engineering, Chinese Academy of Sciences, Beijing, China
{guoping,huyue,liyunpeng}@iie.ac.cn
[2] School of Cyber Security, University of Chinese Academy of Sciences, Beijing, China

Abstract. Word Sense Disambiguation (WSD) is a core task in NLP fields and has many potential applications. Traditional supervised methods still have obstacles, such as the problem of variable size of label candidates and the lack of annotated corpora. Although attempts are made to integrate gloss information to the model, no existing models have paid attention to the divergences among glosses. In this paper, we propose a Multi-Glosses BERT (MG-BERT) model with two main advantages for WSD task. Our model jointly encodes the context and multi-glosses of the target word. We show that our Context with Multi-Glosses mechanism can find out and emphasize the divergences among glosses and generate nearly orthogonal gloss embeddings, which makes it more accuracy to match the context with the correct gloss. We design three classification algorithms, Gloss Matrix Classifier (GMC), General Gloss Matrix Classifier (GGMC) and Space Transforming Classifier (STC), all of which can disambiguate words with full-coverage of WordNet. In GMC and GGMC, we utilize gloss embeddings as weight matrix. For STC, we transform different label space to a same label space. Experiment shows that our MG-BERT model achieves new state-of-the-art performance on all WSD benchmarks.

Keywords: MG-BERT · Context with Multi-Glosses · Space Transforming Classifier

1 Introduction

Word sense disambiguation (WSD) is to assign a word in context with a correct sense, which is a core task of Natural Language Processing (NLP)[12]. Table 1 shows an example of WSD task: given a context with a target word, our goal is to find out the correct sense of the target word in context. Although some breakthroughs have been made, WSD task remains a long-standing challenge in the field. In general, WSD methods can be divided into two distinct categories: knowledge-based methods and supervised methods. Knowledge-based methods aim to extract lexical knowledge from resources like WordNet [9] to infer the

© Springer Nature Switzerland AG 2020
G. Li et al. (Eds.): KSEM 2020, LNAI 12275, pp. 263–275, 2020.
https://doi.org/10.1007/978-3-030-55393-7_24

senses of the word in context. Supervised methods, on the other hands, rely on annotated semantic corpora for training. Even though supervised methods have surpassed knowledge-based methods, supervised methods still suffer from two major challenges.

Table 1. An example of WSD task (target word and label is highlighted with Bold font).

Context:
Visits and interviews with key personnel were also **undertaken**
Candidates:
1. undertake%2:41:01::
Gloss: accecpt as a challenge
2. undertake%2:41:00::
Gloss: accecpt as a charge
3. undertake%2:36:00::
Gloss: enter upon an activity or enterprise
4. undertake%2:32:01::
Gloss: enter into a contractual arrangement
5. undertake%2:32:00::
Gloss: promise to do or accomplish

First, WSD has the problem of variable size of label candidates, because different words have totally different label candidates. Traditional supervised methods usually trained a dedicated classifier for each word, such as IMS [18], which makes it impossible to scale up to all-words WSD, for the words not in training corpus have no suitable classifier. With the breakthroughs of the pre-trained contextual word representation, recent methods focus on creating sense-level representations and use a simple Nearest Neighbors (k-NN) method for classification, an example of this method is LMMS [6]. One disadvantage of these methods is that the algorithm must compute the distance and sort all the candidate senses at each prediction, which means we need to have access to all training data any time we want to disambiguate a word.

The other bottle-neck of supervised WSD is the lack of the annotated corpora. And the production of annotated data in WSD is even more complicated than other NLP tasks, for semantic annotation needs expert annotators. And so far, the most common manually annotated corpus remains to be SemCor [10]. To deal with the data hungry problem, different approaches use various kinds of external knowledge.

One popular method is to utilize gloss (sense definition) to represent sense, match context with each gloss, and take the sense corresponding to the best-matching gloss as the prediction results. GAS [8] and HCAN [7] incorporates gloss information (a sentence that extensionally explains a word meaning). LMMS takes advantage of the semantic relations in WordNet (synset, hypernym and lexname) to extend annotated corpus. GlossBERT [3] constructs context-gloss pairs and fine-tune the pre-trained BERT model. Although these methods integrate gloss information, they may not make full use of gloss knowledge. All of the models above encode gloss information separately with no consideration of other glosses. And as Table 1 shows, two quite different senses (the first two in the table) have nearly identical glosses. How to use different gloss representations to denote different senses remains to be a tricky problem.

Recently, the pre-trained language models, such as BERT [2] and XLNET [17], encode several linguistic information and produce contextual word representations. They have achieved excellent results in several NLP fields. In this paper, we propose Multi-Glosses BERT (MG-BERT) model that jointly encodes context and multi-glosses into embeddings. We first preprocess raw data into Context with Multi-Glosses format and then feed them into a BERT base model. Instead of training a dedicated classifier for each individual word, we propose a new method that we treat the gloss embeddings as the classifier weight matrix. Thus, as long as we have a word's glosses, we have the dedicated classifier for the word. What's more, we also design a Space Transforming Classifier that can convert different label space to a same one. Our contributions are summarized below:

1. We propose Multi-Glosses BERT (MG-BERT) model which pays attention to the divergences among glosses. Our model can generate nearly orthogonal gloss embeddings which are more suitable for WSD tasks.
2. We design three classification algorithms, all of which can solve WSD task with full-coverage of WordNet. We adopt gloss embeddings as weight matrix for the first two algorithm, and design a label space transforming algorithm for the last one.
3. We conduct several experiments and experimental results show that our MG-BERT model achieves new state-of-the-art performance on all English all-words WSD benchmark datasets.

2   Related Work

Although some restrictions are on the supervised methods, the best performances in WSD still belongs to the supervised ones. Many of the approaches have learnt from the knowledge-based methods to use the lexical resource to enhance their sentence representations. Some use glosses (sense definitions) and others even take the graph structure, hypernym and lexname etc., to try to solve WSD on full-coverage of WordNet without MFS (Most Frequent Sense) fallbacks.

GAS and HCAN combine gloss information in WordNet with the context in training data. The models use GloVe embeddings [13] as their sentence embeddings, and through two Bi-LSTM models for generating contextual embeddings of the context and the gloss respectively. Both of these works rely on MFS fallbacks.

Static word embeddings, such as GloVe, suffers from the meaning conflation around word types. And start with ELMo [14], the implementation of the contextual embeddings has gradually gained wide attention. The contextual embeddings is context-sensitive, which means that the same word, given different context, will be represented differently. Soon, BERT came out, and achieved new state-of-the-art results on 11 NLP tasks. BERT model uses multi-layer bidirectional Transformer encoder, and is pretrained on a large corpus. For better performance, BERT model has to be fine-tuned on the specific tasks.

Contextual Word embeddings have proven to encode the context into them and this gives an important cue for the word disambiguation. Indeed, with the contextual embeddings generated by BERT, a simple k-NN method can surpass the performance of all the previous methods. However, this method still relies on MFS fallbacks. LMMS created sense-level embeddings with full-coverage of WordNet. LMMS used synsets, hypernyms, lexnames and the aggregation of them to generate the missing sense-level embeddings. Instead of adopting k-NN method, GlossBERT treats WSD tasks as a sentence-pair classification problem. For a target word that has N senses, GlossBERT constructs N context-gloss pairs, and choose the sense which has the highest probability.

Although attempts are made to integrate gloss information into the models, the gloss embeddings they generated are usually similar to each other and may influence the results of their model. In this paper, we present MG-BERT model, a BERT-based model that simultaneously encodes the context of the target word and all its glosses into embeddings. We prove that gloss embeddings can be applied as the weight matrix of classifier. Furthermore, we also design a Space Transforming Classifier, which transform different label space to a same one.

3   Multi-glosses BERT Model

Our model can be divided into two parts: Embedding Generator and Space Transforming Classifier as we illustrate in Fig. 1. We formulate raw data to the Context with Multi-Glosses shape and then through our Embedding Generator to generate target word embedding and its corresponding sense embeddings. Then, our Space Transforming Classifier transforms the original label space to a unified label space and calculates the probability distribution over all the possible senses of the target word. We will describe each part thoroughly in this section.

3.1   Notations

In this paper, we denote the target word as x, its context sentence of length l as C_x. We use c_x^i to other tokens in the context sentence C_x where i to index tokens.

$$C_x = (c_x^1, ..., x, ..., c_x^l) \tag{1}$$

For each target word x, its candidate senses are dented as $\{s_x^1, s_x^2, ..., s_x^N\}$, where N is the total number of the candidate senses. All candidate senses form a label space denoted as Ω_x^S.

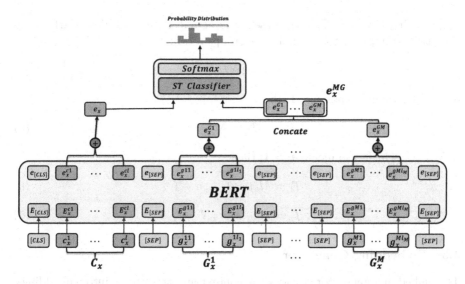

Fig. 1. A high-level view of our MG-BERT model, the embedding generator encodes the context and M glosses into embeddings simultaneously while the Space Transforming (ST) Classifier outputs the probability distribution over all sense candidates

$$s_x^i \in \Omega_x^S, i = 1, 2, ..., N \tag{2}$$

And each sense s_x^i has a sense definition called gloss, denoted as G_x^i. Like the context sentence, we use g_x^{ij} to the jth word in the ith gloss of the target word x. The length of each gloss G_x^i is l_i.

$$G_x^i = (g_x^{i1}, ..., g_x^{ij}, ..., g_x^{il_i}), i = 1, 2, ..., N \text{ and } j = 1, 2, ..., l_i \tag{3}$$

Thus, our goal is to find the most closely related sense s_x^i from Ω_x^S given the target word x and the context C_x.

3.2   Context with Multi-glosses

To better leverage gloss information, we reshape the raw data to form Context with Multi-Glosses. An example of Context with Multi-Glosses is shown in Table 2. We use M, a hyperparameter, to control the upper limit of glosses we combine together with the context each time. To emphasize which word the target word is, we add double quotation mark around the target word x in the context C_x. Then we concatenate the context and all its glosses together with $[CLS]$ and $[SEP]$ marks. For the words that have more senses than M, $N > M$, we divide its senses into $\lceil N/M \rceil$ groups, where $\lceil \rceil$ means the ceiling of the number. We keep the first $\lceil N/M \rceil - 1$ groups M senses and the last group $\leq M$ senses. In each group, we extract the corresponding glosses and concatenate them with context.

Formally, if we use T_i to represent the actual number of glosses ith group has, use n to represent the number of groups a target word needs. We have:

$$n = \lceil N/M \rceil \tag{4}$$

If $N \leq M$, which means $n = 1$, then:

$$T_1 = N \tag{5}$$

If $N > M$, which also means $n > 1$, then:

$$T_i = M, i = 1, ..., n - 1 \tag{6}$$

$$T_n = N - M * (N - 1) \tag{7}$$

3.3 Embedding Generator

The embedding generator encodes the context and the glosses into embeddings. To better extract the semantic relationship among context and glosses, we adopt BERT base model for producing the embeddings. We modify BERT model so that it can match our Context with Multi-Glosses. As we have multi-sentences input, we use 0 and 1 as segment tokens alternately as shown in Table 2. We use the same truncating methods (always truncating the longest sentence) as BERT. We take the last hidden layer of BERT as our embedding. Intuitively, all words in the context C_x have influence on the disambiguation task, thus, we sum up the tokens corresponding to the context C_x as our target word embedding e_x. Similarly, we sum up the tokens corresponding to each gloss G_x^i as each gloss embedding e_x^{Gi}.

$$e_x = \sum_{C_x} \text{BERT}(C_x + \sum_{i=1}^{T} G_x^i) \tag{8}$$

$$e_x^{Gi} = \sum_{G_x^i} \text{BERT}(C_x + \sum_{i=1}^{T} G_x^i), i = 1, 2, ..., T \tag{9}$$

where $\sum_X$ means to sum up the tokens corresponding to the X part, $\text{BERT}(\cdot)$ means the last layer representations of BERT model, and $C_x + \sum_{i=1}^{M} G_x^i$ denotes one group of our Context with Multi-Glosses. Finally, we concatenate the sense embeddings together to form a gloss matrix e_x^{MG}. As mentioned above, different groups have different values of T. We pad the gloss matrix with $\mathbf{0}$ vector to match the matrix dimension to a fix number M.

If $T = M$:

$$e_x^{MG} = [e_x^{G1} : ... : e_x^{GT}] \tag{10}$$

If $T < M$:

$$e_x^{MG} = [e_x^{G1} : ... : e_x^{GT} : \mathbf{0} : ...] \tag{11}$$

where: means concatenation operator.

Table 2. An example of Context with Multi-Glosses, target word in context is emphasized with double quotation. We apply 0 and 1 alternately as segment ids.

Context with Multi-Glosses:

[CLS] Visits and interviews with key personnel were also "undertaken".
[SEP] accept as a challenge [SEP] accept as a charge [SEP] enter upon an activity or enterprise [SEP] enter into a contractual arrangement [SEP] promise to do or accomplish [SEP]

Segment Ids:

$0_{[CLS]}$ ... $0_{context}$ ... $0_{[SEP]}$ ... 1_{gloss} ... $1_{[SEP]}$ ... 0_{gloss} ... $0_{[SEP]}$ ... 1_{gloss} ... $1_{[SEP]}$ ... 0_{gloss} ... $0_{[SEP]}$ ... 1_{gloss} ... $1_{[SEP]}$

3.4 Classification Algorithms

As mentioned in Sect. 3.1, the label space Ω_x^S varies with x, which makes it impossible to use a traditional unified classifier for all words. Traditional classifiers can be formulated as:

$$Y = W_x \times x \tag{12}$$

where W_x is the classifier weight matrix, which also varies with the target word and need dedicated training. Thus, each word needs to train a dedicated weight matrix W_x, which makes traditional methods only capable with the small fraction of WordNet senses covered by the training corpus.

To solve the problem of disambiguating words with only small fraction of WordNet, we propose three classification algorithms, Gloss Matrix Classifier (GMC), General Gloss Matrix Classifier (GGMC) and Space Transforming Classifier (STC). GMC and GGMC are two methods suitable for WSD task, which we use the gloss matrix as weight matrix W_x. Different from the two algorithms above, STC maps different label space Ω_x^S to a same space. Thus, STC is not restricted in WSD task, it can be applied to other tasks which also have different label space problem.

Gloss Matrix Classifier. In traditional classifier, each row in W_x multiplies with our input x and the outcome serves as the score of how much our input matches this class. So, each row in W_x can be seen as the feature of each class. In WSD task, we can seem the gloss embeddings, the sense definitions, as the feature of senses, which means that our gloss embeddings can serve the same purpose as W_c. So, we use e_x^{MG} as the weight matrix, and the classifier can be rewritten as:

$$Y = e_x^{MG^T} \times e_x \tag{13}$$

Our gloss matrix can also be learnt through training in our Embedding Generator.

General Gloss Matrix Classifier. In order to improve our model expressive ability, we try to add another weight matrix W_a to our classifier, the function can be formulated as:

$$Y = e_x^{MG^T} \times W_a \times e_x \tag{14}$$

where $W_a \in \mathbb{R}^{H \times H}$, H is the hidden size of our embedding.

Space Transforming Classifier. As the first two classifiers use different gloss matrix e_x^{MG} as classifier weight matrix, each word still can be seen having a dedicated classifier. Differently, our Space Transforming Classifier tries to map different label space Ω_x^S to a same label space. Though target word x and its corresponding label space Ω_x^S changes simultaneously, the relationship between them remain invariant. So instead of taking the target word embedding e_x as input, our Space Transforming Classifier takes the relationship between target word x and its corresponding label space Ω_x^S as input, and aims to find out the sense with the highest relationship score. Our Space Transforming Classifier replicates the target word embedding e_x M times, and then concatenates it with the gloss matrix. Then we use a single-layer perceptron such that:

$$Y = W_a[(e_x)^M : e_x^{MG}]^T \tag{15}$$

We scaled our classifier function, $Y \in \mathbb{R}^{1 \times M}$, with the scaling factor $\frac{1}{\sqrt{H}}$. Then, we use softmax function to generate probability distribution over sense candidates, which formulates as:

$$P(s_x^i | C_x) \sim softmax(\frac{Y}{\sqrt{H}}) \tag{16}$$

where $s_x^i \in \Omega_x^S$. We select the sense with the highest probability score as our prediction. During training, all model parameters are jointly learned by minimizing a standard cross-entropy loss between $P(s_x^i | C_x)$ and true label.

4 Experiment and Evaluation

4.1 Training Dataset

We use SemCor as our training set, which is the largest corpus manually annotated with WordNet senses. It has 352 documents for a total of 226,036 sense annotations and is the most popular corpus used for training WSD systems.

4.2 Evaluation Dataset

We evaluate our model on several English all-words fine-grained WSD datasets. To be fair, we use the same benchmark datasets proposed by Raganato [16]. The datasets are from the Senseval and Semeval competitions, which are Senseval-2 (**SE2**), Senseval-3 task 1 (**SE3**), SemEval-07 task 17 (**SE07**), SemEval-13 task 1 (**SE13**) and SemEval-15 task 13 (**SE15**). Following previous methods [15], we use SE07 as our validation set to find the optimal settings of our framework. Other four datasets are used as testing set.

4.3   Implementation Details

In this paper, we use BERT base model for fine-tuning, which has 12 layers of transformer blocks with 12 self-attention heads each. The hidden size is 768. To balance sequence length and batch size, we set M to 10, sequence length to 256 and batch size to 16. Noteworthy, with $M = 10$, all words need only two turns to get the final prediction. We trained our model on a GeForce RTX 2080 Ti for 4 epochs with learning rate as 2e−5.

For the words that have multi-groups inputs, as mentioned above, during training, if the correct sense is not in group, we randomly choose a sense and replace it with the correct sense. Thus, we make sure that each group has a correct label and use it as supervised signals during training. While during testing, as we do not have label data in advance, we select the sense corresponding to the highest probability from each group. And instead of simply taking the highest probability as our final result, we use these senses to construct context with multi-glosses again, and feed back into our model for another turn prediction. We repeat this procedure until the words have only one group input and we choose the highest probability sense.

4.4   Results and Discussion

We show the performance of some prior models on the English all-words WSD benchmark datasets in Table 3, and compare the performance of our proposed model with them.

The first block shows the MFS baseline. To disambiguate word, it simply chooses the most frequent sense in the training corpus.

We select two knowledge-based models for comparison. One is $Lesk_{ext+emb}$ [1], which is a variation of Lesk algorithm [5]. use word embedding to calculate gloss-context overlap. The other is Babelfy [11], which builds a graph-based architecture based on the semantic network structure from Babelnet.

Supervised models have remained surprisingly competitive for quite a long time. We compare our approach with two traditional supervised methods. IMS [18] method trains SVM classifiers and use semantic local features. Besides all the features IMS used, IMS_{emb} [4] utilized word embeddings as its feature and improved IMS performance. Neural-based model aims to train an end-to-end unified model for WSD tasks. $Bi\text{-}LSTM_{att.+LEX+POS}$ [15] used a neural sequence learning task, with sequence of words as input and sequence of senses as output. GAS_{ext} [8] and HCAN [7] applied gloss information to the neural model and use co-attention mechanism.

Recently, BERT has shown great performance on NLP tasks, and dozens of BERT-based WSD models have come out. LMMS [6] created sense-level embeddings and utilize a simple K-NN method for classification. GlossBERT [3] treats WSD as a sentence-pair classification problem by constructing context-gloss pair and fine-tunes the pre-trained BERT model.

The final blocks in Table shows our results for the all-word English WSD tasks. To be mentioned, the order of the glosses does have a slightly influence of

Table 3. English all-words task results in F1-score(%). **Bold** font means new state-of-the-art (SOTA) system

Model	SE07	SE13	SE15	SE2	SE3	ALL
MFS baseline	54.5	63.8	67.1	65.6	66.0	64.8
Lesk$_{ext+emb}$	56.7	66.2	64.6	63.0	63.7	64.2
Babelfy	51.6	66.4	70.3	67.0	63.5	66.4
IMS	61.3	65.3	69.5	63.0	63.7	68.9
IMS$_{emb}$	62.6	65.9	71.5	72.2	70.4	70.1
Bi-LSTM$_{att.+LEX+POS}$	64.8	66.9	71.5	72.0	69.1	69.9
GAS$_{ext}$	-	67.2	72.6	72.2	70.5	70.6
HCAN	-	68.5	72.8	72.8	70.3	71.1
LMMS$_{2348}$	68.1	75.1	77.0	76.3	75.6	75.4
GlossBERT (Token-CLS)	71.9	74.6	79.3	77.0	75.4	76.3
GlossBERT (Sent-CLS-WS)	72.5	76.1	80.4	77.7	75.2	77.0
SG-BERT$_{Baseline}$	67.9	72.2	78.1	77.3	74.8	75.0
MG-BERT$_{GMC}$	74.1	78.5	80.7	79.8	77.8	78.8
MG-BERT$_{GGMC}$	72.5	**78.8**	81.0	79.8	77.2	78.6
MG-BERT$_{STC}$	**74.3**	78.2	**81.3**	**80.1**	**78.0**	**78.9**

the outcome. We shuffle the glosses and conduct our experiment three times. The fluctuation varies in an acceptable range and the outcomes show in Table 3 is the average of our three experiments. SG-BERT$_{Baseline}$ is our baseline, which we use context with single gloss(SG) as model input, just like GlossBERT without the weak supervised signals. In SG-BERT$_{Baseline}$, we adopt the same Gloss Matrix Classifier as MG-BERT$_{GMC}$. Compared with MG-BERT$_{GMC}$, MG-BERT$_{GMC}$ outperforms SG-BERT$_{Baseline}$ by a substantial margin. By jointly encoding M glosses, our model obtains the divergences among glosses. What's more, as we use gloss matrix as weight matrix, the gradient can not only flow through single gloss embeddings but also flow among glosses. Compared our MG-BERT$_{GMC}$ with previous model, using gloss matrix as classifier weight matrix gains marked improvement. Although the total number of parameters in MG-BERT$_{GGMC}$ are more than those in MG-BERT$_{GMC}$, the overall performance decreases slightly. This is probably because the randomly initialized matrix destroyed the class feature contain in the gloss matrix. Moreover, MG-BERT$_{STC}$ performs better than the other two models. Although no one systems always performs better on all datasets, our MG-BERT$_{STC}$ and MG-BERT$_{GGMC}$ achieve new state-of-the-art performance on all datasets including evaluation set. Our model proves that simultaneously encodes all gloss embeddings corresponding to the target word, the WSD task can be improved by at most **2.2%** F1-score.

Table 4. Between-glosses similarities in our SG-BERT and MG-BERT models.

Model	NOUN	VERB	ADJ	ADV	Avg.
SG-BERT$_{Baseline}$	0.4360	0.4989	0.5408	0.4763	0.4687
MG-BERT$_{GMC}$	0.0137	0.0153	0.0169	0.0175	0.0146

Orthogonal Gloss Embeddings. To better explain why our context with multi-glosses mechanism can help improve the performance, we conduct an experiment which calculates the cos-similarities between candidate gloss embeddings which belong to the same target word from SG-BERT$_{Baseline}$ and MG-BERT$_{GMC}$. We separate our results with parts-of-speech (POSs) as shown in Table.

From Table 4, we can find out that after we adopt context with multi-glosses into our model, the gloss embeddings similarities greatly decreased, from 0.46 to almost 0 (orthogonal). This means that in our MG-BERT model, the model can compare the glosses with each other and emphasize the divergences among glosses. As our model is to calculate the similarities between the context embedding and the gloss embedding to do the classification, having nearly orthogonal gloss embeddings will greatly improve the performance. What's more, we can also find that in both models, VERBs and ADJs have more similar gloss embeddings than NOUNs and ADVs, which is consistent with our understanding that VERBs and ADJs are usually more difficult to disambiguate than NOUNs and ADVs.

5   Conclusion

In this paper, we propose MG-BERT model for WSD task. Our model greatly improves the performance on all WSD task because of two main advantages. First, our model encodes all glosses together simultaneously. We proved that by doing so, our model can learn about the divergences among glosses. As a result, the gloss embeddings our model generated are nearly orthogonal, which greatly improve our model performance. Second, we design three classification algorithms, Gloss Matrix Classifier (GMC), General Gloss Matrix Classifier (GGMC) and Space Transforming Classifier (STC), for WSD task with full-coverage of WordNet. In the first two, we try to use gloss embeddings as weight matrix of our classifier, which achieves great improvement. For the last, we transform different label space to a same one. This algorithm can be extended to other fields with the same problem. Results show that our model outperforms all existing models and achieve new state-of-the-art performance on all WSD benchmarks.

Acknowledgements. We thank the reviewers for their insightful comments. We also thank Effyic Intelligent Technology (beijing) for their computing resource support. This work was supported by in part by the National Key Research and Development Program of China under Grant No. 2016YFB0801003.

References

1. Basile, P., Caputo, A., Semeraro, G.: An enhanced lesk word sense disambiguation algorithm through a distributional semantic model. In: Proceedings of COLING 2014, the 25th International Conference on Computational Linguistics: Technical Papers, pp. 1591–1600 (2014)
2. Devlin, J., Chang, M.W., Lee, K., Toutanova, K.: Bert: Pre-training of deep bidirectional transformers for language understanding. arXiv preprint arXiv:1810.04805 (2018)
3. Huang, L., Sun, C., Qiu, X., Huang, X.: Glossbert: Bert for word sense disambiguation with gloss knowledge. arXiv preprint arXiv:1908.07245 (2019)
4. Iacobacci, I., Pilehvar, M.T., Navigli, R.: Embeddings for word sense disambiguation: an evaluation study. In: Proceedings of the 54th Annual Meeting of the Association for Computational Linguistics (Volume 1: Long Papers), pp. 897–907 (2016)
5. Lesk, M.: Automatic sense disambiguation using machine readable dictionaries: how to tell a pine cone from an ice cream cone. In: Proceedings of the 5th Annual International Conference on Systems Documentation, pp. 24–26 (1986)
6. Loureiro, D., Jorge, A.: Language modelling makes sense: Propagating representations through wordnet for full-coverage word sense disambiguation. arXiv preprint arXiv:1906.10007 (2019)
7. Luo, F., Liu, T., He, Z., Xia, Q., Sui, Z., Chang, B.: Leveraging gloss knowledge in neural word sense disambiguation by hierarchical co-attention. In: Proceedings of the 2018 Conference on Empirical Methods in Natural Language Processing, pp. 1402–1411 (2018)
8. Luo, F., Liu, T., Xia, Q., Chang, B., Sui, Z.: Incorporating glosses into neural word sense disambiguation. arXiv preprint arXiv:1805.08028 (2018)
9. Miller, G.A.: WordNet: An Electronic Lexical Database. MIT press, Cambridge (1998)
10. Miller, G.A., Chodorow, M., Landes, S., Leacock, C., Thomas, R.G.: Using a semantic concordance for sense identification. In: Proceedings of the workshop on Human Language Technology, pp. 240–243. Association for Computational Linguistics (1994)
11. Moro, A., Raganato, A., Navigli, R.: Entity linking meets word sense disambiguation: a unified approach. Trans. Assoc. Comput. Linguist. **2**, 231–244 (2014)
12. Navigli, R.: Word sense disambiguation: a survey. ACM Comput. Surv. (CSUR) **41**(2), 1–69 (2009)
13. Pennington, J., Socher, R., Manning, C.D.: Glove: global vectors for word representation. In: Proceedings of the 2014 Conference on Empirical Methods in Natural Language Processing (EMNLP), pp. 1532–1543 (2014)
14. Peters, M.E., et al.: Deep contextualized word representations. arXiv preprint arXiv:1802.05365 (2018)
15. Raganato, A., Bovi, C.D., Navigli, R.: Neural sequence learning models for word sense disambiguation. In: Proceedings of the 2017 Conference on Empirical Methods in Natural Language Processing, pp. 1156–1167 (2017)

16. Raganato, A., Camacho-Collados, J., Navigli, R.: Word sense disambiguation: a unified evaluation framework and empirical comparison. In: Proceedings of the 15th Conference of the European Chapter of the Association for Computational Linguistics: Volume 1, Long Papers, pp. 99–110 (2017)
17. Yang, Z., Dai, Z., Yang, Y., Carbonell, J., Salakhutdinov, R.R., Le, Q.V.: XLNet: generalized autoregressive pretraining for language understanding. In: Advances in Neural Information Processing Systems, pp. 5754–5764 (2019)
18. Zhong, Z., Ng, H.T.: It makes sense: a wide-coverage word sense disambiguation system for free text. In: Proceedings of the ACL 2010 System Demonstrations, pp. 78–83 (2010)

Top Personalized Reviews Set Selection Based on Subject Aspect Modeling

Muhmmad Al-Khiza'ay[1]([⊠]), Noora Alallaq[1], Firas Qays Kamal[2],
Tamather Naji Alshimmari[3], and Tianshi Liu[4]

[1] School of Information Technology, Deakin University, Geelong, Australia
{malkhiza,nalallaq}@deakin.edu.au
[2] Osmania University, Main Road, Amberpet, Hyderabad 500007, Telangana, India
Faris_qais2001@yahoo.com
[3] College of Engineering, University of al-Qadisiyah, Al Diwaniyah, Iraq
Tamatheralshimmari1@gmail.com
[4] School of Computer, Xi'an Shiyou University, Xi'an, China
232228949@qq.com

Abstract. People are checking different sites before doing their business to either purchase any item online or select any service or product. Many commercial sites rely on the reviews to evaluate their product and services. Other sites are especially designed for the users and reviews to e valuate any product or service. However, select the best review is still a big challenge for the user to select. Many works have been proposed to select the best reviews but with contain redundant information. The best personalized review that is really related to the main topic that the user is searching on is very important. For this reason, in this work, a new personalized reviews' selection is proposed. We based on the idea of that different point of view for the user causes different evaluation and revering. For this reason, searching on the best reviews in a specific subject gives more accurate and significant selection results. In this paper, design a new approach for the best personalized reviews' selection that is based on two stages. The first one in the predict the subject aspect modeling (distribution) based on using the A latent Dirichlet allocation (LDA) model. Second, we design a new weighted personalized reviews selection based subject aspect scoring function to select the top personalized reviews. The experimental results show our method selects reviews that are more focusing on the product or service subject aspect. The reviews that are more emphasizing on a different subject are selected.

Keywords: Personalized reviews · Subject aspect · LDA · Top personalized reviews

1 Introduction

Users who regularly make purchases via e-commerce websites such as CNet, Shopping and Amazon usually post their experiences on the website in the form of reviews. Consumers are permitted to write their reviews which reflect their opinion on the purchased

G. Li et al. (Eds.): KSEM 2020, LNAI 12275, pp. 276–287, 2020.
https://doi.org/10.1007/978-3-030-55393-7_25

goods or services. Consumer reviews have become a significant form of content generated by the customer which serves as a reliable source of information. The posted data can benefit a wide range of entities, including product designers, product manufacturers, potential users, and owners of e-commerce websites.

Online reviews have become a significant type of customer-produced content It is typical of online users to be eager to read all reviews of a specific product that is of interest to them [13, 16] this is particularly true for mobile apps where the display screen is small and resources are limited. A five-inch screen, for example, can display only two restaurant reviews from Yelp.com. Most of the time, most users just read the reviews at the top and use it to make a purchase. Thus, selecting and displaying only a set of best N reviews to a customer, a likely massive number of product reviews, is a very serious challenge. Many online portals solved this problem using inappropriate and unethical methods, since they chose only the reviews with the highest ratings [1].

In general, various previous works have focused on the selection of the best quality reviews [10–14]. Some major aspects of a product being reviewed may be excluded because some of these methods used by it are not accountable for redundancy in the content of the review. Recently conducted studies to offer a comprehensive analysis of the item (i.e., product aspects) [2, 3, 16] They have concentrated on the systematic inclusion of aspects of the company as a detailed rating of the top reviews.

Based on literature evaluation of previous approaches, studies have investigated how the various aspects of products are relevant to users [1, 6, 9], Although they failed to use this element to increase the rating of reviews. Therefore, previous techniques continued to maintain the same group of top reviews to different users without taking into account their varying interests (e.g., U1 and U2). Against this context, our approach focuses on improving the type of top reviews that are shown to customers as they search for product details and regard user expectations as customization. The enhancement would allow the user to pick the aspects of a product which are more important to themselves. Similar to the previous algorithms, the proposed method is capable of selecting high-quality reviews covering various aspects of a product, with a focus on the user-focused product aspects.

After identifying which aspects are valuable for the customer, one more issue will be how to choose the reviews which will be presented to the user based on the aspect of the product that is of interest to them. The proposed framework for a product to retrieve the customized sub-set of reviews based on aspect analysis (TPRS) therefore involves two steps. The first one in the modeling (distribution) of the subject aspect based on the use of the model A latent Dirichlet Allocation (LDA). Furthermore, design a new weighted personalized review scoring function for the selection of top custom reviews based on the subject aspect. Since the reviews cover essential aspects of the product, our approach is always mindful of their feelings. The findings of the present study show that the proposed model will summarize feedback and add a limited subset of feedback that takes into account user expectations and preferences.

The remaining part of the paper is categorized according to the following. Section 1 shows preceding studies on methods of selection review and other related techniques. Section 2 Will provide various definitions, formularization of problems. Section 3 System for the design of topic aspects. Section 4 describes the methodology suggested here.

Section 5 Provides the experimental findings and comments on the study issues and performance. Section 6 sheds light on the conclusions drawn and on suggestions for future sample selection study.

2 Literature Review

Extensive knowledge, the proposed Subject Aspect Modeling, and Top Personalized Review Selection(TPRS) issues were not evaluated in an earlier study however there are a number of related works discussed in this section.

In previous studies carried out by [1, 3, 4, 10], Latent Aspect Rating Analysis (LARA) that is an opinionated text data analysis was proposed. The LARA analyses any opinion that is given of a product in a social network review in terms of relevant subjects for the discovery of each reviewers' latent opinion on the various subjects, and the subjective emphasis on the various aspects when an overall judgment of the entity is given. No attention has been paid to personalized reviews and the aspects that are of interest to users.

In [2, 6, 8] a system for automatic summarization was proposed and developed. In the proposed system, the use of topic representation approaches is employed first of all obtaining an intermediate representation of the text that covers the discussed topics in the input. However, no attention was paid to the preferences of the costumers.

The study carried out by [9, 12] showed how consumers' subjective preferences for various product features can be learned using textual data, and how textual data can be used for predictive modeling of future changes in sales.

In [11], the authors investigated divergent aspects of review text like readability, the usefulness of information, subjectivity, and linguistic correctness in reviews as factors that influence sales and perceived usefulness. The study focused on the extraction of different aspects but paid no attention to the preferences and interests of customers.

An analysis of the emotional content of a huge amount of online product reviews was carried out by [13–15] using (NLP) methods. Their main aim was the reduction of data dimensionality, rather than focusing on the interests of customers.

In order to analyze the sentiments of people, their emotions, attitudes, opinions, etc., [11, 16] proposed (SA) system. The system is intended to perform the analysis in terms of elements like topics, individuals, products, services, organizations, etc.

A set of design guidelines for aspect-based opinion mining was presented by [9, 15]. In their studies, they discussed different complex LDA models. They also proposed a novel methodology that can be used in the identification of review factors which shoppers can utilize in evaluating the usefulness of reviews.

In a study conducted by [13], a system was proposed for the extraction of reviews of product attributes and the polarity of the attributes.in their study, customers are presented with a graphic representation of the better version of two products using some criteria which include the date of review, start rating, the polarity of reviews and helpfulness of the review score. The review of previous studies has shown that no studies have been carried on the selection of personalized review base on aspect and sentiment analysis and putting into consideration the preferences of the customers.

3 Problem Formulation

3.1 Notation

In terms of defining our problem, we assume that the set of users are defined as $\mathcal{U} = \{u_1, u_2, \ldots, u_n\}$, and set of products are defined also as $\mathcal{P} = \{p_1, p_2, \ldots, p_m\}$ where (n) and (m) are the users and products dimensional set. Basically, each product in the products set $(\mathcal{P})$ belongs to the same domain (m). Usually, each sub-set of products are sharing the same subject aspects such as hotel, restaurant, etc. In this case, we define a set of subject aspect features that each subset of product is sharing with such that as $\mathcal{A} = \{a_1, a_2, \ldots, a_d\}$ where (d) is the dimensionality of the subject aspect feature set. For instance, $(\mathcal{U})$ includes set of users form the airlines services website, $(\mathcal{P})$ also includes different airlines companies such as American airlines, United etc. $(\mathcal{A})$ includes different subject aspects that the users $(\mathcal{U})$ are concerned about each product (Airlines company) in $(\mathcal{P})$ such as the "service type", "delayed flight", "cancelled flight", and/or "customer services".

Typically, each user $(u_{j \in n})$ from the defined users set $\mathcal{U}^{U \in \mathbb{R}}$ can write review about any product $(p_{j \in m})$ from the defined product list (any product website set) $\mathcal{P}^{P \in \mathbb{R}}$. In this case, we denoted the (r_i^j) to be a review that is written by (u_i) to accomplish his/her opinion, concern, or idea about a certain product (p_j). Different review form that the user can obtain his opinion about the product such as text review or certain recommendation tag such e.g. number of starts that the user (u_i) can give to the product (p_j).

3.2 Reviews Objectives

The main objective of writing a review (r_i^j) form a certain user (u_i) about any product or service (p_j) is to evaluate that product. I n another word a set of reviewers are written by set of users to describe and evaluate a product. In this case, we want to discuss the variety of different review forms that the users are used to describe their opinion. As been discussed, there are many forms that the users can used to describe their opinion about the product. One of them such as starts recommendation form does not provide plenty of space for the user to describe his/her idea. Rather than that the other user can not have much information about the reason (lack number of starts) about that product. Another formulation is the text review. This form provides more space for the user to describe his/her idea about the product. However, this formula of the product evaluation has some difficulties about what the ground truth of product's reviews. In another words, what you will care about the product/service. This causes some issue since there is no limitation for the user to write his/her review. For instance, user (u_1) can write his/her opinion (text) about a product (p_1) to have a review (r_1^1). In another way, another user (u_2) can write the same review about the same product (p_1) but using different vocabulary in different subject which causes having another review (r_2^1). This issue can enlarge our domain which causes some difficulties to select which review is the best.

To overcome this issue, we propose a knowledge representation reduction in which what (u_i) was caring about (p_j) when he/she wrote the review (r_i^j). In another words, we will try to re-modeling the set of reviewers $(\mathcal{A})$ and redwing them based on their subject aspect we called subject aspect modeling. The subject aspect modeling is a distribution of what the main subject that a set of users having the same review about the same product or service based on a certain subject. For instance, set of users ($u_i \in \mathcal{U}$) are having the same subject when they have evaluated the same product ($p_j \in \mathcal{P}$) but using m-dimensional vector of vocabulary to describe the same subject $\Phi^i_j = \left(\phi^{i,j}_i, \phi^{i,j}_2, \ldots, \phi^{i,j}_m \right)$ based on using the same set of reviews $\mathcal{A} = \{a_1, a_2, \ldots, a_m\}$. Each factor $\phi^{i,j}_l$ in Φ^i_j represent a certain subject aspect of (a_l) of product or service (p_l) to a user (u_l). In our proposal, we try to extract the main subject aspect (distribution) (Φ^i_j) based on the same set of reviews (r^j_i). In another words, if a set of users (u_i) have written the different reviews to evaluate a product (p_j) then (Φ^i_j) can subject them based on different aspects.

3.3 Top Personalized Reviews Set Selection

Personalized reviews' selection is defined as the select a set of tops of users give the best review about the certain product. Let assume that the $\left(\mathcal{R}^p_i \right)$ is a set of reviews that are written by set of users $\left(\mathcal{R}^u_i \right)$. In this case, the top personalized reviewers' selection is defined a best set of reviews that the algorithm will retrieve to the user based the requested key. For example, if a user (u_i) request a review about a certain product or service (p_j) using a user defined key (k), the top-reviewers section will retrieve a $k - sized$ of users' subset $\left(\hat{\mathcal{R}}_{i,j} \right)$ that match the product $\left(\mathcal{R}^p_i \right)$. The main criteria of selecting the best reviews is based on predict the best subject aspect (Φ^i_j) that surmised and grouped the users (u_i) and then select the best top review sets of a product that are same for different users where $\left(\hat{\mathcal{R}}_{x,y} = \hat{\mathcal{R}}_{i,j} \right)$ for any user ($u_x, u_x \in \mathcal{U}, \ x \neq y$).

4 Subject Aspect Modeling (SAM) Framework

Subject aspect model is defined as a collection of $\mathcal{U}$ reviewers with set of certain subject aspect ($\mathcal{A}$). Subject aspect model is a mixtures model of set of different subjects $\theta_1, \ldots, \theta_D$, over different K of topics that characterized by vectors of word probabilities $\varphi_1, \ldots, \varphi_K$. The model assumes that the subject aspect model mixtures $\theta_1, \ldots, \theta_D$, and the topics $\varphi_1, \ldots, \varphi_K$ follow a Dirichlet distribution with concentration parameters α and β respectively[4].

The subject aspect model $\theta_1, \ldots, \theta_D$ based mixture model produces set of probability vectors of length K, where K is the number of topics. The entry θ_{di} in the subject aspect modeling is the probability of topic i appearing in the dth document. The subject aspect model mixtures correspond to the rows of the Document Topic Probabilities (DTP). The topics $\varphi_1, \ldots, \varphi_K$ are probability vectors of length V in each subject, where V is the number of words in the topic vocabulary. The entry φ_{iv} corresponds to the probability of the vth word of the vocabulary appearing in the ith topic. The topics $\varphi_1, \ldots, \varphi_K$ correspond to the columns of the Topic Word Probabilities property (TWP) in each subject [5].

A latent Dirichlet allocation (LDA) model is a text document topic model which discovers underlying topics in a collection of documents and infers word probabilities in subject aspects (topics). The following flowchart illustrates the personalized best review selection based subject aspect topic modeling based LDA as a probabilistic graphical model. Given the subjects $\varphi_1, \ldots, \varphi_K$, the Dirichlet prior α on the topic mixtures, LDA assumes the following generative process for a text reviews to subject them. Under this generative process, the joint distribution of a document with words $w_1, \ldots, w_N$, with topic mixture θ, and with topic indices $z_1, \ldots, z_N$ is given by Eq. (1) [6, 7]:

$$p(\theta, z, w | \alpha, \varphi) = p(\theta | \alpha) \prod_{n=1}^{N} p(z_n | \theta) p(w_n | z_n.\varphi) \tag{1}$$

where N is the number of words in the document. Summing the joint distribution over z and then integrating over θ yields the marginal distribution of a text document w [8]:

$$p(w | \alpha, \varphi) = \int_{\theta}^{a} p(\theta | \alpha) \prod_{n=1}^{N} \sum_{z_n} p(z_n | \theta) p(w_n | z_n.\varphi) d\theta \tag{2}$$

The Dirichlet distribution is a continuous generalization of the multinomial distribution. Given the number of categories $K \geq 2$, and concentration parameter α, where α is a vector of positive reals of length K, the probability density function of the Dirichlet distribution is given by [9, 10]:

$$p(\theta | \alpha) = \frac{1}{\beta(\alpha)} \prod_{i=1}^{K} \theta_i^{\alpha_i - 1} \tag{3}$$

where β denotes the multivariate Beta function given by

$$\beta(\alpha) = \frac{\prod_{i=1}^{K} \Gamma(\alpha_i)}{\Gamma\left(\sum_{i=1}^{K} \alpha_i\right)} \tag{4}$$

A special case of the Dirichlet distribution is the symmetric Dirichlet distribution. The symmetric Dirichlet distribution is characterized by the concentration parameter α, where all the elements of α are the same [8–10].

5 Proposed System: Top Personalized Reviews Selection (TPRS)

5.1 Predicting Personalized Subject Aspect- Distribution

Based on the idea of the main standard approach for recommendation system by applying the collaborative filtering [11–13] that bases basically of the preference aggregating to find the similar user's target u_i.

To predict the important subject aspect θ_i that belongs to the user's target u_i to select the best review p_j, we first aggregate the whole subject aspect distribution model Φ_j^v as been showing "Personalized Reviewers Selection based Subject Aspect Modeling". Each user u_v who has comments on the p_j is included and aggregated in the subject aspect importance distribution Φ_j^v. In this case, each Φ_j^v is weighted and scored based on the correlation score and similarity between the user's subject aspect profile (aspect profile) which is denoted as λ^i. That means two reviews are close to each other and similar if they agree with each other and write a review in the same subject aspect which is denoted as $\mathcal{P}$.

Mathematically, illustrate that by assuming that weighted average of the similarity score between each review is drawn based on the mean average score of the subject aspect importance distribution as it shown in Eq. (5) [14]:

$$w^{i,j} = cost\left(\lambda^i, \lambda^j\right) \tag{5}$$

Where $w^{i,j}$ is the similarity score between two reviews based on the cosine similarity score between two reviews' profiles λ^i, λ^j which in other word means the subject aspect importance distribution.

Each review's profile such as λ^i and λ^j in the subject aspect importance distribution Φ_j^v is drawn based on the prediction score of each review and different subject aspect Φ_j^i as it shown in the Eq. (6) [15]:

$$\Phi_j^i = \frac{\sum_{u_v \in U_j} w^{i,v} \Phi_j^v}{\sum_{u_v \in U_j} w^{i,y}} \tag{6}$$

Where U_j illustrates reviews that comments on the same subject aspect p_j. Among set of different reviews u_i on different product p_j we define a best personalized review scoring selection function $F_{ps}\left(\hat{\mathcal{R}}\right)$ as the Eq. (7) illustrates below:

$$F_{ps}\left(\hat{\mathcal{R}}\right) = \sum_{a_l \in A} \left(\Phi_j^i + \delta\right) f\left(\hat{\mathcal{R}}, a_l\right) \tag{7}$$

Where Φ_j^i is the estimated subject aspect importance distribution between each review and each subject as is illustrated as a pair of $\left(u_i, p_j\right)$ as the Eq. (8) shows below:

$$\Phi_j^i = \phi_1^{i,j}, \phi_2^{i,j}, \ldots, \phi_m^{i,j} \tag{8}$$

Where m is the descent distribution number of different subject.

By adding the δ to the main equation of the best personalized review selection (8) which is set-up to 10^{-4} in case of impact the importance of the prediction subject aspect. In this case, and after adding the δ to the estimated subject aspect importance distribution Φ_j^i the personalized weighted average of each review based each subject aspect in the final prediction score become non-zero. In this case, the best personalized reviews' selection and prediction function gives a chance to each subject aspect to be selected which becomes less sensitive and more accurate for all reviews the written on any product based different subject aspect viewpoint.

5.2   Top Personalized Selection Based Subject Aspect Modeling

In order to select or retrieve the best top personalized reviews $\hat{\mathcal{R}}_{i,j}$ after estimated the subject aspect modeling of the reviews set $\Phi_j^i = \left(\phi_1^{i,j}, \phi_2^{i,j}, \ldots, \phi_m^{i,j}\right)$ for different users (u_i) to the same product or service (p_j) by (u_i, p_j), the selection reviews process is needed. Simple, the reviews are ranking based on their rating and the best one is selected [16]. In this case, the main issue is that many redundancies are accrues sine two or more users rating of reviewing the same product using different text. To overcome this issue, the comprehensiveness way to retrieve the best reviews based the subject aspect is proposed in this work. To select the best review $\hat{\mathcal{R}}_{i,j}$ the selection process should go over many subject related to the user query to ensure that is related to the same service/product (p_j) within the same subject aspect as possible. In our subject aspect selection process, first every review is scoring by using the quality function (q). The quality function maps every review (r) to a real score (number) $q(r)$ as is it given in Eq. (9) [13].

$$F(\hat{\mathcal{R}}) = \sum_{a_l \in \mathcal{A}} f(\hat{\mathcal{R}}, a_l | p_j) \tag{9}$$

where $f(\hat{\mathcal{R}}, a_l)$ is denoted as that how much $(\hat{\mathcal{R}})$ is contributing on providing information on subject (a_l) when the product (p_j) is evaluated or reviewed.

Basically, to find the best review, the scoring function is maximized by the best review selection based subject aspect as is defined as is shown in Eq. (10) [11, 16]:

$$f(\hat{\mathcal{R}}, a_l) = \max_{r \in \hat{\mathcal{R}}_{a_l}} q(r) \tag{10}$$

where $\hat{\mathcal{R}}_{a_l}$ is denoted as a set of reviews in $\hat{\mathcal{R}}$ that is mentioned before at a_l.

This value is used as a key to be retrieved from the data through the selection process. In this case, one review may have different scoring value related with different subject aspect. At the end the highest score will be selected to retrieve that review for the related subject.

Moreover, to consider the best personalized reviews among different subject aspect, the scoring function $F(\hat{\mathcal{R}})$ is adjusted based on each product/service that is related to that subject. We assume that the top personalized reviews' selection is retrieving different reviews (a_l) form different users (u_i) on product/service (p_j), a personalized subject aspect scoring $F_p(\hat{\mathcal{R}})$ is define to ensure that the best score is selected regarding to the main related subject aspect $(\hat{\mathcal{R}})$ as is shown in Eq. (11):

$$F_p(\hat{\mathcal{R}}) = \sum_{a_l \in \mathcal{A}} \left(\tilde{\phi}_l^{i,j} + \delta\right) f(\hat{\mathcal{R}}, a_l | p_j) \tag{11}$$

Where $\tilde{\phi}_l^{i,j}$ is denoted as the predicted related subject aspect of the user (a_l) to the product/service (p_j), and δ is the weighted factor (normalization) that is setup to very small value. The top personalized review's selection based subject aspect modeling is described below in Algorithm 1.

Algorithm 1 Top Personalized Reviews Selection Based Subject Aspect Modeling

---

Input: Set of reviews on the target product $p_j : \mathcal{R}_j^p$.

    Set of users that evaluated product $p_j : U_j$.

    Set of product subject aspects modeling: $\mathcal{A}$;

    Integer subject aspect domain size K:

    Scoring function: F.

Output: Top personalized reviews $\hat{\mathcal{R}}_{i,j}$

---

1.   **Calculate** the cosine similarity score between two reviews' profiles λ^i and λ^j for each user $u_v \in U_j$.

2.   **Calculate** the weighted average of the similarity score $w^{i,v}$ for each user $u_v \in U_j$.

3.   **Predict** the subject aspect modeling Φ_j^v.

4.   Set the top personalized reviews selection list to empty $\tilde{\mathcal{R}}^0 = \emptyset$.

5.   **for all** $i = 1, .., k$ **do**

6.   **for all** $r \in \mathcal{R}_j \backslash \tilde{\mathcal{R}}^{i-1}$ **do**

        Selects the review that achieves the maximum incremental gain $\Delta_{i-1}(r) = F_p\left(\tilde{\mathcal{R}}^{i-1} \cup \{r\}\right) - F_p(\tilde{\mathcal{R}}^{i-1})\right)$.

7.   **end for**

8. $r_i = \underset{r \in \mathcal{R}_p \backslash \tilde{\mathcal{R}}^{i-1}}{\operatorname{argmax}} \Delta_{i-1}(r)$.

9.   **Select** the reviewer $\tilde{\mathcal{R}}^i = \tilde{\mathcal{R}}^i \cup \{r\}$.

10. **end for**

11. Set $\tilde{\mathcal{R}}^k$ as the top k reviews showing to the user (u_i)

---

6 Experimental Results

6.1 Reviews Analyzing Based Subject Aspect Model

In this experiment, estimated the subject aspect distribution $\left(\Phi_j^i\right)$ for each review on a product (p_j) by user (u_i). First, fit an LDA model with 4 Subject Aspects (an initial subject aspect number k). In order to predict the $\left(\Phi_j^i\right)$, a Latent Dirichlet Allocation (LDA) model is used for subject aspect modeling which discovers underlying the subject aspect (topics) in a collection of review's comment text and infers word probabilities in subject aspect (topics).

Randomly select two reviews and visualized personalized subject aspect scoring $F_p\left(\hat{\mathcal{R}}\right)$. For example, in Fig. 1(a) shows the first reviews has 40% personalized similarity score $F_p\left(\hat{\mathcal{R}}\right)$ that is related to first subject, while it has about 20%, 22%, and 17% personalized similarity scores that are related to subject 2,3, and 4 respectively. Also, Fig. 1(b) shows the second reviews has 42% personalized similarity score $F_p\left(\hat{\mathcal{R}}\right)$ that is related to second subject, while it has about 23%, 16%, and 15% personalized similarity scores that are related to subject 1, 3, and 4 respectively.

The top 10 words (bag of words) of each subject aspect distribution $\left(\Phi_j^i\right)$ from each review on a product (p_j) by user (u_i) is extracted and visualized.

Finally, the LDA model is used to predict the subject aspect for each review by maximize the personalized scoring function $F\left(\hat{\mathcal{R}}\right)$ for the whole reviews.

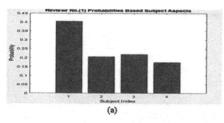
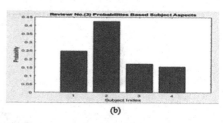

(a)   (b)

Fig. 1. Personalized reviewers' similarity scores visualization

6.2 Top Personalized Review Set Selection Based Subject Aspect

First have to select which subject that customer want to request in. for instance, the second subject (UNIT SUBJECT), the main reviews set that are related to the unit subject are retrieved and visualized. Have three criteria of the reviews (negative, neutral, and positive), the reviews are order based on their personalized similarity score to this subject respect to their response (negative as shown in blue plotting, neutral the orange plotting, and positive reviews the yellow plotting). The individual distribution of each criterion based over the whole reviews in the UNIT subject is shown in Fig. (2) below.

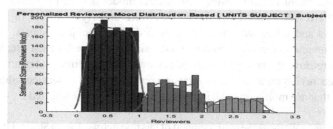

Fig. 2. Individual distribution of each criterion (negative, neutral, and positive) based second selected subject

Finally, the top personalized reviews' selection is retrieving different reviews (a_l) form different users (u_i) on product/service (p_j), a personalized subject aspect scoring $F_p\left(\hat{\mathcal{R}}\right)$.

Moreover, and the best weighted similarity scores $\left(\tilde{\phi}_l^{i,j} + \delta\right)$ for the top selected personalized reviews for each predicted related subject aspect $\tilde{\phi}_l^{i,j}$ to the user (a_l) to the product/service (p_j) where the initial value of δ is selected integer value based selected subject aspect (Fig. 3).

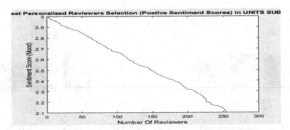

Fig. 3. Ordered top weighted personalized similarity scores for the top personalized reviews selection

7 Conclusion

In this paper, a new personalized review's selection approach based subject aspect modeling is proposed and implemented. By relying on the idea of different users (reviews) may have impacted on different subject when they wrote their evaluation (different subject aspect) for the same product or services. In this case, retrieve the top personalized reviews based on selective subject aspect gives much accurate and significant results. This approach requires to predict different subject aspect models (distribution). For this reason, the latent Dirichlet allocation (LDA) model is used to predict the subject aspects for the reviews text. LDA model discovers underlying topics in a collection of text reviews and infers word probabilities in different subject aspects. Finally, top personalized review's selection based weighted simulate score for each user (review) is calculated to retrieve the best personalized reviews. The experimental results show our method selects reviews that are more focusing on the product or service subject aspect. The reviews that are more emphasizing of different subject are selected.

Acknowledgment. We thank the partial support of this research work from Shannxi Province Key R&D Plan (2018KW-010) and National Natural Science Fund of China (Project No. 71871090).

References

1. Vu, H.Q., et al.: Exploring tourist dining preferences based on restaurant reviews. J. Travel Res. **58**(1), 149–167 (2019)
2. Wang, X., Li, G., Jiang, G., Shi, Z.: Semantic trajectory-based event detection and event pattern mining. Knowl. Inf. Syst. **37**(2), 305–329 (2011). https://doi.org/10.1007/s10115-011-0471-8
3. Niu, W., et al.: Multi-granularity context model for dynamic Web service composition. J. Netw. Comput. Appl. **34**(1), 312–326 (2011)
4. Lu, Y., Mei, Q., Zhai, C.: Investigating task performance of probabilistic topic models: an empirical study of PLSA and LDA. Inf. Retrieval **14**(2), 178–203 (2011)
5. Zhao, W.X., et al.: Comparing twitter and traditional media using topic models. In: Clough, P., et al. (eds.) ECIR 2011. LNCS, vol. 6611, pp. 338–349. Springer, Heidelberg (2011). https://doi.org/10.1007/978-3-642-20161-5_34
6. Blei, D.M., Ng, A.Y., Jordan, M.I.: Latent Dirichlet allocation. J. Mach. Learn. Res. **3**(Jan), 993–1022 (2003)

7. Foulds, J., et al.: Stochastic collapsed variational Bayesian inference for latent Dirichlet allocation. In: Proceedings of the 19th ACM SIGKDD International Conference on Knowledge Discovery and Data Mining (2013)

8. Moro, S., Cortez, P., Rita, P.: Business intelligence in banking: a literature analysis from 2002 to 2013 using text mining and latent Dirichlet allocation. Expert Syst. Appl. **42**(3), 1314–1324 (2015)

9. Jelodar, H., et al.: Latent Dirichlet allocation (LDA) and topic modeling: models, applications, a survey. Multimed. Tools Appl. **78**(11), 15169–15211 (2019)

10. Moghaddam, S., Jamali, M., Ester, M.: Review recommendation: personalized prediction of the quality of online reviews. In: Proceedings of the 20th ACM International Conference on Information and Knowledge Management (2011)

11. Chen, G., Chen, L.: Augmenting service recommender systems by incorporating contextual opinions from user reviews. User Model. User-Adap. Inter. **25**(3), 295–329 (2015). https://doi.org/10.1007/s11257-015-9157-3

12. Alallaq, N., Al-Mansoori, A., Al-Sudani, A.R.: Personalized reviews based on aspect analysis and polarity. In: 2019 8th International Conference on Modeling Simulation and Applied Optimization (ICMSAO), pp. 1–6. IEEE, April 2019

13. Manke, S.N., Shivale, N.: A review on: opinion mining and sentiment analysis based on natural language processing. Int. J. Comput. Appl. **109**(4) (2015)

14. Zhang, L., Liu, B.: Aspect and entity extraction for opinion mining. In: Chu, Wesley W. (ed.) Data Mining and Knowledge Discovery for Big Data. SBD, vol. 1, pp. 1–40. Springer, Heidelberg (2014). https://doi.org/10.1007/978-3-642-40837-3_1

15. Zha, Z.-J., et al.: Product aspect ranking and its applications. IEEE Trans. Knowl. Data Eng. **26**(5), 1211–1224 (2013)

16. Al-khiza'ay, M., Alallaq, N., Alanoz, Q., Al-Azzawi, A., Maheswari, N.: Personalize review selection using PeRView. In: Liu, W., Giunchiglia, F., Yang, B. (eds.) KSEM 2018. LNCS (LNAI), vol. 11061, pp. 238–249. Springer, Cham (2018). https://doi.org/10.1007/978-3-319-99365-2_21

SCX-SD: Semi-supervised Method for Contextual Sarcasm Detection

Meimei Li[1,2], Chen Lang[1,2], Min Yu[1,2(✉)], Yue Lu[1,2], Chao Liu[1], Jianguo Jiang[1], and Weiqing Huang[1]

[1] Institute of Information Engineering, Chinese Academy of Sciences, Beijing, China
yumin@iie.ac.cn
[2] School of Cyber Security, University of Chinese Academy of Sciences, Beijing, China

Abstract. Sarcasm detection is to identify the text with the author's sarcastic attitude. Verbal sarcasm is one main error sources of sentiment analysis tasks. However, labeled sarcastic samples are expensive to obtain. Previous approaches, e.g., model user and topic embedding from multiple perspectives together with large-scale network training, are not suitable for real business scenarios that expect low cost and high speed. In this paper, we propose a semi-supervised method for contextual sarcasm detection in online discussion forums. We adopt author and topic sarcastic prior preference as context embedding that supply simple but representative background knowledge. Then we introduce a sarcasm-unlabeled learning method to utilize a few labeled sarcastic samples and model the classification boundary. Experiments are conducted on real-world data from Reddit, and the results indicate the outperformance over existing methods.

Keywords: Semi-supervised learning · Sarcasm detection · Context modeling

1 Introduction

Sarcasm is a special rhetorical expression, often expressed in a positive literal sense to express negative emotions, and it is omnipresent in daily conversations as well as in written communication. It can change or reverse the true meaning of the sentence. Automatic sarcasm detection is essential in different application domains, such as political ideology analysis, sentiment analysis, and product research. There is evidence that mining negative estimation from twitter can be utilized to generate revenue for clients. In some natural language processing tasks, such as sentiment analysis, failure to consider sarcasm is one of the main causes of system errors. Contextual sarcasm detection is important to correct the sentiment analysis system.

The difficulty of sarcasm detection lies in the inconsistency between real emotion and literal emotion. Existing methods are mainly based on supervised learning and strongly depend on the quality and quantity of labeled training data. Most methods [1, 2] train neural networks with large scale data to find vocabulary and pragmatic clues from the content. Porwal et al. [3] used a recurrent neural network (RNN) model along with long short-term memory (LSTM) cells for sarcasm detection because it automatically

© Springer Nature Switzerland AG 2020
G. Li et al. (Eds.): KSEM 2020, LNAI 12275, pp. 288–299, 2020.
https://doi.org/10.1007/978-3-030-55393-7_26

captures grammatical and semantic information over Twitter tweets. Samonte et al. [4] extracted features of the tweets such as pragmatic and punctuation, and built the classification model by using a series of supervised learning algorithms such as Naïve Bayesian and Maximum Entropy. But sarcasm is a contextual phenomenon as sarcasm comprehension needs to be based on contextual information such as who is speaking and in which subject. Researchers began to integrate context embedding to improve classification performance. Oprea et al. [5] utilized multiple neural network models to learn the historical tweets posted by the author, and obtained representations as author context. Context-based approaches show significant gains in detection accuracy. These methods [6, 7] also prove the validity of the context. However, all of the above detection methods rely on large amounts of labeled data to model complex feature representations. These methods have the following two limitations: (1) They require the design and implementation of complex features, whose feature engineering is labor-intensive and dependent on external tools and resources; (2) The models they built require a large amount of labeled data and complicated deep learning networks.

In real scenarios, labeled sarcasm samples are hard to get and cost expensive human labor. We consider that even though the mixed data is unlabeled, its feature distribution contains some information that should be exploited. Instead of using large amounts of labeled data, we can model the boundaries of their distribution with a small number of sarcastic samples and a large number of unknown samples. In this paper, we carried out the semi-supervised method for sarcasm detection which does not require abundant labeled data and extensive feature engineering. We utilize the convolutional neural network (CNN) to learn content representation and input the sarcasm-unlabeled learning together with the satirical preferences as context representation to detect sarcasm with a small amount of positive data (sarcastic data). Our model achieves the same or slightly better results than the current advanced models with a small amount of labeled data and a combination of uncomplicated features. In summary, the contributions are as follows:

(1) We propose the sarcasm detection method based on semi-supervised learning, which can achieve 78% accuracy with only 10% (or less) labeled satiric samples.
(2) Without complex feature extraction and deep learning network modeling, concise content and context embeddings are selected and the features are proved to be effective by ablation experiment.

The remainder of this paper is organized as follows. The related work is introduced in Sect. 2. Section 3 describes the details of our proposed method. Section 4 demonstrates the effectiveness of the proposed method with experiments. Finally, concluding remarks are offered in Sect. 5.

2   Related Work

Social media users are inclined to adopt a creative language such as sarcasm to express the true meaning implicitly. In the past, researchers favored the use of supervised learning to model sarcasm representations. Some methods [8, 9] try to obtain sentence representation with a combination of neural networks. Ghosh et al. [10] first identified satirical and non-satirical tweets using a combination of convolutional neural networks, recursive neural networks (long-term and short-term memory), and deep neural networks. Majumder et al. [11] obtained sentence representation using gated recurrent unit with attention mechanism and presented a multi-task deep learning framework. They demonstrated improvements in deep learning architectures. With the development of research, some scholars began to add auxiliary information. Shimura et al. [12] leveraged the main sense of words in different domains as auxiliary information to improve the overall performance of text categorization. Liu et al. [13] performed a three-layer neural network and conducted a statistical analysis to select the appropriate auxiliary variables. Their model exceeded the state-of-the-art methods in three of the four datasets. Auxiliary information, such as punctuation and emojis, has been shown to be effective in helping the model understand the sarcastic samples. In recent years, researchers began to explore more context representation, not just in sentiment-related field [14–16]. Hazarika et al. [17] learned stylometric features, personality features, forum and content embedding from historical comments, and utilized a neural network to classify sarcasm. The user embedding model has proven most informative in a sarcasm detection pipeline so far. Ghosh and Veale[18] presented a combination CNN/LSTM (long short-term memory RNN) architecture, which uses the user affect inferred from recent tweets, the text of the tweet, and that of the parent tweet as input. They showed when the speaker's emotion can be inferred at the time, detection accuracy will be significantly improved. Kolchinski et al. [19] showed augmenting a bidirectional RNN with author representation improves performance. The supervised learning method shows satisfactory results with sufficient marking data, and the addition of multiple contextual features is indeed beneficial to the detection of sarcasm. However, the heavy work of data annotation and complex model construction make it difficult to apply automatic satire detection to the industry.

Recently, researchers have introduced the semi-supervised learning model for text classification [20], sentiment analysis [21], semantic role labeling [22], and so on, which used a small number of labeled samples to reduce dependence on data. Semi-supervised learning is crucial for alleviating labeling burdens in people-centric sensing. Xu et al. [23] proposed the semi-supervised sequential variational autoencoder which maximizes the lower bound of variational evidence for data likelihood, and implicitly derived the underlying label distribution for the unlabeled data. They used only 2.5K labeled instances on the Internet Movie Database data set and surpassed the supervised learning trained with 20K labeled samples. Lee et al. [24] built a base sentiment dictionary from a small training dataset by using a lasso-based ensemble model and used an adaptive instance-based learning model to estimate the score of words outside the training data set. The performance of this method can be compared with the supervised learning models trained on large datasets. Semi-Supervised learning is favored by researchers because it requires as little manual labor as possible and can provide relatively high accuracy. This is in line with our intention to build a simple, lightweight sarcasm detector that does not require complex data tagging and modeling work.

3   Methods

In this section, we present the technical details of our proposed method. First, we raise the problem definition and notations. Then we cover the overview of the Semi-supervised method for ConteXtual Sarcasm Detection model (SCX-SD). Finally, we detail each component of SCX-SD.

3.1 An Overview of SCX-SD

The task is to detect sarcastic text in topic forums, such as Reddit. Let $U = \{u_1, \ldots, u_N\}$ be a set of users. Each user posts comments on one or more topic forums. $T = \{t_1, \ldots, t_M\}$ is the topic set. Our goal is to find a model that predicts whether comment C_{ij} made by $u_i \in U$ on the forum $t_j \in T$ is sarcastic. In this paper, content refers to the comment, and contextual information refers to the background information related to the comment, which can assist in understanding the real sentiment of the comment. It includes the information related to the author u_i and the forum t_i.

The overall architecture of SCX-SD is shown in Fig. 1. The model can be divided into two parts: text representation (left) and semi-supervised classification (right). For a comment C_{ij}, we first generate the text representation (Sect. 3.2) which contains content and context embedding. We choose CNN to learn max-pooling vectors as content representation and then extract its context representation composed of the author's satirical preference and topic satirical preference. The author embedding and topic embedding are obtained through the sarcasm analysis of the authors' statements and statements on different topics in the training set. From the training data, the context representations of various known users and topics are generated. When predicting the test data, directly take out the embedding of the corresponding author and topic, and enter it together with the content representation into the classifier built with the training data. Finally, we concatenate the all embedding into the semi-supervised classifier (Sect. 3.3) to assign the label "sarcastic" or "non-sarcastic".

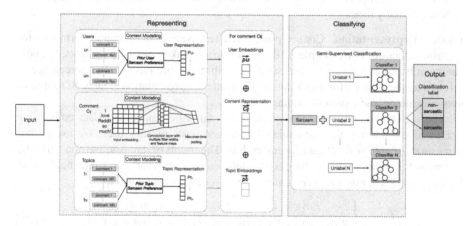

Fig. 1. The overall architecture of SCX-SD.

3.2 Text Representation

Content Representation. CNNs extract location-invariant local patterns to generate text representations. Vectors capture both the syntactic and semantic forms which are helpful for our task. CNN contains the input layer, convolutional layer, pooling layer, fully-connected layer, and output layer. Since we don't use CNN for supervised learning to directly obtain the classification results, but utilize the text representation generated by CNN as the input of the semi-supervised classifier, so we select the pooling layer vector instead of the dense layer behind.

Given a user's comment C_{ij}, which is a sentence composed of n words $S = [x_1, x_2, \cdots, x_n]$. The input sequence consists of word embedding $x_i \in R_{em}$ using the pre-trained Fast Text embedding [25] for each word and R_{em} is the size of word embedding. We use the single-layered CNN [26] to model the input sequence. First, a convolutional layer is applied having three filters $W[1, 2, 3] \in R_{em*h[1,2,3]}$ of filter sizes $h[1, 2, 3]$. CNN extracts the local information of each part of the input sequence by sliding the convolution kernel with different sizes on the sequence. Three filters create three feature maps. For example, a feature c_i is generated from a window of words $x_{i:i+h_k-1}$ by filter $k \in \{1, 2, 3\}$ as:

$$c_{ki} = f\left(W_k \cdot x_{i:i+h_k-1} + b_k\right) \tag{1}$$

Here, $b_k \in R$ is the bias term for filter W_k and $f(\cdot)$ is a non-linear function.

Each filter creates M feature maps, so we get 3 M feature maps as total. Max pooling operation can capture the most important feature for each feature map—one with the highest value. We then apply a max-pooling operation over the feature maps and take the maximum value $\hat{c}_k = max\{c_k\}$ as the feature of the filter W_k. Finally, the content representation $\hat{c} = [\hat{c}_1 \oplus \hat{c}_2 \oplus \hat{c}_3]$ is generated by concatenating each $\hat{c}_k$. We choose the max-over-time vectors as the content representation which can well reflect the local correlation in the sentence. And since we want to simplify the feature engineering, we only run one epoch to get the max-pooling vectors, without having to adjust the parameters many times.

Context Representation. Context provides rich and varied background information for sarcasm detection. For example, if the author often makes sarcastic remarks, his next comment is more likely to be sarcastic. The political forum has a higher percentage of satires than the weather forum and so on. The context representation in SCX-SD contains user embedding and topic embedding.

The user embedding tries to capture users' sarcastic tendencies. User embedding is created considering the accumulated historical posts of each user. We use *author sarcasm prior preference Pu* as user embedding. The counts of sarcastic and non-sarcastic comment for authors seen in the training data, serve a prior sarcasm frequency. $Pu_i = \{(sarc\_preference, nonsarc\_preference)|i = 1, \cdots, n\}$ is calculated as:

$$Pu_i = \left\{ \frac{Pu\{y = 1\}}{Pu\{y = 1 \text{ and } y = 0\}}, 1 - \frac{Pu\{y = 1\}}{Pu\{y = 1 \text{ and } y = 0\}} \right\} \tag{2}$$

The topic embedding is also extracted from the comments within each discussion forum. The probability of users' sarcastic comments is different under different forums, and controversial topics such as politics and sports are easier to cause sarcastic comments. The same as *Pu*, *topic sarcasm prior preference Pt_i* is:

$$Pt_i = \left\{ \frac{Pt\{y = 1\}}{Pt\{y = 1 \text{ and } y = 0\}}, 1 - \frac{Pt\{y = 1\}}{Pt\{y = 1 \text{ and } y = 0\}} \right\} \tag{3}$$

Here, y is the sarcasm label for training data. For unknown users or topics, set it to (0, 0). Subsequent experiments prove that they effectively improve the detection effect.

3.3 Sarcasm-Unlabeled Classifying

For the problem of fewer sarcastic (positive) samples and more unlabeled samples, we treat it as the Sarcasm-Unlabeled problem. Our goal is training a classifier that can distinguish between positive and negative, with only a few labeled sarcasm data and easily accessible unlabeled data. Following content and context modeling, we concatenate several embeddings as the final vectors $\vec{q} = [\hat{c} \oplus Pu \oplus Pt]$. We adopt the sarcasm-unlabeled bagging method to generate several base classifiers and predict test data by calculating the mean score of them. Here are the steps and the algorithm of the sarcasm-unlabeled bagging method.

- Create a training set consisting of all sarcastic data and K samples randomly selected from unlabeled data, with replacement. K is the number of sarcastic data.
- Build a base classifier from these "bootstrap" samples, where sarcastic and unlabeled data points are treated as positives (label = 1) and negatives (label = 0) respectively.
- Apply the base classifier to all unlabeled samples except the selected k samples in the unlabeled dataset – hereafter called OOB (out of the bag) points – and record their scores.
- Repeat the above three steps T times, get the prediction probability by averaging the results of T classifiers and obtain the average OOB score of each sample.

Algorithm 1 Sarcasm Unlabeled bagging algorithm

INPUT: *sarcasm set* P, *unlabeled set* U, K: size of sarcasm samples, T : number of bootstraps

OUTPUT: a function $f: X \rightarrow R$, a score $s: U \rightarrow R$

Initialize $\forall x \in U, n(x) \leftarrow 0, f(x) \leftarrow 0$

For t =1 to T **do**

    Draw a subset u_t of size K from U

    Train a classifier f_t to discriminate P against u_t

    For any $x \in U \backslash u_t$, **update**:

$$f(x) \leftarrow f(x) + f_t(x),$$
$$n(x) \leftarrow n(x) + 1,$$

end for

Return

$$f = \frac{1}{T}\Sigma_{t=1}^{T} f_t, \quad s(x) = \left. f(x) \middle/ n(x) \right. \ for \ x \in U$$

The bagging strategy is used to construct multiple base classifiers for the sampled training data set, and the average probability given by each classifier is calculated to label test data "1" or "0". OOB scores can estimate the generalization error. When the labeled samples, especially the labeled sarcasm samples are lacking, the method performs effectively than the supervised method.

4 Experiments

In this section, we introduce the dataset and training details at first, and then analyze the experimental results.

4.1 Dataset

There are some collected datasets for sarcasm detection. The sources used to get sarcastic and non-sarcastic statements mainly contain Reddit, Twitter, Internet Argument Corpus (IAC), and so on. Twitter has been the most common source for sarcasm detection in previous corpora. However, tweets are often written in abbreviated language and the frequent appearance of hashtagged tokens in the statement easily caused confusion. IAC has also been used as a source of sarcastic comments, but it just contains arguments. Reddit is a social media site, users post on topic-specific forums known as subreddits and comment on submissions.

With the release of the Self-Annotated Reddit Corpus (SARC[1]), Khodak et al. [27] created an advantage for context-based sarcasm detection. SARC includes unprecedented 533 M comments. If the author marked the comment with the "/s" tag, then it is considered sarcastic. Each comment came with a sarcasm label, author, the subreddit it appeared in, the comment voting score, the published date, and identifiers linking back to the original dataset of all comments.

We choose SARC for several reasons. Firstly, it is larger than other Reddit datasets and suitable for training complex models. Secondly, it has user and subreddit information which is a strong guarantee for context embedding in our model. Thirdly, Reddit comments are not constrained by length and its self-annotated labels are in a more standardized form because they are largely anonymous and do not rely on a shared context to communicate sarcasm. In order to make a comparison with other models like CASCADE, we also consider the three variants of the SARC dataset.

- **Main balanced:** It is the primary dataset that contains an equal amount of sarcastic and non-sarcastic comments. It contains comments from 1246058 users (118940 in training set and 56118 in testing set) distributed across 6534 forums (3868 in training set and 2666 in testing set).
- **Main unbalanced:** To simulate the proportion of sarcastic samples in real scenarios less than that of non-sarcastic, we constructed the unbalanced dataset with sarcastic: non-sarcastic = 1:3.
- **Pol balanced:** To assessment our context embedding effect, we choose the subset of main corresponding to comments in topic/*r/politics*. Table 1 shows the dataset details used in our experiment.

4.2 Training Details

For comments, each word is represented as word embedding using the pre-trained Fast Text embedding and *embedding_dim* is 300. In CNNs, each input is either restricted or

---

[1] http://nlp.cs.princeton.edu/SARC/2.0/.

Table 1. Details of SARC dataset.

		Train set		Test set	
		Sarcastic	Non-sarcastic	Sarcastic	Non-sarcastic
Main	Balanced	77351	77351	32333	32333
	Unbalanced	25784	77351	10778	32333
Pol balanced		6834	6834	1703	1703

padded to 100 words for uniformity. Other parameters are: $num\_filters = 128, filter\_size = \{3, 4, 5\}, batch\_size = 4096$, and $f = ReLU$.

To fit the sarcasm-unlabeled bagging method, we set a *hidden_size* of sarcasm data in the training set and *n_estimators* which stands for the number of base classifiers. We select the decision tree as the base classifier. *Hidden_size* means how many sarcasm samples be treated as unlabeled, and in this way we can simulate an experimental environment with few positive samples. The parameters and features we set for different datasets are shown in Table 2. Positive/Total is the percentage of positive examples after hiding to total data.

Table 2. The Sarcasm-Unlabeled bagging parameters of three datasets.

	Hidden_size	Positive/Total	Feature	n_estimators
Main balanced	62000	15351/154702	$\hat{c}, Pu, Pt$	100
Main unbalanced	15000	10784/103135	$\hat{c}, Pu, Pt$	100
Pol balanced	5500	1334/13668	$\hat{c}, Pu$	100

We treat sarcastic samples as positive samples and non-sarcastic samples as negative samples. The evaluation metrics for the method are *Accuracy, Precision, Recall*, and *F1-score. Accuracy* reflects the overall classification effect of the model, while the other three can respectively represent the recognition of a certain category.

4.3 Results and Analysis

We compare our model with the following baseline models, and all of them base on the SARC dataset. In Table 3 we report the means result of ten runs with several evaluation metrics, unlike previous work only reports accuracy or F1-score. We consider the comprehensive evaluation model of multiple indicators, not only the overall accuracy of the model, but also the accuracy of the two categories of 0 and 1.

- **CASCADE [17]:** CASCADE extracts stylometric and personality features of users and discourse embedding through historical comments, and models a supervised classification for hybrid training with content presentation.

- **Khodak et al.** [27]: Khodak et al. get the strongest baseline numbers by Bag-of-Bigrams.
- **Kolchinski et al.** [19]: They use bidirectional RNN with GRU cells as a baseline and add Bayesian prior probabilities representing user behavior.

It can be observed that when we use only ten percent sarcasm samples of the entire dataset as labeled data (Sect. 4.2), the detection effect of SCX-SD can compete with the current advanced model, and sometimes even better. In the main-balanced dataset, SCX-SD achieves the best effect, no matter for class 1, class 0 or the average of two classes. In the main-unbalanced dataset, SCX-SD performs a bit behind CASCADE. As for pol-balanced dataset, it presents as the same level as others.

Table 3. Several evaluation metrics for the comparison methods.

Models		Main								Pol balanced			
		Balanced				Unbalanced				Acc	F1	Pre	Rec
		Acc	F1	Pre	Rec	Acc	F1	Pre	Rec				
CASCADE		77.0	77.0	–	–	79.0	86.0	–	–	74.0	75.0	–	–
Khodak et al.		–	75.0	–	–	–	–	–	–	–	76.5	–	–
Kolchinski et al.		–	74.0	–	–	–	–	–	–	–	77.6	–	–
SCX-SD	avg	**78.5**	**78.2**	**79.9**	**78.5**	73.3	75.1	81.7	73.4	72.6	72.3	73.8	72.7
	0		75.8	86.4	67.5		79.8	92.9	69.9		69.4	78.9	61.9
	1		**80.6**	**73.3**	**89.5**		61.2	48.2	83.9		**75.3**	68.6	83.4

Surprisingly, our features were remarkably simple, without a complex analysis of user personalities. Our content representation is extracted once through CNN's max-pooling layer, and the context representation is derived by calculating user and topic sarcastic preferences. SCX-SD in both cases allows more flexibility on the task. We have also tried user stylometric features, user personality features, and topic embedding as posted in CASCADE to enrich context representation. However, it does not seem to improve the detection model. Besides, we use the text CNN to extract the representation of ancestors' comments which reply to the comment and hope to increase the accuracy, but it has little effect. Hence one can see that SCX-SD uses as few features and labeled data as possible to achieve a satisfactory detection effect, which is more suitable for the actual scenes.

In the real world application sceneries, speed is an important criterion in this line of research. SCX-SD is divided into two parts: building decision tree and predicting the test set. The relatively time-consuming part is the training phase, so researchers can store the trained tree in advance and use it directly to categorize the test data while the time to predict a test set of 60,000 is 2.85 s. This speed is acceptable and competitive.

About the parameters hidden_size and n_estimators, it is concluded that when hidden_size continues to increase, the detection accuracy hardly decreases, and when n_estimators decreases, the model effect will not deteriorate until n < 10. It demonstrates the powerful generalization of the sarcasm-unlabeled bagging method and the strong representational nature of our features. Text CNN can be replaced by Gated Recurrent Unit (GRU) or BiGRU and it may enhance the effect, but we don't talk about it in our paper.

4.4 Ablation Study

To explore the role of our contextual embedding, first we test performance for the content-based CNN (Table 4. row 1). This setting provides the worst relative performance with almost 20% lower accuracy than optimal ones. Next, we add contextual features to this network. With the addition of *author sarcasm prior probability Pu* (row 2), the performance of the model on three datasets has improved 20% significantly. The *topic sarcasm prior probability Pt* (row 3) further strengthens the effect of the model. As mentioned above, both the users and the forums have their sarcasm habits, which can provide a prior knowledge for judging sarcasm, and it can help us to distinguish between satirical and non-satirical comments.

Table 4. Comparison with variants of SCX-SD.

Models		Main								Pol balanced			
		Balanced				Unbalanced				Acc	F1	Pre	Rec
		Acc	F1	Pre	Rec	Acc	F1	Pre	Rec				
Content only	avg	54.6	54.3	54.8	54.6	54.2	57.3	66.5	54.2	53.2	53.2	53.2	53.2
	0		50.4	55.6	46.1		63.5	79.0	53.1		53.2	53.3	53.2
	1		58.2	53.9	63.1		38.7	29.1	57.7		53.3	53.3	53.2
Content + Pu	avg	77.4	77.1	77.3	77.3	71.2	73.1	81.6	71.2	72.6	72.3	73.8	72.7
	0		74.5	85.2	66.2		77.5	93.5	66.2		69.4	78.9	61.9
	1		79.6	72.4	88.5		62.1	46.7	64.8		75.3	68.6	83.4
Content + Pu + Pt	avg	78.5	78.2	79.9	78.5	73.3	75.1	81.7	73.4	–	–		
	0		75.8	86.4	67.5		79.8	92.9	69.9		–		
	1		80.6	73.3	89.5		61.2	48.2	83.9		–		

5 Conclusion

Verbal sarcasm is one of the main error sources in sentiment analysis tasks. The sheer volume of labeling and the complexity of feature engineering has kept researchers away. In this paper, we proposed the sarcasm-unlabeled method for contextual sarcasm detection, with the concatenation of content representation based on CNN and sarcastic preference embedding. Experiment results show that SCX-SD (our method) achieves more accurate detection results with only ten percent labeled data, in an uncomplex neural network. We will explore the interpretability of sarcastic features in the future.

Acknowledgments. This work is supported by the National Natural Science Foundation of China (No.71871090).

References

1. Madasu, A., Rao, V.A.: Sequential learning of convolutional features for effective text classification. In: Proceedings of the 2019 Conference on Empirical Methods in Natural Language Processing and the 9th International Joint Conference on Natural Language Processing (EMNLP-IJCNLP), pp. 5662–5671 (2019)
2. Zou, Y., Gui, T., Zhang, Q., et al.: A lexicon-based supervised attention model for neural sentiment analysis. In: Proceedings of the 27th International Conference on Computational Linguistics, pp. 868–877 (2018)
3. Porwal, S., Ostwal, G., Phadtare, A., et al.: Sarcasm detection using recurrent neural network. In: 2018 Second International Conference on Intelligent Computing and Control Systems, pp. 746–748. IEEE (2018)
4. Samonte, M.J.C., Dollete, C.J.T., Capanas, P.M.M., et al.: Sentence-level sarcasm detection in English and Filipino tweets. In: Proceedings of the 4th International Conference on Industrial and Business Engineering, pp. 181–186 (2018)
5. Oprea, S., Magdy, W.: Exploring author context for detecting intended vs perceived sarcasm. In: Proceedings of the 57th Annual Meeting of the Association for Computational Linguistics, pp. 2854–2859 (2019)
6. Iida, R., Kruengkrai, C., Ishida, R., et al.: Exploiting Background Knowledge in Compact Answer Generation for Why-Questions. In: Proceedings of the AAAI Conference on Artificial Intelligence, vol. 33, pp. 142–151 (2019)
7. Li, Q., Zhang, Q., Si, L.: Rumor detection by exploiting user credibility information, attention and multi-task learning. In: Proceedings of the 57th Annual Meeting of the Association for Computational Linguistics, pp. 1173–1179 (2019)
8. Huang, T., Shen, G., Deng, Z.H.: Leap-LSTM: enhancing long short-term memory for text categorization. In: Proceedings of the 28th International Joint Conference on Artificial Intelligence, pp. 5017–5023. AAAI Press (2019)
9. Chen, X., Zhang, Y., Qin, Z.: Dynamic explainable recommendation based on neural attentive models. In: Proceedings of the AAAI Conference on Artificial Intelligence, vol. 33, pp. 53–60 (2019)
10. Ghosh, A., Veale, T.: Fracking sarcasm using neural network. In: Proceedings of the 7th Workshop on Computational Approaches to Subjectivity, Sentiment and Social Media Analysis, pp. 161–169 (2016)
11. Majumder, N., Poria, S., Peng, H., et al.: Sentiment and sarcasm classification with multitask learning. IEEE Intell. Syst. **34**(3), 38–43 (2019)
12. Shimura, K., Li, J., Fukumoto, F.: Text categorization by learning predominant sense of words as auxiliary task. In: Proceedings of the 57th Annual Meeting of the Association for Computational Linguistics, pp. 1109–1119 (2019)
13. Liu, L., Priestley, J.L., Zhou, Y., et al.: A2Text-Net: a novel deep neural network for sarcasm detection. In: 2019 IEEE First International Conference on Cognitive Machine Intelligence (CogMI), pp. 118–126. IEEE (2019)

14. Niu, W., Lei, J., Tong, E., et al.: Context-aware service ranking in wireless sensor networks. J. Netw. Syst. Manage. **22**(1), 50–74 (2014)
15. Liu, W., Salzmann, M., Fua, P.: Context-aware crowd counting. In: Proceedings of the IEEE Conference on Computer Vision and Pattern Recognition, pp. 5099–5108 (2019)
16. Niu, W., Li, G., Tang, H., et al.: CARSA: a context-aware reasoning-based service agent model for AI planning of web service composition. J. Netw. Comput. Appl. **34**(5), 1757–1770 (2011)
17. Hazarika, D., Poria, S., Gorantla, S., et al.: CASCADE: contextual sarcasm detection in online discussion forums. In: Proceedings of the 27th International Conference on Computational Linguistics, pp. 1837–1848 (2018)
18. Ghosh, A., Veale, T.: Magnets for sarcasm: Making sarcasm detection timely, contextual and very personal. In: Proceedings of the 2017 Conference on Empirical Methods in Natural Language Processing, pp. 482–491 (2017)
19. Kolchinski, Y.A., Potts, C.: Representing social media users for sarcasm detection. In: Proceedings of the 2018 Conference on Empirical Methods in Natural Language Processing, pp. 1115–1121 (2018)
20. Kiyono, S., Suzuki, J., Inui, K.: Mixture of expert/imitator networks: scalable semi-supervised learning framework. In: Proceedings of the AAAI Conference on Artificial Intelligence, vol. 33, pp. 4073–4081 (2019)
21. Zhang, Z., Zhai, S.: Semisupervised autoencoder for sentiment analysis: U.S. Patent Application 15/838,000[P], 14 June 2018
22. Cai, R., Lapata, M.: Semi-supervised semantic role labeling with cross-view training. In: Proceedings of the 2019 Conference on Empirical Methods in Natural Language Processing and the 9th International Joint Conference on Natural Language Processing (EMNLP-IJCNLP), pp. 1017–1026 (2019)
23. Xu, W., Tan, Y.: Semisupervised text classification by variational autoencoder. IEEE Trans. Neural Netw. Learn. Syst. **31**(1), 295–308 (2019)
24. Lee, G.T., Kim, C.O., Song, M.: Semisupervised sentiment analysis method for online text reviews. J. Inform. Sci. (2020)
25. Bojanowski, P., Grave, E., Joulin, A., et al.: Enriching word vectors with subword information. Trans. Assoc. Comput. Linguist. **5**, 135–146 (2017)
26. Kim, Y.: Convolutional neural networks for sentence classification. In: Proceedings of the 2014 Conference on Empirical Methods in Natural Language Processing (EMNLP), pp. 1746–1751 (2014)
27. Khodak, M., Saunshi, N., Vodrahalli, K.: A large self-annotated corpus for sarcasm. In: Proceedings of the Eleventh International Conference on Language Resources and Evaluation (LREC 2018) (2018)

End-to-End Multi-task Learning for Allusion Detection in Ancient Chinese Poems

Lei Liu$^{(\boxtimes)}$, Xiaoyang Chen , and Ben He$^{(\boxtimes)}$

School of Computer Science and Technology, University of Chinese Academy
of Science, Beijing, China
{liulei175,chenxiaoyang19}@mails.ucas.ac.cn, benhe@ucas.ac.cn

Abstract. Much efforts have been devoted to research about ancient
Chinese poems. However, tasks around allusions, a fundamental element of ancient Chinese poetry, has received little attention. To mitigate
this gap, we introduce three allusion tasks: allusion entity recognition
(AER), allusion source identification (ASI), and allusion entity classification (AEC). For each task, we create a large corpus extracted from allusion dictionary. We explore the performance of two learning strategies:
single-task model and allusion hierarchical multi-task learning (AHMTL)
model. Compared with the single-task model, experimental results show
that the AHMTL model improves each task's overall performance by
formulating relationship between tasks. In addition, poem readability, a
downstream task of allusion tasks, is combined to gain improvement in
the F1-score by 1.4%.

Keywords: Allusion detection · Allusion entity recognition · Allusion
classification · Allusion source identification · Multi-task learning

1 Introduction

An allusion is usually viewed as quotations of ancient stories or phrases with
origins. Unlike ordinary words, allusion words link to an external reference to
events, stories, or myths. Using allusion in poetry is a typical writing skill adored
by poets to express rich emotions in short words. However, it also brings more
complications, leading to that identifying and understanding allusion becomes
a challenging work for common readers. Relying on auxiliary materials such as
a professional allusions dictionary helps readers grasp the meaning of allusion
words. Nevertheless, this way is time-consuming and hard to expand reading due
to the limited auxiliary materials. Therefore, developing an allusion application
has practical value by automatically providing low-cost but high-quality hints
and interpretation of allusion as poem reading.

Recently, poetry analysis has emerged as a hot research topic, covering poetry
generation [20] and ancient poetry sentiment analysis [13]. These studies greatly

© Springer Nature Switzerland AG 2020
G. Li et al. (Eds.): KSEM 2020, LNAI 12275, pp. 300–311, 2020.
https://doi.org/10.1007/978-3-030-55393-7_27

enrich expressive form of classical poetry in modern literature. However, there are still few studies on allusion, an important linguistic component of poems.

In this work, we conduct pilot study on allusions by introducing four tasks, including allusion entity recognition (AER), allusion entities classification (AEC), allusion source-identifying (ASI), as well as poetry readability (PR) assessment, a downstream task of the allusion related tasks. These four tasks cover a complete procedure from detecting allusion entities to finding out its source text, and finally, integrated into a downstream PR task.

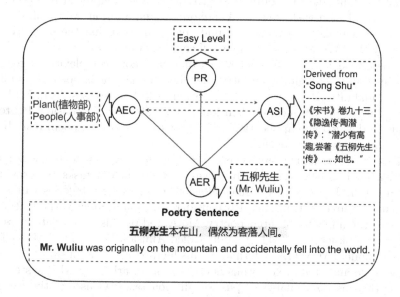

Fig. 1. Example of Allusion 'Mr. Wuliu'

Here we give a brief description of each task and the relationship between them by giving an example shown in Fig. 1. The goal of allusion entities recognition is to identify whether a poem sentence contains an allusion entity and locate the beginning and ending position of allusion. For example, there exists an allusion 'Mr. Wuliu' in Fig. 1, representing a famous poet call TaoYuanming, which is used to express the willingness of adoring rural life and tiredness of officialdom. Instead of annotated by experts, the AER model can identify this allusion automatically. Based on allusion identified by AER, we can continue to make a further profound analysis of allusion in semantics by subsequent AEC and ASI tasks.

Based on detected allusion words, we further group them into different classes according to their similarity in semantics or topic called allusion entity classification (AEC). AEC makes it available to manage categories of allusions appearing in the poem library and distinguish them according to tags, which provides a handy way to look up specific classes of allusion for poetry lovers during poetry composition. It also presents a sharp sense for readers of poetry in a short time without understanding details. Here, 'Mr. Wuliu' is put in the 'Plant' and 'People' class. It should be noted that an allusion has multiple tags, possibly due to the diversity in meaning.

Another follow-up application is allusion source-identifying (ASI), which aims to determine the origins or source of allusion among numerous external literature, which is ace in the hole to reduce a heavy reading burden of poetry. Since the interpretation of allusion words only derive from source text instead of literal meaning, ASI effectively bridges the knowledge gap between detecting allusion and understanding it. The source text gives a full and comprehensive story represented by allusion with context. Therefore, taking allusion 'Mr. Wuliu' as an example, readers easily perceive the underlying emotion that the poet tried to express by 'Mr. Wuliu' after learning full content of its source text.

As described above, using allusion can bring more complexity in reading poem. In other words, allusion entities play an essential role in poem readability. Poetry readability (PR) is a text classification task that assesses the difficulty of ancient Chinese poetry in comprehension proposed by [21]. We consider PR to be an allusion downstream application and tempt to measure the poem readability by involving features of allusion.

Apart from discussing single-task, Fig. 1 presents heavy topological inter-dependencies between tasks. AER serves as fundamental tasks by providing allusion entity. Then, AEC classifies allusion into specific categories while ASI finds out its source text. Finally, PR considers allusion as an essential feature in measuring the difficulty of input poetry. Motivated by this close relatedness, we build multi-task learning (MTL) model called Allusion Hierarchical Multi-Task Learning model (AHMTL) to formalize this hierarchical relationship.

One large and creditable corpus is crucial for an arbitrary NLP system. At present, there is no available corpus for allusion tasks. Consequently, we construct a high-quality corpus[1] of allusions from scratch by extracting example sentences of dictionaries and make it public to the general serving as further researches.

To summarize, we proposed four allusion tasks that aim to build an allusion system for aiding in reading ancient Chinese poetry. We construct a large allusion library preparing for training single model and explore how to model the inter-dependencies between tasks by the AHMTL framework. The paper is organized as follows: In Sect. 2, we present the related works. In Sect. 3, we introduce our model architecture. In Sect. 4, we describe the datasets and experimental settings. In Sect. 5, we resolve and analyze our results and make a further conclusion in Sect. 6.

2 Related Work

In this section, we briefly introduce the previous works related to our tasks, covering pre-trained models (PTMs), sequential labeling model, and multi-task learning (MTL).

---

[1] https://github.com/lailoo/Allusion_detection.

PTMs have shown a significant effect in improving the performance of many natural language processing tasks such as text classification or sequential labeling task [27]. By pre-training on a large corpus of unlabeled text, PTMs are capable of learning universal syntactic and semantic representations and transfers it to a new NLP task, which avoids training a new model from scratch and alleviates the overfitting problem when training numerous parameters on a small corpus. Currently, the development of PTMs has been advanced from shallow global word embedding such as Word2Vec [22] and GloVe [24] to deep contextualized word representation of ELMo [25], GPT [28] and BERT [8]. Compared with many PTMs, BERT model performs better by utilizing bidirectional encoder from transformer [32] to learn word representation based on its left and right context jointly.

Besides, BERT also provides a simple architecture for text classification with taking a degenerate text-$\varnothing$ pair as input and feeding the $[CLS]$ representation into an output layer to make a prediction.

For AER, we regard it as a sequential labeling task similar to conventional NER to recognize allusion words step-by-step. There exist vast prior works to apply neural sequential labeling model in NER task. [18] introduced and neural architecture with bidirectional LSTMs (BiLSTM) and conditional random fields (CRF) to make tagging decisions without resorting to external resources and hand-crafted features. [30] addressed issues of NER on the Portuguese language by employing a pre-trained BERT with CRF layer architecture and explored the effect of two training strategies of feature-based and fine-tuning. As for poetry, [16] extended the scope of conventional NER [26] from normal text into English poetry. By applying BiLSTM encoder, they build a sequential labeling model trained on poetry corpus, achieving better results comparing the poorly performance of off-the-shelf taggers like SpaCy [6], proving the power of a sequential model when processing poems text.

MTL is used to share knowledge between related tasks [9]. [31] argues that a low-level layer is used to encode general information while the high-level layers focus on task-oriented semantics. [7] collected 12 biomedical entity datasets and applied MTL to augment the overall scale of entities set by mixing those datasets. [29] proposed a hierarchical MTL architecture to combines four related tasks with different corpus, achieving STOA results.

3   Model

This section will give a full description of our model, including four single-model for individual task and a hierarchical architecture, namely AHMTL, for learning the relationship between four tasks. As shown in Fig. 2, our model consists of three primary components: input encoder layer, middle encoder layer, and task-specific layer. Pre-trained BERT shared by all tasks is served as input encoder which is responsible for mapping the input text into contextualized representations. Middle encoder shared by tasks except for AER is used to attain sentence or document representation for whole input text by attention mechanism or

BiLSTM encoder. The selection of middle encoder leads to differences in model complexity. Finally, the task-specific layer makes predictions for each task.

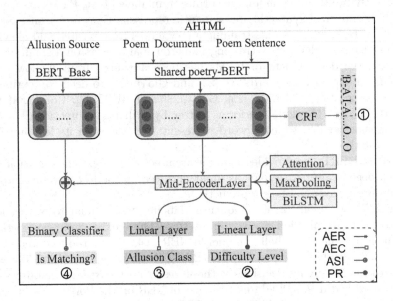

Fig. 2. The architecture of AHMTL

It is noted that Fig. 2 lists four paths of data flow. Practically, each path denotes a single model for a specific task. In the following subsection, we will discuss the details of the models.

3.1 PTM: Poetry-BERT Model

Existing BERT-Base model provided by [3] is trained on modern Chinese Wikipedia, lacking syntax and semantics from ancient Chinese. It can be enhanced by incorporating domain knowledge.

[15] has already trained BERT for ancient Chinese and applied it in segment sentence task. Unfortunately, the model is not publicly available. Inspired by their work, we conducted a further unsupervised pre-training to tailor poetry-BERT on the corpus with size of 3.1G drawn from SiKuQuanShu [11] and ancient Chinese poetry library [1, 4, 33] with more than 800,000 poem documents.

Further training is regarded as a fine-tuning procedure with parameters initialized by BERT-Base [3] model, keeping that target tasks of training still are 'masked LM' and 'next sentence prediction.' We also keep default settings of text processing in line with BERT-Base [8]. All raw texts are pre-processed into a signal file following standard format with a sentence each line and a blinking line between document. Finally, we run the pre-training script[2] lasting one million steps to tune our domain BERT, i.e., poetry-BERT.

---

[2] https://github.com/google-research/bert.

3.2 Single-Model for Allusion Entity Recognition (AER)

We formalize AER as a sequential labeling task to identify the location of an allusion entity instance. Therefore, the goal of AER is to detect span of allusion entity $A = (B - A, I - A, ...I - A)$ in a poem sentence $S, S = [w_1; w_2; ::::; w_d]$. $B - A, I - A$ denote the beginning and ending of allusion words respectively. The remaining words are classified into non-allusion tag: O.

BERT-CRF architecture is adopted to formalize the AER task following work [30] as shown in the path of data flow (No. 1) in Fig. 2. As a sentence-to-vector encoder, poetry-BERT takes poem sentence S as input and outputs contextual representation $T_i \in \mathbb{R}^h$ for i^{th} token by extracting only the hidden vector of the final layer where h represents the hidden dimension. Finally, the sequence of token vector T_i are fed into Linear-Chain CRF layer, which serves as a sequential classifier to make allusion prediction for each token.

3.3 Single-Model for Allusion Entities Classification (AEC) and Poem Readability (PR)

AEC aims at grouping allusion identified by AER into different classes in terms of semantics. However, currently AER model performs rather poorly, contributing to worse results if AEC made predictions directly based on the output of AER on word-level. Considering that allusion expresses the core idea of a poem sentence, we hypothesize that the semantics of poem sentences are equivalent to allusion. In this way, we convert AEC task from word-level classification into sentence-level.

Poem readability is another classification task but on document-level. Despite existing subtle differences in granularity with AEC, we still adopt the same architecture for both of them.

Both paths, No. 2 and No. 3, in Fig. 2 shows framework of single-model for AEC and PR. It is composed of poetry-BERT, following middle encoder and upper linear prediction layer. Firstly, the poem sentence S_s for AEC or poem document S_d for PR is fed into poetry-BERT to attain vector T_i for each token w_i. Then, all of the vectors are delivered to the middle encoder to generate sentence or document representation $D_v, D_v \in \mathbb{R}^k$. Finally, the last task-specific layer makes a prediction by projecting D_v into class space, i.e., from $\mathbb{R}^k$ to $\mathbb{R}^c$ where c denotes the number of class.

3.4 Single-Model for Allusion Source Identification (ASI)

Similar to the assumption in AEC, we simplify ASI into text matching tasks intending to identify whether a poem sentence is relevant to the original text in semantics.

ASI model derives from prior work [28] as shown by Path No. 4 in Fig. 2. Unlike other tasks, the input is text pair (S_r, S_s), where S_s denotes poem sentence, and S_r represents allusion's external reference. Owing to expressed in model Chinese, S_r is fed into BERT-Base model while S_s is processed by poetry-BERT. Then, both kinds of BERT generate embedding (D_r, D_s) for input text with same dimension jointly. Instead of utilizing middle encoder like S_s, we simply extract representation of token $[CLS]$ as final D_r. Please refer to paper [8] to learn function of token $[CLS]$. Afterward, the upper linear layer makes binary classification to determine whether the semantics of a poem sentence matches source text according to the concatenation of (D_r, D_s).

3.5   AHMTL: Allusion Hierarchical Multi-task Learning

Inspired by close inter-dependencies between tasks, we orchestrate a unified framework deriving from previous work [29] to combine four single tasks together in a hierarchical way.

Due to all single models designed based on BERT, it is easier to gather them together by setting BERT as a shared layer. Undoubtedly, AER lies the bottom layer for its low-level semantics. Then ascending layers are supervised by labels of high-level tasks, aiming to furthermore post-process allusion. In the end, the task-specific layer is arranged at the top of AHMTL.

Because there are no unified datasets with label suited for all tasks, asynchronous parameter updating is applied to adjust an inductive bias. AHMTL randomly selects one target task and updates parameters of the shared layer and corresponding task-specific layer while keeping the parameters of other task-specific layers frozen. Therefore, the loss function of AHMTL can be deemed as a weighted sum of every single model where the weight is measured by the scale of their corpus.

4   Experiments

4.1   Corpus

Two allusion dictionary: 'QuanTangShiDianGuCiDian' [10] and 'ShiDianXin-Bian' [5] serve as our data source for extracting allusion annotated by experts, guaranteeing the quality of the final dataset. The first dictionary contains 3451 allusions covering almost allusions used in the Tang poem. Each allusion is accompanied by numerous example sentences, interpretation, and source text which are well organized. Hence it is convenient to collect plenty of training instances for both AER and ASI. Regrettably, allusion from [10] is unannotated with tag, which is unable to train model for AEC. Therefore, we select the latter dictionary [5] with 11 categories of allusion to generate training dataset suited for AEC task.

The structuralization of raw text makes it possible to construct a large allusion corpus automatically. Firstly, we need to extract poem sentences with an allusion from the dictionary with the help of regular expression and filter out useless contents with messy code or duplication. During extraction, much efforts are devoted to compiling lots of scripts to adapt diversity of patterns such as single examples and multi-example. After obtaining the initial version of allusion library, we continue to examine every case to ensure the correctness.

Based on example sentences, we make post-processing furthermore to construct corpus for AER, AEC, and ASI, respectively. For AER, the BIO tagging scheme[26] is adopted to generate a label sequence. For each sentence, we generate a tag sequence by exactly matching words in sentences with allusion words. Then, we extend AER corpus to suit ASI by matching corresponding source text for each allusion example sentence to compose text pair as a positive instance. To enhance the generation of models, we also append the negative pair for each instance into ASI corpus. The proportion of positive and negative instances is 1:1 in training while 1:9 in test set. Unlike AER and ASI, AEC corpus only consists of poem sentences with an allusion class, whose scale is rather small. For poem readability, we choose APRD corpus proposed by [21] as our training dataset. APRD contains 1915 samples which are divided into three difficulty level: easy-level, middle-level, difficulty-level. Table 1 lists statistical information of corpus for each task.

Table 1. The details of corpus for each task

Corpus	Case form	Size	Class num
AER corpus	Sentence	21,594	–
ASI corpus	Sentence	215,940	1
AEC corpus	Sentence	2,666	11
APRD	Poem	1,915	3

4.2 Metrics for Each Task

For tasks of AER, AEC and PR, Precision, Recall, and F1 score are leveraged as primary metrics. Although those metrics are the same in both sequential labeling tasks and text classification, the actual object processed exist differences. The metrics take into account output on entity-level exactly matched for AER [19] as opposed to instance-level for AEC and poem readability [21].

As for ASI, we argue that ranking position of positive instances indicate effectiveness of model. Therefore, Mean Reciprocal Rank (MRR) [23] is applied to evaluate the performance of ASI model for single allusion.

4.3 Method Setting

The AHMTL model involves several vital components. We attempt to explore the effectiveness of them by a list of method settings as follows.

– **AER, AEC, ASI, PR:** single-model for four tasks individually as discussed in model section.

- **AER+AEC, AER+ASI, AER+PR, AER+AEC+ASI:** denotes grouped tasks based on AER model.
- **AHMTL+BERT-Base:** replaces the poetry-BERT model of default AHMTL with BERT-Base
- **AHMTL:** default AHMTL with poetry-BERT model.

4.4 Implementation Details

As described in the pre-trained model section, the poetry-BERT is the output of incremental training using ancient Chinese corpus based on BERT-Base, which is initialized with the pre-trained weights of Chinese-L-12-H-768-A-12 model. Here we keep most hyper-parameters of model as default value such as $H = 768$, character-based tokenization.

AHMTL is based on interfaces designed by Kashgari [2], a friendly NLP framework based on TensorFlow1.x. To achieve better reproducibility of our work, we reimplement the Kashgari framework in TensorFlow2.0 and extend the text-matching module. HMTL [29] model implemented by PyTorch resolved the issue of multi-input and multi-output appearing in MTL by task selector. We also apply this solution to our AHMTL model. Moreover, owing to the shared poetry-BERT as the first layer of the model, adapter [14] provided by keras-bert [12], a light BERT implementation, is utilized as a fine-tuning strategy.

We use a batch size of 64. Learning rates with initial value 1e−3 dynamically declines when the metrics stop improving lasting 5 epochs. The maximum numbers of epochs are 50, but early-stopping is employed to prevent overfitting. Besides, we use the customized Adam [17] as our optimizer. The more detailed experimental settings are listed in our allusion project published early. Besides, we report the average results by 5-fold cross-validation to attain more precise evaluation.

5   Results

Table 2 summarizes the results of our experiments. In a word, AHMTL achieves promising performance on three tasks, namely AEC (+1.3 in F1 score), ASI (+2.0 in MRR) and poem readability (+1.4 in F1 score). It suggests that the shared poetry-BERT layer can learn universal information across tasks, aiding in transferring knowledge to improve generalization. To explore more factors that affect the output of model, we conduct comparative trials on different scenarios.

Single Models vs. AHMTL. Table 2 presents the performance of AHMTL, which promotes the improvement of AEC, ASI, and PR which trains model on the small corpus. However, the results of AER appear little fluctuation. One reason is that model is easily impacted by imbalance of the corpus scale between tasks. During training, parameters of AHMTL is primarily dominated by AER, leading to less sensitivity to knowledge from other tasks. By contrast, the better results of the remaining task show that AER is beneficial to its downstream application.

Table 2. Results (%) of AHMTL and single model

Model	AER_t			AEC_t			ASI_t	Poem-readability (PR_t)		
	P	R	F1	P	R	F1	MRR	P	R	F1
AER	35.4	25.8	29.5	–	–	–	–	–	–	–
AEC	–	–	–	**35.6**	26.7	27.6	–	–	–	–
ASI	–	–	–	–	–	–	48.9	–	–	–
PR	–	–	–	–	–	–	–	70.1	65.2	66.7
AER+AEC	45.5	30.3	36.2	14.8	12.0	9.9	–	–	–	–
AER+ASI	40.7	**33.6**	**36.6**	–	–	–	**55.0**	–	–	–
AER+PR	41.0	28.7	32.1	–	–	–	–	66.3	61.3	60.0
AER+AEC+ASI	**51.8**	24.8	32.1	19.7	45.5	23	37.7	–	–	–
AHMTL+BERT-Base	42.3	28	33.5	21.8	**36.7**	16.7	43	63.4	59.9	57.9
AHMTL	31.4	26.7	28.7	33.3	29.3	**29.0**	51.1	**70.4**	**66.7**	**68**

Progressive Tasks Group Based on AER. To further analyze the contribution of AER, we combine AER with other tasks into mini-group. Benefit from outstanding expansibility AHMTL model, mini-AHMTL is easy to implement by removing the needless task via configuring the task selector. Table 2 also reports the experiment results for each group. More surprisingly, there is a significant improvement in AER and ASI, as opposed to a considerable reduction to AEC and PR. The mutual-improvement of AER and ASI suggests that the source text is helpful for allusion detecting and vice versa. However, the severe imbalance between AER and AEC or PR will bring more noise into models.

Effect of Pre-trained Domain BERT Model. We perform a comparing experiment between domain BERT and universal BERT. As reported in Table 2, applying the poetry-BERT model leads a better result in most tasks such as PR and AEC, especially when the training dataset is small. Nevertheless, the effect on AER task is unexpected for its better performance. One possible reason is that although poetry-BERT has learned domain knowledge, the impact it brings to model will weaken gradually as the scale of corpus growing.

Although AHMTL has promoted the results of most tasks, all tasks' overall performance is still weak. For example, the span F1 score of single-model on AER reaches unsatisfied value by 0.29, meaning that most of the allusion is still undetected or wrongly identified. The poor performance of single-model indicates the weak power of existing NLP architectures that works well on normal text when taking poems as input.

6   Conclusion

We proposed four tasks focused on the allusion of poetry, covering a complete procedure from allusion identification to finding out its source text. Those studies have a positive effect on understanding allusion. We also construct individual corpus drawn from allusion dictionaries for each allusion task and make it public to the general, which is beneficial to further research. In addition, based on the relationship between tasks, we design a hierarchical MTL model AHMTL to combine them together, contributing to better results than single-model. Although AHMTL can bring improvement in prediction, the overall performance is unsatisfied, even training models based on poetry-BERT. On the one hand, it indicates the complexity of allusion tasks. On the other hand, studying a specific model according to the feature of allusion is necessary, which is our future work.

Acknowledgments. This work is supported in part by the National Natural Science Foundation of China (61472391) and the University of Chinese Academy of Sciences.

References

1. Chinese-poetry/Chinese-poetry. https://github.com/chinese-poetry/chinese-poetry
2. GitHub - BrikerMan/Kashgari: Kashgari is a production-level NLP Transfer learning framework built on top of tf.keras for text-labeling and text-classification, includes Word2Vec, BERT, and GPT2 Language Embedding. https://github.com/BrikerMan/Kashgari
3. Google-research/bert. https://github.com/google-research/bert
4. GuShiWenWang. https://www.gushiwen.org/
5. ShiCiDianGuCiDian. http://ch.eywedu.com/Story/
6. spaCy. Industrial-strength Natural Language Processing in Python. https://spacy.io/
7. Crichton, G., Pyysalo, S., Chiu, B., Korhonen, A.: A neural network multi-task learning approach to biomedical named entity recognition **18**(1), 368. https://doi.org/10.1186/s12859-017-1776-8. http://bmcbioinformatics.biomedcentral.com/articles/10.1186/s12859-017-1776-8
8. Devlin, J., Chang, M.W., Lee, K., Toutanova, K.: BERT: pre-training of deep bidirectional transformers for language understanding. http://arxiv.org/abs/1810.04805
9. Evgeniou, T., Pontil, M.: Regularized multi-task learning. In: Proceedings of the 2004 ACM SIGKDD International Conference on Knowledge Discovery and Data Mining - KDD 2004. p. 109. ACM Press. https://doi.org/10.1145/1014052.1014067. http://portal.acm.org/citation.cfm?doid=1014052.1014067
10. Fan, Z.: QuanTangShiDianGuCiDian
11. garychowcmu: Garychowcmu/daizhigev20. https://github.com/garychowcmu/daizhigev20
12. HG, Z.: CyberZHG/keras-bert. https://github.com/CyberZHG/keras-bert
13. Hou, Y., Frank, A.: Analyzing sentiment in classical Chinese poetry. In: Proceedings of the 9th SIGHUM Workshop on Language Technology for Cultural Heritage, Social Sciences, and Humanities (LaTeCH), pp. 15–24. Association for Computational Linguistics. https://doi.org/10.18653/v1/W15-3703. http://aclweb.org/anthology/W15-3703
14. Houlsby, N., et al.: Parameter-Efficient Transfer Learning for NLP. http://arxiv.org/abs/1902.00751
15. Hu, R., Li, S.: Knowledge representation and sentence segmentation of ancient Chinese based on deep language models. CCL

16. Iv, J.J.F.: Poetry: Identification, Entity Recognition, and Retrieval p. 131
17. Kingma, D.P., Ba, J.: Adam: a method for stochastic optimization. http://arxiv.org/abs/1412.6980
18. Lample, G., Ballesteros, M., Subramanian, S., Kawakami, K., Dyer, C.: Neural architectures for named entity recognition. In: NACCL, pp. 260–270. https://doi.org/10.18653/v1/N16-1030. http://aclweb.org/anthology/N16-1030
19. Li, J., Sun, A., Han, J., Li, C.: A survey on deep learning for named entity recognition. http://arxiv.org/abs/1812.09449
20. Liao, Y., Wang, Y., Liu, Q., Jiang, X.: GPT-based generation for classical Chinese poetry. http://arxiv.org/abs/1907.00151
21. Liu, L.: An annotated dataset for ancient Chinese poetry readability. http://cips-cl.org/static/anthology/CCL-2019/CCL-19-070.pdf
22. Mikolov, T., Sutskever, I., Chen, K., Corrado, G.S., Dean, J.: Distributed Representations of Words and Phrases and their Compositionality, p. 9
23. Minaee, S., Kalchbrenner, N., Cambria, E., Nikzad, N., Chenaghlu, M., Gao, J.: Deep learning based text classification: a comprehensive review 1(1), 42
24. Pennington, J., Socher, R., Manning, C.: Glove: global vectors for word representation. In: Proceedings of the 2014 Conference on Empirical Methods in Natural Language Processing (EMNLP), pp. 1532–1543. Association for Computational Linguistics. https://doi.org/10.3115/v1/D14-1162. http://aclweb.org/anthology/D14-1162
25. Peters, M.E., et al.: Deep contextualized word representations. http://arxiv.org/abs/1802.05365
26. Pradhan, S., Moschitti, A., Xue, N., Uryupina, O., Zhang, Y.: CoNLL-2012 Shared Task: Modeling Multilingual Unrestricted Coreference in OntoNotes p. 40
27. Qiu, X., Sun, T., Xu, Y., Shao, Y., Dai, N., Huang, X.: Pre-trained Models for Natural Language Processing: A Survey p. 28
28. Radford, A., Narasimhan, K., Salimans, T., Sutskever, I.: Improving Language Understanding by Generative Pre-Training p. 12
29. Sanh, V., Wolf, T., Ruder, S.: A hierarchical multi-task approach for learning embeddings from semantic tasks. http://arxiv.org/abs/1811.06031
30. Souza, F., Nogueira, R., Lotufo, R.: Portuguese Named Entity Recognition using BERT-CRF. http://arxiv.org/abs/1909.10649
31. Thung, K.H., Wee, C.Y.: A brief review on multi-task learning 77(22), 29705–29725. https://doi.org/10.1007/s11042-018-6463-x. http://link.springer.com/10.1007/s11042-018-6463-x
32. Vaswani, A., et al.: Attention is all you need. http://arxiv.org/abs/1706.03762
33. Werner: Werneror/Poetry. https://github.com/Werneror/Poetry

Defense of Word-Level Adversarial Attacks via Random Substitution Encoding

Zhaoyang Wang and Hongtao Wang[✉]

School of Control and Computer Engineering,
North China Electric Power University, Baoding, China
luckychizuo@gmail.com, wanght@ncepu.edu.cn
https://github.com/Raibows/RSE-Adversarial-Defense

Abstract. The adversarial attacks against deep neural networks on computer vision tasks have spawned many new technologies that help protect models from avoiding false predictions. Recently, word-level adversarial attacks on deep models of Natural Language Processing (NLP) tasks have also demonstrated strong power, e.g., fooling a sentiment classification neural network to make wrong decisions. Unfortunately, few previous literatures have discussed the defense of such word-level synonym substitution based attacks since they are hard to be perceived and detected. In this paper, we shed light on this problem and propose a novel defense framework called Random Substitution Encoding (RSE), which introduces a random substitution encoder into the training process of original neural networks. Extensive experiments on text classification tasks demonstrate the effectiveness of our framework on defense of word-level adversarial attacks, under various base and attack models.

1 Introduction

Deep Neural Network (DNN) has become one of the most popular frameworks to harvest knowledge from big data. Despite their success, the robustness of DNNs has ushered a serious problem, which has prompted the adversarial attacks on them. Adversarial attack refers to generating imperceptible perturbed examples to fool a well-trained DNN model making wrong decisions. In the Computer Vision (CV) domain, adversarial attacks against many famous DNN models have been shown to be an indisputable threat.

Recently, adversarial attacks on DNN models for Natural Language Processing (NLP) tasks have also received significant attentions. Existing attack methods can be classified into two categories: character-level attacks and word-level attacks. For character-level attacks, attackers can modify several characters of an original text to manipulate the target neural network. While character-level attacks are simple and effective, it is easy to defend when deploying a spell check and proofread algorithm before feeding the inputs into DNNs [12]. Word-level attacks substitute a set of words in original examples by their synonyms, and

© Springer Nature Switzerland AG 2020
G. Li et al. (Eds.): KSEM 2020, LNAI 12275, pp. 312–324, 2020.
https://doi.org/10.1007/978-3-030-55393-7_28

thus can preserve semantic coherence to some extent. The adversarial examples, created by word-level attackers, are more imperceptible for humans and more difficult for DNNs to defend.

Until now, there are few works on defense of adversarial attacks against NLP tasks, e.g., text classification. Most efforts had gone into increasing the model robustness by adding perturbations on word embeddings, e.g., adversarial training [5] or defensive distillation [11]. Although these approaches exhibit superior performance than base models, they assume there are no malicious attackers and could not resist word-level adversarial attacks [15]. The only work against word-level synonym adversarial attacks is [15]. It proposed a Synonym Encoding Method (SEM) which maps synonyms into the same word embeddings before training the deep models. As a result, the deep models are trained only on these examples with only fixed synonym substitutions. The reason why SEM based deep models can defend word-level attacks is that it can transform many unseen or even adversarial examples 'move' towards 'normal' examples that base models have seen. While SEM can effectively defend current best synonym adversarial attacks, it is too restrictive when the distances are large between transformed test examples and the limited training examples.

This paper takes a straightforward yet promising way towards this goal. Unlike modifying word embeddings before the training process, we put the synonyms substitutions into the training process in order to fabricate and feed models with more examples. To this end, we proposed a dynamic random synonym substitution based framework that introduces Random Substitution Encoding(RSE) between the input and the embedding layer. We also present a Random Synonym Substitution Algorithm for the training process with RSE. The RSE encodes input examples with randomly selected synonyms so as to make enough labeled neighborhood data to train a robust DNN. Note that the RSE works in both training and testing procedure, just like a dark glasses dressed on the original DNN model.

We perform extensive experiments on three benchmark datasets on text classification tasks based on three DNN base models, i.e., Word-CNN, LSTM and Bi-LSTM. The experiment results demonstrate that the proposed RSE can effectively defend word-level synonym adversarial attacks. The accuracy of these DNN models under RSE framework achieves better performance under popular word-level adversarial attacks, and is close to the accuracy on benign tests.

2   Related Work

Adversarial attack and defense are two active topics recently. In natural language processing, many tasks are facing the threat of adversarial attack, e.g., Text Classification [3,4,9], Machine Translation [2], Question & Answer [14], etc. Among them, text classification models are more vulnerable and become the targets of malicious adversaries. The state-of-the-art adversarial attacks to text classification in literatures can be categorized into the following types:

- Character-level attacks. Attackers can modify a few characters of an original text to manipulate the target neural network. Gao et al. [4] proposed Deep-WordBug, an approach which adds small character perturbations to generate adversarial examples against DNN classifiers. Ebrahimi et al. [3] proposed an efficient method, named by Hotflip, to generate white-box adversarial texts to trick a character-level neural network. In [9], text adversarial samples were crafted in both white-box and black-box scenarios. However, these approaches are easy to defend by placing a word recognition model before feeding the inputs into neural network [12].
- Word-level attacks. Word-level attacks substitute words in original texts by their synonyms so they can preserve semantic coherence. Liang et al. [9] designed three perturbation strategies to generate adversarial samples against deep text classification models. Alzantot et al. [1] proposed a genetic based optimization algorithm to generate semantically similar adversarial examples to fool a well-trained DNN classifier. To decrease the computational cost of attacks, Ren et al. [13] proposed a greedy algorithm, namely PWWS, for text adversarial attack. Word-level adversarial examples are more imperceptible for humans and more difficult for DNNs to defend.

There exists very few works on defending word-level text adversarial attacks. To the best of our knowledge, [15] is the only work on defenses against synonym substitution based adversarial attacks. They proposed Synonym Encoding Method (SEM) that encodes synonyms into the same word embeddings to eliminate adversarial perturbations. However, it needs an extra encoding stage before the normal training process and is limited on the fixed synonym substitution. Our framework adopts a unified training process and provides a flexible synonym substitution encoding scheme.

3   Preliminaries

In this section, we firstly present the problem of adversarial attack and defense in text classification tasks. Next we provide preliminaries about attack models: several typical word-level synonym adversarial attacks.

3.1   Problem Definition

Given a trained text classifier $F : \mathcal{X} \to \mathcal{Y}$, $\mathcal{X}$ and $\mathcal{Y}$ denote the input and the output space respectively. Suppose there is an input text $x \in \mathcal{X}$, the classifier can give a predicted true label y_{true} based on a posterior probability P.

$$\underset{y_i \in \mathcal{Y}}{\operatorname{argmax}} P(y_i|x) = y_{true} \tag{1}$$

An **adversarial attack** on classifier F is defined that the adversary can generate an adversarial example x' by adding an imperceptible perturbation

Δx, such that:

$$\underset{y_i \in \mathcal{Y}}{\operatorname{argmax}} P(y_i|x') \neq y_{true}$$

$$s.t. \quad x' = x + \Delta x, \quad \|\Delta x\|_p < \epsilon \tag{2}$$

where $\|\cdot\|_p$ denotes the p-norm and ϵ controls the small perturbation so that the crafted example is imperceptible to humans.

The **defense** against adversarial attack requires to train an enhanced text classifier F^* over F. A successful defense means that for a given input text example x, the attacker failed to craft an adversarial example, or the generated adversarial example x' could not fool the classifier F^*.

$$\underset{y_i \in \mathcal{Y}}{\operatorname{argmax}} P(y_i|x') = \underset{y_i \in \mathcal{Y}}{\operatorname{argmax}} P(y_i|x) = y_{true} \tag{3}$$

3.2 Synonym Adversarial Attacks

To ensure the perturbation small enough, the adversarial examples need to satisfy semantic coherence constraints. An intuitive way to craft adversarial examples is to replace several words in the input example by their synonyms. Let $x = w_1, \cdots, w_n$ denote an input example, where $w_i \in W$ denotes a word. Each word w_i has a synonym candidate set $\mathcal{S}_i$. For a synonym adversarial attack, adversary can substitute K words denoted by $\mathcal{C}_x$, to craft an adversarial example $x' = w'_1 \cdots w'_n$:

$$w'_i = \begin{cases} w_i & \text{if } w_i \notin \mathcal{C}_x \\ s_i^j & \text{if } w_i \in \mathcal{C}_x \end{cases} \tag{4}$$

where s_i^j denotes the jth substitution candidate word in $\mathcal{S}_i$.

Existing synonym substitution based adversarial attacks had gone into proposing fast searching algorithms, such as Greedy Search Algorithm(GSA) [7] and Genetic Algorithm (GA) [1]. [13] proposed a fast state-of-the-art method called Probability Weighted Word Saliency(PWWS) which considers the word saliency and the classification confidence.

4   The Proposed Framework

In this section, we first present our motivation, and then demonstrate the detailed defense framework.

4.1   Motivation

There are many possible reasons why DNNs have vulnerabilities to adversarial attacks. One of the pivotal factors comes from the internal robustness of neural networks. Given a normal example x, suppose x is within the decision boundary in which the classifier can make a correct prediction, as seen in Fig. 1(a).

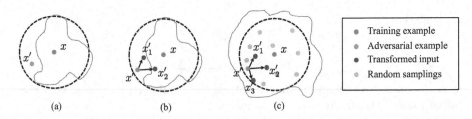

Fig. 1. Decision boundary around normal example.

However, attackers can craft an adversarial example x' in the neighborhood of x such that the classifier will make a wrong prediction on x'.

For word-level adversarial attacks, adversarial examples within neighborhood of normal examples are generally created by substituting parts of words in the text by their synonyms. Therefore a judicious solution is to encode synonyms into the same embeddings [15], and use the modified embeddings to train neural networks as well as testing. However, this method does not enlarge the decision boundary too much because the number of training data is limited. Thus under this encoding, a carefully crafted adversarial example may or may not go through the decision boundary. From Fig. 1(b) we can see that under such encoding, adversarial example x' may be mapped to x'_1 (defense fail) or x'_2 (defense success).

In this paper, we apply a different way to generate more robust word embeddings. We randomly involve neighborhood examples of all training data into the model training process. The neighborhood examples come from random synonym substitutions and they share the same label as the original example. Thus the decision boundary of one example may be expanded to cover most unseen neighborhood examples including adversarial examples, as shown in Fig. 1(c). Note that we did not generate a large number of neighborhood examples for a training data because of the expensive training time.

To address this challenge, we adopt a dynamic synonym substitution strategy in the training process and the number of training data remains unchanged. As presented in Fig. 1(c), a neighborhood example(a green circle) replaces the original example(the blue circle) to involve in the training process in an epoch. Thus different neighborhood examples are generated and work in different epochs. In the test process, testing examples are also required to randomly substitute by their synonyms. As a result, no matter an unseen example(may be adversarial) x' is mapped to x'_1, x'_2 or x'_3, the model can also give the correct prediction. We give the details of our framework in the next subsection.

4.2 Framework Specification

Given a set of training examples $\{x_i, y_i\}_N$, a text classification model $\mathcal{M}$ with parameter θ, the objective of M is to minimize the negative log-likelihood:

$$\min_{\theta} \left\{ \mathcal{L}(\theta) := -\sum_{i}^{N} \log P(y_i|x_i; \theta) \right\} \tag{5}$$

Input   Encoder   Embedding Model

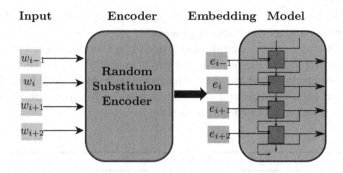

Fig. 2. The proposed RSE framework

To make the decision boundary more refined and ease the training load, our approach do not generate many labeled examples in advance. We dynamically generate neighborhood examples instead of original examples in every epoch of the training process. To this end, we proposed a dynamic random synonym substitution based framework RSE that introduces a Random Substitution Encoder between the input and the embedding layer. Then the training objective is to minimize:

$$\min_{\theta} \left\{ - \sum_{i}^{N} \operatorname*{rand}_{\|\delta_i\| < \epsilon} \log P(y_i | x_i + \delta_i; \theta) \right\} \tag{6}$$

where rand denotes the random synonym substitution operation, and $\|\delta_i\| < \epsilon$ guarantees that the generated example $x_i + \delta_i$ stays in the neighborhood of x_i.

Since rand operation does not need an optimization, we can fuse random synonym substitution and encodes the inputs into a new embeddings in real time in the training process of model $\mathcal{M}$. Figure 2 illustrates the representation of the proposed framework.

From Fig. 2, we can see that RSE reads an input text and encodes it to an embedding using a random synonym substitution algorithm. For example, given an original example $x = w_1 \cdots w_n$, RSE outputs a neighborhood x' and feeds the embedding $x'_e = e_1 \cdots e_n$ into the subsequent model $\mathcal{M}$. Then model $\mathcal{M}$ is trained on the perturbed examples. Here $\mathcal{M}$ can be one of any specific DNN models in NLP tasks and in this paper we focused on text classification model such as CNN and LSTM.

4.3   Random Synonym Substitution Algorithm

Next we introduce the details of RSE and the training process under the proposed framework. In practice, to satisfy the constraints $\|\delta_i\| < \epsilon$, we adopt a substitution rate sr instead of neighborhood radius ϵ. There are three steps to generate a neighborhood x' for original example x. Firstly, we select a substitution rate sr between a minimal rate r_{min} and a maximal rate r_{max}. Then we randomly sample a candidate words set $\mathcal{C}$ in which will be substituted. Finally we randomly chose synonyms for all words in $\mathcal{C}$. Algorithm 1 presents the details of these steps as well as the training process.

In the test stage, a test example also needs to be encoded under the proposed RSE in order to mitigate the possible adversarial noise. For example, given a test example x, we firstly transform it to a representation of its neighborhood x' by performing an algorithm(Lines 4–6 in Algorithm 1). Then the embedding is fed into the well-trained model to give a prediction.

Algorithm 1: Training for the RSE framework

Input: Training data $D = \{x_i, y_i\}_N$, model prameter θ, minimal rate r_{min} and maximal rate r_{max}

1 **for** *epoch* $= 1 \cdots N_{ep}$ **do**
2 | **for** *minibatch* $B \subset D$ **do**
3 | | **for** *original example* $x \in B$ **do**
4 | | | Sample a substitution rate sr between r_{min} and r_{max};
5 | | | Randomly sample candidate words set $\mathcal{C}$
6 | | | Sample synonyms for all words in $\mathcal{C}$ to generate x';
7 | | | Replace x with x' in B.
8 | | **end**
9 | | Update θ using gradient ascent of log-likelihood 6 on *minibatch* B.
10 | **end**
11 **end**

---

5   Experiments

We evaluate the performance of the proposed RSE framework experimentally in this section. We firstly present the experiment setup, and then report the experiment results on three real-world datasets. The results show that models under our RSE framework could achieve much better performance in defending adversarial examples.

5.1   Experiment Setup

In this subsection we give an overview of the datasets, target models, attack models and baselines used in our experiments.

Datasets. We test our RSE framework on three benchmark datasets: *IMDB*, *AG's News* and *Yahoo! Answers*.

IMDB [10] is a dataset for binary sentiment classification containing 25,000 highly polarized movie reviews for training and 25,000 for testing.

AG's News [16] is extracted from news articles using only the title and description fields. It contains 4 classes, and each class includes 30,000 training samples and 1900 testing examples.

Yahoo! Answers [16] is a topic classification dataset with 10 classes, which contains 4,483,032 questions and corresponding answers. We sampled 150,000 training data and 5,000 testing data from the original 1,400,000 training data and 60,000 testing data for the following experiments. Each class contains 15,000 training data and 500 testing data respectively.

We also used padding length when preprocessing the input text. The padding length is decided by each datasets' average sentence length. Table 1 lists the detailed description of the aforementioned datasets.

Table 1. The statistic and preprocessing settings for each dataset

Dataset	# of training samples	# of testing samples	# of vocab words	Padding length
IMDB	25,000	25,000	80,000	300
AG's News	120,000	7,600	80,000	50
Yahoo! Answers	150,000	5,000	80,000	100

Base Models. We used three main classic deep neural networks as base models in our RSE framework for text classification task: *LSTM*, *Bi-LSTM* and *Word-CNN*.

LSTM has a 100-dimension embedding layer, two LSTM layers where each LSTM cell has 100 hidden units and a fully-connected layer.

Bi-LSTM also has a 100-dimension embedding layer, two bi-directional LSTM layers and a fully-connected layer. Each LSTM cell has 100 hidden units.

Word-CNN [6] has two embedding layers, one is static for pretrained word vectors, and another is non-static for training, three convolutional layers with filter size of 3, 4, and 5 respectively, one 1D-max-pooling layer and a fully-connected layer.

Attack Models. We adopt three synonym substitution adversarial attack models to evaluate the effectiveness of defense methods. We suppose attackers can obtain all testing examples of three datasets (IMDB, AG's News, Yahoo! Answers) and can call prediction interfaces of any models at any time.

Random. We first randomly choose a set of candidate words that has synonyms. Then keep replacing the original word in the candidate set with a randomly synonym until the target model predicts wrong.

Textfool [8] uses sorted word candidates based on the word similarity rank to replace with the synonym and keep perform until the target model predicts wrong. We will not use the typos substitution because we pay most attention to synonyms replacements attacks.

PWWS [14] is a greedy synonym replacement algorithm called Probability Weighted Word Saliency (PWWS) that considers the word saliency as well as the classification probability. As the same, we only use synonym replacements but not specific named entities replacements.

Baseline. We take *NT, AT* and *SEM* as three baselines. *NT* is a normal training framework without taking any defense methods. *AT* [5] is an adversarial training framework, where extra adversarial examples are generated to train a robust model. We adopt the same adversarial training configurations as in [15], which uses *PWWS* to generate 10% adversarial examples from each dataset for every normal trained neural network. Then the adversarial examples and original training examples are mixed for the training process. *SEM* [15] is an adversarial defense framework which inserts a fixed synonym substitution encoder before the input layer of the model. We evaluate our framework and baseline frameworks by using *LSTM, Bi-LSTM* and *Word-CNN* as base models respectively.

5.2 Evaluations

We evenly sampled each class from the origin test data to form 1,000 clean testing examples for every datasets. Then these examples are used to generate adversarial examples by the above attack models, which will take *NT, AT, SEM* and the proposed *RSE* as the victim targets.

Table 2. The evaluation results under different settings.

Dataset	Attack model	LSTM (%)			Bi-LSTM (%)			Word-CNN (%)		
		NT	AT	RSE	NT	AT	RSE	NT	AT	RSE
IMDB	No attack	**88.8**	87.9	87.0	**89.5**	88.7	86.5	87.6	86.3	**87.8**
	Random	80.6	77.8	**83.1**	81.5	79.2	**81.9**	77.5	75.0	**83.0**
	Textfool	75.4	76.2	**84.2**	74.6	77.0	**83.7**	71.2	71.0	**83.1**
	PWWS	26.3	29.3	**82.2**	27.3	28.1	**79.3**	13.5	10.5	**81.2**
AGs News	No attack	90.5	92.6	**92.9**	**96.4**	94.5	94.1	95.9	**96.9**	94.8
	Random	84.7	87.9	**89.2**	91.4	89.7	**92.2**	91.6	92.6	**93.1**
	Textfool	79.6	85.3	**88.7**	86.3	89.1	**90.6**	88.0	**92.6**	92.2
	PWWS	63.0	72.8	**84.2**	70.4	78.0	**88.3**	67.5	77.1	**89.9**
Yahoo! Answers	No attack	72.5	**73.1**	72.1	**73.2**	72.7	71.8	**71.2**	66.0	70.1
	Random	60.5	65.1	**68.6**	61.2	64.2	**68.9**	58.9	54.3	**67.3**
	Textfool	58.9	63.4	**67.4**	60.4	63.6	**67.1**	57.9	56.2	**66.4**
	PWWS	29.1	39.3	**64.3**	26.3	39.2	**64.6**	28.8	26.8	**62.6**

The key metrics to evaluate the performance of different defense frameworks in this paper are *Accuracy, Accuracy Shift* and *Attack-Success Rate. Accuracy* refers to the ratio that the number of correctly predicted examples against the total number of testing examples. *Accuracy Shift* refers to the reduced accuracy before and after the attack. *Attack-Success Rate* is defined by the number of successfully attacked examples by attack models against the number of correctly predicted examples with no attack. It can be computed by $(Accuracy\ Shift)/(No\ attack\ Accuracy)$. The better defense performance the target model has, the lower Attack-Success Rate the attacker gets.

Table 2 shows the accuracy results of base models(*LSTM, Bi-LSTM* and *Word-CNN*) against various attack models (*Random, Textfool, PWWS*) under

NT, *AT*, and the proposed *RSE* defense framework. For each base model with each dataset, we highlight the highest classification accuracy for different defense frameworks in **bold** to indicate the best defense performance.

From Table 2, we can see the following observations when looking at each box to find the best accuracy result:

1. When there is no attack, either NT or AT usually has the best accuracy. But under other attack models, our RSE framework can get the best accuracy.
2. For each column in each box, target models have the lowest accuracy under PWWS attack, which demonstrates PWWS is the most effective attack model. The accuracy of NT and AT drop significantly under PWWS attack. But RSE has the best defense performance since the accuracy loss is very small compared with 'No attack'.
3. Under different settings (various datasets, attack models and base models), our RSE framework has a better performance with few accuracy decrease. This demonstrates the generalization of RSE framework to strengthen a robust deep neural network against synonym adversarial attacks.

Table 3. SEM VS. RSE under PWWS attack model.

Metric %	Base model	IMDB		AGs News		Yahoo! Answers	
		SEM	RSE	SEM	RSE	SEM	RSE
Before-attack accuracy	LSTM	86.8	87.0	90.9	92.9	69.0	72.1
	Bi-LSTM	87.6	86.5	90.1	94.1	70.2	71.8
	Word-CNN	86.8	87.8	88.7	94.8	65.8	70.1
After-attack accuracy	LSTM	77.3	82.2	85.0	84.2	54.9	64.3
	Bi-LSTM	76.1	79.3	81.1	88.3	57.2	64.6
	Word-CNN	71.1	81.2	67.6	89.9	52.6	62.6
Accuracy shift	LSTM	9.5	**4.8**	**5.9**	8.7	14.1	**7.8**
	Bi-LSTM	11.5	**7.2**	9.0	**5.8**	13.0	**7.2**
	Word-CNN	15.7	**6.6**	21.1	**4.9**	13.2	**7.5**
Attack-success rate	LSTM	10.94	**5.52**	**6.49**	9.36	20.43	**10.82**
	Bi-LSTM	13.13	**8.32**	9.99	**6.16**	18.52	**10.03**
	Word-CNN	18.09	**7.52**	23.79	**5.17**	20.06	**10.70**

RSE vs. SEM. We also compared SEM with our RSE as shown in Table 3. Please note that we evaluate the performance of RSE and SEM only under PWWS attack model on three datasets because: (1) PWWS has the strongest attacking efficacy; and (2) SEM has no opened source codes yet and we directly cite the results in [15] under the same experimental settings.

It can be seen from Table 3 that the average After-Attack Accuracy of RSE is higher than SEM for about 5%–10%. We also compared the Accuracy Shift of SEM and RSE since the parameters of each base model may be different. We find out that except for AG's News dataset with LSTM model, models under RSE have smaller Accuracy Shift, and the shifts are stable with only 5% decrease in average. But for SEM the decrease is about 10% in average. For AG's News dataset with Word-CNN model, the Accuracy Shift reaches 21.1% and thus the performance of the model is unacceptable. We can also see that our RSE has lower Attack-Success Rate for nearly all settings. This means that it is more difficult for PWWS attacker to craft adversarial examples under RSE framework.

Substitution Rate. When crafting an adversarial example, it is better to add smaller perturbations. Thus noisy rate is an important metric in adversarial attack. It means that the crafted examples may not be imperceptible if the noisy rate is high. On the contrary, the defense mechanism is better if it causes the attacker have to add more noise to success. Thus in this paper we introduce Substitution Rate as a metric, which is defined as the number of substituted words against the sentence length. The better performance the defense framework has, the more the substituted words the attack model costs.

Table 4. Performance on substitution rate.

Dataset	Attacker	LSTM (%)			Bi-LSTM (%)			Word-CNN (%)		
		NT	AT	RSE	NT	AT	RSE	NT	AT	RSE
IMDB	Textfool	17.98	18.22	**20.09**	17.63	18.26	**20.06**	17.57	17.35	**20.03**
	PWWS	10.54	11.23	**19.13**	10.55	11.25	**18.12**	6.41	5.36	**17.99**
AGs News	Textfool	20.93	21.27	**22.18**	21.18	21.39	**22.11**	22.09	21.59	**22.17**
	PWWS	19.07	20.88	**21.66**	20.09	21.44	**22.20**	19.35	20.75	**23.08**
Yahoo! Answers	Textfool	13.69	**14.19**	13.25	13.59	**14.29**	13.38	13.88	**13.71**	13.50
	PWWS	10.28	12.47	**16.09**	9.79	12.85	**16.30**	10.30	9.74	**16.21**

The Table 4 shows the Substitution Rate of each base model without (NT) or with defend frameworks (AT and RSE). We could not list the results of SEM since they did not report in [15]. From Table 4 it could be seen that the Substitution Rate for attacking the models with RSE is over 20% in most cases, better than NT and AT. So we can safely conclude that RSE makes the attackers pay more cost for perturbing origin sentences.

6   Conclusion and Future Work

In this paper, we propose a defense framework called RSE to protect text classification models against word-level adversarial attacks. With this framework, a random synonym substitution encoder is fused into the deep neural network to endow base models with robustness to adversarial examples. And a corresponding training algorithm is also proposed. Extensive experiments on three popular real-world datasets demonstrate the effectiveness of our framework on defense of word-level adversarial attacks. In the future, we will explore how the parameters of our algorithm impact the performance and transfer our RSE framework into other typical NLP tasks, e.g., Machine Translation and Question & Answer, to protect deep models from word-level adversarial attacks.

Acknowledgments. This work was supported by the National Natural Science Foundation of China (Grant No. 61802124).

References

1. Alzantot, M., Sharma, Y., Elgohary, A., Ho, B.-J., Srivastava, M.B., Chang, K.-W.: Generating natural language adversarial examples. In: Proceedings of the 2018 Conference on Empirical Methods in Natural Language Processing, pp. 2890–2896 (2018)
2. Ebrahimi, J., Lowd, D., Dou, D.: On adversarial examples for character-level neural machine translation. In: Proceedings of the 27th International Conference on Computational Linguistics, pp. 653–663 (2018)
3. Ebrahimi, J., Rao, A., Lowd, D., Dou, D.: HotFlip: white-box adversarial examples for text classification. In: Proceedings of the 56th Annual Meeting of the Association for Computational Linguistics (ACL 2018), pp. 31–36 (2018)
4. Gao, J., Lanchantin, J., Soffa, M.L., Qi, Y.: Black-box generation of adversarial text sequences to evade deep learning classifiers. In: 2018 IEEE Security and Privacy Workshops (SP Workshops 2018), pp. 50–56 (2018)
5. Goodfellow, I.J., Shlens, J., Szegedy, C.: Explaining and harnessing adversarial examples. In: 3rd International Conference on Learning Representations (ICLR 2015) (2015)
6. Kim, Y.: Convolutional neural networks for sentence classification. In: Proceedings of the 2014 Conference on Empirical Methods in Natural Language Processing (EMNLP 2014), pp. 1746–1751 (2014)
7. Kuleshov, V., Thakoor, S., Lau, T., Ermon, S.: Adversarial examples for natural language classification problems (2018)
8. Kulynych, B., Hayes, J., Samarin, N., Troncoso, C.: Evading classifiers in discrete domains with provable optimality guarantees. arXiv preprint arXiv:1810.10939 (2018)
9. Liang, B., Li, H., Su, M., Bian, P., Li, X., Shi, W.: Deep text classification can be fooled. In: Proceedings of the Twenty-Seventh International Joint Conference on Artificial Intelligence, pp. 4208–4215 (2018)
10. Maas, A.L., Daly, R.E., Pham, P.T., Huang, D., Ng, A.Y., Potts, C.: Learning word vectors for sentiment analysis. In: The 49th Annual Meeting of the Association for Computational Linguistics: Human Language Technologies, pp. 142–150 (2011)
11. Papernot, N., McDaniel, P.D., Wu, X., Jha, S., Swami, A.: Distillation as a defense to adversarial perturbations against deep neural networks. In: IEEE Symposium on Security and Privacy, pp. 582–597 (2016)
12. Pruthi, D., Dhingra, B., Lipton, Z.C.: Combating adversarial misspellings with robust word recognition. In: Proceedings of the 57th Conference of the Association for Computational Linguistics (ACL 2019), pp. 5582–5591 (2019)
13. Ren, S., Deng, Y., He, K., Che, W.: Generating natural language adversarial examples through probability weighted word saliency. In: Proceedings of the 57th Conference of the Association for Computational Linguistics (ACL 2019), pp. 1085–1097 (2019)

14. Ren, Y., Du, Y., Wang, D.: Tackling adversarial examples in QA via answer sentence selection. In: Proceedings of the Workshop on Machine Reading for Question Answering (ACL 2018), pp. 31–36 (2018)
15. Wang, X., Jin, H., He, K.: Natural language adversarial attacks and defenses in word level. arXiv preprint arXiv:1909.06723 (2019)
16. Zhang, X., Zhao, J.J., LeCun, Y.: Character-level convolutional networks for text classification. In: 2015 Annual Conference on Neural Information Processing Systems, pp. 649–657 (2015)

Document-Improved Hierarchical Modular Attention for Event Detection

Yiwei Ni[ID], Qingfeng Du[✉][ID], and Jincheng Xu[ID]

School of Software Engineering, Tongji University, Shanghai, China
{nee_11235,du_cloud,xujincheng}@tongji.edu.cn

Abstract. The task of event detection aims to find the event trigger for a sentence and identify the correct event type. Contextual information is crucial to event detection, which helps a model to identify the triggers better. Existing models utilizing contextual information only take document information as additional features for deep learning model, without considering the specific contribution of document information to trigger classification. In this paper, we propose a *Document-Improved Hierarchical Modular Event Detection* (DIHMED) model to extract hierarchical contextual information. Specifically, considering the relevance between event types, we build independent modules that combine the document-level information to express this relevance. Given the fact that events from the same document are often related, these modules can make better use of document-level information. We conduct several experiments on a Chinese political event dataset and the results show that our model can outperform the state-of-the-art models.

Keywords: Event detection · Neural network · Contextual information

1 Introduction

Event detection is a very important subtask of event extraction, which selects event triggers from a sentence and judges the event types. Here is an example in ACE-2005 guideline [17]:

"In Baghdad, a cameraman died when an American tank fired on the Palestine hotel."

an ED model should find the trigger words *"died"* and *"fired"*, and identify the corresponding event type *"Die"* and *"Attack"*. This task is challenging because an event can be identified by different triggers while a trigger word can be classified into different event types according to different contexts.

Most researchers regard event detection as a multi-classification task. In the past, feature-based methods were often taken [5,8], and representation-based methods are more widely used recently [2,14,15]. In general, most of them

This work is supported by the National Natural Science Foundation of China (Grant No. U1934212).

only pay attention to single sentences, ignoring the contextual information, and obviously, the information from a single sentence is insufficient. Sometimes it is impossible to infer the actual meaning of words from a single sentence. For instance, here are two sentences from the same document in our dataset: *"军方将在海面上空进行试飞。(The military will have a test over the sea.)... 发射的导弹是由另一种型号改良而来。(The missile launched was improved from another type.)"*

With only one trigger word *"试飞(test)"*, we can not determine the meaning of the sentence due to the ambiguity. It may describe an event about a test flight of an airplane, or a test of a missile. Only with contextual information can we know what the event is about and identify the event type *"Exhibit force posture"*.

Meanwhile, the events from the same document are often related. In the Chinese political event dataset, every document has a topic, and all sentences in the document expand on this topic to describe more details. As a result, the events in the same document have similar event types. We analyze the dataset statistically to show it more clearly. There are twenty event types in the dataset. We combine related event types and get six superordinate concepts. For example, both event type *"Assault"* and event type *"Fight"* describe a hostile relationship between two sides so that we combine these two event types to compose a superordinate concept. According to our analysis, the proportion of events types having the same superordinate concept in the same document can reach 59.9%. In other words, when there are multiple events in a document, more than half of them are related to each other. This indicates that the context can be utilized to understand the meaning of a sentence, and such sufficient information in the context can help us identify the event types.

Therefore, some researchers combine contextual information in event detection [3,20]. However, after transforming documents into vectors, they only take these vectors as additional features for all the sentences in this document. In this situation, the same contextual information is attached to all the sentences in the document. This method allows all the sentences in a document to share the same feature but ignores the specific contribution of the document. As a result, it is of limited help to event detection.

In this paper, we propose a *Document-Improved Hierarchical Modular Event Detection* (DIHMED) model, which detects events from a sentence with its contextual information. The attention mechanism is proved effective in computing the contribution of each word in the sentence, but the traditional attention mechanism can not handle hierarchical classification. Referring to previous work on hierarchical classification [1,16,18], we build a hierarchical modular attention component for each superordinate concept so that we can extract and keep the corresponding superordinate concept information at the sentence level. Besides, to process the information of sentence and context separately, we set up another component for document-level superordinate concept information. Finally, we merge them to obtain document-improved hierarchical modular information.

In summary, our contributions are as follows: (1) We analyze the problem of existing event detection models utilizing contextual information, then build hierarchical modular attention components to tackle it. (2) We balance the weight of contribution scores at different levels to improve performance. (3) We experiment on a Chinese political event dataset with several baseline models, and the result shows that our model outperforms these state-of-the-art models.

2   Related Work

Event detection is a sub-task of event extraction, which is a challenging and promising task in natural language processing. In the past, feature-based methods have been widely used to solve this problem, while with the development of deep learning, more researchers adopt representation-based methods. The existing researches of event extraction can be mainly divided into two categories: the pipelined approach and the joint approach.

The pipelined approach divides event extraction into two subtasks: event detection and argument extraction. An event detection model needs to find the trigger word and identify the event type in the given sentence [3,15,20], while an argument extraction model should find arguments such as time and location [11,18]. As a pipeline, the performance of argument extraction relies on the output of event detection, so the improvement of event detection is crucial to event extraction.

The joint approach extracts the triggers with their arguments simultaneously [9,14,19]. It is theoretically helpful for both event detection and argument extraction because the dependencies between triggers and arguments are captured. However, existing experiments only show obvious improvement in argument extraction.

3   Proposed Model

In this paper, event detection is regarded as a multi-classification task. For each token in a sentence, the model needs to determine its event type. There are 20 event types and 300 subtypes in the Chinese political event dataset, and we regard these 300 subtypes as the output event type T of our model. Formally, given a document $d = s_1, s_2, ..., s_l$, each sentence of the document s_i ($0 \leq i \leq l$) consists of words $w_{i1}, w_{i2}, ..., w_{im}$, and the goal of the model is to assign an event type $t \in T$ or $None$ to $\forall w_{ij} (1 \leq j \leq m)$.

As shown in Fig. 1, the model is composed of three parts: sentence-level information extraction, document-level information extraction, and a context-improved classifier. We have encoders in each extraction step to extract basic information. A hierarchical modular attention (HMA) component (shown in Fig. 2) is applied then to compute attention for every superordinate concept. After computing attention at both the sentence level and the document level, we merge them to obtain the document-improved hierarchical modular information and feed it to the classifier.

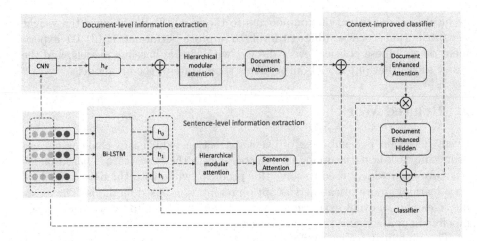

Fig. 1. The architecture of our proposed DIHMED model.

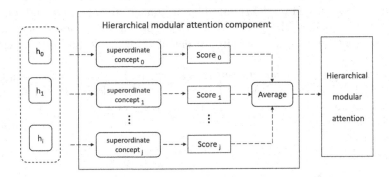

Fig. 2. The structure of the hierarchical modular attention component.

3.1   Sentence-Level Information Extraction

In this phase, we focus on extracting sentence-level information. Bi-directional recurrent neural network (RNN) is proved to be suitable for processing sequential inputs. We choose long short-term memory (LSTM) [4], a variant of RNN, to encode our sentence input. By concatenating the word embedding g_{ij} and entity embedding e_{ij} of a token w_{ij}, each token is transformed into a vector x_{ij} [15], and then a bi-directional LSTM is adopted to gain the hidden state of the sentence:

$$h_{ij} = [\overrightarrow{LSTM}(x_{ij}), \overleftarrow{LSTM}(x_{ij})] \tag{1}$$

To explore the relevance between words and superordinate concepts, we employ an HMA component. Referring to the previous work [12,18], we build a sentence-level Hierarchical Modular Attention(HMA). Specifically, for super-ordinate concepts $c_1, c_2, ..., c_n$, a vector $u_k(1 \leq k \leq n)$ is prepared to express the hidden information of each concept. A multi-layer perceptron is adopted to compute the hidden state, and then obtain the attention score:

$$f_{ij}^k = tanh(W_a[h_{ij}; u_k]) \tag{2}$$

$$a_{ij}^k = \frac{exp(W_b f_{ij}^k)}{\sum_{j'=1}^m exp(W_b f_{ij'}^k)} \tag{3}$$

We average the attention scores of all n concepts, and then receive the sentence-level attention:

$$a_{ij} = \frac{1}{n} \sum_{k=1}^n a_{ij}^k \tag{4}$$

3.2   Document-Level Information Extraction

Generally, one document has a specific topic which the events in the document are related to. In previous work [3,20], document-level information is extracted by the encoder and is directly used as the additional feature of each sentence, which ignores the relationship between the document and the corresponding topic. Similarly to the sentence-level information extraction, we need hierarchical modular attention to make better use of the context. After extracting the document-level information with an encoder, we employ an HMA component to compute the contribution of documents to each superordinate concept.

Convolutional Neural Network (CNN) is often adopted to capture semantic and syntactic features in natural language processing tasks [2,6,7], so we choose CNN as the encoder of document-level information. For document $d = \{w_{00}, w_{01}, ..., w_{0m}, w_{10}, ..., w_{l0}, ...w_{lm}\}$, we concatenate word embedding g_{ij} of each word w_{ij}, and obtain the representation of the document $x_d = (g_{00}; g_{01}; ...; g_{0m}; g_{10}; ...; g_{l0}; ...; g_{lm})$ as the input of CNN:

$$h_d' = CNN(x_d) \tag{5}$$

Similarly, we build a document-level hierarchical modular attention component and compute the contribution of the document to a superordinate concept k:

$$f_{ij}^{k'} = tanh(W_a'[h_{ij}; h_d'; u_k']) \tag{6}$$

$$a_{ij}^{k'} = \frac{exp(W_b' f_{ij}^{k'})}{\sum_{j'=1}^m exp(W_b' f_{ij}^{k'})} \tag{7}$$

We also average these attention scores for the document-level attention:

$$a_{ij}' = \frac{1}{n} \sum_{k'=1}^n a_{ij}^{k'} \tag{8}$$

3.3   Context-Improved Classifier

After calculating sentence-level attention and document-level attention respectively, to obtain the document-improved hierarchical modular attention, we combine them with hyperparameters λ and μ:

$$a_{ij}^{''} = \lambda\, a_{ij} + \mu\, a_{ij}^{'} \tag{9}$$

By utilizing this document-improved attention, we receive the weighted hidden:

$$r_{ij} = a_{ij}^{k''} h_{ij} \tag{10}$$

We concatenate x_{ij} r_{ij} and $h_d^{'}$ as the input of softmax and get the probability vector O. The t-th element O_t of O indicates the confidence score of w_{ij} for event type t. Finally, we compute the score of w_{ij} for all event types:

$$p(t|w_{ij}, \theta) = \frac{e^{O_t}}{\sum_{t' \in T} e^{O_{t'}}} \tag{11}$$

where θ is the hyperparameter of the model. Given all the instances (w_{ij}, y_{ij}) in the dataset, we define the following objective function:

$$J(\theta) = -\sum_{i=1}^{l} \sum_{j=1}^{m} \log p(y_{ij}|w_{ij}, \theta) \tag{12}$$

4   Experiment

4.1   Dataset and Experiment Setup

We conduct experiments on a Chinese political event dataset. Besides our proposed model, we reproduce several state-of-the-art models for comparisons and perform the experiments with the same dataset. This dataset is constructed based on the syntax tree and trigger word dictionary in our previous work, and it contains 8012 documents. We randomly choose 1600 documents of them as the test set, 800 as the validation set, and the remaining 5612 as the training set. For each predicted trigger, we judge its correctness by checking if its event subtype and offsets match those of a reference trigger. We take $Precision(P)$, $Recall(R)$, $F1 - Score(F1)$ as the evaluation metrics.

The hyperparameters of the reproduced models are set according to the original researches. For our proposed model, the word embedding dimension and the entity embedding dimension are set to 300 and 50 respectively. We initialize the word embedding with a pre-trained one [13]. The LSTM has one layer and its hidden size is 300. The window size of CNN is 3 and its number of feature maps is 100. We set the dimension of HMA to 300 for both sentence-level and document-level. The dropout is set to 0.5. Adam algorithm is applied during the training to minimize the loss.

4.2   Baseline Models

In order to evaluate our proposed DIHMED model, we choose several state-of-the-art models for comparisons on the same dataset. It is proved that the representation-based models outperform the feature-based models, so we only choose the former as our baseline.

Sentence-Level Models:

DMCNN [2]: This model adopts a dynamic multi-pooling layer according to event triggers and arguments, to reserve more crucial information.

JRNN [14]: This model exploits the inter-dependencies between event triggers and argument roles via discrete structures.

JMEE [10]: This model jointly extracts multiple event triggers and arguments by introducing syntactic shortcut arcs to enhance information flow and attention-based graph convolution networks to model graph information.

Document-Level Models:

DLRNN [3]: This model automatically extracts cross-sentence clues to improve sentence-level event detection without designing complex reasoning rules.

DEEB [20]: This model learns the event detection oriented embeddings of documents, and uses the learned document embedding to enhance another bidirectional RNN model to identify event triggers and their types in sentences.

4.3   Effect of Different Attention Mechanism

Compared with the basic attention mechanism, our proposed DIHMED model applies hierarchical modular attention at both the sentence level and the document level. To demonstrate the effect of these components, we design several models with different attention mechanism:

Basic model: This model adopts a bi-directional LSTM as the encoder. This model only takes sentences as input and does not adopt any attention mechanism.

Basic model with attention: This model employs a simple attention mechanism based on the basic model.

DIHMED1/2/3: These models apply HMA in different ways. DIHMED1 only takes the sentence-level HMA ($\lambda = 1$ and $\mu = 0$) while DIHMED2 only uses the document-level HMA($\lambda = 0$ and $\mu = 1$). Finally, DIHMED3 combines HMA in both the sentence level and the document level ($\lambda = 0.5$ and $\mu = 0.5$).

In Table 1 we can see that, compared with the basic attention mechanism, hierarchical modular attention can help improve performance. With the HMA component, the model can take advantage of the underlying information and exploit the hierarchical relations. In addition, the model with document-level HMA can achieve a higher recall. This is mainly because with contextual information, we can detect more trigger words, so the result is not restricted by local information. However, this model gets a lower precision because, with the

Table 1. Experimental results of models with different attention mechanism.

Methods	λ	μ	P	R	F1
Basic model	–	–	73.6	75.6	74.6
Basic model with attention	–	–	75.0	78.1	76.5
DIHMED1	1	0	77.5	77.8	77.7
DIHMED2	0	1	73.9	84.5	78.8
DIHMED3	0.5	0.5	80.5	82.1	81.3

increase of candidate trigger words, the model with only document-level attention gains less information from the sentence, and thus misclassifies the trigger words. With the combination of sentence-level and document-level attention, our proposed DIHMED model can achieve better performance.

4.4 Overall Model Performance

Table 2 shows the overall performance of our proposed model and other baseline models. All the models we propose with different HMA components can achieve a comparable result with state-of-the-art models, and the DIHMED model outperforms these baselines at all metrics. The result can be explained as follows: (1) Sentence-level methods (DMCNN, JRNN, JMEE) pay all their attention to the sentence, and they can achieve a good result to a certain extent. However, only with context can we know the meaning of some specific trigger words. For this reason, our model combines contextual information and can better identify trigger words. (2) Document-level methods (DLRNN, DEEB) only regard document vector as additional features, and can not exploit the deep relation between documents and events. With hierarchical modular attention, we can extract the contribution of the input for event types. We combine the contributions of both

Table 2. Overall model performance on the test set. † indicates that the model employs the document-level information.

Methods	P	R	F1
Chen's DMCNN [2]	73.6	80.5	76.9
Ngyuen's JRNN [14]	75.1	78.6	76.8
Liu's JMEE [10]	78.7	80.3	79.5
Duan's DLRNN [3]†	74.4	78.8	76.5
Zhao's DEEB [20]†	74.7	82.0	78.2
DIHMED1	77.5	77.8	77.7
DIHMED2†	73.9	**84.5**	78.8
DIHMED3†	**80.5**	82.1	**81.3**

the sentence level and the document level to make better use of the local and global information and thus get better performance.

4.5 Qualitative Analysis

To further demonstrate the effect of our proposed model, we conduct a qualitative analysis. By visualizing the attention score, we can intuitively see the contribution of each word computed by attention modules. We select a sentence from the dataset and compute the scores of sentence-level attention and document-level attention respectively. In Fig. 3 we can see that attention at different levels are interested in different words. In this case, the place of the event "there" and the result of the event "injured" get higher scores in sentence-level attention, while the trigger word "explosion" gets a higher score in document-level attention. This document describes some details of this event, such as the time of this explosion and the reason for it. So the document-level attention extracts more information rather than single sentences, and help improve event detection.

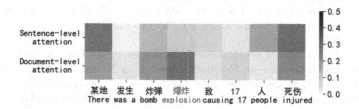

Fig. 3. The heatmap of the sample sentence. The word in red is the trigger word of the event.

5 Conclusions

In this paper, we propose a document-improved hierarchical modular event detection model. We conduct several experiments on a Chinese political event dataset and find that with hierarchical modular attention, contextual information can better help event detection. In the future, we will further seek the balance between the sentence-level and the document-level information and try to adjust the weights dynamically.

References

1. Andreas, J., Rohrbach, M., Darrell, T., Klein, D.: Neural module networks. In: 2016 IEEE Conference on Computer Vision and Pattern Recognition (CVPR 2016), Las Vegas, NV, USA, 27–30 June 2016, pp. 39–48 (2016). https://doi.org/10.1109/CVPR.2016.12

2. Chen, Y., Xu, L., Liu, K., Zeng, D., Zhao, J.: Event extraction via dynamic multi-pooling convolutional neural networks. In: Proceedings of the 53rd Annual Meeting of the Association for Computational Linguistics and the 7th International Joint Conference on Natural Language Processing of the Asian Federation of Natural Language Processing (ACL 2015), 26–31 July 2015, Beijing, China, Volume 1: Long Papers, pp. 167–176 (2015). https://www.aclweb.org/anthology/P15-1017/

3. Duan, S., He, R., Zhao, W.: Exploiting document level information to improve event detection via recurrent neural networks. In: Proceedings of the Eighth International Joint Conference on Natural Language Processing (IJCNLP 2017), Taipei, Taiwan, 27 November–1 December 2017 - Volume 1: Long Papers, pp. 352–361 (2017). https://www.aclweb.org/anthology/I17-1036/

4. Hochreiter, S., Schmidhuber, J.: Long short-term memory. Neural Comput. **9**(8), 1735–1780 (1997). https://doi.org/10.1162/neco.1997.9.8.1735

5. Hong, Y., Zhang, J., Ma, B., Yao, J., Zhou, G., Zhu, Q.: Using cross-entity inference to improve event extraction. In: The 49th Annual Meeting of the Association for Computational Linguistics: Human Language Technologies, Proceedings of the Conference, 19–24 June 2011, Portland, Oregon, USA, pp. 1127–1136 (2011). https://www.aclweb.org/anthology/P11-1113/

6. Kalchbrenner, N., Grefenstette, E., Blunsom, P.: A convolutional neural network for modelling sentences. In: Proceedings of the 52nd Annual Meeting of the Association for Computational Linguistics (ACL 2014), 22–27 June 2014, Baltimore, MD, USA, Volume 1: Long Papers, pp. 655–665 (2014). https://doi.org/10.3115/v1/p14-1062

7. Kim, Y.: Convolutional neural networks for sentence classification. In: Proceedings of the 2014 Conference on Empirical Methods in Natural Language Processing (EMNLP 2014), 25–29 October 2014, Doha, Qatar, A meeting of SIGDAT, a Special Interest Group of the ACL, pp. 1746–1751 (2014). https://doi.org/10.3115/v1/d14-1181

8. Li, Q., Ji, H., Hong, Y., Li, S.: Constructing information networks using one single model. In: Proceedings of the 2014 Conference on Empirical Methods in Natural Language Processing (EMNLP 2014), 25–29 October 2014, Doha, Qatar, A meeting of SIGDAT, a Special Interest Group of the ACL, pp. 1846–1851 (2014). https://www.aclweb.org/anthology/D14-1198/

9. Li, Q., Ji, H., Huang, L.: Joint event extraction via structured prediction with global features. In: Proceedings of the 51st Annual Meeting of the Association for Computational Linguistics (ACL 2013), 4–9 August 2013, Sofia, Bulgaria, Volume 1: Long Papers, pp. 73–82 (2013). https://www.aclweb.org/anthology/P13-1008/

10. Liu, X., Luo, Z., Huang, H.: Jointly multiple events extraction via attention-based graph information aggregation. In: Proceedings of the 2018 Conference on Empirical Methods in Natural Language Processing, Brussels, Belgium, 31 October–4 November 2018, pp. 1247–1256 (2018). https://doi.org/10.18653/v1/d18-1156

11. Luo, Z., Sui, G., Zhao, H., Li, X.: A shallow semantic parsing framework for event argument extraction. In: Douligeris, C., Karagiannis, D., Apostolou, D. (eds.) KSEM 2019. LNCS (LNAI), vol. 11776, pp. 88–96. Springer, Cham (2019). https://doi.org/10.1007/978-3-030-29563-9_9

12. Luong, T., Pham, H., Manning, C.D.: Effective approaches to attention-based neural machine translation. In: Proceedings of the 2015 Conference on Empirical Methods in Natural Language Processing (EMNLP 2015), Lisbon, Portugal, 17–21 September 2015, pp. 1412–1421 (2015). https://www.aclweb.org/anthology/D15-1166/

13. Mikolov, T., Sutskever, I., Chen, K., Corrado, G.S., Dean, J.: Distributed representations of words and phrases and their compositionality. In: Advances in Neural Information Processing Systems 26: 27th Annual Conference on Neural Information Processing Systems 2013. Proceedings of a meeting held 5–8 December 2013, Lake Tahoe, Nevada, United States, pp. 3111–3119 (2013). http://papers.nips.cc/paper/5021-distributed-representations-of-words-and-phrases-and-their-compositionality

14. Nguyen, T.H., Cho, K., Grishman, R.: Joint event extraction via recurrent neural networks. In: NAACL HLT 2016, The 2016 Conference of the North American Chapter of the Association for Computational Linguistics: Human Language Technologies, San Diego California, USA, 12–17 June 2016, pp. 300–309 (2016). https://www.aclweb.org/anthology/N16-1034/

15. Nguyen, T.H., Grishman, R.: Event detection and domain adaptation with convolutional neural networks. In: Proceedings of the 53rd Annual Meeting of the Association for Computational Linguistics and the 7th International Joint Conference on Natural Language Processing of the Asian Federation of Natural Language Processing (ACL 2015), 26–31 July 2015, Beijing, China, Volume 2: Short Papers, pp. 365–371 (2015). https://www.aclweb.org/anthology/P15-2060/

16. Shimura, K., Li, J., Fukumoto, F.: HFT-CNN: learning hierarchical category structure for multi-label short text categorization. In: Proceedings of the 2018 Conference on Empirical Methods in Natural Language Processing, Brussels, Belgium, 31 October–4 November 2018, pp. 811–816 (2018). https://www.aclweb.org/anthology/D18-1093/

17. Walker, C., Strassel, S., Medero, J., Maeda, K.: ACE 2005 multilingual training corpus. Linguistic Data Consortium, Philadelphia 57 (2006)

18. Wang, X., et al.: HMEAE: hierarchical modular event argument extraction. In: Proceedings of the 2019 Conference on Empirical Methods in Natural Language Processing and the 9th International Joint Conference on Natural Language Processing (EMNLP-IJCNLP 2019), Hong Kong, China, 3–7 November 2019, pp. 5776–5782 (2019). https://doi.org/10.18653/v1/D19-1584

19. Zhang, T., Ji, H., Sil, A.: Joint entity and event extraction with generative adversarial imitation learning. Data Intell. 1(2), 99–120 (2019). https://doi.org/10.1162/dint_a_00014

20. Zhao, Y., Jin, X., Wang, Y., Cheng, X.: Document embedding enhanced event detection with hierarchical and supervised attention. In: Proceedings of the 56th Annual Meeting of the Association for Computational Linguistics (ACL 2018), Melbourne, Australia, 15–20 July 2018, Volume 2: Short Papers, pp. 414–419 (2018). https://doi.org/10.18653/v1/P18-2066

Fine-Tuned Transformer Model
for Sentiment Analysis

Sishun Liu[1], Pengju Shuai[1], Xiaowu Zhang[1], Shuang Chen[1], Li Li[1(✉)],
and Ming Liu[2]

[1] School of Computer and Information Science, Southwest University,
Chongqing, China
liusishun5@gmail.com,
{shuaipengju,xiaohuangren,chen60423351}@email.swu.edu.cn,
lily@swu.edu.cn
[2] Chongqing College of Electronic Engineering, Chongqing, China
liuming@cqcet.edu.cn

Abstract. Sentiment analysis is one of important natural language processing (NLP) tasks. The transformer-based language models have become new baseline in sentiment analysis tasks in recent years. These models trained by efficient unsupervised learning methods have pushed accuracy of sentiment analysis tasks to a higher level. In this paper, we propose a set of novel transformation methods to tackle a sentiment analysis task. Specifically, we design four novel transformations in our transformer-models. For comparison purpose, the new data manipulation method is implemented on two language models (BERT and RoBERTa) separately. The nonlinear transformation method is evaluated on the large-scale real world dataset (Weibo sentiment analysis dataset (https://www.datafountain.cn/competitions/423)). The experiment results demonstrate the superiority of the proposed method.

Keywords: Natural language process · Sentiment analysis · Language model

1 Introduction

Nowadays, pretrained transformer language models, such as BERT [1], RoBERTa [2], ALBERT [3], Electra [4], have achieved remarkable results on plenty of natural language process tasks, for instance in the field of machine translation and sentiment analysis [5]. These models can maintain more latent information because of their self-attention [6] architecture and large unlabelled language corpus. By using transfer learning, lots of NLP downstream tasks can simply add several simple information extraction transformations on above language models.

Compared with negative sampling word embedding models, such as Word2Vec [7], the key difference is unsupervised learning tasks. Negative sampling models assume latent information is maintained in a fixed number of context words and these models' task is maintaining latent semantic information by

© Springer Nature Switzerland AG 2020
G. Li et al. (Eds.): KSEM 2020, LNAI 12275, pp. 336–343, 2020.
https://doi.org/10.1007/978-3-030-55393-7_30

maximizing context words probability and minimizing negative words probability. Transformer language models expand such idea and can collect information from whole sentences using self-attention mechanism. They are trained using more efficient unsupervised tasks such as MLM (Masked Language Model) [1] and RTD (Replaced Token Detection) [4].

In this paper, we design four novel extraction transformation methods catching not only latent information in all language model outputs, but information in hidden layers for they are less affected by downstream tasks. We test two transformer language model, BERT [1] and RoBERTa [2] on Weibo dataset. By using these extraction transformations our models outperform baselines in this task, and show latent information existing in hidden layers.

The rest of this paper is organized as follows. The related work is introduced in Sect. 2, followed by the detailed descriptions of proposed methods in Sect. 4. Experiment results are shown in Sect. 5. Our conclusion is in Sect. 6.

2   Related Work

Among most natural language process models, we can divide these models into two parts: part one establishes a transformation from individual words into word or word-like token embedding, while part two extracts information from the obtained embedding and returns the expected result.

Before transformer language models are created, the most popular embedding method is one-hot methods. However, this embedding method does not contain context information, and it is what Word2Vec [7] solves. Compared with the previous one-hot word representation, Word2Vec can maintain latent information of each words. Besides it, there are another word embedding methods like fastText [8] and GloVe [9].

In information extraction parts, since natural languages are sequential. It is very natural to use sequential model to get the information from input, that's why lots of researchers use RNN to extract information from natural languages, like RNN decoder-encoder model for statistical machine translation [10].

2.1   Transformer Language Model

In order to solve drawbacks in traditional sequential models, Devlin et al. decided using pure self-attention architecture [6] to construct a model which can learn semantic information from both previous tokens and following tokens without the influences of distance, as known as BERT [1]. BERT proves self-attention is better than almost all previous language model architectures, and now all state-of-the-art language models are based on self-attention architecture, like XLNet [11], RoBERTa [2], ALBERT [3], and Electra [4]. The differences between them are the language learning tasks and input embedding methods. Models except Electra select a small subset of tokens in unlabelled input sequences, mask small part of these tokens (typically 15%), or shuffle the given sequences and attention to all another tokens (XLNet [11]), then train the models to recover the original sequence input. While Electra uses a GAN-like replaced token detection task. Experiment result shows this task is very efficient: Electra is better than BERT [1] and DistilBERT [12] when the size of model is very small.

2.2 Sentiment Analysis

Sentiment analysis is an important part of natural language processing, and transformer-based language models also achieves lots of state-of-the-art result on multiple datasets. Colin et al. reported a transformer-based language model called Text-to-Text Transfer Transformer, or T5 [13]. This model treats every NLP problem as a "text-to-text" problem, so this model's training task would fit every NLP tasks. This language model achieves state-of-the-art on SST-2 Binary Classification task [5]. Xie et al. introduced a novelty semi-supervised learning methods called Unsupervised Data Augmentation(UDA) [14]. By using this method and BERT model, they achieves state-of-the-art on several datasets with surprising small amount of available labels.

3   Problem Definition

The input data has a comment set C and a corresponding target label set T. Comment set C contains several non-equal length texts $c^s = \{w_1^s, w_2^s, ..., w_k^s\}$ where w_i^s refers to a single token defined in language model vocabulary in comment c^s. The max length of every comment c^s is not bigger than 140. Target label set T has three labels: -1 means negative mood, 0 means neutral mood and 1 means positive mood.

After tokenziation and language models, each comment c^s should be first replaced by a number string $\{t_1^s, t_2^s, ..., t_k^s\}$ where t_i^s is a non-negative value, then become k token embedding $\{v_1^s, v_2^s, ..., v_k^s\}$. Each comment has only one corresponding label l^s where $l^s \in \{-1, 0, 1\}$. In this paper, we aim to select proper label for each comment in test dataset. The method is: for each comment we have three probabilities p_{-1}, p_0, p_1 after our transformation methods. Label which has the maximum probability is the final result.

4   Framework Description

First, the input data c^s becomes token embedding v^s after transformer language models. Second, our four transformation methods will pick up output v^s or hidden state vectors $v_{hidden\_i}^s$ from hidden layer i. Every transformation method gives probabilities of three labels p_{-1}, p_0, p_1. Finally, after ensemble layer we can get proper labels by finding maximum probability.

4.1   Self-attention Architecture

Self-attention is developed based on the attention mechanism which is widely used in computer vision and sequential learning task [15,16]. Vaswani et al. [6] was inspired by attention mechanism and proposed a self attention architecture to completely replace RNN into full attention layers.

The central part of self attention is replacing traditional linear model by pure attention. An attention module will calculate a set of weight values and map it with original query, then output the weighted sum of the values. If we assume the value of query is packed into a matrix Q, the key value is packed into another matrix K, and matrix V is the value matrix, the dimension of a single key value is d_k, the attention function can be described as [6]:

$$Attention(Q, K, V) = softmax(\frac{QK^T}{\sqrt{d_k}})V \tag{1}$$

Based on this small but powerful module, Vaswani et al. created several self-attention based transformer models and outperformed state-of-the-art models on machine translation and English Constituency Parsing tasks [6].

4.2   Context-Aware Semantic Information Transformation

There are lots of ways to extract data from token embedding, like full connection layer, LSTM [17] and GRU [10]. In order to find the differences between different context-aware information transformations, we create four different independent transformations:

1. Full connection layer. The input is a concatenation of first vector in transformer language model output and that in last two hidden layers outputs. The reason is hidden layers outputs are less affected by downstream tasks, meaning more original latent information can be obtained here.
2. Full connection layer. The input also is a concatenation of normal output and hidden output described in Transformation 1, but this method will catch one more hidden layer output to preserve more latent information.
3. Bidirectional GRU layer. The input of GRU is the full output of transformer language model.
4. GRU and LSTM layer. The input of LSTM is the full output of transformer language model, then the hidden state and output vectors from LSTM containing memory information are transferred into another GRU module.

For the first and second type transformations, the output of language model will contain two parts: one is the first vector of output embedding and the another part is the first vector of last two or three hidden layers. These two transformations can be described as:

$$v_1^s = LM_{total\_layer\_number}(c^s) \tag{2}$$
$$v_{hidden\_i,1}^s = LM_{total\_layer\_number-i}(c^s) \tag{3}$$
$$output_1 = argmax(f(v_1^s, v_{hidden\_1,1}^s, v_{hidden\_2,1}^s)) \tag{4}$$
$$output_2 = argmax(f(v_1^s, v_{hidden\_1,1}^s, v_{hidden\_2,1}^s, v_{hidden\_3,1}^s)) \tag{5}$$

where $LM_{layer\_number}(\cdot)$ returns the first output vector after a given number of self-attention layers, $argmax(\cdot)$ retrieves the index of the largest number in final vector and transfers it into proper sentiment label, $f(\cdot)$ is a full connection layer.

The third and fourth transformations use a linear module to extract the latent data. These transformations does not need the hidden output from these huge language models. They can be described as:

$$\{v_1^s, v_2^s, ..., v_k^s\} = LMA(c^s) \tag{6}$$
$$h_1, h_2 = BiGRU(\{v_1^s, v_2^s, ..., v_k^s\}) \tag{7}$$
$$h_1 = LSTMGRU(\{v_1^s, v_2^s, ..., v_k^s\}) \tag{8}$$
$$output_3 = argmax(f(h_1, h_2)) \tag{9}$$
$$output_4 = argmax(f(h_1, v_1^s)) \tag{10}$$

where $LMA(\cdot)$ returns the full model output of a single input tokens, $BiGRU(\cdot)$ means the BiGRU module in transformation 3 and $LSTMGRU(\cdot)$ means LST-MGRU module in transformation 4.

5   Experiment

5.1   Datasets

We evaluate these models on a given Weibo dataset. Weibo is a Chinese short text messages exchange platform. The final result is a label referring to the emotion of this weibo. This label has three values: 1 means positive mood, 0 means neural mood, and -1 means negative mood. Comment time of these weibo begins on Jan 1, 2020 and ends on Feb. 19, 2020. There is 100,000 labelled training data, 900,000 unlablled data and 10,000 test data.

Because there are lots of unlabelled training dataset, and in order to introduce the distribution of test dataset, we use a popular semi-supervised training methods called pseudo-label. In the following experiment part, we will investigate the influence of these pseudo-labels by training these models on a combination of original training datasets and 1,000 pseudo-labels. These pseudo labels are created by a single BERT-wwm [18] model with a bidirectional GRU layer.

5.2   Baselines and Evaluation Metrics

In this part we will introduce two different language models and used evaluation metric.

BERT: BERT [1] is a pure self attention based language model created in 2018. Unlike ELMo, BERT can learn deep language representations using unlabelled data, making it easy to fine-tune a state-of-the-art model by just adding a simple additional transformation.

RoBERTa: RoBERTa [2] is a robust BERT by using far more unlabelled training dataset and computation resources because Liu et al. investigated the original BERT training methods and found it is underfitting. After longer training processes, RoBERTa performs better than BERT in several tasks [18].

In this paper, we use pretrained BERT-wwm [18] and RoBERTa_zh4[1] models.
All of these language models use four different context-aware semantic transformations. Because it is a multi-classification task, the evaluation metric is
macro-F1 value. The original F1 value is the harmonic mean of accuracy and
recall, and macro-F1 is the average number of F1 value on given different types.

5.3 Results and Discussions

The main goal of this experiment is checking the effectiveness of different models
and optimization method. The parameters of the training process are fixed and
the optimizer of this model is an improved Adam algorithm named AdamW [19].
The detailed training parameters are listed in Table 1.

Table 1. Implementation details

Learning rate	$1e^{-5}$	Warmup steps	200
Loss function	Cross entropy	Accumulation steps	4
Total step	20,000	Batch size	8

The accumulation steps number means a gradient optimization step every
four batches, it can help increase the batch size efficiently without consuming
more computation resources like memory. Our baseline model uses the same
language models and its context-aware transformation is a single full connection
layer. Training results are generated using training dataset with training labels
excluded, and test results are generated using normal test dataset. The ensemble
results and baseline results are listed in Table 2. "Original" means this model are
trained using pure training dataset while "Pseudo-labelled" means this model
are trained using training dataset and pseudo-label dataset.

Table 2. Experiment result (F1 value)

Models	Training dataset		Test dataset	
	RoBERTa	BERT	RoBERTa	BERT
Baseline	–	–	0.7330	0.7317
Original	0.7571	0.7447	0.7359	**0.7390**
Pseudo-labelled	**0.7575**	**0.7460**	**0.7413**	0.7384

For RoBERTa part, because the model size of RoBERTa is larger than BERT,
meaning it can stow more latent information, the training results and test results

---

[1] https://github.com/brightmart/roberta_zh.

are better than BERT. However, the gap in test results are smaller. As for BERT part, the model size of itself limits the performance. Not only train results, but test results are lower than RoBERTa. After investigating each single model, BERT or RoBERTa with transformation 2 and transformation 3 are always better than another two models.

As for pseudo-labelled training dataset, because RoBERTa is much larger than BERT, it is expected that RoBERTa is much better than BERT. As for test dataset things are a little different. RoBERTa shows that our model outperforms baseline and do better using pesudo-label dataset. But in BERT model using original dataset are slightly better than that with pseudo-label dataset. After investigation we find BERT may be affected by several controversial labels existing in pseudo-label dataset.

6   Conclusion

In this paper, we investigated four different context-aware information transformations on BERT and RoBERTa. The result shows the simple full connection layers and concatenation of output and hidden state is the overall best method in this task, meaning there is latent information hidden in these transformer layers and it is less affected by fine-tuning training dataset.

Acknowledgement. This research was supported by National Natural Science Foundation of China (Grants No. 61877051), and Natural Science Foundation Project of CQ, China (Grants No. cstc2018jscx-msyb1042, and cstc2018jscx-msybX0273). The authors would like to thank anonymous reviewers for their constructive comments.

References

1. Devlin, J., Chang, W.M., Lee, K., Toutanova, K.: BERT: pre-training of deep bidirectional transformers for language understanding. arXiv preprint arXiv:1810.04805 (2018)
2. Liu, Y., et al.: A robustly optimized BERT pretraining approach. arXiv preprint arXiv:1907.11692 (2019)
3. Lan, Z., et al.: Albert: a lite BERT for self-supervised learning of language representations. In: International Conference on Learning Representations (2020)
4. Clark, K., Luong, M.T., Le, Q.V., Manning, C.D.: ELECTRA: pre-training text encoders as discriminators rather than generators. In: International Conference on Learning Representations (2020)
5. Socher, R., et al.: Recursive deep models for semantic compositionality over a sentiment treebank. In: Proceedings of the 2013 Conference on Empirical Methods in Natural Language Processing, pp. 1631–1642 (2013)
6. Vaswani, A., et al.: Attention is all you need. In: Advances in Neural Information Processing Systems (2017)
7. Mikolov, T., Chen, K., Corrado, G., Dean, J.: Efficient estimation of word representations in vector space. arXiv preprint arXiv:1301.3781 (2013)
8. Bojanowski, P., Grave, E., Joulin, A., Mikolov, T.: Enriching word vectors with subword information. Trans. Assoc. Comput. Linguist. **5**, 135–146 (2017)

9. Pennington, J., Socher, R., Manning, C.D.: GloVe: global vectors for word representation. In: Proceedings of the 2014 Conference on Empirical Methods in Natural Language Processing (EMNLP), pp. 1532–1543 (2014)
10. Cho, K., et al.: Learning phrase representations using RNN encoder-decoder for statistical machine translation. In: Proceedings of the 2014 Conference on Empirical Methods in Natural Language Processing (EMNLP) (2014)
11. Yang, Z., et al.: XLNet: generalized autoregressive pretraining for language understanding. In: Advances in Neural Information Processing Systems 32 (2019)
12. Sanh, V., Debut, L., Chaumond, J., Wolf, T.: DistilBERT, a distilled version of BERT: smaller, faster, cheaper and lighter. arXiv preprint arXiv:1910.01108 (2019)
13. Raffel, C., et al.: Exploring the limits of transfer learning with a unified text-to-text transformer. arXiv preprint arXiv:1910.10683 (2019)
14. Xie, Q., et al.: Unsupervised data augmentation for consistency training. arXiv preprint arXiv:1904.12848 (2019)
15. Chen, Y., et al.: A2-Nets: double attention networks. In: Advances in Neural Information Processing Systems 32 (2018)
16. Zhao, G., et al.: MUSE: parallel multi-scale attention for sequence to sequence learning. arXiv preprint arXiv:1911.09483 (2019)
17. Sak, H., Senior, A., Beaufays, F.: Long short-term memory based recurrent neural network architectures for large vocabulary speech recognition. arXiv preprint arXiv:1402.1128 (2014)
18. Cui, Y., et al.: Pre-training with whole word masking for Chinese BERT. arXiv preprint arXiv:1906.08101 (2019)
19. Loshchilov, I., Hutter, F.: Decoupled weight decay regularization. In: International Conference on Learning Representations (2017)

An Algorithm for Emotion Evaluation and Analysis Based on CBOW

JiaXin Guo and ZhengYou Xia[✉]

College of Computer Science and Technology,
Nanjing University of Aeronautics and Astronautics, Nanjing, China
jiaxin_guo_nuaa@163.com, zhengyou_xia@nuaa.edu.cn

Abstract. As a tool to study people's views and opinions on things and events around them, sentiment analysis is widely used in the analysis and processing of mass evaluation information. Traditional emotion analysis is generally based on emotion dictionary, but the construction of emotion dictionary needs a lot of artificial time, and in different application fields need to establish different emotion dictionaries. Meanwhile, emotion dictionary can't contain the semantic information of words and it also ignore the role of non-emotional words in the expression of emotion. This paper proposes a new emotion analysis algorithm (CBOW-PE) based on word embedding and part of speech. First, in the stage of pre-processed experiment, we use part of speech to preprocess the experimental data, fully consider the role of non-emotional words on emotion analysis. And then we add emotional information to assist word vector training, so that the word embedding of words related to emotion has both semantic information and emotional information. Finally, this paper makes a lot of experimental analysis and comparison of this paper and its related algorithms. The results show that the algorithm is effective and efficient in Chinese emotion analysis.

Keywords: Word embedding · Part-of-speech · Emotion dictionary · CBOW · CBOW-PE

1 Introduction

In recent years, with the popularity and continuous development of the Internet, people enjoy more and more convenient e-commerce, network social, electronic entertainment and other services. It has become an important research topic to conduct sentiment analysis on a large number of evaluation texts on the Internet to facilitate new users to browse and refer.

More attention has been paid to the sentiment analysis of short texts. Most of the existing methods follow the direction of Peng et al. [1], using the sentiment polarity of manual tagging to construct sentiment classification. The main work is to design effective features to obtain better performance [2–5]. By learning documents, obtaining dense word representations or word embeddings [6,7]

G. Li et al. (Eds.): KSEM 2020, LNAI 12275, pp. 344–351, 2020.
https://doi.org/10.1007/978-3-030-55393-7_31

has become a new direction for people to study. GloVe is another unsupervised learning word embedding model which based on global word frequency statistics, considering co-occurrence information between words and achieves better results than Word2Vec [8]. Tang et al. proposed three neural network structures to introduce emotional information into the loss function to learn word embeddings with specific emotions, and achieved good results in Twitter English sentiment analysis [9]. Jacob et al. Put forward the Bert model [10], through adjusting the information in all layers and contexts, pre training the deep two-way representation from the marked text to get the word vector.

In the previous related sentiment analysis methods at home and abroad, mainly based on dictionary analysis method and machine learning method. Taking a dictionary-based approach requires a lot of manual processes, and because of the different application scenarios, a single dictionary is not sufficiently versatile, and often requires the establishment of different dictionaries. Machine learning based methods rely directly on a large number of data to learn, but lack of supervision, learning results are difficult to interpret. This paper tries to add part of speech as a reference in the pretreatment stage, and add some non-emotional words with emotional information outside the affective words to form the enhanced learning part of sentiment analysis. Introducing emotion as supervision in the process of word embedding learning, adding a layer of hidden layer in the process of word embedding learning, so that the word embedding can learn both semantic information and emotional information in the process of learning, So as to get embedded words that are more conducive to Chinese text sentiment analysis.

The remainder of this paper is as follows: Sect. 2 introduces the improved word embedding learning model CBOW-PE. The Sect. 3 is the relevant experimental analysis and discussion. Finally, a concise conclusion is given in Sect. 4.

2   CBOW Combined with Part-of-Speech and Emotional Information

In this paper, the CBOW algorithm is improved to obtain word embeddings containing a mixture of semantic and emotional information. In the CBOW algorithm, window movement is used, and the target word and its context are continuously used as input to obtain unsupervised word embeddings. Each word embedding is obtained by weighting its context word embedding, making the target word and the context relevant, So that the word embedding has semantic information. In order to make the obtained word embedding better applicable to the sentiment classification task, we added a loop of learning content on the basis of the original CBOW model to train the sentiment information of the word embedding. For a sentence containing emotional polarity, CBOW simply applies a window and moves on the basis of the sentence, and the CBOW-PE model in this article adds an emotional learning loop on this basis, and associates each sentence with the emotional expression. In previous studies on English, the accuracy of sentiment analysis can be improved by including part of speech as one of the criteria for feature extraction [11]. In Chinese, words of different parts of speech have different effects on emotional expression, among which verbs and adjectives often contain more emotional information.

2.1   CBOW-PE Model Framework

Compared with the traditional sentiment analysis task, this paper adds part-of-speech as one of the considerations in the data preprocessing stage, and improves the algorithm flow for the goal of sentiment analysis.

First, in the preprocessing process, the paper further processed the experimental data by using the part-of-speech to obtain the text used in the traditional word embedding learning. At the same time, it also obtained the emotion-related words containing emotional information and filtering based on part-of-speech. In the process of word embedding learning, the improved CBOW-PE model is used for training. Finally, the obtained word embedding is applied to the emotion classification task. The specific CBOW-PE framework is as follows (Fig. 1).

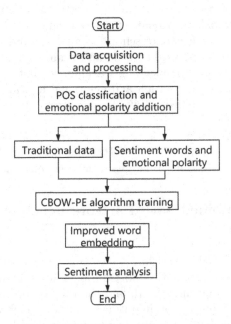

Fig. 1. CBOW-PE model framework

2.2   CBOW-PE Model

For the corpus, words are divided into emotionally related words and non-emotional related words. For non-affective related words, the learning model is trained using the CBOW model. For emotion-related words, the CBOW-PE model is used for training. The sentiment-related words in the first hidden layer and context form a set of $\{x_{t-c}, ..., x_t, ..., x_{t+c}\}$ training semantic information.

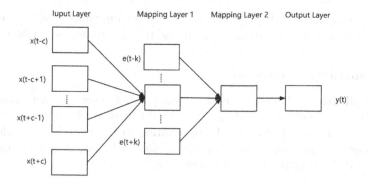

Fig. 2. CBOW-PE model framework

CBOW reduces the algorithm time by negative sampling, and its loss function is as formula (1) and (2) (Fig. 2):

$$L = -\log(\sigma(v_{x_t}'^T h)) - \sum_{i=1}^{k} \log(\sigma(-v_{x_i}'^T h)) \tag{1}$$

$$h = \frac{1}{2c} \sum_{i=1}^{c} (v_{t+i} + v_{t-i}) \tag{2}$$

v is the vector form of the word x. k is the number of negative samples. Improve the prediction accuracy by minimizing the loss function L. In the second hidden layer, there is a set of $\{e_{t-l}, ..., e_t, ..., e_{t+l}\}$ related to the emotion-related words that have appeared in its context, and we calculate the correlation between the set and the emotion expressed by the sentence. The Emotional loss is as obtain formula (3) and (4):

$$L = -\log(\sigma(e_{true}'^T h_e)) - log(\sigma(e_{false}'^T h_e)) \tag{3}$$

$$h_e = \frac{1}{2l} \sum_{i=1}^{l} (v_{t+i} + v_{t-i}) \tag{4}$$

v is the vector form of the emotion-related word e. e_{true} is the correct sentiment of this review sentence, e_{false} is the wrong sentiment. For emotional words, two layers of learning must be performed simultaneously during the learning process, so the final word vector loss function is the sum of semantic information loss and emotional information loss, as in formula (5):

$$L_{cp} = L + L_e \tag{5}$$

During the experiment, Adam's optimization algorithm was selected to adjust the word embedding to reduce the loss function.

3 Experiment

In the experimental analysis and discussion, it is mainly divided into three parts. First, we introduce the experimental preparation. Then, we compare the

CBOW-PE model and the CBOW model after adding part-of-speech and emotion. Finally, we compare our model with traditional sentiment analysis methods.

3.1 Experiment Preparation

Data Collection. The data used in this paper is user product evaluations obtained from Taobao. This data set is 11.3M in size, with a total of 62,774 sentences, containing 66,872 words. In addition to the main body of the text, each review has its domain identification (movies, books, etc.) and emotional polarity (31,728 positive and 31,046 negative).

Data Processing. First, we first remove the special symbols in the text. Then the text is segmented in Chinese and its part of speech is marked for training. Finally, delete the stop words which have no practical influence on the expression of the sentence.

3.2 Influence of Part-of-Speech and Sentiment

This section compares the initial word embedding (CBOW), word embedding and part-of-speech (CBOW-P), word embedding and emotion (CBOW-E), and improved word embedding model (CBOW-PE). Among them, CBOW-P is a training layer that adds an additional layer of emotion-related part-of-speech word embeddings to the original CBOW, and CBOW-E adds emotion as an independent "word" to CBOW training for training. According to Mikolov's article, the optimal window size when using the word embedding algorithm is 5, the number of iterations is 5, and the word embedding dimension is 200.

The experimental results are shown in Table 1. It can be seen that after adding part-of-speech and emotion to the CBOW algorithm, the word embedding obtained by the CBOW-PE algorithm has a better effect of sentiment analysis. Adding part-of-speech alone as a reference is slightly improved, but it is better to add part-of-speech and emotion together, and adding emotional information alone may make the sentiment analysis of the word embedding trained worse.

3.3 Comparison with Traditional Related Emotion Classification Algorithms

After parameter tuning (process omitted), the set window size of CBOW and CBOW-PE in this experiment is 5, the word embedding dimension is 350, and the number of iterations is 15 times. Choose to use the SVM classifier for sentiment analysis. At the same time, the sentiment analysis method based on sentiment dictionary is added as a reference in this experiment. It can be seen from the results that the sentiment analysis effect of the word embedding combined with the SVM classifier trained by the CBOW-PE model is better than the original method (Table 2).

Table 1. Influence of part-of-speech and sentiment

	Accuracy	Recall	Precision	F-score
CBOW	79.95%	80.90%	81.59%	81.22%
CBOW-E	80.45%	79.51%	83.32%	81.33%
CBOW-P	78.92%	79.66%	80.81%	80.19%
CBOW-PE	80.69%	80.27%	82.19%	82.04%

Table 2. Comparison with traditional related emotion classification algorithms

	Accuracy	Recall	Precision	F-score
Sentiment dictionary	67.06%	80.05%	66.41%	72.60%
Glove+SVM	79.48%	82.10%	80.12%	81.10%
BERT+SVM	81.98%	82.66%	83.55%	83.10%
CBOW+SVM	82.47%	83.57%	83.70%	83.63%
CBOW-PE+SVM	84.48%	85.41%	85.60%	85.50%

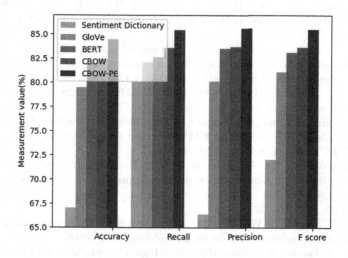

Fig. 3. Comparison with traditional related emotion classification algorithms

4   Conclusion

This paper is mainly improved based on the CBOW model. By adding a layer of hidden layers, the word embedding is trained simultaneously on semantic information and emotional information during the learning process, thereby obtaining a word embedding containing emotional information. The experimental results show that the word embedding sentiment classification result obtained by the improved CBOW-PE algorithm is better than the traditional CBOW method, and it can complete the sentiment classification task more accurately. However, the model in this paper is still inadequate. On the one hand, it depends on the size of the corpus, and it does not handle new words; on the other hand, it does not consider the case of polysemy. In the following research, further improvements are needed to address the shortcomings of the model (Fig. 3).

References

1. Pang, B., Lee, L., Vaithyanathan, S.: Thumbs up?: sentiment classification using machine learning techniques. In: Proceedings of the ACL-2002 Conference on Empirical Methods in Natural Language Processing, vol. 10, pp. 79–86. Association for Computational Linguistics (2002)
2. Jain, A.K., Pandey, K.: Analysis and implementation of sentiment classification using lexical POS markers. Int. J. **2**(1), 36–40 (2013)
3. Mohammad, S.M., Kiritchenko, S., Zhu, X.: NRC-Canada: building the state-of-the-art in sentiment analysis of tweets. arXiv preprint arXiv:1308.6242 (2013)
4. Bengio, Y., Courville, A., Vincent, P.: Representation learning: a review and new perspectives. IEEE Trans. Pattern Anal. Mach. Intell. **35**(8), 1798–1828 (2013)
5. Bengio, Y.: Deep learning of representations: looking forward. In: Dediu, A.H., Martín-Vide, C., Mitkov, R., Truthe, B. (eds.) SLSP 2013. LNCS (LNAI), vol. 7978, pp. 1–37. Springer, Heidelberg (2013). https://doi.org/10.1007/978-3-642-39593-2_1
6. Mikolov, T., Sutskever, I., Chen, K., Corrado, G.S., Dean, J.: Distributed representations of words and phrases and their compositionality. In: Advances in Neural Information Processing Systems, pp. 3111–3119 (2013)
7. Mikolov, T., Chen, K., Corrado, G., Dean, J.: Efficient estimation of word representations in vector space. arXiv preprint arXiv:1301.3781 (2013)
8. Pennington, J., Socher, R., Manning, C.D.: Glove: global vectors for word representation. In: Proceedings of the 2014 Conference on Empirical Methods in Natural Language Processing (EMNLP) (2014)
9. Tang, D., Wei, F., Qin, B., Zhou, M., Liu, T.: Building large-scale twitter-specific sentiment lexicon: a representation learning approach. In: Proceedings of Coling 2014, the 25th International Conference on Computational Linguistics: Technical Papers, pp. 172–182 (2014)
10. Devlin, J., Chang, M.W., Lee, K., Toutanova, K.: BERT: pre-training of deep bidirectional transformers for language understanding (2018)
11. Badr, B.M, Fatima, S.S.: Using skipgrams, bigrams, and part of speech features for sentiment classification of twitter messages. In: Sharma, D.M, Sangal, R., Sherly, E. (eds.) Proceedings of the 12th International Conference on Natural Language Processing, ICON 2015, Trivandrum, India, 11–14 December 2015, pp. 268–275. NLP Association of India (2015)
12. Jiang, L., Yu, M., Zhou, M., Liu, X., Zhao, T.: Target-dependent twitter sentiment classification. In: Proceedings of the 49th Annual Meeting of the Association for Computational Linguistics: Human Language Technologies, vol. 1, pp, 151–160. Association for Computational Linguistics (2011)
13. Hu, X., Tang, J., Gao, H., Liu, H.: Unsupervised sentiment analysis with emotional signals. In: Proceedings of the 22nd International Conference on World Wide Web, pp. 607–618 (2013)
14. Li, X., Bing, L., Lam, W., Shi, B.: Transformation networks for target-oriented sentiment classification. arXiv preprint arXiv:1805.01086 (2018)

15. Chen, P., Sun, Z., Bing, L., Yang, W.: Recurrent attention network on memory for aspect sentiment analysis. In: Proceedings of the 2017 Conference on Empirical Methods in Natural Language Processing, pp. 452–461 (2017)

16. Wang, Y., Huang, M., Zhu, X., Zhao, L.: Attention-based LSTM for aspect-level sentiment classification. In: Proceedings of the 2016 Conference on Empirical Methods in Natural Language Processing, pp. 606–615 (2016)

17. Turney, P.D., Littman, M.L.: Measuring praise and criticism: inference of semantic orientation from association. ACM Trans. Inf. Syst. (TOIS) 21(4), 315–346 (2003)

18. Hu, M., Liu, B.: Mining and summarizing customer reviews. In: Proceedings of the tenth ACM SIGKDD International Conference on Knowledge Discovery and Data Mining, pp. 168–177 (2004)

19. Joulin, A., Grave, E., Bojanowski, P., Mikolov, T.: Bag of tricks for efficient text classification. arXiv preprint arXiv:1607.01759 (2016)

20. Bollegala, D., Mu, T., Goulermas, J.Y.: Cross-domain sentiment classification using sentiment sensitive embeddings. IEEE Trans. Knowl. Data Eng. 28(2), 398–410 (2015)

21. Kim, Y.: Convolutional neural networks for sentence classification. arXiv preprint arXiv:1408.5882 (2014)

22. Zheng, X., Chen, H., Xu, T.: Deep learning for Chinese word segmentation and POS tagging. In: Proceedings of the 2013 Conference on Empirical Methods in Natural Language Processing, pp. 647–657 (2013)

23. Liu, T., Cui, Y., Yin, Q., Zhang, W., Wang, S., Hu, G.: Generating and exploiting large-scale pseudo training data for zero pronoun resolution. arXiv preprint arXiv:1606.01603 (2016)

24. Narasimhan, K., Yala, A., Barzilay, R.: Improving information extraction by acquiring external evidence with reinforcement learning. arXiv preprint arXiv:1603.07954 (2016)

25. Goldberg, Y., Levy, O.: Word2vec explained: deriving Mikolov et al.'s negative-sampling word-embedding method. CoRR, abs/1402.3722 (2014)

Predicting Crowdsourcing Worker Performance with Knowledge Tracing

Zizhe Wang[1,2]([✉]), Hailong Sun[1,2], and Tao Han[1,2]

[1] SKLSDE Lab, School of Computer Science and Engineering,
Beihang University, Beijing 100191, China
{wangzz,hantao}@act.buaa.edu.cn,
sunhl@buaa.edu.cn
[2] Beijing Advanced Innovation Center for Big Data and Brain Computing,
Beihang University, Beijing 100191, China

Abstract. Knowledge-intensive crowdsourcing (KI-C) plays an important role in today's knowledge economy. And competitive knowledge-intensive crowdsourcing (CKI-C) is a kind of KI-C in which tasks are released in the form of competitions. The worker performance prediction is important for CKI-C platforms to recommend tasks to proper workers. Traditional worker performance prediction methods do not consider the complex properties of tasks and worker skills, thus they do not function in CKI-C. In this work, we design the KT4Crowd framework to incorporate knowledge tracing, used effectively in intelligent tutoring systems (ITS), into CKI-C for predicting worker performance. The experimental results on a large-scale Topcoder dataset show the effectiveness of our framework and the DKVMN model with our framework achieves the best performance among the compared state-of-the-art methods.

Keywords: Crowdsourcing · Knowledge tracing · Performance prediction

1 Introduction

Knowledge-intensive crowdsourcing (KI-C) is one of the most promising areas of next generation crowdsourcing [1,6], for its essential role in today's knowledge economy[1]. Existing crowdsourcing applications mainly focus on simple micro-tasks (e.g. image annotation) while KI-C often targets more complex tasks that require workers to master advanced knowledge, such as crowdsourced software development, product design and knowledge creation. In this paper, we only discuss the KI-C in which each crowd worker or team finishes the KI-C task alone and submits the work (such as Kaggle, Topcoder) and we call it competitive KI-C (CKI-C). Collaborative KI-C like Wikipedia is not considered.

Task assignment is the fundamental process of crowdsourcing, and a precise worker performance prediction model is the precondition of an efficient task

---

[1] https://en.wikipedia.org/wiki/Knowledge_economy.

© Springer Nature Switzerland AG 2020
G. Li et al. (Eds.): KSEM 2020, LNAI 12275, pp. 352–359, 2020.
https://doi.org/10.1007/978-3-030-55393-7_32

assignment method. In traditional crowdsourcing, this problem has been studied thoroughly [2,5]. But most of these methods are inapplicable in CKI-C, since CKI-C tasks have multiple features and the answer set is infinite, which does not match the settings of these methods. Another weakness of traditional methods are that worker's ability is usually denoted as a constant or a fixed confusion matrix. While some researches show that worker's ability changes when he performs complex tasks [8]. So we need a worker performance prediction model which can dynamically and precisely estimate worker's ability in CKI-C.

Predicting worker performance in CKI-C shares similar challenges faced by intelligent tutoring system (ITS). And knowledge tracing (KT) is a kind of method that traces the evolving knowledge state of students as they engage a sequence of learning activities [3]. KT has a big progress these years after it is combined with deep learning [7,9]. We propose a Knowledge Tracing for crowdsourcing (KT4Crowd) framework to use the knowledge tracing models to solve the worker performance prediction problem in CKI-C.

The framework consists of three parts, data preprocessing, training and predicting. We use task decomposition and results transformation to preprocess the CKI-C data. The training process is similar to that of a KT model. And the predicting process uses the trained KT model to predict the subtasks of one task, and use majority voting to get the final prediction of the original task. Finally we conducted two experiments to evaluate the performance of KT models under our KT4Crowd framework. Both of them show the efficiency of our framework.

Our main contributions are summarized as follows:

- We propose a general framework to adapt KT models to predict worker performance in CKI-C. To the best of our knowledge, this is the first work using KT models to solve worker performance prediction problem in CKI-C.
- We propose two methods to improve traditional KT methods to work with the multi-feature, no-standard-answer CKI-C data.
- We have conducted two kind of experiments to validate the effectiveness of our framework and get impressive results. Experiments show that DKVMN with our framework has the best performance.

2 Knowledge Tracing for Crowdsourcing

In this section, we try to apply knowledge tracing methods in CKI-C for worker performance prediction. First, we give a formalization of worker performance prediction problem in CKI-C. Then we introduce our knowledge tracing for CKI-C (KT4Crowd) framework in detail. The framework consists of two parts, training and predicting.

2.1 Worker Performance Prediction in CKI-C

We already describe the concept of CKI-C in our introduction. In this section, we will first give a more clear definition to it. At first we need to define the concepts of task, worker and result.

Task. *A CKI-C task* t *on platform* P *usually requires multiple skills. We call them features of the task and denote them as* $F^t = \{f_1^t, f_2^t, ..., f_N^t\}$. *It is a subset of the features set* $\{f_1, f_2, ..., f_L\}$, *which contains all the features that appear in tasks on* P.

Worker. *The worker set of task* t *is denoted as* $\{w_1^t, w_2^t, ..., w_M^t\}$. *Each worker* w *has expertise in some skills.*

Result. *The* M *workers of task* t *submit their works* $\{s_1^t, s_2^t, ..., s_M^t\}$. *The submissions are evaluated by an organization or individual except the workers. The evaluation results set is* $\{a_1^t, a_2^t, ..., a_M^t\}$, *which are usually scores. The results can be ranked by a certain rule, and the top* K *workers will win the rewards. And the final results is a binary set* $\{c_1^t, c_2^t, ..., c_M^t\}$, *in which* c_i^t *means whether worker* i *win the rewards of* t.

The multiple features and the submissions evaluation process show that the tasks are knowledge-intensive. And ranking the results indicate they are competitive. The feature amount of task t can be 1, same as the worker number M_t.

With the definition above, we can formalize the worker performance prediction problem. And there are two ways to predict worker's performance.

Sequence Prediction. *Let* I_w *be the series numbers set of the tasks that worker* w *has done. We define* $\{(F^i, a_w^i)\}_{i \in I_w}$ *as the answer sequence of worker* w. *Sequence prediction predicts* c_w^k, *the final result of worker* w *under task* $k(k \in I_w)$, *with its feature set* F^k, *previous tasks' feature sets* $\{F^i\}_{i \in I_w, i < k}$ *and previous tasks' answers* $\{a_i\}_{i \in I_w, i < k}$.

Winner Prediction. *When given a new task* t, *its feature set* F^t *and a worker set* W, *we denote all the tasks they have done as a set* T. *the winner prediction process predicts the final results or rank of the workers under task* t *with the workers' historical data including their historical results* $\{c_w^{t'}\}_{w \in W, t' \in T}$ *and former tasks' feature sets* $\{F^{t'}\}_{t' \in T}$.

2.2 Knowledge Tracing Model

Before introducing our framework, we need to give a brief definition to knowledge tracing model first.

Knowledge Tracing Model. *A KT model contains a classifier* $K(x; \theta)$ *and its training algorithm* U. *The input data* x *is a result sequence. For example, the result sequence of worker* w *is* $\{F^i, c_w^i\}_{i \in I_w}$. *In a KT model,* F^i *has only one feature, and the value of* c_w^i *is 0 or 1.* θ *is the parameter vector of the classifier. The output of* K *is a prediction sequence* $\{p_w^i\}_{i \in I_w}$, p_w^i *is the prediction result of worker* w *under task* i. *The training algorithm* U *is an algorithm that updates the parameter vector* θ *of* K *using the training data.*

Then we will introduce the two main differences between ITS and CKI-C.

First, tasks on CKI-C are more complex than those on ITS. Since the ITS's purpose is training the students, the exercises on it are simpler, usually contain only one knowledge point each. But the tasks in CKI-C are usually released for practical applications, which means more skills are required.

Second, there is no standard answer for the tasks in CKI-C. After a task is finished, what we have is the evaluations of workers' submissions. The evaluations of submissions are usually real-number scores, not binary data. And the evaluations may be not objective enough. But in ITS, the system will automatically judge whether the answer is correct with the standard answers. Due to the two differences, we need to preprocess the CKI-C data.

2.3 KT4Crowd Framework

Data Processing Methods. The framework consists of two main data processing methods as follow.

Task Decomposition. A task with multiple features can be approximately regarded as multiple tasks with one feature. So we decompose each task t with N_t features into N_t subtasks with one feature.

Results Transformation. For a task t in CKI-C, the result $a_{w,t}$ of a worker is 1 only when her rank is higher than R and score is higher than S. Otherwise, $a_{w,t} = 0$. R and S is determined by the characteristic of dataset. One important thing is that, the R and S here is not two hyperparameters but a criterion to judge whether a worker has a good performance. The value of them depends on the dataset.

With these two methods, the processed data can meet the requirement of KT models.

When solving the second research question, we need to convert the workers' continuous scores into a binary form. And the binary data should represent whether the developer has a good performance. The question is how to determine that a developer performs well. So we give a definition to good performance.

Good Performance. A worker w performs well on task t only when his score $s_{w,t}$ is higher than a threshold S and his rank $r_{w,t}$ is higher than a threshold R.

With this definition, we design the results transformation method above.

Then we will introduce the training and predicting processes. Suppose we have a CKI-C dataset and a KT model $\{K(\boldsymbol{x}; \boldsymbol{\theta})\}, U$, we need to train a model to predict the performance of the CKI-C worker.

Training. In the training process, we reorganize the dataset into a set of sequences. Each sequence is an answer sequence of a worker, which is in the form of $\{(F^i, a_w^i)\}_{i \in I_w}$. The tasks in a sequence are ranged in a chronological order. For each sequence, firstly, we use the results transformation method to transform the answers $\{a_w^i\}$ into binary results $\{c_w^i\}$. Then we use the task decomposition method to decompose the tuple (F^i, c_w^i) into multiple tuples $\{(f_1^i, c_w^i), (f_2^i, c_w^i), ..., (f_N^i, c_w^i)\}$, in which $f_j^i (1 \leq j \leq N)$ is the features in the feature set F^i. After that, we get a new answer sequence of subtasks. Then we can use the new answer sequences and the updating algorithm U to train the classifier $K(\boldsymbol{x}; \boldsymbol{\theta})$. And finally we will get the predicting model $K(\boldsymbol{x}; \hat{\boldsymbol{\theta}})$.

Predicting. When a new task t is coming, we can predict whether worker w will perform it well in the following way. At first we get the historical tasks that worker w has done and his answers, and reorganize them into an answer sequence $\{(F^i, a_w^i)\}_{i \in I_w}$ chronologically. Same as the training process, we use the results transformation and task decomposition methods to transform the answer sequence into a new answer sequence S. After that, we suppose the result of t is 0 and decompose it into multiple subtasks with the task decomposition method. For each subtask t_j, we add the tuple $(f_j^i, 0)$ into the tail of the new sequence and get S_j. Then we input S_j into $K(x; \hat{\theta})$ and get the prediction sequence o_j. The last element of o_j is the prediction of subtask with feature f_j. After we get the predictions of all the subtasks, we use majority voting to get the prediction of task t.

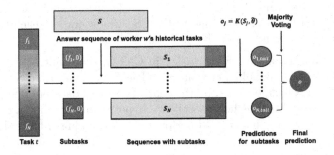

Fig. 1. The predicting process of KT4Crowd framework.

Figure 1 shows how we use the trained KT model $K(x; \hat{\theta})$ to predict worker w's performance doing a task with N features.

3 Experiments

We perform two kinds of experiments that introduced in 2.1.

3.1 Sequence Prediction

Data Preparation. We use a Topcoder dataset for experiment. We crawled submission records of developer on Topcoder from 2006 to 2019. The invalid statistics are deleted first. Finally we get 50625 valid submissions. These submissions involve 21522 tasks and 8584 developers. The statistics of the dataset can be seen in Table 1.

We transform the submission records into 8584 features-answer sequences, and randomly split the 8584 sequences into train dataset, validation dataset and test dataset by 60%, 20%, 20%. Then we put the data into four models and get the results. We set $R = 1$, $S = 90$ as the threshold, which ensure that there are enough positive samples for training and the positive sample refer to a good performance.

Table 1. The statistics of Top-coder dataset

Statics	Original	Pruned
Submissions	83566	50625
Developers	11562	8584
Tasks	57596	21522
Features	275	223

Table 2. Performance of four methods.

Metrics	DKT-S	DKT-O	DKVMN-O	DKVMN-S
AUC	0.7467	0.7166	0.8266	**0.8412**
Accuracy	0.7125	0.6980	0.7738	**0.8254**
Precision	0.6161	0.5803	0.6729	**0.7273**
Recall	0.5041	0.5224	0.6212	**0.7146**
F1	0.4807	0.5498	0.6460	**0.7540**

Methods and Metrics. We design a baseline framework called one-feature framework.

One-Feature. In this framework, we choose one feature for each task with multiple features. We first count the occurrence number of all the features in all the tasks. For the task which has multiple features, we choose the feature with largest occurrence number as the key feature. The form of worker results is the same as that in KT4Crowd framework.

We compare four methods in this experiment.

- **DKT-O:** DKT with one-feature framework [7].
- **DKT-S:** DKT with KT4Crowd framework.
- **DKVMN-O:** DKVMN with one-feature framework [9].
- **DKVMN-S:** DKVMN with KT4Crowd framework.

The metrics we choose are area under curve (AUC), accuracy, precision, recall and F1 score. We use all the predicted final results and the real results in the test dataset to calculate these metrics. How to transform the evaluation results is given in Sect. 2.

Results. Since the five metrics all range from 0 to 1, we put them into one figure. Table 2 shows the results. In Table 2, we can see that DKVMN-S gets the best performance on all metrics. The two DKVMN methods are both better than the two DKT methods. We can also find that the KT4Crowd framework is better than the baseline framework.

3.2 Winner Prediction

We first propose two methods to use the rating systems for worker performance prediction. Then we conduct an experiment to compare the prediction results between KT models and rating systems.

One Dimension: In this method, features of task are not considered. We use the results of tasks only to calculate the ratings of workers. Each developer's rating is a number r, which changes only when he loses or wins in a task, no matter how the task is. We calculate the new ratings after all submissions, and use them for prediction. For a task in the test dataset, we find the most recent rating records of the participants, and get the predicted rank by ranking the ratings.

Multi-features Linear Regression: In this method, the features of tasks are considered. We use a 223 dimensions vector $\mathbf{r} = \{r_1, r_2, ..., r_{223}\}$ to represent a developer's ability. The 223 dimensions $\{r_i\}_{i=1}^{223}$ are the developer's ratings over 223 features. When a task with features $\{f_1, f_2, ..., f_n\}$ is finished, $\{r_{f_1}, r_{f_2}, ..., r_{f_n}\}$ will be updated by the rating system. Similarly, we record the new $\mathbf{r}$ after each submission. Then we use linear regression to calculate the expression $F(\cdot)$ between $\mathbf{r}$ and the score in corresponding submission. Then we use the expression $F(\cdot)$ to predict the results of tasks in test dataset.

Data Preparation. Since the rating systems need all the historical data to calculate the ratings, we need to split the dataset by time. We first split the tasks into three parts by the time stamp, the proportions are 60%, 20%, 20%. Then we split the submission records by the tasks, and the three datasets of submissions are respectively train dataset, validation dataset, test dataset.

Models and Metrics. We choose 6 models for this experiment:

- **Glicko-2-M:** Glicko-2 with multi-feature linear regression method [4].
- **Glicko-2-O:** Glicko-2 with one dimension method.
- **Topcoder-M:** Topcoder rating with multi-feature linear regression method.
- **Topcoder-O:** Topcoder rating with one dimension method.
- **DKT-S:** DKT with KT4Crowd framework.
- **DKVMN-S:** DKVMN with KT4Crowd framework.

We choose 6 metrics in this experiment: AUC, accuracy, precision, recall and F1 score.

Table 3. Comparisons between KT and rating system.

Metrics	Glicko-2-M	Glicko-2-O	Topcoder-M	Topcoder-O	DKT-S	DKVMN-S
AUC	–	–	–	–	0.6332	**0.7817**
Accuracy	0.6359	0.7465	0.6967	0.7625	0.6043	**0.7771**
Precision	0.4449	0.6150	0.5348	0.6488	0.5067	**0.7139**
Recall	0.4129	0.5796	0.5037	0.6126	0.6107	**0.6962**
F1	0.4199	0.5877	0.5096	0.6210	0.5538	**0.7594**

Results. The results are shown in Table 3. From Table 3, we can see that DKVMN with KT4Crowd framework outperforms all the other methods. And we can conclude that DKVMN with KT4Crowd framework has a better performance than traditional rating systems in winner prediction in CKI-C.

4   Conclusion

In this work, we study how to apply KT methods for worker performance prediction on CKI-C tasks. We propose a KT4Crowd framework and design two experiments to estimate its effectiveness. The DKVMN model with KT4Crowd framework not only outperforms other KT models in the answer sequence prediction experiment, but also outperforms the traditional rating systems in the winner prediction experiment. So we can conclude that the KT models can be used in CKI-C, they have a good performance, especially the state-of-the-art model.

Acknowledgements. This work was supported partly by National Key Research and Development Program of China under Grant No. 2019YFB1705902, partly by National Natural Science Foundation under Grant No. (61932007, 61972013, 61421003).

References

1. Basu Roy, S., Lykourentzou, I., Thirumuruganathan, S., Amer-Yahia, S., Das, G.: Task assignment optimization in knowledge-intensive crowdsourcing. VLDB J. **24**(4), 467–491 (2015). https://doi.org/10.1007/s00778-015-0385-2
2. Bragg, J., EDU, W., Weld, D.S.: Learning on the job: optimal instruction for crowdsourcing. In: ICML Workshop on Crowdsourcing and Machine Learning (2015)
3. Corbett, A.T., Anderson, J.R.: Knowledge tracing: modeling the acquisition of procedural knowledge. User Model. User-Adap. Inter. **4**(4), 253–278 (1994). https://doi.org/10.1007/BF01099821
4. Glickman, M.E.: Dynamic paired comparison models with stochastic variances. J. Appl. Stat. **28**(6), 673–689 (2001)
5. Hassan, U., Curry, E.: A capability requirements approach for predicting worker performance in crowdsourcing. In: 9th IEEE International Conference on Collaborative Computing: Networking, Applications and Worksharing, pp. 429–437. IEEE (2013)
6. Kittur, A., et al.: The future of crowd work. In: Proceedings of the 2013 Conference on Computer Supported Cooperative Work, pp. 1301–1318. ACM (2013)
7. Piech, C., et al.: Deep knowledge tracing. In: Advances in Neural Information Processing Systems, pp. 505–513 (2015)
8. Wang, Z., Sun, H., Fu, Y., Ye, L.: Recommending crowdsourced software developers in consideration of skill improvement. In: Proceedings of the 32nd IEEE/ACM International Conference on Automated Software Engineering, pp. 717–722. IEEE Press (2017)
9. Zhang, J., Shi, X., King, I., Yeung, D.Y.: Dynamic key-value memory networks for knowledge tracing. In: Proceedings of the 26th International Conference on World Wide Web, pp. 765–774. International World Wide Web Conferences Steering Committee (2017)

Deep Learning

Watermarking Neural Network with Compensation Mechanism

Le Feng$^{(\boxtimes)}$ (iD) and Xinpeng Zhang

School of Computer Science and Technology, Fudan University, Shanghai, China
{19110240030,zhangxinpeng}@fudan.edu.cn

Abstract. In recent years, the rapid development of neural networks has also brought his intellectual property (IP) protection. Embedding a watermark in a neural network is an effective scheme to protect its IP. In this paper, we propose a new watermark embedding scheme with compensation mechanism that is different from the previous regularization embedding. First, we select the weights of the watermark to be embedded pseudo-randomly. Then, we perform an orthogonal transformation on the selected weights, and embed the watermark by the binarization method in the obtained coefficients, and use the inverse orthogonal transformation on the watermarked coefficients to obtain the watermarked weights. Finally, we propose a model fine-tuning scheme with compensation mechanism, which can eliminate the slight accuracy degradation caused by binarization without destroying the watermark in the model. In our scheme, due to the concealment of watermark embedding location, it can overcome the defects of previous schemes which cannot resist watermark overwriting attack. Moreover, compared with the regularization embedding method, our scheme uses the fine-tuning with compensation mechanism, which requires less embedding cost and is more stable. In addition, it has achieved favorable performance in resisting weight pruning attack, weight fine-tuning and fidelity evaluation.

Keywords: Watermarking · Neural network · Binarization · Compensation mechanism

1 Introduction

The development of deep learning has brought huge changes to our lifestyle. Its success in computer vision, natural language processing and other fields has fully verified its effectiveness. However, DNN (Deep Neural Network) training requires a large amount of data and computing resources, which is often difficult for ordinary developers. An effective solution is that since the neural network model posseses favorable transfer learning characteristics, the trained model can be directly transferred to the problem to be solved. However, at the same time, this has also created a new problem, that is, the IP protection of the neural network model. The owner may not want their IP of the model to be violated,

© Springer Nature Switzerland AG 2020
G. Li et al. (Eds.): KSEM 2020, LNAI 12275, pp. 363–375, 2020.
https://doi.org/10.1007/978-3-030-55393-7_33

or even it is hoped that these IP can bring certain legal benefits to themselves. To this end, researchers [1–4, 8, 10, 11] have proposed some schemes of embedding watermarks in models to protect their intellectual property rights. Among them, IP protection in black box scenarios is mainly achieved by embedding backdoor images [1, 4], which is not the focus of our research. This article mainly discusses the issue of IP protection through watermarking neural network model in the more widely used white-box scenario.

The first scheme of watermarking neural network model was proposed by Yusuke Uchida [10]. Its main idea is to embed the watermark information composed of bit strings into the weight distribution of one layer of the middle layer in the form of the regularization. By embedding the watermark information into the weight distribution, the network can be robust to some extent, for example, it can model fine-tuning and weight pruning [6]. However, its disadvantages are also obvious, that is, the scheme cannot resist watermark overwriting attack. Because the layers of the network are limited, an attacker can use overwriting attack on each layer, which will necessarily destroy the original watermark of the owner. In addition, embedding the watermark in this regularized manner may interfere with the normal training of the network, especially for complex networks such as GAN (Generative Adversarial Network).

In response to the above shortcomings, Rouhani [4] tried to propose his improvement scheme. Since the activation of the middle layer of the neural network is approximately subject to Gaussian mixture distribution, the author also selects a certain layer of the middle layer and embeds a watermark in the activation distribution of this layer. The disadvantage of this scheme is that the embedded watermark capacity is small.

In addition to the watermarking scheme, Fan [5] proposed to embed the passports layer in the neural network to protect the IP of the model. This scheme can make it difficult for attackers to forge new passports to achieve the purpose of obfuscating IP. However, its shortcomings are also obvious. To ensure that the passport layer is private, it takes an extra 1 times of training time, which will greatly limit its application in practical scenarios.

In response to these shortcomings, we have proposed our solution. First, We use keys to pseudo-randomly select the weights to be embedded in the watermark. When an attacker wants to destroy the watermark through overwriting attack, it is no longer effective because he cannot grasp the position of the watermark. Of course, the attacker can also embed more watermark to cover the position of the original watermark as much as possible, but the amount of watermarks that the model can carry is limited, which means that the more watermarks are embedded, the higher the cost, and even the original function of the model may be destroyed. Then, in our scheme, compared with watermark overwriting attack, the weight pruning may damage the original watermark more. Therefore, to avoid this, we perform spread-spectrum modulation on the original watermark and dispersedly embed the modulated watermark into the weights of different layers of the model to enhance the robustness of the watermark.

Moreover, it is different from the previous watermarking scheme that changes the weight distribution with regularization during the training process. We combine binarization and compensation mechanism to ensure that watermark embedding can be achieved without impacting the model. Binarization can embed watermarks into the model. The compensation mechanism is implemented to eliminate the slight impact on model accuracy by fine-tuning the model's weights other than the embedded watermark. Since the watermarked weights are relatively small compared to the whole weights of the model, only a few epochs of fine-tuning are required to restore the model accuracy. All in all, the contributions of this article are as follows:

(I) By making the location where the watermark is embedded is hidden, it is impossible for an attacker to grasp the location, and then use watermark overwriting attack to erase the original watermark.

(II) We combine binarization and compensation mechanism to ensure that the watermark is embedded while the model accuracy does not degrade. In doing so, the embedding cost is slight and more stable than the previous regularization embedding schemes.

(III) By spread-spectrum modulating the watermark and embedding it in different layers, this reduced the risk of the watermarked weights being set to zero when weights pruning is required to compress the model.

2   The Proposed Scheme

Here, we embed the watermark into the weights of the intermediate layers of the neural network model, excluding the output layer. Because if it is embedded in the output layer, on the one hand this will have a large impact on the performance of the model, on the other hand, the users will often use a new output layer instead of the original output layer. Since our scheme has nothing to do with the structure of the neural network, our scheme is effective for convolution layer, fully connected layer and recurrent layer. For the convenience of description, we still take the most widely used classification neural network model as an example. This model is denoted by M, which contains intermediate layers L of depth d, where each intermediate layer can be represented as $l_i(i \in [0, d-1])$, and we use $\text{Len}(l_i)$ to represent the number of weights in the layer l_i, and $\text{W}(l_i)$ to represent the weights of layer l_i.

Herein, we embed a watermark B of length n into the model M. The overall process is listed in Algorithm 1. Next, we will introduce the detailed embedding process.

2.1   Weights Selection

To make the watermarked weights as dispersed and hidden as possible, we determine to embed the watermark in as many layers as possible. The selection criterion is that the number $\text{Len}(l_i)$ of weights is sufficiently large relative to the length n of watermark B to be embedded. Here, we choose a gap of 10 times, that is, the selected layers $SL = \{l_i | Len(l_i) \geq 10 \times n, i \in [0, d-1]\} = \{sl_j | j \in [0, sd-1]\}$, and we use NL to represent the middle layers without embedding the watermark. This is done for three reasons. One is that if the weight contained in this layer is less than the length of the watermark information to be embedded, then obviously this layer cannot carry all the watermark information. The second is to meet the requirements of undetectability and security. If the layer contains not enough weights, the attacker's brute force solution may destroy the watermark information of the layer. The third is to consider the embedding cost. If the embedded watermark information changes the excessive weights of the layer, the loss of model accuracy will also increase, and the cost of modifying the model by fine-tuning will be relatively increased.

As for which weights in sl_j are selected, we use a pseudo-random number generator with the key K_0 as the initial value to generate the n positions where the watermark is embedded and get the corresponding weights sw_j. The unselected weights in the layer sl_j are represented by snw_j.

2.2   Watermark Generation

We find that if the original watermark B is simply embedded in the n weights, its robustness is poor, especially for pruning attacks. To this end, we need to perform spread-spectrum modulation [12] on the B, so that each bit of it is distributed redundantly in all selected sd layers to enhance the watermark robustness. Here, sd is the chip rate of spread-spectrum modulation. The process is as follows. For the original watermark $B = \{b_u | b_u \in \{-1, 1\}, u \in [0, n-1]\}$, we expand it to $EB = \{eb_v = b_u, u = v \bmod sd, v \in [0, sd \times n - 1]\}$, and then use K_1 as the key of the pseudo-random number generator to generate a pseudo-random noise sequence $P = \{p_v | p_v \in \{-1, 1\}, v \in [0, sd \times n - 1]\}$, and finally use P to modulate EB to obtain the final watermark $SB = \{sb_v | sb_v = p_v \times eb_v, v \in [0, sd \times n - 1]\}$. The purpose of using P for modulation is to ensure that the spread-spectrum watermark SB possesses noise characteristics, which makes it difficult for attackers to locate. For example, if $B = \{-1, 1, 1, -1\}$, $sd = 3$, $P = \{1, 1, -1, 1, -1, -1, 1, -1, 1, -1, 1, 1\}$, Fig. 1 shows the specific spread-spectrum modulation process.

2.3   Embedding Watermark

In this section, we will embed the watermark SB into the weight SW. To make the embedded watermark as dispersed as possible, we divide SB evenly into sd groups, and each group is represented as $se_j (j \in [0, sd-1])$. Then, we embed each group se_j of watermark information into the corresponding weight sw_j in the same binarization method. The following is a specific binarization embedding method.

(I) Transforming weights: The purpose of this step is to ensure that the watermarked weights will not be detected by the attacker on the one hand,

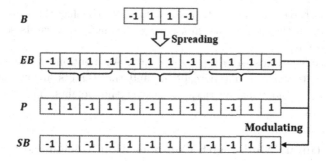

Fig. 1. Example of spread-spectrum modulating B to SB

and on the other hand, to make the change of the weights caused by binarization more evenly distributed in different weights, so as not to tilt the weights to impact the accuracy of the model. The specific method is that we use the orthogonal matrix ϑ of shape $n \times n$ to transform the weight sw_j to obtain $sw^{(1)}_j$, that is, $sw^{(1)}_j = \vartheta \times sw_j$.

(II) Binarizing to embed watermark: To ensure that the embedding of the watermark does not cause excessive degradation of the model accuracy, and also to ensure the concealment of the embedding position of the watermark so that the statistical features will not be discovered by the attacker, we require that after embedding the watermark, the overall energy of watermark weights remains unchanged, that is, their second norm is unchanged. Therefore, we will employ the following binarization scheme.

First, we need to calculate the MSE σ_j of the weights of each selected layer. Since the weights of each layer of the neural network approximately follow $(0, \sigma_j)$-Gaussian distribution [9] with a mean value of 0, the sw_j composed of the randomly selected weights from which also conforms to this distribution, and its second norm is $\|sw_j\|_2 = \sqrt{\sum_{u=0}^{n-1} sw_j[u]^2} = \sqrt{n \times (\frac{1}{n}\sum_{u=0}^{n-1}(sw_j[u] - 0)^2)} \approx \sqrt{n} \times \sigma_j$. Orthogonal transformation has the property that the second norm is invariant, so the second norm of the coefficient $sw^{(1)}_j$ obtained after the orthogonal transformation in (I) is also $\sqrt{n} \times \sigma_j$. Next, we embed the watermark by a scheme of binarizing the coefficient $sw^{(1)}_j$ using σ_j, as shown in Eq. (1).

$$sw^{(1)}_j[u] = \begin{cases} -\sigma_j & \text{if } se_j[u] = -1 \\ \sigma_j & \text{if } se_j[u] = 1 \end{cases} \tag{1}$$

where, $u \in [0, n-1]$. It can be seen that if the uth bit of the embedded watermark $se_j[u]$ is -1, then we set $sw^{(1)}_j[u]$ to negative σ_j, and if the uth bit of the embedded watermark $se_j[u]$ is 1, then we set $sw^{(1)}_j[u]$ to positive σ_j. In this way, we can verify that, after embedding watermark, the second norm of coefficient $sw^{(1)}_j$ is still $\sqrt{n} \times \sigma_j$, thus ensuring that the weight recovered after the orthogonal inverse transformation of (III) has the same energy as the weights of the original unembedded watermark, that is, its second norm is also $\sqrt{n} \times \sigma_j$.

(III) Inverse transforming coefficient: As mentioned above, to ensure that the watermark is not perceptible and the weights are not skewed, we need to use orthogonal matrices to recover the weights. That is, $sw^{(2)}{}_j = \vartheta^T \times sw^{(1)}{}_j$.

2.4 Fine-Tuning Model with Compensation Mechanism

Different from the regularization embedding scheme [4,10] that guarantees that the model accuracy does not degrade during the embedding process, we use the mechanism of accuracy compensation after embedding the watermark to eliminate the impact of the watermark embedding on the performance of the model. Specifically, in the process of fine-tuning, the watermarked weights $SW^{(2)}$ are kept unchanged, and only other weights in the model including NL, SNW and output layer are fine-tuned.

Because watermarked weights are very small compared to the weight of the model, and our embedding scheme ensures that the overall energy of the weight before and after the watermark is embedded is constant. Therefore, the model's accuracy degradation is slight, and only a few epochs of fine-tuning can restore the original performance of the model. Section 3.4 prove this. As shown in Fig. 4, after embedding a 256-bit watermark, the accuracy degradation of both Resnet18 and DenseNet is rather slight, the model performance can be restored through few epochs of fine-tuning.

After fine-tuning model with accuracy compensation, we will get the final watermarked model WM.

2.5 Watermark Extraction

Extracting the watermark is divided into three steps. The first step is to find out where the watermark is embedded in different layer of the model using the scheme mentioned in Sect. 2.1 according to the key K_0, and then obtain weights sw_j'. Then use the orthogonal matrix ϑ to sw_j' to get $sw^{(1)}{}_j' = \vartheta \times sw_j'$. In the second step, we need to extract the corresponding watermark se_j' from $sw^{(1)}{}_j'$. The method is listed in Eq. (2).

$$se_j'[u] = \begin{cases} -1 \text{ if } sw^{(1)}{}_j'[u] \leq 0 \\ 1 \text{ if } sw^{(1)}{}_j'[u] > 0 \end{cases} \tag{2}$$

where $u \in [0, n-1]$. Corresponding to the process of embedding watermark, we can determine whether the embedded watermark is 1 or -1 by judging the positive and negative of the coefficient. Then, all the se_j' extracted from different layers are combined into a watermark SB'. In the third step, the key sequence K_1 is used to generate a noise sequence P, and P is used to demodulate SB' to obtain $EB' = SB'/P'$. Finally, according to the method of Eq. (3) ($u \in [0, n-1]$), the watermark B' is extracted and compared with the original watermark B to determine IP of the model.

$$B'[u] = \begin{cases} -1 & \text{if } \sum_{k=0}^{sd-1} eb'_{k \times n + u} \leq 0 \\ 1 & \text{if } \sum_{k=0}^{sd-1} eb'_{k \times n + u} > 0 \end{cases} \tag{3}$$

Algorithm 1. Watermarking neural network model

Input: A model M containing intermediate layers L of depth d, where each intermediate layer can be represented as $l_i (i \in [0, d-1])$, Watermark signature B (length of which is n), *training set* and *testing set*, Secret key K_0 and K_1, Orthogonal matrix ϑ.

Output: Watermarked model WM

1: **(I)** SW, SNW, NL = Selecting_weights (L, n, K_0), where SW are selected from sd layers, NL are the layers that does not satisfy embedding security, SNW are the unselected weights in the layers that satisfy embedding security.
2: **(II)** SB = Generating_watermark (B, sd, K_1)
3: **(III)** $SW^{(1)}$ = Transforming_Weights (SW, ϑ)
4: $SW^{(1)}$ = Binarizing_coefficient ($SW^{(1)}$, SB)
5: $SW^{(2)}$ = Recovering_weights ($SW^{(1)}$, ϑ)
6: WM = Fine-tuning_model_with_compensation (NL, SNW, *training set*, *testing set*)

Table 1. Training strategy of Resnet18-Cifar10, Resnet18-Cifar100 and DenseNet-MNIST

Structure-Dataset	Batch size	Epochs	Optimizer	Learning rate	Learning rate update strategy
Resnet18-Cifar-10	64	70	Adm	0.001	Decrease by 1/10 each 30 epochs
Resnet18-Cifar-100	64	70	Adm	0.001	Decrease by 1/10 each 30 epochs
DenseNet-MNIST	64	30	Adm	0.001	Learning rate is fixed

3 Experimental Evaluation

Next, we will evaluate our watermarking scheme. The network structure tested included Resnet18 [7] and the DenseNet. DenseNet is the neural network we constructed for MNIST and consists of fully connected layers, whose structure is 512(Fc)-Dropout(0.2)-512Fc(relu)-Dropout(0.2)-10Fc(softmax). The other datasets include cifar10 (10 classification), cifar100 (100 classification). The final evaluation model includes three types: Resnet18-Cifar10, Resnet18-Cifar100 and DenseNet-MNIST, whose training strategies are listed in Table 1. As with the [10], unless otherwise specified, we choose to embed a 256-bit watermark (a 256-bit watermark is sufficient for security). In addition, the hardware platform is the graphics card of Nvidia GTX 1080ti, CPU of Intel(R) Xeon(R) e5-2620 v4 with the main frequency of 2.1 GHZ, and the memory of 64 GB.

3.1 Fidelity Evaluation

In our scheme, the number of watermarked weights is much less than the whole weights of the model, so the degradation of model accuracy after embedding the watermark is slight. Taking DenseNet-MNIST as an example, the weight of the middle layer of the model is about 663552, and the number of watermarked weights is about $2 \times 256 = 512$, that is, only the weight of about 0.07% is changed. Then, we fine-tune the model, which can completely guarantee that the accuracy of the model will not degrade, as shown in Table 2.

Table 2. Fidelity evaluation of three sets of watermarked models, we call the original accuracy of the model as baseline accuracy, and call the accuracy of the watermarked model as watermarked accuracy.

Structure-Dataset	Baseline accuracy	Ratio of watermarked weights	Watermarked accuracy
Resnet18-Cifar10	91.40%	0.04%	91.41%
Resnet18-Cifar100	68.59%	0.08%	68.53%
DenseNet-MNIST	98.37%	0.04%	98.42%

3.2 Robustness Evaluation

Resisting Fine-Tuning. In most transfer learning scenarios, users need to fine-tune the model using their own datasets. Therefore, for a watermarked model, the watermark should not be destroyed by the user's fine-tuning of the model. In order to fully consider the robustness of the watermark, we assume that all layers of the model are fine-tuned, and the training strategy for fine-tuning follows the settings in Table 1. We set up three sets of experiments: using Cifar-10 to fine-tune the watermarked DenseNet-MNIST; using MNIST to fine-tune the watermarked Resnet18-Cifar10; using Cifar10 to fine-tune the watermarked Resnet18-Cifar100. Due to the differences between different datasets, we need to make the following adjustments: First, adjust the size of the image in MNIST to be the same as in Cifar10. The second is that since the output layer of Resnet18-Cifar100 includes 100 neurons, when fine-tuning it with Cifar10, the output layer needs to be replaced with 10 neurons. As shown in Fig. 2, even if the model accuracy changes greatly, the watermark is still not destroyed.

Table 3. Loss of original watermark after overwriting a 256-bit or 2560-bit watermark.

Structure-Dataset	Overwriting capacity	
	256	2560
DenseNet-MNIST	0	0
Resnet18-Cifar10	0	0
Resnet18-Cifar100	0	0

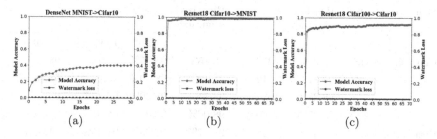

    (a)   (b)   (c)

Fig. 2. Changes in model accuracy and watermark loss in these three watermarked models as the number of fine-tuning epochs increases.

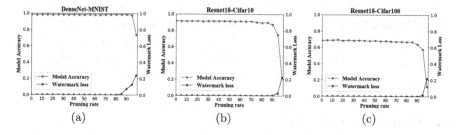

    (a)   (b)   (c)

Fig. 3. Changes in model accuracy and watermark loss with increasing pruning rate.

Resisting Weights Pruning. Due to the huge computing and storage resource consumption of neural networks, they sometimes need to be compressed. Compression includes pruning, quantization, low-rank approximation, and model distillation. Herein, we use the pruning scheme proposed in [6], where the weights of $\alpha\%$ with the smallest absolute value of all layers except the output layer are set to 0. The upper limit of the pruning rate α is set to 99%.

As shown in Fig. 3, when the pruning rate reaches 80%, the watermark loss is still 0. When the pruning rate reaches 99% and the model accuracy is severely degraded, only about 20% of the watermark information is destroyed.

Resisting Watermark Overwriting. In our watermark scheme, the watermark is implicitly embedded in the weights of each layer through the key mechanism. When an attacker uses watermark overwriting attack, since they cannot grasp where the watermark is embedded, it is difficult for them to destroy the original watermark of the owner. Therefore, as shown in Table 1, even if an attacker overwrites a new 2560-bit watermark in the model, the loss of the original watermark is still 0 (Table 3). Of course, the attacker may try to overwrite more watermark, but as shown in Sect. 3.3, the more watermark is embedded, the greater the accuracy loss of the model will eventually make the model unusable.

Table 4. The relation between the length of embedded watermark and model accuracy

Structure-Dataset	Baseline accuracy	Capacity	Binarized accuracy	Fine-tuning accuracy		
				1 epoch	10 epochs	20 epochs
DenseNet-MNIST	98.37%	256	98.15%	98.35%	98.45%	98.46%
	98.37%	512	98.19%	98.39%	98.46%	98.48%
	98.37%	1024	98.11%	98.30%	98.39%	98.43%
	98.37%	2560	98.06%	98.35%	98.41%	98.42%
Resnet18-Cifar10	91.40%	256	90.87%	91.35%	91.35%	91.52%
	91.40%	512	90.29%	90.96%	91.43%	91.43%
	91.40%	1024	88.63%	90.55%	91.17%	91.45%
	91.40%	2560	83.92%	89.53%	90.85%	90.69%
Resnet18-Cifar100	68.59%	256	66.47%	68.37%	68.55%	68.61%
	68.59%	512	65.49%	67.78%	68.45%	68.63%
	68.59%	1024	63.50%	67.55%	68.15%	68.38%
	68.59%	2560	42.56%	64.02%	67.27%	67.49%

3.3   Watermark Capacity

A 256-bit watermark is not the limit of capacity. Our scheme can embed more watermark. Of course, the more watermark, the more likely the original performance of the model degrades. We call the accuracy of the model after binarizing before fine-tuning the binarized accuracy. Table 4 shows the binarized accuracy and the accuracy of fine-tuning 1 epoch, 10 epochs, and 20 epochs when embedding respectively 256, 512, 1024, and 2560-bit watermark in three sets of models. We will draw three conclusions. Take Resnet18-Cifar100 as an example. First, as the amount of embedded watermark increases, the binarized accuracy gradually decreases. After embedding 256, 1024, and 2560-bit watermarks, the binarized accuracy of the model gradually decreases, which are 66.47%, 63.50%, and 42.56%, respectively. The second is that when fine-tuning for 1 epoch, the lost accuracy will be quickly compensated. The accuracy is restored to 68.37%, 67.55%, and 64.02%, which is close to the baseline accuracy of the model. The third is that as the amount of embedded watermark increases, the cost of accuracy loss compensation turn greater. When embedding a 256-bit watermark, 1 epoch of fine-tuning can restore the accuracy to 68.37%, while at embedding 2560-bit watermark, even after 20 epochs of fine-tuning, the accuracy restores to 67.49%, which is lower than the baseline accuracy by about 1%.

3.4 Embedding Cost

In our scheme, the computational cost of embedding the watermark includes two parts, one is the process of selecting the embedding position and binarization embedding, and the other is fine-tuning to ensure that the model accuracy does not degrade. The consumption of the first part is almost negligible compared to the training time of a neural network. Therefore, we mainly analyze the calculation cost of fine-tuning in the second part. From Fig. 4, we can find that after a few epochs of fine-tuning, the accuracy loss caused by watermarking will be made up, which means our scheme has a very low embedding cost.

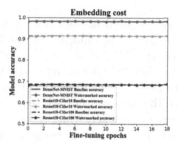

Fig. 4. After watermarking weights, the accuracy of the model gradually recovers as the number of fine-tuning epochs increases.

3.5 Comparison with Previous Schemes

In order to fully illustrate the superiority of our scheme, we compare it with several previous schemes [4,5,10]. The Uchida [10] and Darvish [4] schemes both use watermark to protect the IP of neural network model. Uchida embeds the watermark in the weight of a certain layer in the middle layer, Darvish embeds a watermark in the activation of a middle layer. Fan [5] scheme is to protect the IP of the model by adding a passport layer to the neural network. By comparison, we find that these solutions have favorable fidelity and can resist weight fine-tuning and weight pruning. But they also have defects. The main drawback of Uchida's scheme is that it cannot resist watermark overwriting attack. The disadvantage of Darvish's scheme lies in the lower watermark capacity, and obvious watermark loss has already occurred when 128-bit watermark is embedded. In addition, both Uchida and Darvish schemes embed watermarks in a regularized way during training models. This may make training easier, but it may also make model training more difficult, which means that the embedding cost is unstable.

Fan scheme needs to change the structure of the model, which can prevent the attacker from embedding a new watermark. The author thinks that the current watermarking schemes used for IP protection of neural network models are unable to resist copyright obfuscation attacks, that is, when an attacker embeds a new watermark again, there will be two watermarks in the model: the model owner's original watermark and a watermark embedded by attacker.

Table 5. Comparison of different algorithms (N means failed, Y means pass)

Algorithm	Fidelity	Robust			Capacity	Embedding cost
		Fine-tuning	Pruning	Overwriting		
Ours	Y	Y	Y	Y	High	Fine-tuning few epochs
Uchida [10]	Y	Y	Y	N	High	Unstable
Darvish [4]	Y	Y	Y	Y	Low	Unstable
Fan [5]	Y	Y	Y	Y	–	100% more training time

However, we think that such a problem does not exist. Although our scheme cannot prevent the attacker from embedding a new watermark again, even if the model has two watermarks, the IP of the model can still be determined. The solution is based on the fact that the true owner of the model owns the original model with only one watermark. When IP verification is required, the owner can use his key to extract the same watermark from both the original model and the model mastered by the attacker, while the attacker cannot extract the watermark from the original model of the owner using his key. In this way, the true owner of the model can be identified. Moreover, the drawback of Fan scheme is its huge time consumption. In order to ensure that the passport layer is not detected by the user, it will take more than twice the training time, which will greatly limit its use in real scenarios.

All comparisons are listed in Table 5. We think that our scheme makes up for the shortcomings of these three schemes. The specific has been described in the experimental evaluation and will not be repeated here.

4   Conclusion

In this paper, we propose a new model watermarking scheme, which is irrelevant to the neural network structure. To ensure the robustness of the embedded watermark, we first spread-spectrum modulate the original watermark information, and then embed them pseudo-randomly in different layers of the network through the key. The embedding strategy is to binarize coefficients obtained by orthogonally transforming weights to embed the watermark −1 or 1. Finally fine-tune the model with compensation mechanism to ensure that accuracy does not degrade. Compared with the previous scheme, our scheme can well resist watermark overwriting attacks and has a lower embedding cost. And in other various tests including fidelity, resisting fine-tuning, resisting weights pruning and capacity, all perform well and have reliable practical application significance.

Acknowledgment. This work was supported by the Natural Science Foundation of China (U1936214, U1636206, and 61525203).

References

1. Adi, Y., Baum, C., Cisse, M., Pinkas, B., Keshet, J.: Turning your weakness into a strength: watermarking deep neural networks by backdooring. In: 27th USENIX Security Symposium (USENIX Security 2018), pp. 1615–1631 (2018)
2. Chen, H., Fu, C., Rouhani, B.D., Zhao, J., Koushanfar, F.: DeepAttest: an end-to-end attestation framework for deep neural networks. In: Proceedings of the 46th International Symposium on Computer Architecture, pp. 487–498 (2019)
3. Chen, H., Rouhani, B.D., Fu, C., Zhao, J., Koushanfar, F.: DeepMarks: a secure fingerprinting framework for digital rights management of deep learning models. In: Proceedings of the 2019 on International Conference on Multimedia Retrieval, pp. 105–113 (2019)
4. Darvish Rouhani, B., Chen, H., Koushanfar, F.: DeepSigns: an end-to-end watermarking framework for ownership protection of deep neural networks. In: Proceedings of the Twenty-Fourth International Conference on Architectural Support for Programming Languages and Operating Systems, pp. 485–497 (2019)
5. Fan, L., Ng, K.W., Chan, C.S.: Rethinking deep neural network ownership verification: embedding passports to defeat ambiguity attacks. In: Advances in Neural Information Processing Systems, pp. 4716–4725 (2019)
6. Han, S., Mao, H., Dally, W.J.: Deep compression: compressing deep neural networks with pruning, trained quantization and Huffman coding. arXiv preprint arXiv:1510.00149 (2015)
7. He, K., Zhang, X., Ren, S., Sun, J.: Deep residual learning for image recognition. In: Proceedings of the IEEE Conference on Computer Vision and Pattern Recognition, pp. 770–778 (2016)
8. Li, H., Willson, E., Zheng, H., Zhao, B.Y.: Persistent and unforgeable watermarks for deep neural networks. arXiv preprint arXiv:1910.01226 (2019)
9. Lin, D., Talathi, S., Annapureddy, S.: Fixed point quantization of deep convolutional networks. In: International Conference on Machine Learning, pp. 2849–2858 (2016)
10. Uchida, Y., Nagai, Y., Sakazawa, S., Satoh, S.: Embedding watermarks into deep neural networks. In: Proceedings of the 2017 ACM on International Conference on Multimedia Retrieval, pp. 269–277 (2017)
11. Zhang, J., et al.: Protecting intellectual property of deep neural networks with watermarking. In: Proceedings of the 2018 on Asia Conference on Computer and Communications Security, pp. 159–172 (2018)
12. Zhao, Z., Yu, N.H.: A novel watermark embedding algorithm. In: Proceedings of International Conference on Machine Learning and Cybernetics, vol. 4, pp. 1929–1934. IEEE (2002)

Information Diffusion Prediction with Personalized Graph Neural Networks

Yao Wu, Hong Huang$^{(\boxtimes)}$, and Hai Jin

National Engineering Research Center for Big Data Technology and System, Services
Computing Technology and System Lab, Cluster and Grid Computing Lab,
Huazhong University of Science and Technology, Wuhan, China
{yaowu,honghuang,hjin}@hust.edu.cn

Abstract. Online social networks are crowded with massive information, which is more likely to spread rapidly on a large scale. Therefore, understanding and predicting information diffusion on social networks will be much helpful to improve the performance of marketing and control the dissemination of misinformation. Recently, the deep learning techniques have enhanced the methods for diffusion prediction and provide a new way to model the diffusion process in time and space. However, these models introduce the temporal and structural factors affecting the diffusion with two sequential or parallel steps separately. Moreover, they neglect the whole influence of the diffusion cascade from a global view. Hence, we propose a novel method for diffusion prediction with personalized graph neural networks, namely *infGNN*, to model the interactions between structural and temporal factors. Furthermore, we integrate local and global influence of diffusion cascade for prediction. Experiments results on three datasets show the superiority of the proposed model.

Keywords: Diffusion prediction · Social influence · Graph neural networks

1 Introduction

Various online social networks enable individuals to keep in touch with each other anywhere at any time, accompanied by huge information spreading rapidly on a large scale. However, while promoting the development of the information industry, it also encourages the dissemination of misinformation. Thus, the study of information diffusion has drawn the attention of academia and industry.

Researchers have made efforts to explore information diffusion and study diffusion prediction models. Since the online platforms make the social network visible and diffusion path traceable, a large amount of data about diffusion sequences as well as social network structure are available. Early studies suggest basic diffusion models, such as *Linear Threshold* (LT) and *Independent Cascade* (IC). However, these models are not able to quantify the differences in

© Springer Nature Switzerland AG 2020
G. Li et al. (Eds.): KSEM 2020, LNAI 12275, pp. 376–387, 2020.
https://doi.org/10.1007/978-3-030-55393-7_34

propagation probabilities between users. To make a more accurate prediction, researchers analyze the factors affecting the transmission of information and manually extracted features to measure the diffusion probability, for example, user's characteristics [1], contents [2], structure features [3], and time decay factor [4]. Meanwhile, traditional machine learning methods are used for prediction, like Logistic Regression [3] and Conditional Random Field [4]. But these models rely on feature extraction and need extra domain knowledge.

Recently, with the advent of deep learning techniques, diffusion models have been enhanced by automatically learning features for prediction. [5–7] adopt representation learning to embed features into the latent representations of users or contents. Since the diffusion model is still required for further prediction with the learned embeddings as input, these models fail to model the temporal dynamics of diffusion cascade. Therefore, end-to-end frameworks based on *Recurrent Neural Networks* (RNNs) are proposed to model the diffusion process. Some works [8,9] model diffusion as cascade sequences with RNNs, while other works [10–12] also take the structural dependency on diffusion cascade into account. However, these models integrate temporal and structural factors affecting the diffusion with two sequential or parallel steps separately. Even though TopoLSTM [13] extends basic LSTM with diffusion topologies, it does not differentiate temporal dynamics from structural dependencies on diffusion cascade. Furthermore, previous models neglect the influence of the diffusion cascade from the global view, which makes prediction directly with the last output of RNNs.

To overcome the above limitations, we propose a personalized graph neural networks to model the diffusion process and integrate the influence of diffusion cascade, in a local and global view, for prediction. Given the diffusion logs and social network, the proposed model is designed to predict the following activated users in the cascade. Particularly, we first construct a diffusion graph for each cascade as the input. The personalized GNNs alternatively update the users' hidden state with the neighbors' state on the diffusion graph and previous users' state on diffusion sequence. In this way, it not only considers the interactions between structural dependencies and temporal dynamics on diffusion cascade but also dynamically learns hidden representation for each user on the diffusion graph. Moreover, we generate the influence embedding by combining global and local influence of the diffusion cascade for further prediction. Specifically, an attention pooling layer is followed to learn the embedding of the diffusion graph representing the global influence. Meanwhile, the latest activated user representation preserves the local influence. At last, the activation probability of each user is computed by a softmax layer with the influence embedding. The main contributions include the following three aspects:

- A GNNs based end-to-end framework, namely *infGNN*, is proposed for information diffusion prediction, which models the interactions between the structural dependencies and temporal dynamics in the diffusion cascade.
- The proposed method dynamically learns the users' representations on the diffusion graph and generates the influence embedding of diffusion cascade from both global and local views for prediction.

– Extensive experiments are conducted and the performance gains show the feasibility and superiority of the proposed model.

2   Related Work

Diffusion Prediction aims to predict the next diffusion trends. Early studies [14,15] introduce the SIR model of epidemiology and propose *Independent Cascade* (IC) and *Linear Threshold* (LT) models. However, the probability of the diffusion between users is randomly initialized. In fact, the diffusion probability is related to many factors, such as the user and content features [1,2]. Therefore, researchers extract features to estimate the diffusion probability for more accurate predictions. Traditional methods [3,16] manually extract the required features, whereas, with the help of network embedding, some works [5–7] embed features to the latent representations and have achieved promising results lately. Embedding-IC [5] and inf2vec [6] learn user representations and adopt the IC model for prediction. HUCE [7] projects the users and contents to one vector space and proposes a diffusion model. However, all these methods consist of two steps: 1) learning users or contents embeddings, 2) making a prediction with the diffusion model.

Recently, the end-to-end framework for diffusion prediction has been proposed. [9] and [8] model diffusion sequences directly with *Recurrent Neural Networks* (RNNs). DeepDiffuse [9] predicts the next activated users as well as its infected time and DeepHawkes [8] predicts popularity. Even though the temporal dynamics in the diffusion process are modeled by the RNNs, other works [10–13] also utilize the structural dependencies among users in the diffusion sequence to make a prediction. For example, SNIDSA [11] uses a structure attention mechanism to consider the structural dependencies among nodes before modeling the diffusion process with RNNs. Deepcas [10] randomly selects several walks to represent the diffusion cascade. Then the cascade representations are learned by GRU networks with the attention mechanism on walks to predict the future size of a cascade. FOREST [12] integrates structure information in the social network when modeling cascade with GRUs to make a multi-scale prediction in the micro and macro. TopoLSTM [13] extends basic LSTM for diffusion typologies to predict next activated users. Even though both structural and temporal factors are considered in these models, the interactions between them are ignored when modeling the diffusion process. Additionally, the global influence of the diffusion cascade is also neglected for final prediction.

Graph Neural Networks (**GNNs**) have achieved promising results for graph-based problems, such as the graph classification [17] and traffic forecasting [18,19]. The key advantage of GNNs is to extract features from the graph-structured dataset and learn stable representations for a certain task. For example, Zhang et.al. [17] adopt the idea of message passing to update the node embedding with neighbors' embeddings and propose a graph pooling strategy for graph classification. Li et al. [18] propose a diffusion graph convolutional layer to extract structure features in the traffic network and use the GRU network to process the sequential traffic data.

Recent studies [20–22] also utilize GNNs to improve popularity prediction performance. Meanwhile, DeepInf [23] extracts features of users' ego networks as the input of GNNs for measuring social influence. However, the diffusion model for user-oriented prediction with Graph Neural Networks, which mainly focuses on the user's states, remains to be studied.

3   Frameworks

In this section, the details of the GNNs-based diffusion prediction model are demonstrated. The overall framework of *infGNN* is shown in Fig. 1. Specifically, we first give the problem definition and describe the construction of the diffusion graph from the diffusion logs and the social network. Then personalized graph neural networks is proposed to dynamically learn user representations on the diffusion graph. Given the user representations, we generate influence embedding representing global and local influence of diffusion cascade and calculate the activation probability of each user through a softmax layer. At last, we show the optimization methods with the loss function of cross-entropy.

3.1   Problem Definition

Given the social network and user action logs, we aim to forecast the users affected by a certain action in the near future. Given a directed graph $G = (V, E)$, V and E represent the users and relationships in the social network. For each edge $(u, v) \in E$, if there exists a link between u and v, $e_{u,v} = 1$; otherwise, $e_{u,v} = 0$. Meanwhile, the diffusion cascade L can be collected from the user action logs of the social network. We denote the diffusion of action a_i as $l_i = \{(v_{i,1}, t_{i,1}),$ $(v_{i,2}, t_{i,2}), ..., (v_{i,T}, t_{i,T})\}$, where $v_{i,j}$ is the active user involved in the diffusion of a_i and $t_{i,j}$ records the time of the activation. Thus, given the social network and user action logs, the prediction problem is formulated as predicting active user $v_{i,T+1}$ for action a_i at time step $T + 1$.

Diffusion Graph Construction. For each action a, the cascade sequence $l = \{(v_{l,1}, t_{l,1}), (v_{l,2}, t_{l,2}), (v_{l,T}, t_{l,T})\}$ collects the diffusion path. Given $G = \{V, E\}$, we can construct diffusion graph G_l at each timestamp. At time t_k, $G_l^k = (V_l^k, E_l^k)$ is current diffusion graph, where $V_l^k = \{v_{l,i}|\ where\ t_i <= t_k\}$ and $E_l^k = \{e_{i,j}^l|\ e_{v_{l,j}, v_{l,i}} = 1\ and\ t_j > t_i,\ for\ v_{l,i}, v_{l,j} \in V_l^T\}$. The diffusion graph is a *Directed Acyclic Graph* (DAG). Moreover, A_l^k is denoted as the adjacent matrix of G_l^k. Instead of learning the users embeddings in advance, we use GNNs to dynamically extract features for each user on the diffusion graph and make final prediction with the user representations on diffusion graph.

3.2   User Representation on Diffusion Graph

In this part, we explain how to learn user representation on the diffusion graph with the personalized graph neural networks, which model the interactions between structural dependencies and temporal dynamics on the diffusion graph.

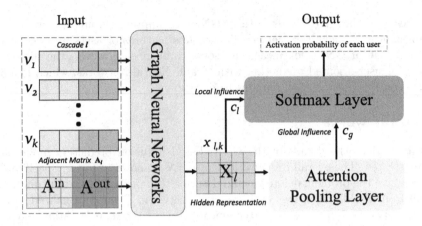

Fig. 1. The framework of the GNNs-based diffusion prediction model

Graph neural networks have achieved promising results to extract features from graph-structured data. Particularly, the model in [24] aims to process the sequential data from a graph by updating the node hidden state with the neighbors' state and the previous nodes' state alternatively. The idea provides a way to model the interactions of structural dependencies and temporal dynamics on diffusion cascade. Inspired by that, we take the diffusion propagation on the diffusion graph as information propagation on the graph and update the users' state with the neighbors' state as well as the previous activated users' state. In this way, the user representation of the diffusion graph can be dynamically learned. Especially, since users on the diffusion graph have different states of sending and receiving information, we differentiate the two states by sender and receiver embeddings respectively. Given a diffusion graph of G_l^k (short for G_l) with k activated users, the users' state can be updated by:

$$x_{l,i}^{(t)} = \begin{cases} x_{l,1}^{(1)}, & t = 1 \\ f(A_l, [x_{l,1}^{(t-1)}, ..., x_{l,k}^{(t-1)}], \overline{X}_l), & t > 1 \end{cases} \tag{1}$$

where $x_{l,i}^{(t)}$ is node vector of $v_{l,i}$ at step t and $x_{l,i}^{(1)}$ is initialized by the receiver embedding of $v_{l,i}$. $A_l = [A_l^{(in)}, A_l^{(out)}]$, which concatenates the incoming and outgoing adjacent matrix of G_l. $[x_{l,1}^{(t-1)}, ..., x_{l,k}^{(t-1)}]$ consists of users representations at step t-1 in G_l and $\overline{X}_l$ are the sender embeddings of users in V_l. Besides, f is the propagation model for diffusion graph and we use Gated mechanism [25] to realize feature propagation on graph. Therefore $x_{l,i}^{(t)}$ can be calculated by:

$$x_{l,i}^{(t)} = (1 - z_{l,i}^{(t)}) \odot x_{l,i}^{(t-1)} + z_{l,i}^{(t)} \odot \widetilde{h_{l,i}^{(t)}} \tag{2}$$

where $z_{l,i}^{(t)}$ is the update gate and $\widetilde{h_{l,i}^{(t)}}$ is the candidate activation, which is derived as follows:

$$\widetilde{h_{l,i}^{(t)}} = \tanh \ (W \ a_{l,i}^{(t)} + U \ (\ r_{l,i}^{(t)} \odot x_{l,i}^{(t-1)}) \tag{3}$$

$$z_{l,i}^{(t)} = \sigma \ (W_z \ a_{l,i}^{(t)} + U_z \ x_{l,i}^{(t-1)}) \tag{4}$$

$$r_{l,i}^{(t)} = \sigma \ (W_r \ a_{l,i}^{(t)} + U_r \ x_{l,i}^{(t-1)}) \tag{5}$$

where W, U, W_z, U_z, W_r, and U_r are parameters and reset gate is computed by Eq. 5. $a_{l,i}^{(t)}$ contains the message passing through both directions on diffusion graph, denoted by:

$$a_{l,i}^{(t)} = [in\_a_{l,i}^{(t)}; out\_a_{l,i}^{(t)}] \tag{6}$$

where $in\_a_{l,i}^{(t)}$ and $out\_a_{l,i}^{(t)}$ merge the features propagated from incoming and outgoing edges respectively, computed by:

$$in\_a_{l,i}^{(t)} = A_{l,i:}^{(in)} \ [x_{l,1}^{(t-1)}, ..., x_{l,k}^{(t-1)}]^\top \ W^{in} \ + b^{in}$$
$$out\_a_{l,i}^{(t)} = A_{l,i:}^{(out)} [\overline{x}_{l,1}, ..., \overline{x}_{l,k}]^\top \ W^{out} + b^{out} \tag{7}$$

where $\overline{x}_{l,i}$ is the sender embedding of $v_{l,i}$ and W^{in}, W^{out}, b^{in}, and b^{out} are the parameters of linear transformer layers.

In this way, we learn the embedding vector $x_{l,i} = x_{l,i}^{(k+1)}$ for each user $v_{l,i}$ on diffusion graph G_l, preserving the structural dependencies and temporal dynamics on diffusion cascade.

3.3   Influence Embedding Generation

Given the users' hidden representations on the diffusion graph, we integrate the global influence and local influence of the diffusion graph for further prediction. The global influence refers to the full impact of the diffusion and combines the influence of all users on the diffusion graph. Since there are differences in the users' influence, an attention layer is used to learn the weights of different users and generates the global influence embedding with users' representations. The global influence embedding c_g is computed by:

$$\alpha_i = \sum_{v_{l,i} \in G_l} W_q \top \sigma(W_1 \ x_{l,i} + W_2 \ x_{l,k}^{(k+1)} + b) \tag{8}$$

$$c_g = \sum_{v_{l,i} \in G_l} \alpha_i \ x_{l,i} \tag{9}$$

Furthermore, considering the influence of users will decline over time, we also introduce the local influence of the diffusion graph to make a more accurate prediction. Specifically, the most recently activated user is taken as the source of local influence and its embedding vectors is the local influence embedding, denoted by $c_i = x_{l,k}$. Then the final influence embedding c_{inf} is obtained by concatenating global and local influence embedding through a linear layer, computed by:

$$c_{inf} = W_g\,[c_g; c_i] \tag{10}$$

Activation Prediction. With the influence embedding c_{inf} of diffusion graph G_l, the user activation score can be calculated by multiplying influence embedding vector with users' receiver embedding matrix X. Then activation probability is computed through a softmax layer, denoted by:

$$p_l^{(k+1)} = softmax(X \cdot c_{inf}) \tag{11}$$

where $p_l^{(k+1)} \in R^{|V|}$ is the probability distribution on users.

Optimization. After getting the activation probability of each user, the diffusion prediction is formulated as a multi-class classification task. So we calculate the cross-entropy for optimization. For a set of cascades $l_j \in L$, the objective function is derived as follow:

$$\mathcal{L} = -\sum_{j=1}^{|L|}\sum_{i=1}^{|l_j|} v_{j,i+1} log\, p_j^{(i+1)} \tag{12}$$

where $v_{j,i+1} \in R^{|V|}$ is the ground truth, denoted by one hot vector.

4   Evaluations

In this section, we test the performance of the proposed model and compare it with several baselines on benchmark datasets. Meanwhile, we study the effects of different diffusion influences on the prediction by the ablation study.

4.1   Datasets

Three datasets are used for evaluation and Table 1 shows the statistics of the following datasets.

- **Twitter** [26] dataset consists of the diffusion path of the tweets containing URLs on Twitter and the following relationships among users.
- **Memes** [27] dataset contains the memes spreading on news websites. We take the website as user and create links between two websites according to the time sequences on any memes.
- **Digg** [28] dataset includes the news stories and its' voters on Digg. The friendship links are also available.

Table 1. The statistics of datasets

Datasets	Twitter	Memes	Digg
User	9,930	4,086	11,579
Edge	619,262	2,716,865	2,617,994
Cascade	3,461	54,847	3,553
Avg. length	40.1	4.8	30.0

4.2 Baselines

We introduce the baselines for comparison and the evaluation metrics for estimation. Then, the experimental settings are described in detail.

- **LSTM** [29] is a common method for sequential prediction problems. We use
  the base LSTM as an anchoring method.
- **DeepDiffuse** [9] is proposed by using RNN networks to predict the next
  activated users and its infected time. We only focus on predicting the next
  activated users.
- **TopoLSTM** [13] integrates the topology of social networks for diffusion prediction. It extends basic LSTM with diffusion topology tailed for diffusion
  prediction, which considers the cascade structure to estimate the probability
  of the inactive user's activation.
- **SNIDSA** [11] considers the structural dependency on diffusion cascade with
  structure attention mechanism and then models the diffusion as a sequence
  with Recurrent Neural Networks. The final output is used for predicting next
  activated users by a softmax layer.

Evaluation Metrics. We adopt two metrics for evaluation: Hits@k and
MRR@k. Hits@k refers to the precision among top-k predicted users ordered
by the value of activation probability. The other metric is *Mean Reciprocal
Rank* (MRR), which evaluates the ranking list of correct predictions.

Experimental Settings. We give the basic settings of experiments. The imple-
mentations of DeepDiffuse and SNIDSA are based on Tensorflow while the
Pytorch is adopted for the rest. Adam optimizer is used for optimization. To
achieve better performance, we also tune all hyper-parameters. The learning
rate is initialized from {0.05, 0.01, 0.001}. Besides, we set the hidden units as
64 for the Twitter dataset and 128 for the remaining datasets. The batch size is
128 and the dropout rate is set as 0.5 if needed. Other parameters of baselines
are the same as the settings in the original paper.

4.3 Overall Results

The overall results of experiments compared with the baselines on three datasets
are demonstrated in Table 2. In conclusion, the proposed method *infGNN*

Table 2. The overall results about *Hits@k* and *MRR@k* on three datasets (%)

Datasets	Methods	Hits@k			MRR@k		
		10	50	100	10	50	100
Twitter	LSTM	11.826	17.301	22.067	8.799	9.033	9.101
	DeepDiffuse	18.436	24.244	26.779	11.810	12.089	12.125
	TopoLSTM	**27.787**	32.912	34.903	16.079	17.198	17.227
	SNIDSA	22.386	31.241	34.637	14.915	15.361	15.407
	infGNN	26.925	**45.224**	**56.175**	**16.545**	**17.389**	**17.544**
Memes	LSTM	34.262	53.064	61.073	16.687	17.613	17.729
	DeepDiffuse	27.645	49.657	59.628	11.717	12.765	12.903
	TopoLSTM	34.319	59.824	68.939	15.742	16.992	17.123
	SNIDSA	21.221	38.861	48.582	9.809	10.636	10.774
	infGNN	**53.809**	**72.438**	**79.364**	**31.151**	**32.074**	**32.170**
Digg	LSTM	9.698	27.941	42.380	3.489	4.265	4.469
	DeepDiffuse	7.248	22.973	34.338	2.471	3.139	3.299
	TopoLSTM	11.651	32.899	46.848	4.346	5.284	5.483
	SNIDSA	8.710	26.845	41.504	2.937	3.712	3.918
	infGNN	**12.802**	**34.832**	**49.271**	**4.726**	**5.690**	**5.890**

achieves the highest performance on three datasets except for the Hits@10 on
Twitter, which is slightly lower than the best. It is also observed that with
infGNN, the Hits@50 and Hits@100 on Twitter are considerably higher than
with all baselines. Moreover, *infGNN* produces more consistent performance
compared with other methods, which embodies that the value of Hits@100 is
over (about) 50% on all datasets and even almost 80% in Memes dataset. Meanwhile, we notice that the methods considering the structural dependencies are
usually more effective than the methods which only take the diffusion as a cascade sequence. However, the SNIDSA is the exception and underperforms on
some datasets, even achieving the worst performance on Memes. Besides, the
base LSTM performs better than DeepDiffuse on two datasets. Overall, the
experiment results verify the feasibility and superiority of *infGNN*.

4.4   Analysis of Global and Local Influence

Since both the global and local influence are considered in *infGNN*, we would
like to explore the effects of different influences on diffusion prediction. Therefore
we estimate the performance gains of *infGNN* and its variants over base GRU
network [25], which is adopted as a reference to quantify the performance gains.
Table 3 shows the performance of the base GRU network, which uses the final
hidden output for prediction. As for the variants, *infGNN-g* and *infGNN-l* only

embed one type of influence to the influence embedding for prediction: global influence and local influence respectively. Figure 2 shows the performance gains achieved by *infGNN* and its variants.

Table 3. The performance of base GRU for diffusion prediction (%)

Datasets	Hits@k			MRR@k		
	10	50	100	10	50	100
Twitter	17.291	27.398	34.747	11.509	11.957	12.060
Memes	40.976	61.602	69.358	21.306	22.312	22.423
Digg	10.393	29.165	43.930	3.774	4.570	4.775

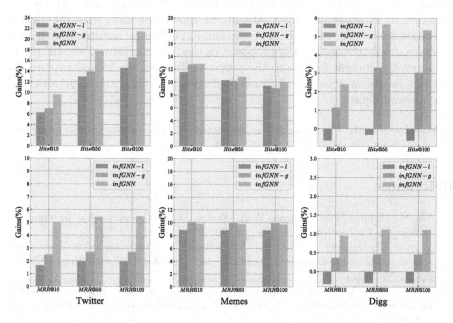

Fig. 2. The performance gains of *infGNN* and its variants over base GRU network (%)

In general, the integration of global or local influence always has a positive impact and the biggest performance gains are achieved by *infGNN*. However, the impact is different in introducing different types of influence. Generally, *infGNN-g* performs better than *infGNN-l*, which even suffers from performance degradation over base GRU on Digg dataset. Moreover, the combination of global and local influence enables more accurate prediction and the improvement is more obvious on Twitter and Digg datasets. Specifically, the Hits@k and MRR@k of *infGNN* rise over 4% and 3% compared with both the variants on Twitter. Besides, it is observed that *infGNN* and its variants boost performance in mostly the same way on Memes dataset. Even though the performance of *infGNN* on Memes remains suboptimal in terms of MRR@k, the performance gap between *infGNN* and *infGNN-g* is small. Furthermore, *infGNN* performs the best in terms of Hit@k. As a result, the integration of global influence can guarantee higher and more stable performance gains, which also indicates the effectiveness of *infGNN*.

5   Conclusion

In the paper, we propose a personalized graph neural networks for information diffusion prediction. We construct the diffusion graph and dynamically learn the hidden representation of each user on the diffusion graph. In this way, we generate influence embedding of diffusion cascade, consisting of global and local influence. Given the influence embedding, the activation probability of each user is computed by a softmax layer. At last, we choose the users with the highest activation probability as the final prediction. Experimental results show the feasibility and effectiveness of the proposed method. In the future, we will try to introduce the content feature to the proposed diffusion model and explore the way to solve the macroscopic diffusion prediction problem with the personalized graph neural networks.

Acknowledgment. The research was supported by National Natural Science Foundation of China (No. 61802140).

References

1. Wu, S., Hofman, J.M., Mason, W.A., Watts, D.J.: Who says what to whom on twitter. In: Proceedings of WWW, pp. 705–714 (2011)
2. Lagnier, C., Denoyer, L., Gaussier, E., Gallinari, P.: Predicting information diffusion in social networks using content and user's profiles. In: Proceedings of ECIR, pp. 74–85 (2013)
3. Suh, B., Hong, L., Pirolli, P., Chi, E.H.: Want to be retweeted? Large scale analytics on factors impacting retweet in twitter network. In: Proceedings of SocialCom, pp. 177–184 (2010)
4. Peng, H.K., Zhu, J., Piao, D., Yan, R., Zhang, Y.: Retweet modeling using conditional random fields. In: Proceedings of ICDM Workshops, pp. 336–343 (2011)
5. Bourigault, S., Lamprier, S., Gallinari, P.: Representation learning for information diffusion through social networks: an embedded cascade model. In: Proceedings of WSDM, pp. 573–582 (2016)
6. Feng, S., Cong, G., Khan, A., Li, X., Liu, Y., Chee, Y.M.: Inf2vec: latent representation model for social influence embedding. In: Proceedings of ICDE, pp. 941–952 (2018)
7. Su, Y., Zhang, X., Wang, S., Fang, B., Zhang, T., Yu, P.S.: Understanding information diffusion via heterogeneous information network embeddings. In: Proceedings of DASFAA, pp. 501–516 (2019)

8. Cao, Q., Shen, H., Cen, K., Ouyang, W., Cheng, X.: Deephawkes: Bridging the gap between prediction and understanding of information cascades. In: Proceedings of CIKM, pp. 1149–1158 (2017)
9. Islam, M.R., Muthiah, S., Adhikari, B., Prakash, B.A., Ramakrishnan, N.: Deep-diffuse: predicting the 'who' and 'when' in cascades. In: Proceedings of ICDM, pp. 1055–1060 (2018)
10. Li, C., Ma, J., Guo, X., Mei, Q.: Deepcas: an end-to-end predictor of information cascades. In: Proceedings of WWW, pp. 577–586 (2017)
11. Wang, Z., Chen, C., Li, W.: A sequential neural information diffusion model with structure attention. In: Proceedings of CIKM, pp. 1795–1798 (2018)
12. Yang, C., Tang, J., Sun, M., Cui, G., Liu, Z.: Multi-scale information diffusion prediction with reinforced recurrent networks. In: Proceedings of IJCAI, pp. 4033–4039 (2019)
13. Wang, J., Zheng, V.W., Liu, Z., Chang, K.C.C.: Topological recurrent neural network for diffusion prediction. In: Proceedings of ICDM, pp. 475–484 (2017)
14. Gruhl, D., Liben-Nowell, D., Guha, R.V., Tomkins, A.: Information diffusion through blogspace. SIGKDD Explor. 6(2), 43–52 (2004)
15. Leskovec, J., McGlohon, M., Faloutsos, C., Glance, N.S., Hurst, M.: Patterns of cascading behavior in large blog graphs. In: Proceedings of SDM, pp. 551–556 (2007)
16. Tsur, O., Rappoport, A.: What's in a hashtag? Content based prediction of the spread of ideas in microblogging communities. In: Proceedings of WSDM, pp. 643–652 (2012)
17. Zhang, M., Cui, Z., Neumann, M., Chen, Y.: An end-to-end deep learning architecture for graph classification. In: Proceedings of AAAI (2018)
18. Li, Y., Yu, R., Shahabi, C., Liu, Y.: Diffusion convolutional recurrent neural network: data-driven traffic forecasting. In: Proceedings of ICLR (2018)
19. Yu, B., Yin, H., Zhu, Z.: Spatio-temporal graph convolutional networks: a deep learning framework for traffic forecasting. In: Proceedings of IJCAI, pp. 3634–3640 (2018)
20. Cao, Q., Shen, H., Gao, J., Wei, B., Cheng, X.: Popularity prediction on social platforms with coupled graph neural networks. In: Proceedings of WSDM, pp. 70–78 (2020)
21. Chen, X., Zhou, F., Zhang, K., Trajcevski, G., Zhong, T., Zhang, F.: Information diffusion prediction via recurrent cascades convolution. In: Proceedings of ICDE, pp. 770–781 (2019)
22. Huang, Z., Wang, Z., Zhang, R.: Cascade2vec: learning dynamic cascade representation by recurrent graph neural networks. IEEE Access 7, 144800–144812 (2019)
23. Qiu, J., Tang, J., Ma, H., Dong, Y., Wang, K., Tang, J.: Deepinf: social influence prediction with deep learning. In: Proceedings of KDD, pp. 2110–2119 (2018)
24. Li, Y., Zemel, R., Brockschmidt, M., Tarlow, D.: Gated graph sequence neural networks. In: Proceedings of ICLR (2016)
25. Cho, K., et al.: Learning phrase representations using RNN encoder-decoder for statistical machine translation. In: Proceedings of EMNLP, pp. 1724–1734 (2014)
26. Hodas, N.O., Lerman, K.: The simple rules of social contagion. Sci. Rep. 4, 4343 (2014)
27. Leskovec, J., Backstrom, L., Kleinberg, J.: Meme-tracking and the dynamics of the news cycle. In: Proceedings of KDD, pp. 497–506 (2009)
28. Hogg, T., Lerman, K.: Social dynamics of digg. EPJ Data Sci. 1(1), 5 (2012)
29. Hochreiter, S., Schmidhuber, J.: Long short-term memory. Neural comput. 9(8), 1735–1780 (1997)

Relationship-Aware Hard Negative Generation in Deep Metric Learning

Jiaqi Huang[1,2], Yong Feng[1,2(✉)] ⓘ, Mingliang Zhou[2,3], and Baohua Qiang[4,5]

[1] College of Computer Science, Chongqing University, Chongqing 400030, China
`fengyong@cqu.edu.cn`
[2] Key Laboratory of Dependable Service Computing in Cyber Physical Society,
Ministry of Education, Chongqing University, Chongqing 400030, China
[3] State Key Lab of Internet of Things for Smart City, University of Macau,
Taipa 999078, Macau, China
[4] Guangxi Key Laboratory of Trusted Software, Guilin University of Electronic
Technology, Guilin 541004, China
[5] Guangxi Key Laboratory of Optoelectronic Information Processing,
Guilin University of Electronic Technology, Guilin 541004, China

Abstract. Data relationships and the impact of synthetic loss have not been concerned by previous sample generation methods, which lead to bias in model training. To address above problem, in this paper, we propose a relationship-aware hard negative generation (RHNG) method. First, we build a global minimum spanning tree for all categories to measure the data distribution, which is used to constrain hard sample generation. Second, we construct a dynamic weight parameter which reflects the convergence of the model to guide the synthetic loss to train the model. Experimental results show that the proposed method outperforms the state-of-the-art methods in terms of retrieval and clustering tasks.

Keywords: Deep metric learning · Sample generation · Distribution quantification · Minimum spanning tree · Relationship preserving

1 Introduction

Deep metric learning (DML) aims at training a deep learning model to learn effective metrics, which measure the similarities between data points accurately and robustly. Deep metric learning has been successfully applied to a variety of tasks, including recommendation [22,26], image retrieval [1,4], person re-identification [10,21], and many others.

Many recent deep metric learning approaches are built on similarity or distance between samples, such as Triplet loss [19], N-pair Loss [15] and Angular Loss [18], etc. Sampling strategy intends to search for samples that profit training most to achieve faster convergence and higher performance, such as Semi-hard Triplet [14] and lifted Structured [11], etc. Recently, sample generation methods

G. Li et al. (Eds.): KSEM 2020, LNAI 12275, pp. 388–400, 2020.
https://doi.org/10.1007/978-3-030-55393-7_35

have been proposed to optimize the network globally by synthetic hard samples. Duan *et al.* [2] utilized a generator with adversarial loss to synthesize potential hard negatives. Zheng *et al.* [25] proposed a hardness-aware method to synthesized hard samples.

However, shortcomings still remain in those sample generation methods. First, they randomly sample different categories of samples to form sample pairs as the basis for synthesizing hard samples, without considering the global geometric distribution of data. Synthetic samples are constrained to be close to the anchor point and remain in different categories, while the distribution of these two categories are of great discrepancy. Although such synthetic sample can generate large gradients for training, it does not conform to the distribution characteristics. Compare to the standard hard sample, it more like an outlier in the data than a hard sample. Training with synthetic outlier samples will make the model learn the wrong features. Second, they did not take the interactions between the training progress and loss of hard synthetic into account. In the early stages of training, the model can easily extract informative samples for training, thus premature training with synthetic samples will prevent the model from learning the characteristics of original data well.

To address the above problems, we propose the Relationship-Aware Hard Negative Generation (RHNG) method. One idea in RHNG is that we construct a continuously updated global minimum spanning tree, which acts as a sampler to screen suitable sample pairs as the basis for synthetic hard samples. Furthermore, we design an adaptive dynamic weight to control the effect of synthetic loss on model training as the training progresses. In this way, the synthetic loss will continuously participate in training to avoid the early impact on training, which will further promote model performance.

In general, the main innovations and contributions of this paper can be summarized as follows:

- We utilize graph structure to learn the data relationship, which is used to maintain the distribution of synthetic samples. As such, synthetic hard negatives are more conducive to promote model training.
- An adaptive dynamic weight is elaborately designed to guide the synthetic loss on model training, which encourages the model fully to learn the original samples and improve the performance by synthetic loss.
- Extensive experimental results demonstrate that the proposed method outperforms the state-of-the-art methods in terms of retrieval and clustering tasks.

The remainder of this paper is organized as follows. Section 2 provides a brief overview of the state-of-the-art related works. Section 3 details our proposed RHNG method. The experimental results and analysis are provided in Sect. 4 and Sect. 5 concludes our work.

2   Related Work

Metric Learning. Metric learning aims at learning effective metrics to measure the similarity of the input data. Many existing methods pay more attention on building a proper loss function to illustrate relation between samples. Contrastive loss [5] is proposed to train on pairwise samples, which gives penalty to negative pairs with distances smaller than a margin. Triplet loss [19] is proposed to construct triplet input samples to ensure the relative distance order between positives and negatives. Furthermore, more classes are involved in N-pair Loss [15]. Instead of comparing samples in euclidean space, distance with angle constraint was measured in Angular Loss [18], which captures additional local structure. Xu *et al.* [20] proposed asymmetric loss for deep metric learning.

Hard Sample Mining. Hard sample mining [6,23] plays an essential role in speeding up convergence and boosting performance of DML model, which continuously mines hard samples that give more information to promote model training. Semi-hard mining [14] is proposed to further utilize the relation among samples by constructing semi-hard triplets within a batch. To take full advantage of the relative relationship, lifted Structured Loss [11] emphasized incorporating all negative samples within a batch. Wang *et al.* [2] focused on applying weights on samples to reflect their importance on optimization. Ge *et al.* [3] proposed Hierarchical Triplet Loss (HTL) that builds a hierarchical tree of all classes.

Hard Negative Generation. Hard negatives usually account for a small part in the training set. Therefore, synthesis methods are proposed to generate potential hard negatives, these synthetics contain more information to fully train the model. Duan *et al.* [2] utilized a generator with adversarial loss to synthesize potential hard negatives that exploit a large number of easy negatives on training. Similarly, Zhao *et al.* [24] proposed an adversarial approach to generate training examples by an adversarial network, which is used to improve triplet-based training. Zheng *et al.* [25] exploited an auto-encoder architecture to synthesize label-preserving hard samples by exploiting existing samples with adaptive hardness-aware augment.

3   Relationship-Aware Hard Negative Generation in Deep Metric Learning

3.1   Overview

Figure 1 shows the overall network architecture of our proposed method, which is comprised of four key components: encoder, graph-based sampler, generator and classifier. The solid arrows represent the data flow and the dotted arrows represent the loss being calculated. The encoder consists of two parts, one is a typical convolutional neural network and the other are several continuous fully-connected layers. Convolutional network acts a feature extractor to

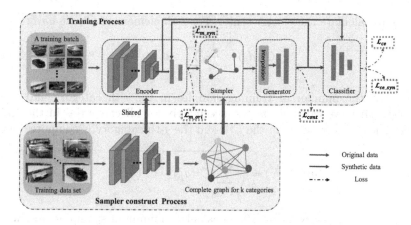

Fig. 1. Overview of the proposed RHNG framework.

extract meaningful image representations, and the fully-connected layers act as the embedding projector following the feature extractor to obtain the embedding. Graph-based sampler is a sample mining process that composes sample pairs based on a minimum spanning tree. The generator takes the sample pairs as input to synthesize hard negatives then maps them back to the feature space which will be exploited for training. However, a generator alone is insufficient to generate reliable synthetic samples, so we use a classifier with cross entropy loss to constrain the generator.

We employ the widely used triplet on our proposed method. Let $\mathcal{X}$ be the training data set, $\mathbf{X} = \{x_i\}_{i=1}^N$ is the mini-batch set from $\mathcal{X}$ and $L = \{l_i\}_{i=1}^N$ is the corresponding labels where $l_i \in \{1 \ldots k\}$. A triplet $<x_i, x_i^+, x_j^->$ is composed by an anchor point x_i, a positive point x_i^+ and a negative point x_j^- with its label $l_j^- \neq l_i$. We denote the feature extractor by F, whose output can be expressed as $\mathbf{Y}(y_i) = F(\mathbf{X})$ where $y_i \in \mathbb{R}^D$ represents a D-dimensional feature. Similarly, $\mathbf{Z}(z_i) = E(\mathbf{Y})$ and $z_i \in \mathbb{R}^M$ express that the embedding projector E projects the features into metric space.

3.2 Relationship-Aware Hard Negative Generation

We employ minimum-spanning tree algorithm to construct the connected subgraphs of all categories in the training set. Categories with edges connected means they are close in the metric space and more reasonable to have hard negatives.

We calculate cluster center for each category to reduce computational complexity. Concretely, we use the encoder to get the encoding for all instances which are denoted as $\mathbf{U} = [\mathbf{u}^1, \mathbf{u}^2, \ldots, \mathbf{u}^k]$ where $\mathbf{u}^i$ represents all instances in i-th category. Denote $C^i = [d_0, d_1, \ldots, d_m]$ as the cluster center of i-th category and d_m is m-th element, we utilize element-wise average for all instances in the category to calculate the cluster center. For p-th element of C^i, we have:

$$C^i(d_p) = \frac{1}{n_i} \sum_{z \in \mathbf{u}^i} z(d_p) \tag{1}$$

where n_i is the number of training samples in the i-th category and z are instances. Then we calculate a distance matrix for all categories in the training set. The distance between the i-th and the j-th category is computed as:

$$d(C^i, C^j) = \sqrt{\sum_{p=1}^{m}(C^i(d_p) - C^j(d_p))^2} \tag{2}$$

Finally, we follow the work in [12] to calculate minimum spanning tree, which acts the sampler to form sample pairs according to the edge connection. The time complexity of the algorithm is $O(n^2)$, in our work n is the number of categories in the dataset. The sampler updates interactively at the certain iterations over the training.

Inspired by the work in [25], we utilize linear interpolation to obtain the hardness of the synthetic samples. The generator takes the embedding of minibatch $\mathbf{X}$ and the sampled triplet sets $\mathcal{T}$ as input. For samples in $\mathbf{X}$, we perform no transformation, i.e. $\tilde{z} = z$, while construct hard embedding $\hat{z}^-$ by linear interpolation based on triplets in $\mathcal{T}$:

$$\hat{z}^- = z^- + \lambda(z - z^-), \quad \lambda = \begin{cases} \frac{d^- - d^+}{d^-} & , d^+ < d^- \\ 1 & , d^+ \geqslant d^- \end{cases} \tag{3}$$

where $d^+ = \|z - z^+\|_2$ and $d^- = \|z - z^-\|_2$ denote the distance between positive pair and negative pair, respectively. The sample embedding obtained by simple interpolation may not meet the sample characteristics, so we apply a network g to map the original embedding $\tilde{z}$ and the synthesize $\hat{z}^-$ to the feature space respectively:

$$\mathbf{Y}(y_i) = g(z_i), \quad (\mathbf{Y}, z_i) \in \{(\tilde{\mathbf{Y}}, \tilde{z}_i), (\hat{\mathbf{Y}}, \hat{z}_i)\} \tag{4}$$

where $\hat{\mathbf{Y}}$ is the synthetic hard negative that will be re-input to the encoder to calculate the synthetic loss, and $\tilde{\mathbf{Y}}$ is used to calculate the reconstruction loss.

3.3 Loss Function

The basic objective function loss used to train the encoder is defined as follows:

$$L_{tri}(<x, x^+, x^->) = [d(x, x^+) - d(x, x^-) + m]_+ \tag{5}$$

where $d(x, x^+) = \|x - x^+\|_2^2$ is the squared Euclidean distance between two embedding vectors, $[\cdot]_+$ represents the loss function only takes the positive component as input and m is the margin. For each batch, we first get its feature $\mathbf{Y}$ and embedding $\mathbf{Z}$ through the encoder. We use the feature set $\mathbf{Y}$ to train the classifier, then use the trained classifier to constrain generator. We train the classifier by minimizing the following cross-entropy loss:

$$\min_{\theta_c} J_{cla} = \mathcal{L}_{ce}(\mathbf{Y}, L) = -log\frac{\exp(y^{l_j})}{\sum_{i=1}^{k}\exp(y^{l_i})} \tag{6}$$

where y^{l_j} represents a feature vector of j-th category and θ_c is the parameter of the classifier.

For the purpose of maintaining the semantic characteristics of the synthesized samples, we formulate the objective function of the generator as follows:

$$
\begin{aligned}
\min_{\theta_g} J_{gen} &= \mathcal{L}_{cont}(\mathbf{Y}, \widetilde{\mathbf{Y}}) + \mathcal{L}_{ce\_syn}(\widehat{\mathbf{Y}}, L) \\
&= \sum_{y\in\mathbf{Y}, \tilde{y}\in\widetilde{\mathbf{Y}}} \|y - \tilde{y}\|^2 + \sum_{\hat{y}\in\widehat{\mathbf{Y}}} \mathcal{L}_{ce}(\hat{y}, L)
\end{aligned}
\tag{7}
$$

where θ_g is the parameter of the generator, $\mathcal{L}_{cont}$ and $\mathcal{L}_{ce\_syn}$ are the content loss and classification loss, respectively.

Finally, we obtain the embedding of synthetic samples $\check{z}$ by $\check{\mathbf{Z}} = E(\widehat{\mathbf{Y}})$. Based on Eq. (5), combining two metric losses calculated from the original and synthetic samples, the objective function to train the encoder can be formulated as:

$$
\begin{aligned}
\min_{\theta_f} J_{metric} &= \mathcal{L}_{m\_ori} + \beta\mathcal{L}_{m\_syn} \\
&= \sum_{i=1}^{N}[d(z_i, z_i^+) - d(z_i, z_j^-) + m]_+ \\
&+ \beta\sum_{i=1}^{N}[d(z_i, z_i^+) - d(z_i, \check{z}_j^-) + m]_+
\end{aligned}
\tag{8}
$$

where θ_f is the parameter of the encoder, and β is a trade-off parameter.

We construct the adaptive dynamic β according to the convergence of $\mathcal{L}_{m\_ori}$. In order to eliminate the training fluctuation, we get a smooth convergence curve by using *Exponential Moving Average* (EMA) [7] for $\mathcal{L}_{m\_ori}$:

$$v_t = \gamma v_{t-1} + (1 - \gamma)\mathcal{L}_{m\_ori}^t \tag{9}$$

where v_t represents the value in t-th iteration and γ is a constant that set to 0.98, which means v_t is calculated from the latest 50 iterations according to the characteristics of the EMA. Then we use the hyperbolic tangent function to map the β between 0 and 1 as:

$$\beta = 1 - tanh(\eta \cdot v_t) \tag{10}$$

where η is a trade-off hyper-parameter.

The necessity of dynamic β lies in two aspects. On the one hand, in the early stages of training, the model can't measure the distance metric of samples well, only using the random triples can generate enough gradients to promote the model convergence. With the training process, the original samples can not support the optimization of the model and synthesize hard negatives as complements to the original ones that helps the model to be further optimized. On the other, we synthesize hard negatives based on the global distribution and the distance of sample pairs, so synthetic hard negatives in the early stage of training process may meaningless that easily damage the metric space structure, resulting the model train in the wrong direction from the beginning.

3.4   Summary of the Proposed Method

The details of the training process of proposed approach are described in Algorithm 1. It is worth noting that even though the methods in [2,24,25] and our method focus on hard sample generation, moreover, some procedures (such as the linear interpolation and autoencoder generation architecture, etc.) are the same as the method in [25]. However, the emphasis is quite different from the following aspects: First, RHNG takes advantage of the graph structure to obtain more precise synthetic hard samples for deep metric learning. Second, previous generation methods lack of considering the impact of synthetic loss on training. However, our method design adaptive dynamic weight to guide the synthetic loss on model training.

Algorithm 1. Training process of proposed RHNG

Input: Training set $\mathcal{X}$; hyper-parameters: η, *update epoch* and the margin m; iteration numbers T

Output: Parameters of the encoder θ_f, the generator θ_g and the classifier θ_c

1: Initialize the graph-based sampler according to Eq. (1) and Eq. (2) by feed-forwarding $\mathcal{X}$ into the encoder
2: **for** $iter = 1$ to T **do**
3:    Sample triplet set $\mathcal{T}$ by the sampler
4:    Generate synthetic hard negatives per Eq. (3) and Eq. (4)
5:    Calculate $\mathcal{L}_{cont}$, $\mathcal{L}_{ce\_syn}$ and $\mathcal{L}_{m\_syn}$ per Eq. (7) and Eq. (8)
6:    Update θ_c by minimising J_{cla} in Eq. (6)
7:    Update θ_g by minimising J_{gen} in Eq. (7)
8:    Update θ_f by minimising J_{metric} in Eq. (8)
9:    Update the minimum spanning tree with current model
10: **end for**
11: **return** θ_c, θ_g, θ_f

---

4   Experiments

4.1   Experimental Settings

Evaluation Metrics and Dataset. We evaluated the proposed method and existing methods on both retrieval and clustering tasks.

Table 1. Experimental results (%) on CUB-200-2011 dataset

Method	R@1	R@2	R@4	R@8	NMI	F_1
Triplet	43.61	55.73	68.93	80.08	55.79	21.54
Semi-hard	45.29	58.25	70.99	80.71	56.37	22.88
N-pair	46.28	59.85	72.21	81.83	58.08	24.48
Lifted	49.72	62.10	74.35	83.86	57.58	26.01
Angular	49.39	61.30	74.48	83.40	58.96	27.58
DAML (Triplet)	37.60	49.30	61.30	74.40	51.30	17.60
HDML (Triplet)	48.51	61.37	74.02	83.65	59.30	26.55
RHNG (Ours)	**50.47**	**63.30**	**75.10**	**84.21**	**59.58**	**27.34**

Table 2. Experimental results (%) on Cars196 dataset

Method	R@1	R@2	R@4	R@8	NMI	F_1
Triplet	55.70	68.20	78.68	86.31	53.82	22.60
Semi-hard	57.72	70.47	80.05	87.84	54.94	23.66
N-pair	60.53	73.28	83.42	89.70	56.82	24.14
Lifted	64.84	76.54	85.49	90.00	57.69	25.10
Angular	68.84	80.23	87.35	92.57	61.31	28.80
DAML (Triplet)	60.60	72.50	82.50	89.90	56.50	22.90
HDML (Triplet)	66.65	78.10	86.17	91.82	59.91	26.61
RHNG (Ours)	**68.68**	**79.28**	**87.45**	**92.89**	**61.64**	**28.74**

For retrieval task, we calculated the percentage of retrieved samples with the same label to the query images in K nearest neighbors as performance metrics, where $K \in \{1, 2, 4, 8\}$, marked as R@K. For clustering task, we employed standard K-means algorithm in test set, which evaluated with normalized mutual information (NMI) and F_1 score. NMI consists of the ratio of mutual information divided by the average entropy of clusters and the average entropy of labels. F_1 score computes the harmonic mean of precision and recalls on determining whether sample attribution to a specific cluster.

We conduct the experiment on widely-used CUB-200-2011 [17] dataset and Cars196 [9] dataset.

- CUB-200-2011 [17] dataset contains 11,788 bird images in 200 bird categories. We exploited the first 100 categories with 5,684 images as training set and the rest 5,924 images in 100 categories for testing.
- Cars196 [9] dataset includes 16,185 car images with 196 categories. We used the first 98 categories with 8,054 images for training and remaining rest 100 categories with 8,131 images for test set.

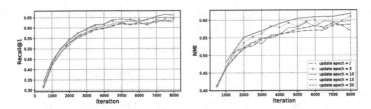

Fig. 2. Comparison of using different update epoch in the clustering (right) and retrieval (left) on Cars196 dataset with $\eta = 0.001$.

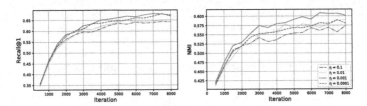

Fig. 3. Comparison of using different values of η in the clustering (right) and retrieval (left) task on Cars196 dataset with update epoch $= 10$.

Implementation Details. We used GoogLeNet [16] architecture as the feature extractor to extract 1024 dimensional features, and embedding projector consists of three full connection layers whose output dimension are 512, 256 and 128 respectively. Meanwhile, we implemented the generator with two fully connected layers whose output dimension are 512 and 1,024 respectively. The classifier is also consisted by three fully-connected layers.

We initialized the GoogLeNet with weights pretrained on ImageNet ILSVRC dataset [13] and all other fully-connected layers with random weights. We set hyper-parameters $\eta = 10^{-3}$, *update epoch* $= 10$ and the margin $m = 1$. We employed the ADAM [8] optimizer with a learning rate of $8e - 4$ on training.

4.2   Comparisons with the State-of-Art Methods

We compared our method with some famous deep metric learning approaches, including the triplet loss [19] and the Semi-hard triplet loss [14], the classical sample mining method N-pair loss [15], the state-of-the-art lifted structure [11] and Angular loss [18], the hard negative generation method DAML [2] and HDML [25]. We followed the settings mentioned in these original papers through the comparison experiments.

Table 1 and Table 2 show the result compared to other baseline methods in two benchmark datasets respectively. We observe our proposed method can achieve competitive performance in terms of two datasets on retrieval and clustering tasks. The comparison with the state-of-the-art DAML and HDML shows the effectiveness of our proposed method in sample generation approaches. For the state-of-the-art Angular loss, we reached a higher performance on the smaller

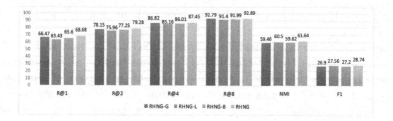

Fig. 4. Results of RHNG and its three sub-models.

CUB-200-2011 dataset, but failed to achieve their performance on the larger Cars196 with a small margin in some metrics. We analyse that the difference comes from the size of the training set, as the increase of samples in each class, we fail to collect sufficient information within a fixed number of samples through simple sampling.

4.3 Hyper-parameters Sensitivity

There are two hyper-parameters affecting our method, which are the "update epoch" and trade-off parameter η, respectively. We discuss the effect of various parameters for the clustering and retrieval tasks in terms of Cars196 dataset.

Figure 2 shows the effect of different update frequencies. Larger "update epoch" means slower update frequency and "update epoch= /" means that we only use the initialized minimum spanning tree and no longer update it. Theoretically, the higher the frequency of updating the minimum spanning tree, the timelier that data distribution can be reflected in different training periods of the model. However, the performance of updating every 10 epochs is better than updating every 5 epochs. We speculate that it is because the model needs sufficient training to learn at different periods, replacing hard negatives too frequently will make the training insufficient.

The impact of η is shown in Fig. 3. We can observe that the performance increases with the increase of η. The proposed method can achieve the best performance when $\eta = 0.001$, and then the performance gradually decreases. We speculate that it is because the constructed dynamic parameter β can better reflect the learning state of the model when $\eta = 0.001$.

4.4 Validation for Single Modules

Due to the deep network architecture, the performance improvements can be caused by many factors. To investigate the impact of different factors on the performance of the proposed method, we conduct a series of validation experiments on the Cars196 dataset. We investigate three RHNG variants: (1) "RHNG-G" is a RHNG variant without graph-based sampler, that is, randomly selecting sample pairs to synthesize hard negatives as the previous generation method. (2) "RHNG-L" is a RHNG variant without linear interpolation, which means

$\hat{z}^- = z^-$. (3) "RHNG-B" is a RHNG variant without dynamic weight β by a constant $\beta = 1$.

Figure 4 shows the experimental results of RHNG and its three variants. From the results, we can see that the full RHNG performs best on all evaluation metrics, which indicates that all of the three components in the model contribute to the final performs. The performance degradation of RHNG-G shows the significance of the graph-based sampler, which enhances the generalization ability of the model by maintaining the data distribution. Furthermore, the performance of RHNG outperforms RHNG-L, which demonstrates that the synthetic samples without enough hard levels cannot promote further training of the model. At the same time, the result of RHNG-B also proves that the strategy of using dynamic parameter to control the weight of synthesis loss is correct.

5   Conclusion

In this paper, we proposed a relationship-aware hard negative generation (RHNG) method in deep metric learning. We utilize global minimum spanning tree to constrain the generation of synthesized hard negatives. Besides, we construct a dynamic weight parameter to guide the synthetic loss to train the model, which prevents synthetic loss from misleading model. Experimental results demonstrate that our RHNG is effective and outperforms some state-of-art methods. In future work, we will focus on precise relationship constraint and efficient synthesis strategy to improve our proposed method.

Acknowledgments. Supported by National Key R&D Program of China (No. 2017YFB1402400), National Nature Science Foundation of China (No. 61762025), Guangxi Key Laboratory of Trusted Software (No. kx202006), Guangxi Key Laboratory of Optoelectroric Information Processing (No. GD18202), and Natural Science Foundation of Guangxi Province, China (No. 2019GXNSFDA185007).

References

1. Deng, C., Yang, E., Liu, T., Li, J., Liu, W., Tao, D.: Unsupervised semantic-preserving adversarial hashing for image search. IEEE Trans. Image Process. **28**(8), 4032–4044 (2019)
2. Duan, Y., Zheng, W., Lin, X., Lu, J., Zhou, J.: Deep adversarial metric learning. In: Proceedings of the IEEE Conference on Computer Vision and Pattern Recognition, pp. 2780–2789 (2018)
3. Ge, W., Huang, W., Dong, D., Scott, M.R.: Deep metric learning with hierarchical triplet loss. In: Ferrari, V., Hebert, M., Sminchisescu, C., Weiss, Y. (eds.) ECCV 2018. LNCS, vol. 11210, pp. 272–288. Springer, Cham (2018). https://doi.org/10.1007/978-3-030-01231-1_17
4. Grabner, A., Roth, P.M., Lepetit, V.: 3D pose estimation and 3D model retrieval for objects in the wild. In: Proceedings of the IEEE Conference on Computer Vision and Pattern Recognition, pp. 3022–3031 (2018)

5. Hadsell, R., Chopra, S., LeCun, Y.: Dimensionality reduction by learning an invariant mapping. In: 2006 IEEE Computer Society Conference on Computer Vision and Pattern Recognition (CVPR 2006), vol. 2, pp. 1735–1742. IEEE (2006)
6. Huang, C., Loy, C.C., Tang, X.: Local similarity-aware deep feature embedding. In: Advances in Neural Information Processing Systems, pp. 1262–1270 (2016)
7. Hunter, J.S.: The exponentially weighted moving average. J. Qual. Technol. **18**(4), 203–210 (1986)
8. Kingma, D.P., Ba, J.: Adam: a method for stochastic optimization. In: ICLR (Poster) (2015)
9. Krause, J., Stark, M., Deng, J., Fei-Fei, L.: 3D object representations for fine-grained categorization. In: 4th International IEEE Workshop on 3D Representation and Recognition (3DRR 2013), Sydney, Australia (2013)
10. Liu, Z., Wang, D., Lu, H.: Stepwise metric promotion for unsupervised video person re-identification. In: Proceedings of the IEEE International Conference on Computer Vision, pp. 2429–2438 (2017)
11. Oh Song, H., Xiang, Y., Jegelka, S., Savarese, S.: Deep metric learning via lifted structured feature embedding. In: Proceedings of the IEEE Conference on Computer Vision and Pattern Recognition, pp. 4004–4012 (2016)
12. Prim, R.C.: Shortest connection networks and some generalizations. Bell Syst. Tech. J. **36**(6), 1389–1401 (1957)
13. Russakovsky, O., et al.: Imagenet large scale visual recognition challenge. Int. J. Comput. Vis. **115**(3), 211–252 (2015)
14. Schroff, F., Kalenichenko, D., Philbin, J.: FaceNet: a unified embedding for face recognition and clustering. In: Proceedings of the IEEE Conference on Computer Vision and Pattern Recognition, pp. 815–823 (2015)
15. Sohn, K.: Improved deep metric learning with multi-class n-pair loss objective. In: Advances in Neural Information Processing Systems, pp. 1857–1865 (2016)
16. Szegedy, C., et al.: Going deeper with convolutions. In: Proceedings of the IEEE Conference on Computer Vision and Pattern Recognition, pp. 1–9 (2015)
17. Wah, C., Branson, S., Welinder, P., Perona, P., Belongie, S.: The caltech-UCSD birds-200-2011 dataset. Technical report CNS-TR-2011-001, California Institute of Technology (2011)
18. Wang, J., Zhou, F., Wen, S., Liu, X., Lin, Y.: Deep metric learning with angular loss. In: Proceedings of the IEEE International Conference on Computer Vision, pp. 2593–2601 (2017)
19. Weinberger, K.Q., Saul, L.K.: Distance metric learning for large margin nearest neighbor classification. J. Mach. Learn. Res. **10**(Feb), 207–244 (2009)
20. Xu, X., Yang, Y., Deng, C., Zheng, F.: Deep asymmetric metric learning via rich relationship mining. In: Proceedings of the IEEE Conference on Computer Vision and Pattern Recognition, pp. 4076–4085 (2019)
21. Yu, H.X., Wu, A., Zheng, W.S.: Cross-view asymmetric metric learning for unsupervised person re-identification. In: Proceedings of the IEEE International Conference on Computer Vision, pp. 994–1002 (2017)
22. Yu, J., Gao, M., Song, Y., Zhao, Z., Rong, W., Xiong, Q.: Connecting factorization and distance metric learning for social recommendations. In: Li, G., Ge, Y., Zhang, Z., Jin, Z., Blumenstein, M. (eds.) KSEM 2017. LNCS (LNAI), vol. 10412, pp. 389–396. Springer, Cham (2017). https://doi.org/10.1007/978-3-319-63558-3_33
23. Yu, R., Dou, Z., Bai, S., Zhang, Z., Xu, Y., Bai, X.: Hard-aware point-to-set deep metric for person re-identification. In: Ferrari, V., Hebert, M., Sminchisescu, C., Weiss, Y. (eds.) ECCV 2018. LNCS, vol. 11220, pp. 196–212. Springer, Cham (2018). https://doi.org/10.1007/978-3-030-01270-0_12

24. Zhao, Y., Jin, Z., Qi, G., Lu, H., Hua, X.: An adversarial approach to hard triplet generation. In: Ferrari, V., Hebert, M., Sminchisescu, C., Weiss, Y. (eds.) ECCV 2018. LNCS, vol. 11213, pp. 508–524. Springer, Cham (2018). https://doi.org/10.1007/978-3-030-01240-3_31

25. Zheng, W., Chen, Z., Lu, J., Zhou, J.: Hardness-aware deep metric learning. In: Proceedings of the IEEE Conference on Computer Vision and Pattern Recognition, pp. 72–81 (2019)

26. Zuo, X., Wei, X., Yang, B.: Trust-distrust aware recommendation by integrating metric learning with matrix factorization. In: Liu, W., Giunchiglia, F., Yang, B. (eds.) KSEM 2018. LNCS (LNAI), vol. 11062, pp. 361–370. Springer, Cham (2018). https://doi.org/10.1007/978-3-319-99247-1_32

Striking a Balance in Unsupervised Fine-Grained Domain Adaptation Using Adversarial Learning

Han Yu, Rong Jiang, and Aiping Li[✉]

College of Computer National University of Defense Technology, Changsha, China
{yuhan17,jiangrong,liaiping}@nudt.edu.cn

Abstract. Fine-grained domain adaptation is an emerging yet very challenging task in representation learning. In this paper, we analyze a possible reason for the poor performance in fine-grained domain adaptation, which is the difficulty in striking a balance between distribution alignment and fine-grained variations elimination. Furthermore, we propose an adversarial fine-grained domain adaptation framework as a step towards alleviating the underlying conflict between fine-grained variations elimination and domain adaptation. Specifically, our adversarial framework consists of two key modules: a joint label predictor for conditional distribution alignment and a rectifier for fine-grained variations elimination. The key balance can be achieved through the adversarial learning. Besides, experiments on domain adaptation benchmark and fine-grained dataset validate the effectiveness of our framework and show that our framework consistently outperforms the state-of-the-art methods including RTN, MADA, Multi-Task, and DASA.

Keywords: Domain adaptation · Fine-grained classification · Adversarial learning · Distribution alignment

1 Introduction

The goal of unsupervised domain adaptation (UDA) is to classify samples from different target domains, in which the labels are unavailable. The idea is to let the classifier become agnostic to the domain by aligning the conditional distribution of the input samples from the two domains (source domain and target domain, respectively). While UDA has been extensively studied, it remains a challenging problem for UDA in the fine-grained setting, which aims to classify images from target domain into the fine-grained categories. For instance, one may have a training dataset collected from the website, where annotation comes for free and the background is blank, but want to apply the resulting model to the fine-grained classification (FGC) in the wild, such as different bird species or car brands, where the label is probably absent.

The difficulty of fine-grained domain adaptation comes from two aspects. On the one side, the representations learned by the model are required to align

© Springer Nature Switzerland AG 2020
G. Li et al. (Eds.): KSEM 2020, LNAI 12275, pp. 401–413, 2020.
https://doi.org/10.1007/978-3-030-55393-7_36

the conditional distribution between domains. On the other side, these representations should be robust enough to eliminate fine-grained variations from the wild, such as view, pose, and background variations. Specifically, it is quite difficult to strike a balance between distribution alignment and fine-grained variations elimination, since the conditional distribution of different fine-grained category (like bird species) may be forcibly aligned due to the variations elimination. For instance, when intuitively combining techniques of UDA and FGC, the fine-grained classifier wants to disentangle the fine-grained variations, while the domain-invariant representation generator is encouraged to narrow the domain difference of the same sub-class, which may strengthen the common fine-grained variations across domains. Such deviation may have a negative impact on the classification performance.

Currently, there are several works [4,7,24] on fine-grained domain adaptation, all of which shed new light on this emerging field with sound improvement. These pioneers give priority on fitting UDA into the FGC setting by using advanced fine-grained techniques, such as self-attention [11,25], attributes information leveraging [23], deep metric learning [5], and pre-trained network transferring [26]. However, to some extent, they ignore the contradiction between fine-grained variations elimination and domain adaptation. Classic approaches [11,25] to address fine-grained domain adaptation imply an assumption that a well-performance domain adaptation model plus good fine-grained classifier will result in sound fine-grained domain adaptation model. Intuitively, such assumption is quite solid, but it probably does not hold in practice, since the intervention between each other, just as mentioned above.

In this work, we propose an adversarial fine-grained domain adaptation framework as a step towards alleviating the underlying conflict between fine-grained variations elimination and domain adaptation. Specifically, our adversarial framework consists of five modules, including representation generator, joint label predictor, class label predictor and rectifier, and discriminator. The representation generator and discriminator are combined, similar to the classic generative adversarial network (GAN). The idea of joint label predictor is to impose a joint adversarial loss instead of the traditional adversarial loss, which is not only category-aware but also domain-aware. In this way, we can align the conditional distribution and handle fine-grained domain adaptation directly (cf. Sect. 3.2). The rectifier is used to rectify view and pose variations, and jointly learn discriminative features for FGC via adversarial training (cf. Sect. 3.3). The class label predictor enforces samples with the same fine-grained label to have similar feature representations that are further used for classification (cf. Sect. 3.4). All modules are jointly and adversarial optimized together, so that they are balanced to contribute to a good representation that may overcoming the underlying conflict. Experimental results on three popular datasets show that our model consistently outperforms the state-of-the-art methods including RTN [16], MADA [18], Multi-Task [7], DASA [24].

The main contributions of this paper are:

- To the best of our knowledge, we are the first to study the underlying conflict between fine-grained classification and domain adaptation. In addition, we propose an adversarial fine-grained domain adaptation framework that strike a balance between FGC and UDA.
- Experimental results on real-world datasets under fine-grained scenario show that our framework obtains better prediction results than the state-of-the-art fine-grained domain adaptation models.

2   Related Work

2.1   Fine-Grained Classification

With the rapid development on representation learning, the FGC becomes a well studied problem [1–3]. There are many FGC techniques could be used in UDA, such as self-attention [11,25], attributes information leveraging [23], deep metric learning [5], and pre-trained network transferring [26]. Generally, these techniques can be divided into two category. The first category is to leverage useful fine-grained information into the end-to-end framework, such as pixels [29], parts [27], and attribute [23], and web images. The second category is totally unsupervised without augmented information, such self-attention [11,25] and pre-trained network fine-tuning and high-resolution images training [4]. However, since these FGC techniques either rely on augmented information and high-resolution images, they are not general approaches for fine-grained domain adaptation.

Our framework differ from [4,7,24] in the following. **Firstly**, our framework do not need augmented information or high-resolution images. By taking the fine-grained variation from the wild into considerations, our framework can capture more fine-grained features. In this way, our framework can eliminate large variations in samples between the wild world and training datasets. **Secondly**, our framework is proposed for the general fine-grained domain adaptation, and not based on the pre-trained network. In addition, our framework is an end-to-end framework. These features are not supported in [4,7]. **Thirdly**, our framework adopt the idea of adversarial training and can strike a balance between fine-grained variations elimination and domain adaptation, which is ignored in [7,24].

2.2   Unsupervised Domain Adaptation

The work on UDA has been working on aligning the distribution between different domains. An effective way is to learn a domain-invariant classifier. Simultaneous Deep Transfer Nets (DC) [21] adopts a discriminator with fully-connected layer to infer which domain the input belongs and impose a soft label constraint to encourage domain-invariant representations. Besides, [14] minimizes the intra-class dispersion to avoid the misalignments, which is a typical way to align conditional distribution. Further, some researcher extract more domain-invariant features via adversarial traininng, i.e., CoGAN [13].

Our method is related to the recent methods [13,14,21] in some aspects. However, **Firstly**, there are some differences between our method and others. Our approach allows to not only align the conditional distributions, but also encourage them to be disjoint. **Secondly**, our framework could be extended to the multi-domain scenario, by a multi classification task on domains. **Thirdly**, our framework can give more priority on the fine-grained features, which boost the classification performance.

3   Our Framework

3.1   Overview

Figure1 shows the architecture of our framework, which consists of five modules: representation generator, joint label predictor, class label predictor, rectifier, and discriminator. We denote the class label predictor with $H_c(\cdot)$ which is used for fine-grained recognition. The representation generator $G(\cdot)$ can be regarded as a generation module, which is used to generate feature representation for the input image. The discrimination module consists of the joint label predictor $H_j(\cdot)$, the rectifier $H_r(\cdot)$, and the discriminator $H_d(\cdot)$. To address the problem of FGC alignment during the migration domain adaptation, our joint label predictor using not a binary adversarial loss on the domain alignment, but a $2K$-way adversarial loss: The first K elements are the known source classes, and

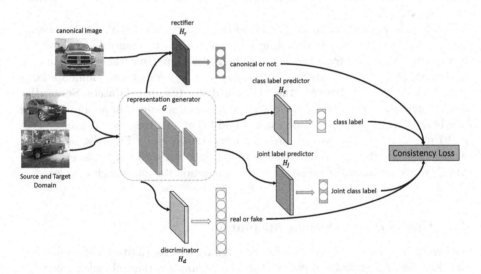

Fig. 1. The network architecture of our framework. Where G is the representation generator, H_j is the joint label predictor, H_c is the class label predictor, H_c is the rectifier, and H_d is the discriminator. We use the source domain, the target domain, and the canonical image as inputs to the generator, and the feature representations outputted by the generator are fed to other modules. We fit the loss functions of each module together for jointly training.

the second K elements are the unknown target classes, where K is the number of fine-grained classes. Besides, to eliminate the impact caused by the difference of the view and poses, we add a rectifier to normalize the feature representation. We also apply a discriminator to determine the real or fake of the image. In the process of the training, we perform alternate min-max optimizations on the objective function to update the parameters of the generation module and the discrimination module. We will detail the objective function of each module and adversarial training below.

3.2   Class Label Predictor

In our framework, the class label predictor is the main module for fine-grained image inference. Given the feature representation generated by $G(\cdot)$, it outputs the K-dimensional probability, which indicates the fine-grained classes of the input image. We use the labeled source domain data to train the class label predictor through cross-entropy loss, and its objective function can be written as:

$$\mathcal{L}_c = \mathbb{E}_{(x_s,y_s)\sim\mathcal{P}_S}\ell_{CE}(H_c(G(x_s)),y_s). \tag{1}$$

where $\mathcal{P}_S$ denotes the data distribution of source domain, x_s is the input image drawn from the source domain with the corresponding label of y_s. During the training process, we updated the representation generator and class label predictor by minimizing this loss function simultaneously.

3.3   Joint Label Predictor

In order to solve the problem of fine-grained class alignment, that is, one class in the source domain is aligned to another class in the target domain, we introduced the joint label predictor. The joint label predictor is trained in conjunction with the generator while aligning the domain conditional distribution and fine-grained class. To train the joint label predictor, we use data from both the source and target domains. When training with source domain data, we use cross-entropy loss to minimize the objective function defined below:

$$\mathcal{L}_{js} = \mathbb{E}_{(x_s,y_s)\sim\mathcal{P}_S}\ell_{CE}(H_j(G(x_s)),[y_s,\mathbf{0}]). \tag{2}$$

where $\mathbf{0}$ represents a K-dimensional vector, which means that for the objective function using the source domain data, the probability of determining the last K (that is, the classes in the target domain) is set to 0. For the target domain data, the objective function is:

$$\mathcal{L}_{jt} = \mathbb{E}_{(x_t,\tilde{y})\sim\mathcal{P}_T}\ell_{CE}(H_j(G(x_t)),[\mathbf{0},\tilde{y}]). \tag{3}$$

where $\mathcal{P}_T$ denotes the data distribution of target domain. It should be noted that the labels of the target domain are unknown, so we need to use the output of the class label predictor trained with the source domain data to serve as pseudo-labels. We use $\tilde{y}$ to represent the pseudo-labels. We assume that an effective classifier can be trained that works on the target domain only from the source domain data (the classification effect is better than random). Therefore, during the training process, our class label predictor is first trained with the source domain data for a period of time and then used to teach the joint label predictor. The overall objective function of the joint label predictor can be written as:

$$\mathcal{L}_j = \mathcal{L}_{js} + \mathcal{L}_{jt}. \tag{4}$$

When updating the joint label predictor, we freeze the parameters of the representation generator $G(\cdot)$ and only update the parameters of the joint label predictor $H_j(\cdot)$.

Similarly, when training the representation generators with source domain data, we minimize the following objective functions:

$$\mathcal{L}_{jsa} = \mathbb{E}_{(x_s, y_s) \sim \mathcal{P}_S} \ell_{CE}(H_j(G(x_s)), [\mathbf{0}, y_s]). \tag{5}$$

The objective function for training the generator with the target domain is defined as follows:

$$\mathcal{L}_{jta} = \mathbb{E}_{(x_t, \tilde{y}) \sim \mathcal{P}_T} \ell_{CE}(H_j(G(x_t)), [\tilde{y}, \mathbf{0}]). \tag{6}$$

We keep the parameters of the joint label predictor unchanged and change joint label from $[y_s, \mathbf{0}]$ to $[\mathbf{0}, y_s]$, and $[\mathbf{0}, \tilde{y}]$ to $[\tilde{y}, \mathbf{0}]$. We freeze the parameters of $H_j(\cdot)$ and minimize the following loss to update the parameters of $G(\cdot)$:

$$\mathcal{L}_{ja} = \mathcal{L}_{js} + \mathcal{L}_{jt}. \tag{7}$$

3.4   Rectifier

One of the challenges in FGC is that objects in images may appear in the high variation of views or poses. In our framework, the rectifier is used to discriminate the difference between the generated feature representation x_z and the canonical image x_c. The input of the rectifier is x and x_c, where x can be the generated representation x_z or the canonical image x_c. If x is the canonical image x_c, we define the data pair x and x_c as a positive sample. Otherwise, it is a negative sample. We define the $\mathcal{P}_{data}$ as the distribution of real images from the source or target domain. The objective function of the rectifier as following:

$$\mathcal{L}_r = \mathbb{E}_{x \sim \mathcal{P}_{data}} \ell_{CE}[(H_r(x, x_c), 1) + (H_r(G(x), x_c), 0)]. \tag{8}$$

When training the rectifier, we set the output probability of canonical image be 1, and set the output probability of the generated representation be 0. We maintain the generator parameters unchanged, so that the rectifier can evaluate the similarity of x and x_c.

When updating the parameters of the representation generator, we minimize the following objective functions:

$$\mathcal{L}_{ra} = \mathbb{E}_{x \sim \mathcal{P}_{data}} \ell_{CE}(H_r(G(x), x_c), 1). \tag{9}$$

We set the output probability of generated representation be 0, and fixed the parameters of the rectifier to make the feature representation x_z generated by $G(\cdot)$ close to x_c.

Algorithm 1. Adversarial training

Input: A batch with labeled source domain samples $\{x_s, y_s\}$, unlabeled target domain samples x_t with corresponding pseudo-labels $\{\tilde{y}\}$, and canonical image x_c
  Output: Updated network parameters θ, ϕ_c, ϕ_j, ϕ_r, ϕ_d
  Initialization: Initialize parameters θ, ϕ_c, ϕ_j, ϕ_r, ϕ_d
  Optimization:
1: **while** not reach an equilibrium **do**
2:      Compute class label predictor loss $\mathcal{L}_c$ by (1).
3:      Update the network parameters ϕ_c:
4:      $\phi_c \leftarrow \phi_c - \frac{\partial \mathcal{L}_c}{\partial \phi_c}$
5:      Compute joint label predictor loss $\mathcal{L}_j$ by (4).
6:      Update network parameters ϕ_j:
7:      $\phi_j \leftarrow \phi_j - \frac{\partial \mathcal{L}_j}{\partial \phi_j}$
8:      Compute rectifier loss $\mathcal{L}_r$ by (8).
9:      Update network parameters ϕ_r:
10:     $\phi_r \leftarrow \phi_r - \frac{\partial \mathcal{L}_r}{\partial \phi_r}$
11:     Compute discriminator loss $\mathcal{L}_d$ by (10).
12:     Update network parameters ϕ_d:
13:     $\phi_d \leftarrow \phi_d - \frac{\partial \mathcal{L}_d}{\partial \phi_d}$
14:     Compute the joint label predictor adversarial loss $\mathcal{L}_{ja}$ by (7).
15:     Compute the rectifier adversarial loss $\mathcal{L}_{ra}$ by (9).
16:     Compute the discriminator adversarial loss $\mathcal{L}_{da}$ by (11).
17:     Combined loss $\mathcal{L}_{adv} = L_c + L_{ja} + \mathcal{L}_{ra} + \mathcal{L}_{da}$
18:     Update the network parameters θ:
19:     $\theta \leftarrow \theta - \frac{\partial \mathcal{L}_{adv}}{\partial \theta}$

3.5 Discriminator

The output of the discriminator is a scalar probability indicating whether the given input is the real image or the generated image representation. The objective function of discriminator is defined as [8]:

$$\mathcal{L}_d = \mathbb{E}_{x \sim \mathcal{P}_{data}} \ell_{CE}[(H_d(x_r), 1) + (H_d(G(x)), 0)]. \tag{10}$$

The discriminator tries to distinguish the generated feature representation from the real image, and the representation generator tries to confuse the discriminator. We optimize the discriminator by minimizing the objective function, while the parameters of the generator remain unchanged. The output probability of real image is set to 1, and the output probability of the generated representation is set to 0.

For the update of the representation generator, we fix the parameters of the discriminator, set the output probability of the generated representation be 1, and minimize the following objective function:

$$\mathcal{L}_{da} = \mathbb{E}_{x \sim \mathcal{P}_{data}} \ell_{CE}(H_d(G(x)), 1). \tag{11}$$

3.6 Adversarial Training

To optimize our framework, we need to perform the adversarial training between the representation generator $G(\cdot)$, the class label predictor $H_c(\cdot)$, and the discrimination module which consists of $H_j(\cdot)$, $H_r(\cdot)$, and $H_d(\cdot)$. Firstly, we train a classifier with labeled source domain data to provide pseudo-labels to the joint label predictor in the subsequent training process. Subsequently, following the adversarial training procedure in [8], we alternately optimized the representation generator and the discrimination module. We define θ as the parameters of representation generator, and ϕ_c, ϕ_j, ϕ_r, ϕ_d are the parameters of class label predictor, joint label predictor, rectifier, and discriminator respectively. The process of adversarial learning is shown in Algorithm 1. For the training of the discrimination module, we freeze the parameters of the generator and only update the relevant parameters of the discrimination module. Correspondingly, when updating the generator and classifier, we maintain the parameters of the discrimination module unchanged.

4   Experiments

4.1   Dataset

GSV Cars dataset introduced in [7] consists of 1095021 images of 2657 categories of cars from 4 sources: craigslist.com, cars.com, edmunds.com and Google Street View. We call the images from craigslist.com, cars.com, and edmunds.com as Web images and the images from Google Street View as GSV images. Cars in Web images are usually clear and have a simple background, while cars in GSV are blurry and have a complex background. All images from the Web images and GSV images have various poses and views. Following the [7], we perform our experiments on a subset consisting of 170 of the most common classes in the dataset, particularly those with at least 100 target images per class.

DA-Retail [24] is a dataset under the retail application which consists of 52,011 images of 263 fine-grained classes from 3 domains include SKU, Shelf, and Web. The images of the SKU are all high-quality pictures taken at a stable angle in a standard environment. The images of Shelf are hard to be classified with low resolution and complex backgrounds. And the Web is the biggest domain consisting of images crawled from the Internet with different resolutions and qualities. In our experiments, we use Web and SKU as the source domain and Shelf as the target domain.

Office dataset [19] is a widely used benchmark dataset, including 31 types of objects in the office (such as backpacks, computers, lamps, and scissors). For each of these objects, images can be obtained from three domains: Amazon (A), WebCam (W), and DSLR (D).

4.2   Performance on Fine-Grained Datasets

We evaluate our proposed framework using two fine-grained datasets GSV Cars and DA-Retail. Following the common setting in the UDA, we use the pre-trained Alexnet [12] and pre-trained ResNet [9] architecture for learning deep representations and implement our framework by PyTorch [17]. We train the model with stochastic gradient descent (SGD) and the learning rate is 0.0001.

We set the mini-batch size is 32, the momentum is 0.5, and the leak slope of LeakyReLU to 0.2. For comparison, we use seven domain adaptation methods DC [21], DANN [6], DCC [22], DeepCoral [20], CoGAN [14], iCAN [28], MADA [18], and two fine-grained domain adaptation methods Multi-Task[7], DASA [24].

We give a brief introduction of the comparison methods. DC uses joint CNN architecture for domain and task transfer. DCC proposes a new CNN architecture which introduces an adaptation layer and an additional domain confusion loss. DeepCoral aligns the second-order statistics of the source and target distributions with a linear transformation. DANN uses adversarial learning to eliminate the distribution gap between the source domain and the target domain. CoGAN enforces weight-sharing constraint that limits the network capacity and favors a joint distribution solution over a product of marginal distributions one. iCAN adds several domain classifiers on multiple CNN feature extraction blocks to learn domain informative representations. MADA captures multimode structures to enable fine-grained alignment of different data distributions by using multiple domain discriminators. Multi-Task uses an attribute based method adaptation loss to increase performance. DASA consists of two modules, DA is responsible for aligning the domain and SA is responsible for extracting features. Among them, DASA is a state-of-the-art of fine-grained domain adaptation method.

Table 1. Results on GSV cars and DA-Retail in UDA: "S", "R" and "W" refer the SKU, Shelf and Web domain in DA-Retail.

Method	GSV cars	Retail ($W \rightarrow R$)	Retail ($S \rightarrow R$)
AlexNet	41.37	9.72	31.06
DC (CaffeNet)	14.98	36.76	41.98
Multi-Task (CaffeNet)	19.05	38.13	46.84
DDC (AlexNet)	15.86	42.81	43.26
DeepCoral (AlexNet)	16.62	34.93	47.57
DASA (CaffeNet)	22.61	47.83	53.44
Ours (AlexNet)	**23.83**	**48.21**	**53.71**
CoGAN	19.19	39.64	45.72
iCAN (ResNet)	26.61	47.81	52.00
MADA (ResNet)	27.34	52.93	49.65
DASA (ResNet)	29.71	53.16	56.10
Ours (ResNet)	**30.62**	**53.42**	**56.27**

We follow standard evaluation protocols for UDA [15]. For all two datasets, we use all labeled source examples and all unlabeled target examples. We compared the average classification accuracy of each method in three random experiments, the final experimental results are shown in Table 1. Since our model is superior to other methods, we can observe performance improvements. For the GSV Cars dataset, it can be seen from Table 1 that our method improves the accuracy rate compared to DASA using both CaffeNet [10] and ResNet structure, which improves 1.22 and 0.91 respectively. For Retail dataset, We achieved better results than DASA in both $W \rightarrow R$ and $S \rightarrow R$ tasks, which improves 0.32 and 0.22 on average.

Analyzing the results, the validity of our framework is mainly due to the function of multiple discrimination modules. The joint label predictor can effectively align fine-grained categories, and the rectifier can normalize feature representations to eliminate the influence of views and poses. DASA also use the method of adversarial learning, and their performance is better than the other adversarial learning model. DASA uses multiple discriminators for domain alignment, which is equivalent to our role of joint label predictor. DASA focuses on specific regions of the image, and we focus on canonical view representations of the image. Compared with DASA, we made more improvements in the GVS Cars dataset than DASA, the main reason may be that the gap between the source domain and the target domain in the GSV Cars is larger. Our framework works well because it eliminates differences in poses, views, backgrounds to generate normalized feature representations. And the representations learned by our framework can align the conditional distribution between domains. In that, we strike a balance between distribution alignment and fine-grained variations elimination.

4.3   Performance on UDA Benchmark

Although our adversarial framework is set for fine-grained, we have also experimented with a generic dataset to show compatibility. We construct our experiments with unsupervised settings as we did in the fine-grained datasets and evaluate all methods across six transfer tasks $A \rightarrow W$, $D \rightarrow W$, $W \rightarrow D$, $A \rightarrow D$, $D \rightarrow A$ and $W \rightarrow A$. Similarly, we build our framework based on AlexNet and Resnet. We compared with DAN [15], DANN [6], RTN [16], MADA, Multi-Task and DASA, and the comparison results are shown in Table 2. As shown in the results, our framework not achieved the best results in simple tasks like $D \rightarrow W$ and $W \rightarrow D$ which the source and target domains are similar. But for $A \rightarrow D$, $D \rightarrow A$ and $W \rightarrow A$ tasks, we achieved better performance. It may be because the source and target domains of $A \rightarrow D$, $D \rightarrow A$ and $W \rightarrow A$ are quite different. Our rectifier can be seen as regularization of the generated representation, thereby reducing this gap. The results proving that our framework also works well on the generic dataset. Also, our framework outperforms the fine-grained domain adaption models Multi-Task and DASA.

Table 2. Results on the generic image dataset: "A", "W" and "D" refer the Amazon, WebCam and DSLR in Office dataset.

Method	$A \rightarrow W$	$D \rightarrow W$	$W \rightarrow D$	$A \rightarrow D$	$D \rightarrow A$	$W \rightarrow A$
AlexNet	60.6	95.4	99.0	64.2	45.5	48.3
DAN (AlexNet)	68.5	96.0	99.0	66.8	50.0	49.8
RTN (AlexNet)	73.3	96.8	99.6	71.0	50.5	51.0
MADA (AlexNet)	78.5	**99.8**	**100.0**	74.1	56.0	54.5
Multi-Task(CaffeNet)	62.4	–	–	–	–	–
DASA (CaffeNet)	64.5	–	–	–	–	–
Ours (AlexNet)	**80.7**	99.6	99.8	**78.3**	**57.1**	**58.6**
ResNet	68.4	96.7	99.3	68.9	62.5	60.7
DAN (ResNet)	83.8	96.8	99.5	78.4	66.7	62.7
DANN (ResNet)	82.0	96.9	99.1	79.7	68.2	67.4
RTN (ResNet)	84.5	96.8	99.4	77.5	66.2	64.8
MADA (ResNet)	90	**97.4**	**99.6**	87.8	70.3	66.4
DASA (ResNet)	85.0	–	–	–	–	–
Ours (ResNet)	**92.8**	97.2	99.1	**89.6**	**70.4**	**69.4**

5 Conclusion and Future Work

In this paper, we presented a novel framework for cross-domain fine-grained recognition, outperforming existing methods on three different datasets. Our framework minimized the discrepancy between different domains, making it more robust under different application views. In the future, it is promising to exploit multiple discriminative fine-grained parts in cross-domain scenarios to further boost the recognition performance.

References

1. Berg, T., Belhumeur, P.: Poof: part-based one-vs.-one features for fine-grained categorization, face verification, and attribute estimation. In: Proceedings of the IEEE Conference on Computer Vision and Pattern Recognition, pp. 955–962 (2013)
2. Branson, S., Van Horn, G., Belongie, S., Perona, P.: Bird species categorization using pose normalized deep convolutional nets. arXiv preprint arXiv:1406.2952 (2014)
3. Chai, Y., Lempitsky, V., Zisserman, A.: Bicos: A bi-level co-segmentation method for image classification. In: 2011 International Conference on Computer Vision, pp. 2579–2586. IEEE (2011)
4. Cui, Y., Song, Y., Sun, C., Howard, A., Belongie, S.: Large scale fine-grained categorization and domain-specific transfer learning. In: Proceedings of the IEEE Conference on Computer Vision and Pattern Recognition, pp. 4109–4118 (2018)

5. Cui, Y., Zhou, F., Lin, Y., Belongie, S.: Fine-grained categorization and dataset bootstrapping using deep metric learning with humans in the loop. In: Proceedings of the IEEE Conference on Computer Vision and Pattern Recognition, pp. 1153–1162 (2016)
6. Ganin, Y., Lempitsky, V.: Unsupervised domain adaptation by backpropagation. arXiv preprint arXiv:1409.7495 (2014)
7. Gebru, T., Hoffman, J., Fei-Fei, L.: Fine-grained recognition in the wild: a multi-task domain adaptation approach. In: Proceedings of the IEEE International Conference on Computer Vision, pp. 1349–1358 (2017)
8. Goodfellow, I., et al.: Generative adversarial nets. In: Advances in Neural Information Processing Systems, pp. 2672–2680 (2014)
9. He, K., Zhang, X., Ren, S., Sun, J.: Deep residual learning for image recognition. In: Proceedings of the IEEE Conference on Computer Vision and Pattern Recognition, pp. 770–778 (2016)
10. Jia, Y., et al.: Caffe: convolutional architecture for fast feature embedding, pp. 675–678 (2014)
11. Kang, G., Zheng, L., Yan, Y., Yang, Y.: Deep adversarial attention alignment for unsupervised domain adaptation: the benefit of target expectation maximization. In: Proceedings of the European Conference on Computer Vision (ECCV), pp. 401–416 (2018)
12. Krizhevsky, A., Sutskever, I., Hinton, G.E.: Imagenet classification with deep convolutional neural networks. In: Advances in Neural Information Processing Systems, pp. 1097–1105 (2012)
13. Li, S., Song, S., Huang, G., Ding, Z., Wu, C.: Domain invariant and class discriminative feature learning for visual domain adaptation. IEEE Trans. Image Process. **27**(9), 4260–4273 (2018)
14. Liu, M.Y., Tuzel, O.: Coupled generative adversarial networks. In: Advances in Neural Information Processing Systems, pp. 469–477 (2016)
15. Long, M., Cao, Y., Wang, J., Jordan, M.I.: Learning transferable features with deep adaptation networks. arXiv preprint arXiv:1502.02791 (2015)
16. Long, M., Zhu, H., Wang, J., Jordan, M.I.: Unsupervised domain adaptation with residual transfer networks. In: Advances in Neural Information Processing Systems, pp. 136–144 (2016)
17. Paszke, A., et al.: Pytorch: an imperative style, high-performance deep learning library. In: Advances in Neural Information Processing Systems, pp. 8024–8035 (2019)
18. Pei, Z., Cao, Z., Long, M., Wang, J.: Multi-adversarial domain adaptation. In: Thirty-Second AAAI Conference on Artificial Intelligence (2018)
19. Saenko, K., Kulis, B., Fritz, M., Darrell, T.: Adapting visual category models to new domains. In: Daniilidis, K., Maragos, P., Paragios, N. (eds.) ECCV 2010. LNCS, vol. 6314, pp. 213–226. Springer, Heidelberg (2010). https://doi.org/10.1007/978-3-642-15561-1_16
20. Sun, B., Saenko, K.: Deep coral: correlation alignment for deep domain adaptation, pp. 443–450 (2016)
21. Tzeng, E., Hoffman, J., Darrell, T., Saenko, K.: Simultaneous deep transfer across domains and tasks. In: Proceedings of the IEEE International Conference on Computer Vision, pp. 4068–4076 (2015)
22. Tzeng, E., Hoffman, J., Zhang, N., Saenko, K., Darrell, T.: Deep domain confusion: maximizing for domain invariance. Computer Vision and Pattern Recognition (2014)

23. Vedaldi, A., et al.: Understanding objects in detail with fine-grained attributes. In: Proceedings of the IEEE Conference on Computer Vision and Pattern Recognition, pp. 3622–3629 (2014)

24. Wang, Y., Song, R., Wei, X.S., Zhang, L.: An adversarial domain adaptation network for cross-domain fine-grained recognition. In: The IEEE Winter Conference on Applications of Computer Vision, pp. 1228–1236 (2020)

25. Xiao, T., Xu, Y., Yang, K., Zhang, J., Peng, Y., Zhang, Z.: The application of two-level attention models in deep convolutional neural network for fine-grained image classification. In: Proceedings of the IEEE Conference on Computer Vision and Pattern Recognition, pp. 842–850 (2015)

26. Yosinski, J., Clune, J., Bengio, Y., Lipson, H.: How transferable are features in deep neural networks? In: Advances in Neural Information Processing Systems, pp. 3320–3328 (2014)

27. Zhang, N., Donahue, J., Girshick, R., Darrell, T.: Part-based R-CNNs for fine-grained category detection. In: Fleet, D., Pajdla, T., Schiele, B., Tuytelaars, T. (eds.) ECCV 2014. LNCS, vol. 8689, pp. 834–849. Springer, Cham (2014). https://doi.org/10.1007/978-3-319-10590-1_54

28. Zhang, W., Ouyang, W., Li, W., Xu, D.: Collaborative and adversarial network for unsupervised domain adaptation, pp. 3801–3809 (2018)

29. Zheng, H., Fu, J., Mei, T., Luo, J.: Learning multi-attention convolutional neural network for fine-grained image recognition. In: Proceedings of the IEEE International Conference on Computer Vision, pp. 5209–5217 (2017)

Improved Performance of GANs via Integrating Gradient Penalty with Spectral Normalization

Hongwei Tan[1,2] , Linyong Zhou[2], Guodong Wang[1], and Zili Zhang[1,3(✉)]

[1] School of Computer and Information Science,
Southwest University, Chongqing 400715, China
`zhangzl@swu.edu.cn`
[2] School of Mathematics and Statistics,
GuiZhou University of Finance and Economics, Guiyang 550025, China
[3] School of Information Technology,
Deakin University, Locked Bag 20000, Geelong, VIC 3220, Australia

Abstract. Despite the growing prominence of generative adversarial networks (GANs), improving the performance of GANs is still a challenging problem. To this end, a combination method for training GANs is proposed by coupling spectral normalization with a zero-centered gradient penalty technique (the penalty is done on the inner function of Sigmoid function of discriminator). Particularly, the proposed method not only overcomes the limitations of networks convergence and training instability but also alleviates the mode collapse behavior in GANs. Experimentally, the improved method becomes more competitive compared with some of recent methods on several datasets.

Keywords: Generative Adversarial Networks · Gradient penalty · Spectral normalization · Training stability · Networks convergence

1 Introduction

Generative Adversarial Networks (GANs) [10] are powerful deep generative models which can be used to learn complex probability distributions. Especially in image research, GANs have been successfully applied to a variety of tasks, including image generation [21,30], image super-resolution [5,16], image-to-image translation [15], image in-painting [37], domain adaptation [35] and many more.

However, while very powerful, GANs are known to be notoriously hard to train. To improve the performance of GANs, the general strategies for stabilizing training are to carefully design the model, such as by crafting the network architectures [16,30,31], by modifying the objective functions [2,21], by

Supported by the National Natural Science Foundation of China: Managing Uncertainty in Service Based Software with Intelligent Adaptive Architecture (No. 61732019).

changing the gradient update modes [13,23,25] and by implementing the penalty or regularization techniques [11,24,34]. Despite practical advances, the performance of GANs still has plenty of place for improvement, especially in stability and convergence.

In this study, we integrate spectral normalization with a zero-centered gradient penalty technique to improve the performance of GANs, which the coalition can either demonstrably improve the stability and convergence of model or effectively alleviate the mode collapse behavior in GANs. Due to the fact that the update dynamic of discriminator network comes completely from the inner function of Sigmoid function, we find that the penalty is more effectively implemented on the inner function than directly done on the discriminator. Meanwhile, the Lipschitz constant of discriminator is leveraged to prevent the expansion of the model gradient. In addition, a training trick is introduced, clipping the gradient norm of network weights, which is conducive to further boosting the training stability. On the other hand, to achieve better convergence, spectral normalization (SN) [24] is added into the discriminator, along with batch normalization (BN) [14] in the generator. This amounts to implementing the penalty on SNGAN [24]. Unlike the original SNGAN, a modified GAN model is trained with spectral normalization. By doing so, our model captures the optimal networks convergence, particularly in the discriminator, it almost converges to a constant. In the experiments, the overall trend of the gradient variation is introduced to reflect the stability of GANs training. At the same time, the results reveal that our method leads to GANs training stability, good networks convergence and improving the quality of generated samples.

In summary, our contributions are as follows:

- By integrating the zero-centered gradient norm penalty on the inner function of Sigmoid function of discriminator with spectral normalization, a method is crafted to improve the performance of GANs.
- A modified SNGAN is introduced, which can demonstrably boost performance. As a training trick, we find that appropriately clipping the gradient norm of network weights assists in improving the stability of GANs training.
- We leverage the overall trend of the gradient variation to mirror the stability of GANs training, where the gradient variations are computed by the average gradient L_2 norm with a batch size samples in each generator update.

The rest of this paper is organized as follows: Sect. 2 introduces the background and related work. Section 3 provides some theoretical underpinnings for our method and proposes a gradient penalty method on a modified SNGAN model. Section 4 examines the performance of the proposed method via a series of experiments on synthetic and benchmark datasets. Finally, the conclusions are drawn in Sect. 5.

2 Background and Related Work

2.1 Backgroud

GANs [10] form a broad class of generative models in which a min-max two-player game is played between a generative network $G(z)$ and discriminative network $D(x)$ whose purpose, respectively, is to map random noise to samples and discriminate real and generated samples. Formally, the GAN objective [10] involves finding a Nash equilibrium to the following min-max problem:

$$\min_G \max_D \mathbb{E}_{x \sim p_{data}}[\log D(x)] + \mathbb{E}_{z \sim p_z}[\log (1 - D(G(z)))], \tag{1}$$

where p_{data} and p_z denote real data distribution (target distribution) and latent distribution (prior distribution such as $N(0, I)$ or $U[-1, 1]$), respectively. According to (1), the loss functions of GANs discriminator network and generator network are as follows:

$$L_D = -\mathbb{E}_{x \sim p_{data}}[\log D(x)] - \mathbb{E}_{z \sim p_z}[\log (1 - D(G(z)))], \tag{2}$$

$$L_{G_1} = \mathbb{E}_{z \sim p_z}[\log (1 - D(G(z)))] \quad (Saturating), \tag{3}$$

$$L_{G_2} = -\mathbb{E}_{z \sim p_z}[\log D(G(z))] \quad (Non - Saturating), \tag{4}$$

where L_{G_1} and L_{G_2} denote the saturating and non-saturating loss function, respectively. From a more pratical standpoint, L_{G_2} makes network training more stable than L_{G_1} [5,8,10,16,34].

2.2   Related Work

Several recent work have focused on addressing the instability and convergence to improve the performance of GANs, where the gradient penalty-based methods are one of the most effective methods. WGAN-GP [11] first used the gradient norm to penalize the criterion function of WGAN [2], which effectively alleviated the limitation of Lipschitz condition in WGAN and significantly improved the stability of WGAN. However, Mescheder et al. [22] proved that the zero-centered gradient penalty converges more easily than the 1-centered gradient penalty (WGAN-GP is a typical 1-centered gradient penalty). Along this line of research, WGAN-LP [29] (WGAN based on the zero-centered gradient penalty) and GAN-0GP [34] (GAN based on the zero-centered gradient penalty) were proposed, respectively. Our proposed method falls under the same category, hopefully provides some context for understanding some of these methods. Specifically, our penalty is done on the inner function of Sigmoid function of discriminator rather than directly penalized on the discriminator as above mentioned methods.

From the optimization perspective, several normalization techniques commonly applied to deep neural networks training have been applied to GANs, such as batch normalization (BN) [14,30], layer normalization(LN) [11] and spectral normalization(SN) [24]. Generally, the BN and LN are simultaneously operated on discriminator and generator, while the SN is only done on discriminator in GANs. In the study, normalization is executed on a modified model (Table 1) based on SNGAN [24], which its discriminator and generator are normalized with the SN and BN, respectively. More importantly, a combination model, the modified model in cooperation with the zero-centered gradient norm penalty (on the inner function), can make the networks more stable and better convergence.

3   Method

In this section, we will lay out the theoretical groundwork for our proposed method. In addition, a training trick (clipping the gradient norm of network weights) and a modified SNGAN will be introduced. Finally, our method will be formally proposed.

3.1   Controlling Gradient Variation

The instability of GAN training is mainly caused by its gradient update instability, while the gradient information of the generator is transmitted by the discriminator gradient [1]. Thus, controlling the gradient update of discriminator can effectively control the instability of GANs training. In image generation, let $X = Supp(\mathbb{P}_{data}) \cup Supp(\mathbb{P}_g)$, where $\mathbb{P}_{data}$ and $\mathbb{P}_g$ denote the real distribution and generative distribution, respectively. Suppose m, n denote the height and width of input image, respectively, and M is the maximal second order derivative of the loss function L_D (2) in X, and then $|\min(\nabla L_D) - \max(\nabla L_D)| \leq M \cdot L\sqrt{12mn}$, where L is the Lipschitz constant of discriminator network [27]. According to this, controlling the Lipschitz constant can effectively control the gradient variation amplitude of discriminator and make GANs training stable. Despite controlling the Lipschitz constant of network is not easy (in fact, even for the two-layer neural networks, the exact computation of the quantity is NP-hard problem [36]), the following theorem [12] provides a feasible scheme.

Theorem 1. *Let $F : \mathbb{R}^n \longrightarrow \mathbb{R}$ be a C^1-function. Then $F(x)$ is Lipschitz function with the Lipschitz constant L for all $x \in \mathbb{R}^n$ if and only if $\|\ \|\nabla F\|^2\ \|_\infty \leq L^2$.*

Theorem 1 (proof see [12] page 73, Lemma 4.3) provides an approach to determine that $F(x)$ is Lipschitz function with Lipschitz constant L, namely, when the square of maximum gradient norm of $F(x)$ is less than or equal to the square of a constant L, and then $F(x)$ is a Lipschitz function. In addition, the theorem also gives the fact: the Lipschitz constant can control the expansion of gradient. To this end, the loss function of discriminator network (2) is transformed into the following penalty form:

$$L'_D = L_D + \lambda \mathbb{E}_{x \sim p_{data}}[\max\{\|\nabla D(x)\|^2, L^2\}], \tag{5}$$

where λ and L are the penalty coefficient and the Lipschitz constant of the discriminator network. According to Theorem 1, this penalty form can effectively control the variation amplitude of discriminator gradient and make GANs gradient update stable.

3.2   Penalizing the Inner Function of Sigmoid

Let $D(x) = Sigmoid(f(x))$, where $Sigmoid(\cdot)$ is the last layer activation function of discriminator. Accordingly, the min-max optimization problem (1) is transformed into the following form:

$$\min_G \max_f \mathbb{E}_{\boldsymbol{x} \sim p_{data}}[\log Sigmoid(f(\boldsymbol{x}))] + \mathbb{E}_{\boldsymbol{z} \sim p_z}[\log (1 - Sigmoid(f(G(\boldsymbol{z}))))]. \quad (6)$$

The loss functions of discriminator and generator, L_D and L_{G_2}, are also replaced correspondingly, and their gradients can be easily proven that

$$\nabla_\phi L_D = -\mathbb{E}_{\boldsymbol{x} \sim p_{data}}[(1 - D(\boldsymbol{x}))\nabla_\phi f(\boldsymbol{x})] + \mathbb{E}_{\boldsymbol{x} \sim p_g}[D(\boldsymbol{x})\nabla_\phi f(\boldsymbol{x})], \quad (7)$$

$$\nabla_\theta L_G = -\mathbb{E}_{\boldsymbol{z} \sim p_z}[(1 - D(G(\boldsymbol{z})))f(G(\boldsymbol{z}))\nabla_{\boldsymbol{x}} f(G(\boldsymbol{z}))J_\theta G(\boldsymbol{z})], \quad (8)$$

where ϕ, θ, $J_\theta G(\boldsymbol{z})$ denote the discriminator network parameters, the generator network parameters and the Jacobi matrix of $G(\boldsymbol{z})$ with respect to θ, respectively. The gradients (7), (8) imply that the update dynamic for GANs training is completely provided by the gradient of the function $f(\boldsymbol{x})$, namely the gradient of the inner function of Sigmoid function. Due to the Lipschitz constant of Sigmoid function is always 0.25, thus controlling the Lipschitz constant of the function $f(\boldsymbol{x})$ is equivalent to controlling the Lipschitz constant of discriminator $D(\boldsymbol{x})$. In view of this, the penalty form of loss function L_D and the loss function L_{G_2} will be transformed into the following form:

$$L_D^* = L_D + \lambda \mathbb{E}_{\boldsymbol{x}^* \sim p^*}[\max\{\|\nabla f(\boldsymbol{x}^*)\|^2, L^2\}], \quad (9)$$

$$L_{G_2} = -\mathbb{E}_{\boldsymbol{x} \sim p_g}[\log (Sigmoid(f(\boldsymbol{x})))], \quad (10)$$

where $\boldsymbol{x}^* = t\boldsymbol{x} + (1 - t)\boldsymbol{y}, t \sim U(0,1), \boldsymbol{x} \in Supp(\mathbb{P}_{data}), \boldsymbol{y} \in Supp(\mathbb{P}_g)$, p^* is the mixed distribution of real distribution and generated distribution. The loss functions L_D^* (discriminator) and L_{G_2} (generator) are adopted throughout this paper, where L_D^* is the loss function of the zero-centered gradient norm penalty with regard to the inner function of Sigmoid function. It is worth emphasizing that the procedure of our algorithm is similar to the algorithm 1 of [10], except for the loss functions of discriminator and generator.

3.3   Exploring the Optimal Model

In this part, an optimal model will be explored, including model structure, clipping the gradient norm of network weights, optimization method. To do this, we expound some comparative cases via the experimental results on CIFAR10 [18], where the number of updates for the generator are 100k.

A Modified SNGAN. Essentially, SNGAN [24] imposes global regularization on the network, which the discriminator was normalized with the SN, along with the BN in generator. In contrast, SN has an advantage over BN and LN in GANs, such as more efficient gradient flow, more stable optimization [19,24]. However, we experimentally find that a modified SNGAN is more effective than original SNGAN in image generation. The differences between two models can be seen from Table 1. The two network structures have not been changed (see Appendix B), we just modified some hyper parameter settings and optimization method based on SNGAN. The modification is simple yet effective. The experimental *FID* [13] value of SNGAN on CIFAR10 is 28.66 (the original SNGAN is 20.70

with hinge loss [24], the BCE loss is used here), while the modified model is 24.53 (the smaller FID value is better). Also, the proposed penalty technique (9), (10) was implemented on the two models, the former FID is 20.12 and the latter is 12.48. The preliminary results show that the modified SNGAN is more effective than the original SNGAN.

Table 1. The hyper parameter setting and optimization method for two models.

Methods	Init	LR	n-dis	Bias	Optimizer
SNGAN	D-Norm G-Norm	0.1	5	T	Adam
Modified SNGAN	D-Orth G-Xavier	0.2	1	T-F-T	OAdam

"Init" is initialization method, which the discriminator and generator in SNGAN are both initialized by normal random number ($N(0,1)$), while the discriminator and generator in the modified SNGAN is initialized by orthogonal [32] and Xavier normal [9], respectively; "LR" is learning rate that uses to the activation function LeakyReLU of discriminator; "n-dis" denotes the numbers of update of the discriminator per one update of the generator; "Bias = T" denotes all biases are true, whereas "Bias = T-F-T" denotes all of the biases of layer are false except for the first layer and last layer.

Clipping the Gradient Norm of Network Weights. In order to further improve the performance of GANs, a training trick, clipping the gradient norms of network weights, is introduced into model training. The norms are computed over all gradients together, as if they were concatenated into a single vector. This process is computationally light and easy to incorporate into existing models, which the upper limit of the clipped norms is a controllable hyper parameter (max-norm) and the lower limit is zero. Note that the operation is performed in discriminator and generator, respectively. We test the different parameters max-norm on the proposed model with the modified SNGAN. Clearly, the parameter configuration E in Table 2 is the optimal combination. As shown in Table 2, it is conducive to appropriately clip the gradient norms of network weights for boosting the quality of the generated samples (FID). Note that the hyper parameter max-norm of discriminator and generator are α and β, respectively. In this study, the parameter combination E is used in all experiments.

Table 2. Comparison of the quality of the generated samples (FID, the smaller is better) based different hyper parameter settings on CIFAR10.

Settings	A	B	C	D	**E**
α	0.001	0.0001	0.1	1	**0.01**
β	0.02	0.0002	0.2	1	**1**
FID	14.93	16.34	15.33	13.25	**10.74**

The experimental results on CIFAR10 confirmed that the performance of GANs is significantly improved by our proposed zero-gradient norm penalty model on the modified SNGAN. In next section, we will further verify the stability and convergence of the proposed model on several datasets. It is worth emphasizing that we have tried different optimization algorithms (such as Lookahead [39]+Adam [17], Lookahead+OAdam [7]) and regularization methods on generator (such as orthogonality regularization [3], the modified orthogonality regularization [5] and group sparse regularization [33]), but none of them go beyond our method.

4   Experimental Results

In this section, the efficacy of our approach will be tested by investigating network convergence, training stability and FID value. Note that, in contrast with IS score [31], the FID value can more comprehensively measure the quality of the generated samples in GANs [4,13].

In order to verify the performance of our algorithm in the abovementioned three aspects, we conducted a set of extensive experiments of unsupervised image generation on CIFAR10 [18], SVHN (unlabeled subset) [26], STL10 [6], CelebA [20] and LSUN (bedroom) [38] datasets. Note that comparison of the networks convergence and training stability are arranged in 4.2, and the FID value in 4.3. In addition, the experiments on two synthetic datasets (8 Gaussians and 25 Gaussians) were performed to investigate the mode collapse behavior in GANs (in 4.1). We also compared our method with the representative ones. Unless otherwise noted, all of the results were obtained by PyTorch [28], where the numbers of update for GANs generator are 100k for all experiments. All codes can be found in https://github.com/thwgithub/GAN-Integration-GP-SN/.

4.1   Mixture of Gaussians

The mode collapse behavior in GANs can seriously affect the performance of GANs. To illustrate the effect of our method for alleviating the mode collapse phenomenon in GANs, a simple model was trained on two 2D mixture of Gaussians datasets (8 Gaussians arranged in a circle and 25 Gaussians in a square). Some technical details are relegated to Appendix A, including network architectures, hyper parameters and description of datasets. The experimental results are shown in Fig. 1. Compared with the original GAN, the least mode collapse behavior is demonstrated by our method, especially in 8 Gaussians dataset, 8 modes are almost learned in Fig. 1(b).

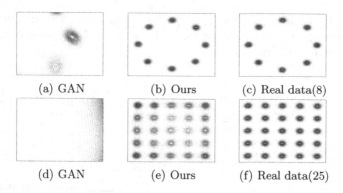

Fig. 1. Testing the mode collapse behavior on two Gaussians datasets ((a) GAN [10] on 8 Guassians; (b) our method on 8 Guassians; (c) real dataset; (d) GAN on 25 Guassians; (e) our method on 25 Guassians; (f) real dataset).

4.2   Results on Benchmark Datasets

In this subsection, we will report the network convergence and training stabiliy of the proposed method on five benchmark datasets(CIFAR10, SVHN, STL10, CelebA and LSUN(bedroom)), which all of the input images were both cropped to $3 \times 32 \times 32$. Due to space limitations, we only exhibit the results on CIFAR10 here and the other results can be found in the supplementary materials (it can be found in https://github.com/thwgithub/Sup-Materials-for-KSEM2020). We also compare against those of other congeneric gradient norm penalty algorithmss, including: WGAN-GP [11], WGAN-LP [29] and GAN-0GP [34].

For the hyper parameters setting, except for using the hyper parameters of the modified SNGAN in Table 1, we set the penalty coefficient λ to 10, as suggested in [11] and set the Lipschitz constant L at 0. The parameter max-norm of clipping weight gradient norm is set to 0.01 in discriminator and 1 in generator (the settings of Table 2). The learning rates of two networks are both 0.0002 with batchsize 64 and latent variable $z \sim N(0, I_{128})$. As for the architecture of the generator and discriminator, we use convolutional neural networks (CNN) that more details is described in Appendix B.

Networks Convergence. As for the network convergence, the results of the experiment on CIFAR10 are illustrated in Fig. 2. Clearly, our method is superior to the other three methods in either discriminator (Fig. 2(a)) or generator (Fig. 2(b)). In fact, we do not care about the convergent value of the loss function (GANs) or the critic function (WGANs), only focus on the amplitude of their oscillations. For the visualization, Fig. 2 is vertically shifted. As shown in Fig. 2, both WGAN-GP and WGAN-LP get stuck in a bad convergence, GAN-0GP is greater improvement for convergence than the former two. While the convergence of our approach (blue curve) significantly outperforms the convergence of the others. It is noted that the update rate of our method is 1:1, that is, the discriminator updates one time per one time update of the generator, whereas the update rate of the others is 1:5.

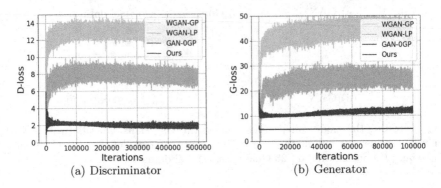

(a) Discriminator   (b) Generator

Fig. 2. Comparison of the convergence among four GANs methods ((a) the convergence of discriminator; (b) the convergence of generator).

Training Stability. In Fig. 3, the stability of GANs training in discriminator (Fig. 3(a)) and generator network (Fig. 3(b)) are exhibited. The overall trend of the gradient variation was used to mirror the stability of training, where the gradient variations were computed by the average gradient L_2 norm with a batch size samples in each generator update. To the best of our knowledge, the method is applied to measure the stability of GANs training for the first time. As observed, WGAN-GP and WGAN-LP both have a large gradient oscillation, this means that the two methods suffer from training instability. Moveover, their gradient variations are similar, it is probably because both algorithms belong to WGANs algorithms. For the GAN-0GP, the gradient behavior is relatively stable at the beginning of the training. However, with the increasing number of training, the method performs poorly. In contrast, our method (blue curve) is even more stable than the other methods with respect to the overall trend gradient variation. Also, we observe a phenomenon from Fig. 3 that the gradient variation trend of the discriminator is similar to the generator. This reveals that the gradient updates of generator and discriminator affect each other. Consequently, only implementing the penalty on the discriminator enable very stable GANs training. These results suggest that the stability of GANs training can be significantly improved by our method.

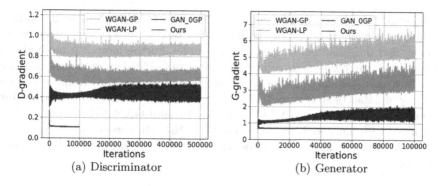

Fig. 3. Comparison of the overall trend of gradient variation among four GANs methods ((a) the gradient variation of discriminator; (b) the gradient variation of generator).

Table 3. Comparison of the generated samples quality (FID, the smaller is better).

Methods	CIFAR-10	SVHN	STL-10	CelebA	LSUN (bedroom)
LSGAN (2017)	22.20	3.84	20.17	5.10	**5.23**
SNGAN (2018)	20.70	4.53	18.11	5.56	12.05
WGAN-GP (2017)	21.89	4.09	18.19	5.01	14.61
WGAN-LP (2018)	21.01	3.62	17.40	5.12	15.21
GAN-0GP (2019)	18.91	6.10	14.49	4.53	7.14
Ours	**10.74**	**3.28**	**11.04**	**4.13**	6.59
Real datasets	0.46	0.24	0.84	0.34	0.55

4.3   Comparison of the Generated Samples Quality

The quality of the generated samples is one of the important indicators to reflect the performance of GANs model. The FID value is used to measure the quality, which the smaller FID value is better. In order to reduce the calculation error, the evaluation of FID is done on 50000 real samples and 50000 fake samples. To be more comprehensive, we compare our approach with five GANs models on five datasets and the results are summarized in Table 3. Clearly, our results (FID) are better than almost all other methods, only LSGAN (5.23) performs slightly better than our approach (5.59) on LSUN (bedroom). Especially, to our knowledge, the FIDs of our method on CIFAR10 and STL10 (10.74 and 11.04) are the state of the art in unsupervised image generation. This indicates that the quality of the generated samples in GANs can be significantly improved by our method. Note that the FIDs of real data are shown at the bottom of Table 3.

5   Conclusions

In this study, we integrated the zero-centered gradient penalty on the inner function of Sigmoid function of discriminator with spectral normalization (the modified SNGAN) to improve the performance of GANs, which is in contrast to the popular algorithms that the integrated method has better convergence, stability and the quality of the generated samples. Furthermore, our method can effectively alleviate the mode collapse behavior in GANs. In the experiments, we have illustrated evidence of improved training with several GANs algorithms on a variety of datasets and the resulting improvements in model performance. Our findings also show that WGAN-GP, WGAN-LP and GAN-0GP do not lead to networks convergence and training stability. In the future work, we would like to further dig into our ideas in more depth and come up with better performance methods.

A   Training Details on Synthetic Datasets

The 8 Gaussians dataset is sampled from a mixture of 8 Gaussians of standard deviation 0.02, this means are equally spaced around a circle of radius 2. 25 Gaussians dataset, like the 8 Gaussians, is sample from a mixture of 25 Gaussians, which is arranged in a square. Two datasets consist of 100 k samples. The discriminator contains three SNLinear layers (bias: True, False and True) with 128 hidden units and LReLU (0.2) activation, and the generator contains three Linear layers (bias: False, False and True) with 256 hidden units, BN and ReLU activation.

As for the hyper parameters setting, both networks are optimized using OAdam with a learning rate of 0.0002 and $\beta_1 = 0.5$, $\beta_2 = 0.9$ (training the original GAN use Adam). The latent variable $z \sim N(0, I_{128})$ and the penalty coefficient $\lambda = 10$ with Lipschitz constant $L = 0$. The batchsize is set to 100.

B   Networks Architecture on Benchmark Datasets

See Tables 4 and 5.

Table 4. Discriminator ($3 \times 32 \times 32$).

SNconv 64 3×3 S $= 1$ P $= 1$ LReLU
SNconv 64 4×4 S $= 2$ P $= 1$ LReLU
SNconv 128 3×3 S $= 1$ P $= 1$ LReLU
SNconv 128 4×4 S $= 2$ P $= 1$ LReLU
SNconv 256 3×3 S $= 1$ P $= 1$ LReLU
SNconv 256 4×4 S $= 2$ P $= 1$ LReLU
SNconv 512 4×4 S $= 1$ P $= 0$ SN
Sigmoid()

Table 5. Generator ($3 \times 32 \times 32$).

dense 512 $\times 4 \times 4$
deconv 512 4×4 S $= 2$ P $= 1$ BN ReLU
deconv 256 4×4 S $= 2$ P $= 1$ BN ReLU
deconv 128 4×4 S $= 2$ P $= 1$ BN ReLU
deconv 3 3×3 S $= 1$ P $= 1$ BN
Tanh()

References

1. Arjovsky, M., Bottou, L.: Towards principled methods for training generative adversarial networks. arxiv e-prints, art. arXiv preprint arXiv:1701.04862 (2017)
2. Arjovsky, M., Chintala, S., Bottou, L.: Wasserstein gan. arXiv preprint arXiv:1701.07875 (2017)
3. Bansal, N., Chen, X., Wang, Z.: Can we gain more from orthogonality regularizations in training deep CNNS? In: Proceedings of the 32nd International Conference on Neural Information Processing Systems, pp. 4266–4276. Curran Associates Inc. (2018)
4. Barratt, S., Sharma, R.: A note on the inception score. arXiv preprint arXiv:1801.01973 (2018)
5. Brock, A., Donahue, J., Simonyan, K.: Large scale gan training for high fidelity natural image synthesis. arXiv preprint arXiv:1809.11096 (2018)
6. Coates, A., Ng, A., Lee, H.: An analysis of single-layer networks in unsupervised feature learning. In: Proceedings of the Fourteenth International Conference on Artificial Intelligence and Statistics, pp. 215–223 (2011)
7. Daskalakis, C., Ilyas, A., Syrgkanis, V., Zeng, H.: Training gans with optimism. arXiv preprint arXiv:1711.00141 (2017)
8. Fedus, W., Rosca, M., Lakshminarayanan, B., Dai, A.M., Mohamed, S., Goodfellow, I.: Many paths to equilibrium: Gans do not need to decrease a divergence at every step. arXiv preprint arXiv:1710.08446 (2017)
9. Glorot, X., Bengio, Y.: Understanding the difficulty of training deep feedforward neural networks. In: Proceedings of the Thirteenth International Conference on Artificial Intelligence and Statistics, pp. 249–256 (2010)
10. Goodfellow, I., et al.: Generative adversarial nets. In: Advances in Neural Information Processing Systems, pp. 2672–2680 (2014)
11. Gulrajani, I., Ahmed, F., Arjovsky, M., Dumoulin, V., Courville, A.C.: Improved training of wasserstein gans. In: Advances in Neural Information Processing Systems, pp. 5767–5777 (2017)
12. van Handel, R.: Probability in high dimension. Technical report (2014)
13. Heusel, M., Ramsauer, H., Unterthiner, T., Nessler, B., Hochreiter, S.: Gans trained by a two time-scale update rule converge to a local nash equilibrium. In: Advances in Neural Information Processing Systems, pp. 6626–6637 (2017)
14. Ioffe, S., Szegedy, C.: Batch normalization: Accelerating deep network training by reducing internal covariate shift. arXiv preprint arXiv:1502.03167 (2015)
15. Isola, P., Zhu, J.Y., Zhou, T., Efros, A.A.: Image-to-image translation with conditional adversarial networks. In: Proceedings of the IEEE Conference on Computer Vision and Pattern Recognition, pp. 1125–1134 (2017)
16. Karras, T., Aila, T., Laine, S., Lehtinen, J.: Progressive growing of gans for improved quality, stability, and variation. arXiv preprint arXiv:1710.10196 (2017)
17. Kingma, D.P., Ba, J.: Adam: A method for stochastic optimization. arXiv preprint arXiv:1412.6980 (2014)
18. Krizhevsky, A., Hinton, G., et al.: Learning multiple layers of features from tiny images (2009)
19. Kurach, K., Lucic, M., Zhai, X., Michalski, M., Gelly, S.: A large-scale study on regularization and normalization in gans. arXiv preprint arXiv:1807.04720 (2018)
20. Liu, Z., Luo, P., Wang, X., Tang, X.: Deep learning face attributes in the wild. In: Proceedings of the IEEE International Conference on Computer Vision, pp. 3730–3738 (2015)

21. Mao, X., Li, Q., Xie, H., Lau, R.Y., Wang, Z., Paul Smolley, S.: Least squares generative adversarial networks. In: Proceedings of the IEEE International Conference on Computer Vision, pp. 2794–2802 (2017)
22. Mescheder, L., Geiger, A., Nowozin, S.: Which training methods for gans do actually converge? arXiv preprint arXiv:1801.04406 (2018)
23. Metz, L., Poole, B., Pfau, D., Sohl-Dickstein, J.: Unrolled generative adversarial networks. arXiv preprint arXiv:1611.02163 (2016)
24. Miyato, T., Kataoka, T., Koyama, M., Yoshida, Y.: Spectral normalization for generative adversarial networks. arXiv preprint arXiv:1802.05957 (2018)
25. Nagarajan, V., Kolter, J.Z.: Gradient descent gan optimization is locally stable. In: Advances in Neural Information Processing Systems, pp. 5585–5595 (2017)
26. Netzer, Y., Wang, T., Coates, A., Bissacco, A., Wu, B., Ng, A.Y.: Reading digits in natural images with unsupervised feature learning (2011)
27. Oberman, A.M., Calder, J.: Lipschitz regularized deep neural networks converge and generalize. arXiv preprint arXiv:1808.09540 (2018)
28. Paszke, A., et al.: Automatic differentiation in pytorch (2017)
29. Petzka, H., Fischer, A., Lukovnicov, D.: On the regularization of wasserstein gans. arXiv preprint arXiv:1709.08894 (2017)
30. Radford, A., Metz, L., Chintala, S.: Unsupervised representation learning with deep convolutional generative adversarial networks. arXiv preprint arXiv:1511.06434 (2015)
31. Salimans, T., Goodfellow, I., Zaremba, W., Cheung, V., Radford, A., Chen, X.: Improved techniques for training gans. In: Advances in Neural Information Processing Systems, pp. 2234–2242 (2016)
32. Saxe, A.M., McClelland, J.L., Ganguli, S.: Exact solutions to the nonlinear dynamics of learning in deep linear neural networks. arXiv preprint arXiv:1312.6120 (2013)
33. Scardapane, S., Comminiello, D., Hussain, A., Uncini, A.: Group sparse regularization for deep neural networks. Neurocomputing **241**, 81–89 (2017)
34. Thanh-Tung, H., Tran, T., Venkatesh, S.: Improving generalization and stability of generative adversarial networks. arXiv preprint arXiv:1902.03984 (2019)
35. Tzeng, E., Hoffman, J., Saenko, K., Darrell, T.: Adversarial discriminative domain adaptation. In: Proceedings of the IEEE Conference on Computer Vision and Pattern Recognition, pp. 7167–7176 (2017)
36. Virmaux, A., Scaman, K.: Lipschitz regularity of deep neural networks: analysis and efficient estimation. In: Advances in Neural Information Processing Systems, pp. 3835–3844 (2018)
37. Yeh, R., Chen, C., Lim, T.Y., Hasegawa-Johnson, M., Do, M.N.: Semantic image inpainting with perceptual and contextual losses. arXiv preprint arXiv:1607.07539 2(3) (2016)
38. Yu, F., Seff, A., Zhang, Y., Song, S., Funkhouser, T., Xiao, J.: Lsun: Construction of a large-scale image dataset using deep learning with humans in the loop. arXiv preprint arXiv:1506.03365 (2015)
39. Zhang, M., Lucas, J., Ba, J., Hinton, G.E.: Lookahead optimizer: k steps forward, 1 step back. In: Advances in Neural Information Processing Systems, pp. 9593–9604 (2019)

Evidential Deep Neural Networks
for Uncertain Data Classification

Bin Yuan[1], Xiaodong Yue[1,2(✉)], Ying Lv[1], and Thierry Denoeux[3,4]

[1] School of Computer Engineering and Science, Shanghai University, Shanghai, China
yswantfly@shu.edu.cn
[2] Shanghai Institute for Advanced Communication and Data Science,
Shanghai University, Shanghai, China
[3] Sino-European School of Technology, Shanghai University, Shanghai, China
[4] Université de technologie de Compiègne, Compiègne, France

Abstract. Uncertain data classification makes it possible to reduce the decision risk through abstaining from classifying uncertain cases. Incorporating this idea into the process of computer aided diagnosis can greatly reduce the risk of misdiagnosis. However, for deep neural networks, most existing models lack a strategy to handle uncertain data and thus suffer the costs of serious classification errors. To tackle this problem, we utilize Dempster-Shafer evidence theory to measure the uncertainty of the prediction output by deep neural networks and thereby propose an uncertain data classification method with evidential deep neural networks (EviNet-UC). The proposed method can effectively improve the recall rate of the risky class through involving the evidence adjustment in the learning objective. Experiments on medical images show that the proposed method is effective to identify uncertain data instances and reduce the decision risk.

Keywords: Uncertain data classification · Evidence theory · Deep neural networks

1 Introduction

In data classification tasks, the data instances that are uncertain to be classified form the main cause of prediction error [2,9,23]. Certain classification methods strictly assign a class label to each instance, which may produce farfetched classification results for uncertain instances. Uncertain classification methods aim to measure the uncertainty of data instances and accordingly reject uncertain cases [3,10,15]. The methodology of uncertain classification is helpful to reduce the decision risk and involve domain knowledge in classification process [21,22,24]. For instance, in decision support for cancer diagnosis, filtering out uncertain cases for further cautious identification, may allow us to avoid serious misdiagnosis [1].

Due to their very good performance, deep neural networks have been widely used in the classification of complex data [14,17], such as various kinds of medical

G. Li et al. (Eds.): KSEM 2020, LNAI 12275, pp. 427–437, 2020.
https://doi.org/10.1007/978-3-030-55393-7_38

images. However, most existing deep neural networks lack a strategy to handle uncertain data and may produce serious classification mistakes. For example, classifying CT images using convolutional neural networks without considering uncertainty may lead to overconfident decisions.

Trying to implement uncertain data classification based on deep neural networks, Geifman and El-Yaniv propose a selective classification method with deep neural networks, in which a selection function is constructed to quantify the reliability of predictions [8,11]. The method relies on the quality of the selection function. If the quantification of reliability is not accurate, the identification of uncertain data cannot be guaranteed. Dempster-Shafer (D-S) evidence theory [5] is also used to measure the uncertainty in machine learning models [6,7,20]. Sensoy, Kaplan and Kandemir formulate the uncertainty in deep neural networks from the view of evidence theory [18]. Moreover, evidential neural networks have been constructed and applied for the uncertain classification of medical images [13,19]. However, if the decision costs of different classes are imbalanced, evidential neural networks are not effective to classify the uncertain data instances of the risky class.

To address these problems, we construct a novel evidential deep neural network model and propose an uncertain data classification method. We formalize the uncertainty of the prediction output with evidence theory. A strategy to adjust the uncertainty in classification is also designed to improve the identification of certain and uncertain data instances in the risky class. The contributions of this paper are summarized as follows:

- Propose a novel evidential deep neural networks with the loss objective of both prediction error and evidence adjustment;
- An uncertain data classification method based on evidential deep neural networks (EviNet-UC) is proposed and applied to medical image diagnosis.

The rest of this paper is organized as follows. Section 2 presents the uncertain data classification method with evidential deep neural networks, which includes the model description and the strategy for uncertain data identification. In Sect. 3, we apply the proposed uncertain classification method to medical image data sets and show that the proposed method is effective to reduce the decision costs. Conclusions are given in Sect. 4.

2   Uncertain Data Classification with Evidential Deep Neural Networks

Given a dataset $\mathcal{D} = \{x_i, y_i\}_{i=1}^{N}$ of N labeled data instances where y_i is the class label of the instance x_i, the loss of data classification with the evidential deep neural networks consists of the *prediction error term* L_i^p and the *evidence adjustment term* L_i^e as

$$L = \frac{1}{N} \sum_{i=1}^{N} (L_i^p + \lambda * L_i^e), \tag{1}$$

where $\lambda = \min(1.0, t/10)$ is the annealing coefficient to balance the two terms, t is the index of the current training epoch. At the beginning of model training, $\lambda < 1$ makes the network focus on reducing the prediction error. When $t \geq 10$ the two terms play equal roles in the loss.

2.1 Prediction Error

For the binary classification of x_i, we define the model output e_i^+, e_i^- as the evidence collected by the deep neural network for the positive and negative classes. The sum of the total evidence is $E = e_i^+ + e_i^- + 2$. According to the evidence, we define the belief values of x_i belonging to positive and negative classes as $b_i^+ = e_i^+/E, b_i^- = e_i^-/E$, the uncertainty of classification is defined as $u_i = 1 - b_i^+ - b_i^-$. Similar to the model proposed in [13], we adopt Beta distribution to formulate the distribution of the prediction with the evidences e_i^+, e_i^-. Suppose p_i is the prediction of the instance x_i belonging to the positive class, the probability density function of the prediction is

$$f\left(p_i; \alpha_i, \beta_i\right) = \frac{\Gamma\left(\alpha_i + \beta_i\right)}{\Gamma\left(\alpha_i\right)\Gamma\left(\beta_i\right)} p_i^{\alpha_i - 1}\left(1 - p_i\right)^{\beta_i - 1} dp_i, \tag{2}$$

where the parameters of Beta distribution are $\alpha_i = e_i^+ + 1, \beta_i = e_i^- + 1$ and $\Gamma(\cdot)$ is the gamma function. The prediction of the positive class can be obtained by $p_i = \alpha_i/E$ and $1 - p_i = \beta_i/E$ denotes the prediction of negative class.

Based on the probability density of the prediction, we construct the prediction error term for each data instance x_i as the following expectation of squared error,

$$L_i^p = \int \|p_i - y_i\|^2 f\left(p_i; \alpha_i, \beta_i\right) dp_i \tag{3}$$

$$= \int \|p_i - y_i\|^2 \frac{\Gamma\left(\alpha_i + \beta_i\right)}{\Gamma\left(\alpha_i\right)\Gamma\left(\beta_i\right)} p_i^{\alpha_i - 1}\left(1 - p_i\right)^{\beta_i - 1} dp_i. \tag{4}$$

Referring to the properties of the expectation and variance of Beta distribution, the formula (4) can be derived as

$$L_i^p = \int \|p_i - y_i\|^2 \frac{\Gamma\left(\alpha_i + \beta_i\right)}{\Gamma\left(\alpha_i\right)\Gamma\left(\beta_i\right)} p_i^{\alpha_i - 1}\left(1 - p_i\right)^{\beta_i - 1} dp_i \tag{5}$$

$$= E\left(\|p_i - y_i\|^2\right) \tag{6}$$

$$= E\left(p_i\right)^2 - 2y_i E\left(p_i\right) + y_i^2 + \mathrm{var}\left(p_i\right) \tag{7}$$

$$= \left(E\left(p_i\right) - y_i\right)^2 + \mathrm{var}\left(p_i\right) \tag{8}$$

$$= \left(y_i - \frac{\alpha_i}{\alpha_i + \beta_i}\right)^2 + \left(1 - y_i - \frac{\beta_i}{\alpha_i + \beta_i}\right)^2 + \frac{\alpha_i \beta_i}{\left(\alpha_i + \beta_i\right)^2\left(\alpha_i + \beta_i + 1\right)}. \tag{9}$$

2.2   Evidence Adjustment

Besides the prediction error, the uncertain cases in the classification should also be considered in real application scenarios. Identifying uncertain data instances for abstaining from classification is helpful to reduce the decision risk. In [13], a regularization term is integrated into the objective of neural network to reduce the evidences of uncertain instances. But this strategy ignores the difference of the risks of uncertain instances from different classes. To find out the uncertain instances of risky class effectively, we expect to rescale the data uncertainty u through adjusting the evidence and add an evidence adjustment term into the loss objective. The evidence adjustment term is constructed by the Kullback-Leibler divergence between the distributions of prediction with original and adjusted evidences. We also adopt the Beta distribution for the prediction with adjusted evidences and define $\lambda > 1$ as the *evidence adjustment factor*. The evidence adjustment term is expressed as

$$L_i^e = KL \left(f\left(p_i; \tilde{\alpha}_i, \tilde{\beta}_i\right) | f\left(p_i; 1, \tilde{\lambda}\right)\right), \tag{10}$$

where $(1, \tilde{\lambda}) = (1, y_i\lambda + (1 - y_i))$, $\left(\tilde{\alpha}_i, \tilde{\beta}_i\right) = ((1 - y_i)\alpha_i + y_i, y_i\beta_i + (1 - y_i))$ are the parameters of the Beta distributions of the prediction p_i with adjusted and original evidences.

Let '1' denote the positive class and '0' denote the negative class. When the instance x_i belongs to positive class, $y_i = 1, (1, \tilde{\lambda}) = (1, \lambda)$ and $\left(\tilde{\alpha}_i, \tilde{\beta}_i\right) = (1, \beta_i)$. If x_i belongs to negative class, $y_i = 0, (1, \tilde{\lambda}) = (1, 1)$ and $\left(\tilde{\alpha}_i, \tilde{\beta}_i\right) = (\alpha_i, 1)$. For a negative-class instance, the adjustment term guides the parameter α_i to 1 and thereby reduce the evidence of positive class to 0. For a positive-class instance, the adjustment term guides the parameter β_i to λ. This will force the neural networks to promote the positive-class evidence for certain positive instances to reduce the prediction error.

According to the definition of KL divergence, the evidence adjustment term can be further simplified as

$$L_i^e = \int f\left(p_i; \tilde{\alpha}_i, \tilde{\beta}_i\right) \log \frac{f\left(p_i; \tilde{\alpha}_i, \tilde{\beta}_i\right)}{f\left(p_i; 1, \tilde{\lambda}\right)} dp_i \tag{11}$$

$$= \int f\left(p_i; \tilde{\alpha}_i, \tilde{\beta}_i\right) \log f\left(p_i; \tilde{\alpha}_i, \tilde{\beta}_i\right) dp_i - \int f\left(p_i; \tilde{\alpha}_i, \tilde{\beta}_i\right) \log f\left(p_i; 1, \tilde{\lambda}\right) dp_i \tag{12}$$

$$= E\left(\log f\left(p_i; \tilde{\alpha}_i, \tilde{\beta}_i\right)\right) - E_{B(\tilde{\alpha}_i\tilde{\beta}_i)}\left(\log f\left(p_i; 1, \tilde{\lambda}\right)\right). \tag{13}$$

Referring to the properties of Beta distribution, the expectations in (13) can be further derived for computation as

$$E\left(\log f\left(p_i; \tilde{\alpha}_i, \tilde{\beta}_i\right)\right) \tag{14}$$

$$=E\left(\log \frac{\Gamma\left(\tilde{\alpha}_i + \tilde{\beta}_i\right)}{\Gamma(\tilde{\alpha}_i)\Gamma\left(\tilde{\beta}_i\right)} p_i^{\tilde{\alpha}_i - 1}(1 - p_i)^{\tilde{\beta}_i - 1}\right) \tag{15}$$

$$=E\left(\log \frac{\Gamma\left(\tilde{\alpha}_i + \tilde{\beta}_i\right)}{\Gamma(\tilde{\alpha}_i)\Gamma\left(\tilde{\beta}_i\right)} + (\tilde{\alpha}_i - 1)\log p_i + \left(\tilde{\beta}_i - 1\right)\log(1 - p_i)\right) \tag{16}$$

$$=\log \frac{\Gamma\left(\tilde{\alpha}_i + \tilde{\beta}_i\right)}{\Gamma(\tilde{\alpha}_i)\Gamma\left(\tilde{\beta}_i\right)} + (\tilde{\alpha}_i - 1)E(\log p_i) + \left(\tilde{\beta}_i - 1\right)E(\log(1 - p_i)) \tag{17}$$

$$=\log \frac{\Gamma\left(\tilde{\alpha}_i + \tilde{\beta}_i\right)}{\Gamma(\tilde{\alpha}_i)\Gamma\left(\tilde{\beta}_i\right)} + \left(2 - \tilde{\alpha}_i - \tilde{\beta}_i\right)\psi\left(\tilde{\alpha}_i + \tilde{\beta}_i\right) + (\tilde{\alpha}_i - 1)\psi(\tilde{\alpha}_i) + \left(\tilde{\beta}_i - 1\right)\psi\left(\tilde{\beta}_i\right), \tag{18}$$

and

$$E_{B\left(\tilde{\alpha}_i, \tilde{\beta}_i\right)}\left(\log f\left(p_i; 1, \tilde{\lambda}\right)\right) \tag{19}$$

$$=E_{B\left(\tilde{\alpha}_i, \tilde{\beta}_i\right)}\left(\log \frac{\Gamma(1 + \tilde{\lambda})}{\Gamma(1)\Gamma(\tilde{\lambda})} p_i^{1 - 1}(1 - p_i)^{\tilde{\lambda} - 1}\right) \tag{20}$$

$$=E_{B\left(\tilde{\alpha}_i, \tilde{\beta}_i\right)}\left(\log \frac{\Gamma(1 + \tilde{\lambda})}{\Gamma(1)\Gamma(\tilde{\lambda})} + (\tilde{\lambda} - 1)\log(1 - p_i)\right) \tag{21}$$

$$=\log \frac{\Gamma(1 + \tilde{\lambda})}{\Gamma(1)\Gamma(\tilde{\lambda})} + (\tilde{\lambda} - 1)E_{B\left(\tilde{\alpha}_i, \tilde{\beta}_i\right)}(\log(1 - p_i)) \tag{22}$$

$$=\log \frac{\Gamma(1 + \tilde{\lambda})}{\Gamma(1)\Gamma(\tilde{\lambda})} + (\tilde{\lambda} - 1)\left(\psi\left(\tilde{\beta}_i\right) - \psi\left(\tilde{\alpha}_i + \tilde{\beta}_i\right)\right), \tag{23}$$

in which $\psi(\cdot)$ denotes the digamma function.

2.3   Classification of Uncertain Data

As explained above, based on the belief values of x_i belonging to positive and negative classes $b_i^+ = e_i^+/E$, $b_i^- = e_i^-/E$, we can measure the classification uncertainty of x_i by $u_i = 1 - b_i^+ - b_i^-$. With this uncertainty measure, applying the evidential neural networks to classify data, we can not only assign class labels to instances but also identify the uncertain ones. Through sorting the classified instances according to their uncertainty in ascending order, we select the top k uncertain instances for classification rejection to reduce the prediction risk.

Applying the proposed evidential neural network to the Breast IDC dataset (see the Sect. 3), Fig. 1(a) shows the evidence distribution of all the instances of positive class for multiple values of λ. We can see that the factor λ adjusts the

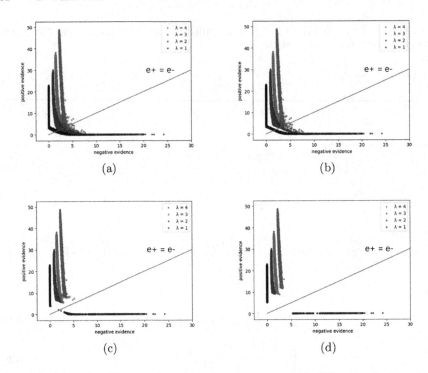

Fig. 1. Evidence distribution of positive-class instances with different rejection rates. (a) rejection rate = 0%, (b) rejection rate = 10%, (c) rejection rate = 20%, (d) rejection rate = 30%.

evidences of instances and the evidences of certain positive instances are promoted. The data instances with low-level evidences for both classes have high uncertainty in classification. Thus the instances located in the bottom-left corner indicate uncertain cases. Figure 1(b–d) display the evidence distribution of data instances after filtering out 10%, 20%, 30% uncertain instances, respectively. Based on the uncertain data identification strategy, we implement the uncertain data classification method with an evidential neural network (EviNet-UC). The effectiveness of the proposed method will be demonstrated in the following section.

3   Experimental Results

To show that the uncertain classification method with evidential neural network is effective to reduce decision costs, we tested the proposed method on the medical datasets Breast IDC [4] and Chest Xray [16]. The Breast IDC dataset consists of the pathological images of patients with infiltrating ductal carcinoma of the breast. The training set has 155314 images and the test set has 36904 images. We set the cancer case as positive class and the normal case as negative class. The Chest Xray dataset has 2838 chest radiographs, in which 427 images are chosen as the test data and the rest are set as the training data. The pneumonia and normal cases are set as positive class and negative class, respectively. For the algorithm implementation, we constructed the evidential deep neural networks based on the resnet18 architecture [14] and we modified the activation function of the output layer to the ReLU function.

To achieve the overall evaluation of the classification methods, we adopt the measures of *accuracy, F1 score, precision, recall rate* and *decision cost*. Suppose the number of the instances of negative class is N and the number of positive-class instances is P, TP and FP denote the numbers of true positive and false positive instances, TN and FN denote the true negative and false negative instances respectively. The measures are defined as

$$accuracy = (TP + TN)/(P + N),$$
$$F1\,score = (2*TP)/(2*TP + FN + FP),$$
$$precision = TP/(TP + FP),$$
$$recall = TP/(TP + FN).$$

Assuming correct classification to have zero cost, $\mathrm{cost}_{NP}, \mathrm{cos}\,t_{PN}$ denote the costs of false-positive classification and false-negative classification, respectively. The average decision cost of classification can be calculated as

$$decision\ cost = \mathrm{cost}_{NP} \cdot \frac{FP}{P+N} + \mathrm{cos}\,t_{PN} \cdot \frac{FN}{P+N}.$$

Based on the measures above, we carried out two experiments to evaluate the performance of the proposed uncertain classification method with evidential neural network (EviNet-UC). The first experiment aims to verify the superiority of the classification of the proposed method. Specifically, we compared the EviNet-UC method with other four uncertain classification methods based on deep neural networks: EvidentialNet [13], SelectiveNet [12], Resnet-pd and Resnet-md [19]. For fair comparison, we implemented all the methods above based on the resnet18 architecture.

We set *rejection rate* = 0 (no rejection), $\mathrm{cost}_{PN} = 5, \mathrm{cos}\,t_{NP=1}$ and applied all the classification methods to the Breast IDC dataset. The comparative experimental results are presented in Fig. 2 and Table 1. We can find that the proposed EviNet-UC method achieves the highest recall rate and the lowest decision cost among all the comparative methods. This means that the proposed method is effective to reduce the misclassification of the risky class (cancer case). Moreover, we changed the rejection rate from 0 to 0.5 to further compare the classification methods. Figure 3 presents the recall rates and the decision costs of different methods with different rejection rates. We can see that the EviNet-UC method achieves the best performance for all rejection rates. Compared to other methods, the proposed method is more effective to reduce the classification risk. The second experiment aims to show that the proposed method is effective to identify uncertain data. Varying the rejection rate in [0, 1] and applying EviNet-UC on the Chest Xray dataset, we obtained the classification results for different numbers of rejected uncertain radiographs. Figure 4 illustrates the evaluation of

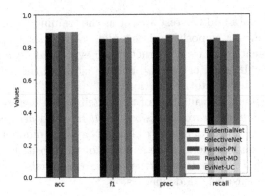

Fig. 2. Comparative experimental results on Breast IDC dataset.

Table 1. Comparative experimental results on Breast IDC dataset.

Methods	Accuracy	F1-score	Precision	Recall	Decision cost
EvidentialNet	0.8874	0.8477	0.8566	0.8389	0.3532
SelectiveNet	0.8874	0.8493	0.8488	0.8498	0.3370
ResNet-PN	0.8912	0.8512	0.8702	0.8330	0.3583
ResNet-MD	0.8911	0.8511	0.8699	0.8330	0.3584
EviNet-UC	**0.8915**	**0.8572**	**0.8432**	**0.8716**	**0.3004**

the classification based on EviNet-UC with varying rejection rates. It can be seen that *accuracy, precision, recall rate* and *F1 score* increase as the *rejection rate* increases. This indicates that the rejected data instances have uncertainty for classification and the EviNet-UC method can improve the classification results through filtering out the uncertain instances.

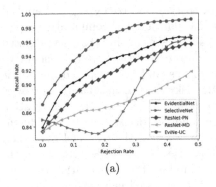
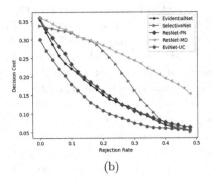

(a)   (b)

Fig. 3. (a) Recall rates of different classification methods with varying rejection rates, (b) decision costs with rejection rates.

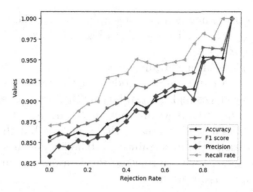

Fig. 4. Classification evaluation of EviNet-UC with varying rejection rates.

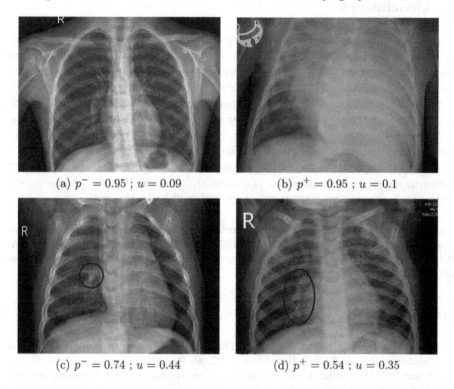

(a) $p^- = 0.95$; $u = 0.09$ (b) $p^+ = 0.95$; $u = 0.1$

(c) $p^- = 0.74$; $u = 0.44$ (d) $p^+ = 0.54$; $u = 0.35$

Fig. 5. (a) certain negative-class instance (normal case), (b) certain positive-class instance (pneumonia), (c) uncertain normal case, (d) uncertain pneumonia case.

When *rejection rate* $= 10\%$, Fig. 5 presents the certain and uncertain instances identified by EviNet-UC. Figure 5 (a) shows a certain negative-class instance of normal radiograph, in which the lung area is very clear. EviNet-UC produces high negative probability $p^- = 0.95$ and low uncertainty $u = 0.09$ to indicate the confident classification. In contrast, Fig. 5 (b) shows a certain positive-class instance of pneumonia radiograph, in which there exist heavy shadows. Correspondingly, EviNet-UC produces high positive probability $p^+ = 0.95$ and low uncertainty $u = 0.1$.

Figure 5 (c) displays an uncertain normal case. In general, the lung area is clear but there exists a dense area of nodule in the right part (marked in red circle). EviNet-UC produces high uncertainty $u = 0.44$ to indicate the judgement is not confident. Figure 5 (d) shows another uncertain case of pneumonia. In the radiograph, there exists a shadow area in the right lung but the symptom is not prominent, which leads to the uncertainty $u = 0.35$ for pneumonia identification. The uncertain radiographs will be rejected for cautious examination to further reduce the cost of misclassification.

4   Conclusions

Certain classification methods with deep neural networks strictly assign a class label to each data instance, which may produce overconfident classification results for uncertain cases. In this paper, we propose an uncertain classification method with evidential neural networks which measures the uncertainty of the data instances with evidence theory. Experiments on medical images validate the effectiveness of the proposed method for uncertain data identification and decision cost reduction. Our method currently focuses on only binary classification problem and the relationship between the decision cost and the evidence adjustment factor also requires theoretical analysis. Exploring the evidence adjustment factor in multi-class classification problems and constructing the precise uncertainty measurement for reducing decision risk will be future works.

Acknowledgment. This work was supported by National Natural Science Foundation of China (Nos. 61976134, 61573235, 61991410, 61991415) and Open Project Foundation of Intelligent Information Processing Key Laboratory of Shanxi Province (No. CICIP2018001).

References

1. Chen, Y., Yue, X., Fujita, H., Fu, S.: Three-way decision support for diagnosis on focal liver lesions. Knowl.-Based Syst. **127**, 85–99 (2017)
2. Chow, C.: On optimum recognition error and reject tradeoff. IEEE Trans. Inf. Theory **16**(1), 41–46 (1970)
3. Cortes, C., DeSalvo, G., Mohri, M.: Boosting with abstention. In: Advances in Neural Information Processing Systems, pp. 1660–1668 (2016)
4. Cruz-Roa, A., et al.: Automatic detection of invasive ductal carcinoma in whole slide images with convolutional neural networks. In: Medical Imaging 2014: Digital Pathology, vol. 9041, p. 904103. International Society for Optics and Photonics (2014)
5. Dempster, A.P.: Upper and lower probabilities induced by a multivalued mapping. In: Classic Works of the Dempster-Shafer Theory of Belief Functions, pp. 57–72. Springer (2008). https://doi.org/10.1007/978-3-540-44792-4_3

6. Denoeux, T.: Maximum likelihood estimation from uncertain data in the belief function framework. IEEE Trans. Knowl. Data Eng. **25**(1), 119–130 (2011)
7. Denoeux, T.: Logistic regression, neural networks and Dempster-Shafer theory: a new perspective. Knowl.-Based Syst. **176**, 54–67 (2019)
8. El-Yaniv, R., Wiener, Y.: On the foundations of noise-free selective classification. J. Mach. Learn. Res. **11**(May), 1605–1641 (2010)
9. Fumera, G., Roli, F.: Support vector machines with embedded reject option. In: Lee, S.-W., Verri, A. (eds.) SVM 2002. LNCS, vol. 2388, pp. 68–82. Springer, Heidelberg (2002). https://doi.org/10.1007/3-540-45665-1_6
10. Fumera, G., Roli, F., Giacinto, G.: Reject option with multiple thresholds. Pattern Recogn. **33**(12), 2099–2101 (2000)
11. Geifman, Y., El-Yaniv, R.: Selective classification for deep neural networks. In: Advances in Neural Information Processing Systems, pp. 4878–4887 (2017)
12. Geifman, Y., El-Yaniv, R.: Selectivenet: a deep neural network with an integrated reject option. arXiv preprint arXiv:1901.09192 (2019)
13. Ghesu, F.C., et al.: Quantifying and leveraging classification uncertainty for chest radiograph assessment. In: Shen, D., et al. (eds.) MICCAI 2019. LNCS, vol. 11769, pp. 676–684. Springer, Cham (2019). https://doi.org/10.1007/978-3-030-32226-7_75
14. He, K., Zhang, X., Ren, S., Sun, J.: Deep residual learning for image recognition. In: Proceedings of the IEEE Conference on Computer Vision and Pattern Recognition, pp. 770–778 (2016)
15. Hellman, M.E.: The nearest neighbor classification rule with a reject option. IEEE Trans. Syst. Sci. Cybern. **6**(3), 179–185 (1970)
16. Kermany, D.S., Goldbaum, M., Cai, W., Valentim, C.C., Liang, H., Baxter, S.L., McKeown, A., Yang, G., Wu, X., Yan, F., et al.: Identifying medical diagnoses and treatable diseases by image-based deep learning. Cell **172**(5), 1122–1131 (2018)
17. LeCun, Y., Bengio, Y., Hinton, G.: Deep learning. Nature **521**(7553), 436–444 (2015)
18. Sensoy, M., Kaplan, L., Kandemir, M.: Evidential deep learning to quantify classification uncertainty. In: Advances in Neural Information Processing Systems, pp. 3179–3189 (2018)
19. Tardy, M., Scheffer, B., Mateus, D.: Uncertainty measurements for the reliable classification of mammograms. In: Shen, D., et al. (eds.) MICCAI 2019. LNCS, vol. 11769, pp. 495–503. Springer, Cham (2019). https://doi.org/10.1007/978-3-030-32226-7_55
20. Tong, Z., Xu, P., Denœux, T.: ConvNet and Dempster-Shafer theory for object recognition. In: Ben Amor, N., Quost, B., Theobald, M. (eds.) SUM 2019. LNCS (LNAI), vol. 11940, pp. 368–381. Springer, Cham (2019). https://doi.org/10.1007/978-3-030-35514-2_27
21. Yao, Y.: Three-way decisions with probabilistic rough sets. Inf. Sci. **180**(3), 341–353 (2010)
22. Yue, X., Chen, Y., Miao, D., Fujita, H.: Fuzzy neighborhood covering for three-way classification. Inf. Sci. **507**, 795–808 (2020)
23. Yue, X., Chen, Y., Miao, D., Qian, J.: Tri-partition neighborhood covering reduction for robust classification. Int. J. Approximate Reasoning **83**, 371–384 (2017)
24. Yue, X., Zhou, J., Yao, Y., Miao, D.: Shadowed neighborhoods based on fuzzy rough transformation for three-way classification. IEEE Trans. Fuzzy Syst. **28**(5), 978–991 (2020)

GDCRN: Global Diffusion Convolutional Residual Network for Traffic Flow Prediction

Liujuan Chen[1] , Kai Han[1(✉)], Qiao Yin[2], and Zongmai Cao[1]

[1] School of Computer Science and Technology/Suzhou Institute for Advanced Study,
University of Science and Technology of China, Hefei, China
hankai@ustc.edu.cn
[2] School of Software Engineering, University of Science and Technology of China,
Hefei, China

Abstract. Traffic flow prediction is a crucial issue for intelligent transportation system. Because of complicated topological structures of road networks and dynamic spatial-temporal patterns of traffic conditions, predicting flows on the road networks is still a challenging task. Most existing approaches focus on the local spatial-temporal correlations, ignoring the global spatial dependences and the global dynamic spatial-temporal correlations. In this paper, we propose a novel deep learning model for traffic flow prediction, called Global Diffusion Convolution Residual Network (GDCRN), which consists of multiple periodic branches with the same structure. Each branch applies global graph convolution layer to capture both local and global spatial dependencies, and further apply GRes to describe global spatial-temporal correlations simultaneously. Extensive experiments on two real-world datasets demonstrate that our model can capture both the global and local spatial-temporal dependencies dynamically. The experimental results show the effectiveness of our method.

Keywords: Traffic prediction · Spatial-temporal network · Graph convolution network

1 Introduction

Intelligent transportation system (ITS) plays an important role in improving efficiency of traffic management and ensuring traffic safety. Predicting traffic conditions is one of the most important tasks in ITS. It can guide traffic management and help drivers avoid congested roads, such that traffic jams can be avoided or alleviated. The traffic prediction on road networks is a typical spatial-temporal data prediction problem, which aims at predicting future traffic flows by making use of historical traffic data and road networks. The complexity of the traffic prediction problem is mainly affected by the following three factors:

K. Han is the corresponding author.

© Springer Nature Switzerland AG 2020
G. Li et al. (Eds.): KSEM 2020, LNAI 12275, pp. 438–449, 2020.
https://doi.org/10.1007/978-3-030-55393-7_39

1. Traffic conditions are different at various times (i.e. morning peak, noon) and also vary greatly from place to place, which shows strong dynamic characteristics in both spatial and temporal dimensions.
2. In the spatial dimension, complicated correlations are observed among different regions. Congestion can affect the traffic conditions of reachable neighbors with different impact. On the other hand, the traffic conditions of places with far distance may also affect each other. In summary, both long-range and short-range spatial relations between places are important in traffic flow prediction.
3. Temporal dependencies follow complicated periodic patterns. Due to people's regular daily life patterns traffic conditions may show some repeated patterns, such as peak flow in mornings. Moreover, just past traffic conditions inevitably have an influence on future traffic flows. Therefore, the temporal patterns of traffic data are not purely periodic.

In recent years, thanks to the development of sensor networks, ITS systems can obtain massive amounts of real traffic network data, which facilitates traffic prediction. Although deep learning methods have brought breakthroughs in traffic prediction, they still have some limitations. 1) The existing studies [11,14,17] assume that spatial dependency relationships only exist among directly connected or very close nodes. 2) RNN-based methods [15,17] are ineffective to learn long-term periodic dependencies, since they have problems such as gradient explosion/disappearances when capturing long sequences. And it is time consuming to train typical chain structured RNN. 3) Current studies [4] do not capture the global spatial-temporal dependencies in the same time. They also reduce the bidirectional traffic network to undirected graphs, which makes these type of methods less effective in practice.

In this paper, we propose a global diffusion convolution residual network (GDCRN) to predict traffic flows, which addresses the three shortcomings we have mentioned above. It contains multiple branches with the same structure for capturing information of different time periods, such as hourly period, daily period and weekly period. We propose global graph convolution (GGC) layer, which integrates a novel graph diffusion convolution unit based on three auxiliary matrices. It contains two local adjacency matrices to capture local spatial correlations of the bidirectional traffic network and a global matrix to capture global spatial dependencies. We further apply the attention mechanism to exploit the most important spatial and temporal dependencies. A global residual (GRes) unit is designed to capture global spatial-temporal information. In this way, our model is able to capture more complicated spatial-temporal correlations with better performance. Our contributions are summarized as follows:

– We propose a novel graph convolution layer which considers dynamicity, local and global spatial dependencies simultaneously. A novel GRes module proposed, which consists of a gated convolution to capture the temporal dependencies and a global residual unit to capture the global spatial-temporal correlations.

- We propose an effective and efficient deep learning framework GDCRN that contains multiple branches for capturing informative features of multiple different periods. Each branch is specially designed to capture spatial-temporal information of this time period.
- We evaluate our model on two real datasets and compare it with six baseline methods by three evalution metrics. Extensive experiments verify the effectiveness of our model.

2   Related Work

Accurate prediction of traffic conditions is essential to data-driven urban management. Researchers have made tremendous efforts in traffic prediction [7,8,13]. Statistical regression methods such as ARIMA and its variants [1,13] are representative models in the early studies on traffic prediction. However, they only study traffic time series for each individual location and fail to consider the spatial correlations. Later, some researchers feed spatial features and other external feature information into the traditional machine learning models [7,12]. But it is still difficult to consider the spatial-temporal correlations of high-dimensional traffic data. The prediction performance of traditional machine learning methods heavily depends on feature engineering.

Recently, deep learning methods have brought tremendous advances in traffic prediction, which outperform many traditional methods. Many models integrate convolution neural network (CNN) and recurrent neural network (RNN) to jointly model complex non-linear spatial and temporal dependences in traffic network, and have achieved inspiring success [8,14]. RNN and its variants [3,5] can effectively use the self-circulation mechanism to learn temporal dependence well. CNN treat city traffic network as images by dividing the traffic network into small grids and use CNN to model the non-linear spatial dependencies. However, the grid structure does not hold the real-word conditions, which makes it unsuitable to capture the spatial dependencies of traffic network effectively. The works in [10,17] propose to capture the structural correlations of traffic network by combing RNN and graph convolution network such as GCN [9] and DCNN [2]. GCN and DCNN models capture the dependence of graphs via operating convolution to pass message between the nodes of graphs. However, RNN-based models are difficult to train, computationally heavy and less effective when capturing long-distance contextual temporal information. To solve these challenges, STGCN [16] apply CNN in the time dimension and GCN in spatial dimension, which enable stable gradient and much faster training speed with fewer parameters. ASTGCN [4] further apply the attention mechanism to adjust spatial-temporal dependence dynamically. Although the schemes mentioned above have improved the accuracy of traffic prediction, they still fail to capture the global and local spatial-temporal dependencies simultaneously in the traffic network.

3   Preliminaries

Definition 1 (Traffic Network). *In this study, the traffic topological network can be defined as a weighted bidirectional graph $G = (V, E, A)$, where V is a set of nodes with limited number ($|V| = N$), E is a set of edges that describe the accessible routes between nodes, and $A \in \mathbb{R}^{N \times N}$ indicates the weighted adjacency matrix of G. Specifically, $a_{ij} \in A$ represents the weight from node v_i to v_j.*

Definition 2 (Traffic Data). *Assuming that the network G has N nodes and traffic data $\mathcal{X}$ contains C features (such as flow, occupy, speed), the traffic data of c-th ($c \in (1, \ldots, C)$) feature on nodes v_i ($i \in (1, \ldots, N)$) at time t can be described as $x_t^{i,c} \in \mathbb{R}$. Then, $X_t^i = (x_t^{i,1}, \ldots, x_t^{i,C}) \in \mathbb{R}^C$ denotes the traffic data with all features on node v_i at time t, and $X_t = (X_t^1, \ldots, X_t^N) \in \mathbb{R}^{N \times C}$ denotes the traffic data with all features and all nodes at time t. The whole historical traffic data can be denoted by $\mathcal{X} = (X_1, \ldots, X_T) \in \mathbb{R}^{N \times C \times T}$.*

Problem 1 (Traffic Flow Prediction). Given a traffic Network G, and its historical signals over past T time slices, i.e. $\mathcal{X} = (X_1, \ldots, X_T) \in \mathbb{R}^{N \times C \times T}$. Our problem is to predict the next T_p horizons traffic flow sequences Y on the whole traffic network. The prediction result Y can be defined as $Y = (Y_1, \ldots, Y_i, \ldots, Y_{T_p}) = (X_{T+1}, \ldots, X_{T+j}, \ldots, X_{T+T_p}) \in \mathbb{R}^{N \times C \times T_p}$, where $0 < j \leq T_p$, $Y_{T+j} = (Y_{T+j}^1, \ldots, Y_{T+j}^N) \in \mathbb{R}^{N \times C}$.

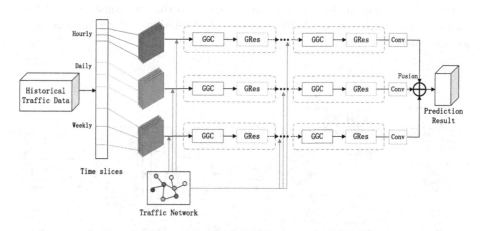

Fig. 1. The architecture of GDCRN.

4   Global Diffusion Convolutional Residual Network

Figure 1 shows the architecture of GDCRN. The inputs of GDCRN are historical traffic data and traffic network, and the outputs are the predictions of the future traffic states. We extract local and global spatial information from topology of traffic network. We set three branches to model the hourly-periodic, daily-periodic and weekly-periodic dependencies. Every branch is able to learn the dynamic spatial-temporal information in every time period by GGC and GRes submodules. A convolution layer is designed in the end to generate prediction results of each branch and keep output shapes consistent. Finally, the outputs of each periodic branch are fused to obtain final prediction results. The detailed mechanism of each module is described in the following subsections.

4.1 Global Graph Convolution Layer

For spatial dimension, the directly connected nodes inevitably affect each other, and those roads which are geographically distant but conveniently reachable are also correlated with each other. So, it is necessary to capture both local and global dependencies. In this paper, we propose a global graph convolution (GGC) unit based on diffusion convolution that simultaneously extract both the local and global spatial dependencies on traffic network. Diffusion convolution [2] is a compositional layers , which smoothes a node's signal by a diffusion process, so that it can directly describe features of bidirectional through multi-dimensional input.

Firstly, three auxiliary matrices are developed to encode spatial information about the topology of the traffic network . For details, we apply forward adjacency matrix A^F and the backward adjacency matrix $A^B = (A^F)^\top$ as local matrices to encode the local spatial proximity. And then we construct a global auxiliary matrix A^G to encode the topological correlations of long distance by a global transform.

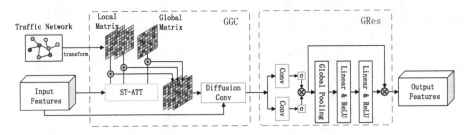

Fig. 2. The substructure of GDCRN module. ST-ATT: Spatial Temporal Attention [4]. Global Pool: Global average pooling layer. Linear & ReLU: Linear Transform and ReLU function

Secondly, traffic conditions of different locations have influence among each other, but not all of these correlations are equally important. Furthermore, correlations between different time horizons are also varying. Therefore, we adopt an attention mechanism (ST-ATT) [4] to generate a spatial-temporal attention matrix β, which can focus on more important spatial-temporal information. Take the temporal attention as an example: $\alpha = \Phi_t \cdot \sigma(((X_{l-1})^T U_1)U_2(U_3 X_{l-1}) + b_t)$, where $X_{l-1} \in \mathbb{R}^{C_{l-1} \times N \times T_{l-1}}$ is the input of the l-th GGC module, σ represents activation function sigmod, $\Phi_t, b_t \in R^{T_{l-1} \times T_{l-1}}, U_1 \in \mathbb{R}^N, U_2 \in \mathbb{R}^{C_{l-1}}, U_3 \in \mathbb{R}^{C_{l-1}}$ are learnable parameters. Each element $\alpha_{i,j}$ represents the strength of the correlation between time i and time j. We apply the normalized temporal attention matrix α' generate dynamic importance-oriented temporal representations $H_t = X_{l-1}\alpha'$. Then, spatial-temporal attention matrix β is generated by the similar attention mechanism based on temporal representations H_t.

Thirdly, we feed the spatial-temporal attention matrix β into following graph convolution layer to adjust correlations between nodes dynamically.

$$\hat{A}^F = (A^F \odot \beta); \hat{A}^B = (A^B \odot \beta); \hat{A}^G = (A^G \odot \beta); \tag{1}$$

where $\odot$ is a Hadamard product. By combining importance-oriented diffusion matrices, our innovative graph convolution layer can be formulated as:

$$H_s = \sigma(\sum_{k=0}^{K} (\hat{A}_k^F X_{l-1}\Theta_{k1} + \hat{A}_k^B X_{l-1}\Theta_{k2} + \hat{A}_k^G X_{l-1}\Theta_{k3})) \tag{2}$$

Where K is the diffusion step, $\Theta_{k1}, \Theta_{k2}, \Theta_{k2}$ is a diagonal matrix of learnable parameters, $\hat{A}_k^F$, $\hat{A}_k^B$ and $\hat{A}_k^G$ are the k-th step diffusion matrices, σ is the activation function of graph convolution layer, X_{l-1} is the input of the l-th GGC unit. By applying diffusion convolution operations with attention mechanisms to the input X_{l-1}, H_s can model dynamic local and global spatial dependencies.

4.2 Global Residual Network

The future traffic conditions have a complex non-linear relationship with the previous traffic conditions. To learn informative temporal correlations, we apply a temporal convolution with a gated mechanism. To capture the interdependencies between spatial and temporal dimensions, we propose a global residual unit which uses global information to selectively emphasise spatial-temporal correlations. As shown in the right part of Fig. 2, Global Residual Network (GRes) combines a gated temporal convolution unit with a global residual unit.

Gated Temporal Convolution Unit. Gated mechanisms have powerful ability to control information. We apply two standard convolution operations with different kernel sizes to learn different hidden representations in the time dimension. Then two different activation functions are applied as output gates to learn complex time features. Given the spatial representations H_s, we can formulate the gated temporal convolution unit as:

$$H_{st} = \sigma_1(Conv_1(H_s)) \odot (\sigma_2(Conv_2(H_s))) \tag{3}$$

where σ_1 and σ_2 are the different non-linear activations, σ_1 is RELU function and σ_2 is tangent hyperbolic function, $\odot$ is element-wise product, $Conv_1$ and

$Conv_2$ are the standard convolution functions. H_{st} can model both spatial and temporal dependencies.

Global Residual Unit. To improve the sensitivity of global spatial-temporal correlations in our model, we design a global residual unit to exploit informative features and suppress less useful ones. Firstly, a global average pooling layer is used to capture global contextual spatial-temporal dependencies directly among all nodes and all time horizons. To limit the model complexity and improve the generalization ability of the model, we use a linear transformation for decreasing dimension and a ReLU function, which can be defined as:

$$f(x) = ReLU(Wx) \tag{4}$$

where $W \in \mathbb{R}^{C_r \times C}$ is learning parameters, C is the input dimension, C_r is the output dimension and $C_r < C$. Different from SElayer [6], we use two same transformations in Fig. 2 instead of different ones, which has been proved by experiments the former performs better. Given local spatial-temporal representations H_{st}, the core of the global residual unit can be defined as:

$$H_o = f(f(GlobalPooling(H_{st}))) \otimes H_{st}, \tag{5}$$

where $\otimes$ is the element-wise product. H_o can further model global dynamic spatial-temporal dependencies. Then, a residual mechanism and LayerNorm are applied to improve generalization performance.

4.3 Fusion Mechanism

To ensure that multiple branches can be effectively merged, we apply a convolution layer at the end of each branch. The output prediction results of the three branches ($\hat{Y}_h, \hat{Y}_d, \hat{Y}_w$) have same shape. Finally, we fuse prediction results of multiple periods to capture global temporal correlations by learning weights and generate the final prediction result $\hat{Y}$, which can be formulated as:

$$\hat{Y} = W_h \odot \hat{Y}_h + W_d \odot \hat{Y}_d + W_w \odot \hat{Y}_w, \tag{6}$$

where $W_h, W_d, W_w \in \mathbb{R}^{N \times T_p}$ are the learning parameters.

5   Experiment Evaluation

5.1   Experiment Settings

DataSet. We verify GDCRN on two large real world highway traffic datasets, PeMSD4 and PeMSD8, released by [4]. PEMSD4 and PeMSD8 records two months of statistics on traffic data on the highways of California. Table 1 presents the details of the two datasets. The traffic data are aggregated every 5 min.

Table 1. The detailed information of the dataset.

Dataset	PeMSD8	PeMSD4
Locations	San Francisco Bay area, California	San Bernardino, California
Detectors	170	307
Time interval	12	12
Time span	01/01/2018-28/02/2018	07/01/2016-31/08/2016

Network Structure and Hyperparameter Settings. We implemented our model in Python with MXNet 1.6.0. We set up three different periodic branches for the model by week, day, and hour. We set the input period length of three branches as: $T_w = 2, T_d = 2, T_h = 1$. Each branch contains two GDCRN blocks. For graph convolution, we construct the global spatial matrix A_G by a random walk with path length $q = 3$ and set graph convolution layers with diffusion step $k = 3$. For gated temporal convolution unit, we set one with 64 filters and the kernel size 3×3, and another with 64 filters and the kernel size 1×1. In the first GDCRN block of branches, we set the strides of temporal convolution as the length of input period (i.e., 2, 2, 1). For the output convolution layer of each branch, we use 12 (prediction horizons) filters with kernel size 1×64. For training phase, the batch size is 16, learning rate is 0.001 and epochs are 50. We split dataset in chronological order with 70% for training, and 20% for testing, and the remaining data for validating.

5.2   Measurement and Baseline Methods

In our experiment, we use three most-widely adopted evaluation metrics, Mean Absolute Error (MAE), Root MeanSquare Error (RMSE), and Mean Absolute Percentage Error (MAPE) to measure our scheme and others. We compare GDCRN with following 6 baseline methods:

- HA: Historical Average model uses the average value of the last 12 time slices as the next prediction value.
- ARIMA [13]: Auto-Regressive Integrated Moving Average method is a widely used time series regression model.
- LSTM [5]: Long Short Term Memory network, which is a spacial RNN model.
- STGCN [16]: Spatio-Temporal Graph Convolutional Network applies purely convolutional structures with a gating mechanism to extract spatial-temporal features simultaneously.
- T-GCN [17]: Temporal Graph Convolutional Network combines with GCN and GRU.
- ASTGCN [4]: Attention based Spatial-Temporal Graph Convolution Network use spatial-temporal mechanism in graph convolution with Chebyshev polynomials approximation.

5.3   Experimental Results

Performance Comparison. We compare the performance of our GDCRN and 6 baseline methods for 15-min, 30-min, 60-min predictions on PEMSD4 and PEMSD8 datasets. Table 2 shows the average results of traffic flow prediction performance on the three prediction intervals. Our GDCRN model obtains superior results on two datasets. It can be seen that GDCRN significantly outperforms the approaches that only take temporal features into account (HA, ARIMA, LSTM).

Table 2. The performance of our model and baselines on different predicting intervals

Dataset	Models	15 min			30 min			60 min		
		MAE	RMSE	MAPE	MAE	RMSE	MAPE	MAE	RMSE	MAPE
PEMSD8	HA	22.92	34.22	14.31	24.97	37.33	15.59	30.03	44.98	18.84
	ARIMA	16.46	25.15	10.06	18.93	29.00	11.54	24.08	35.85	14.83
	LSTM	18.09	28.48	11.86	19.76	30.9	13.85	23.37	36.03	15.5
	T-GCN	17.03	25.06	13.59	17.32	25.74	13.50	18.30	27.29	14.37
	STGCN	14.51	22.58	9.49	15.87	24.58	10.90	18.16	27.08	13.84
	ASTGCN	15.66	23.95	10.22	16.14	24.82	10.51	17.03	26.27	11.17
	GDCRN	**14.03**	**22.03**	**9.31**	**14.59**	**23.11**	**9.74**	**15.39**	**24.59**	**10.40**
PEMSD4	HA	28.64	42.78	19.45	31.02	46.29	21.17	26.83	55.02	25.66
	ARIMA	21.53	34.34	13.78	24.54	38.50	15.82	30.89	47.66	20.41
	LSTM	24.77	39.67	15.94	26.61	42.10	17.06	30.06	46.05	19.54
	T-GCN	22.09	32.90	18.20	22.23	33.34	17.95	23.34	34.97	18.94
	STGCN	18.85	30.00	13.09	20.49	32.19	13.90	23.07	36.92	16.75
	ASTGCN	19.82	31.98	14.33	20.78	32.92	14.8	21.91	34.75	15.81
	GDCRN	**18.65**	**29.91**	**12.98**	**19.37**	**31.13**	**13.46**	**20.40**	**32.81**	**14.25**

Compared to the spatial-temporal models, the prediction results of GDCRN excels RNN-based scheme T-GCN and also performs better than CNN-based schemes STGCN and ASTGCN. As for STGCN, GDCRN achieves bigger enhancement on the 60-min horizons than 15-min horizons. Since GDCRN introduces attention mechanism and GRes module to capture global spatial-temporal correlations, so that our model can better hold the long-term traffic pattern. ASTGCN utilizes GCN to describe spatial dependencies. However, GCN regards the traffic network as a unidirectional graph, which is not practical for real-world traffic network. In contrast, GDCRN adopts global diffusion convolution, which is able to handle bidirectional network and can capture global spatial correlation directly. Therefore, combining with the ability to obtain local and global spatial-temporal correlations on the bidirectional network, GDCRN is able to perform better regarding all the metrics for all predicting horizons.

Figure 3 illustrates the changes of prediction performance of our model and other baseline models as the prediction temporal interval increases. We have two valuable observations which further confirm the superiority of our model.

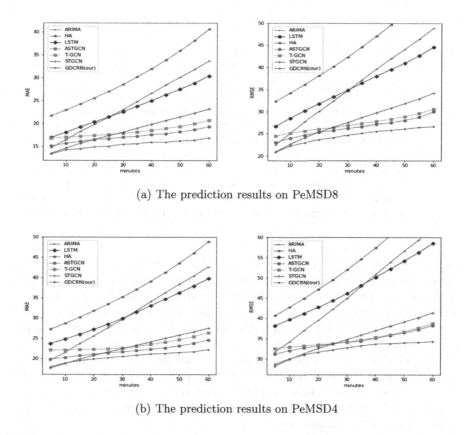

(a) The prediction results on PeMSD8

(b) The prediction results on PeMSD4

Fig. 3. Performance changes of different methods as the predicting horizon increases

Firstly, the growth trends of prediction error of GDCRN are smaller than almost all methods, indicating that our model is insensitive to prediction time interval. Secondly, GDCRN achieves the best forecasting performance in all time dimensions, especially for the long-term prediction. Specifically, the differences between DGCRN and other baseline methods are more significant as the prediction time interval increases, which shows that the scheme of our GDCRN model has advantages not only in short-term predictions, but also in long-term predictions. All the experiment results above demonstrate the advantages of our model in capture spatial-temporal correlation of the highway traffic data.

Ablation Study. In order to verify the effectiveness of every components on our model, we compare the following four variants of our model.

- ChebNet, which replaces diffusion convolution with ChebNet.
- No-GRU, which removes global residual unit in GRes module.
- No-A_G, which removes global spatial matrix in diffusion convolution unit.
- No-Gate, which removes Gate mechanism in temporal convolution unit.

Table 3. Performance of variants of GDCRN on different predicting intervals

Dataset	Models	15 min			30 min			60 min		
		MAE	RMSE	MAPE	MAE	RMSE	MAPE	MAE	RMSE	MAPE
PEMSD8	chebNet	14.83	23.01	10.14	15.65	24.37	10.70	16.83	26.27	11.60
	No-GRU	14.87	23.21	10.28	15.86	25.12	10.68	17.20	27.54	11.68
	No-A_G	14.18	22.21	9.55	14.72	23.26	10.00	15.51	24.71	10.68
	No-Gate	14.22	22.22	9.43	14.79	23.35	9.81	15.65	24.94	10.45
	GDCRN	**14.03**	**22.03**	**9.31**	**14.59**	**23.11**	**9.74**	**15.39**	**24.59**	**10.40**
PEMSD4	chebNet	19.08	30.37	14.01	19.86	31.60	14.82	20.96	33.35	15.88
	No-GRU	19.42	30.73	14.97	20.22	31.98	15.89	21.32	33.73	17.04
	No-A_G	18.73	30.08	13.37	19.52	31.35	14.23	20.53	33.03	15.14
	No-Gate	18.91	30.08	13.82	19.62	31.30	14.27	20.55	32.94	14.82
	GDCRN	**18.65**	**29.91**	**12.98**	**19.37**	**31.13**	**13.46**	**20.40**	**32.81**	**14.25**

Table 3 compares the average performance of every variant over different prediction interval. We can find that GDCRN achieves the best prediction performance. The predicting results of GDCRN excels the ChebNet model, which verifies that capturing bidirectional spatial dependencies is very necessary and useful for prediction tasks on real traffic networks. Compared with the No-GRU model, GDCRN has better prediction precision and is insensitive to prediction interval, which proves that capturing global spatial-temporal features are important for traffic prediction. The GDCRN are superior to No-A_G model, indicating the effectiveness of capture global spatial correlations. In summary, the GDCRN can achieve the best results regardless of the prediction horizon, and each component of our model make sense.

6   Conclusion

In this paper, we propose a novel global diffusion convolution residual network for traffic prediction. Based on the spatial topological structure of the traffic network, we propose a novel graph convolution layer, which leverages global and local information of spatial structure. To exploit informative features, we design a global residual network GRes and combine it with GGC module to capture both global and local spatial-temporal correlations. Experiments on two large-scale real-world datasets verify the effectiveness of our model. Furthermore, GDCRN is a generalized spatial-temporal network prediction framework, and has the potential to be applied to other similar prediction problems such as taxi demand forecasting.

Acknowledgement. This work is partially supported by The National Key R&D Program of China under Grant 2018AAA0101200, National Natural Science Foundation of China (NSFC) under Grant No. 61772491, No. U170921, Anhui Initiative in Quantum Information Technologies AHY150300 and the Fundamental Research Funds for the Central Universities.

References

1. Ahmed, M.S., Cook, A.R.: Analysis of freeway traffic time-series data by usingBox-Jenkins techniques. No. 722 (1979)
2. Atwood, J., Towsley, D.: Diffusion-convolutional neural networks. In: Advances in Neural Information Processing Systems. pp. 1993–2001 (2016)
3. Chung, J., Gulcehre, C., Cho, K., Bengio, Y.: Empirical evaluation of gated recurrent neural networks on sequence modeling. arXiv preprint arXiv:1412.3555 (2014)
4. Guo, S., Lin, Y., Feng, N., Song, C., Wan, H.: Attention based spatial-temporal graph convolutional networks for traffic flow forecasting. Proceedings of the AAAI Conference on Artificial Intelligence. **33**, pp. 922–929 (2019)
5. Hochreiter, S.: Schmidhuber: Long short-term memory. Neural computation **9**(8), 1735–1780 (1997)
6. Hu, J., Shen, L., Sun, G.: Squeeze-and-excitation networks. In: Proceedings of the IEEE conference on computer vision and pattern recognition. pp. 7132–7141 (2018)
7. Jeong, Y.S., Byon, Y.J., Castro-Neto, M.M., Easa, S.M.: Supervised weighting-online learning algorithm for short-term traffic flow prediction. IEEE Transactions on Intelligent Transportation Systems **14**(4), 1700–1707 (2013)
8. Ke, J., Zheng, H., Yang, H., Chen, X.M.: Short-term forecasting of passenger demand under on-demand ride services: A spatio-temporal deep learning approach. Transportation Research Part C: Emerging Technologies **85**, 591–608 (2017)
9. Kipf, T.N., Welling, M.: Semi-supervised classification with graph convolutional networks. arXiv preprint arXiv:1609.02907 (2016)
10. Li, Y., Yu, R., Shahabi, C., Liu, Y.: Diffusion convolutional recurrent neural network: Data-driven traffic forecasting. arXiv preprint arXiv:1707.01926 (2017)
11. Ma, X., Dai, Z., He, Z., Ma, J., Wang, Y., Wang, Y.: Learning traffic as images: a deep convolutional neural network for large-scale transportation network speed prediction. Sensors **17**(4), 818 (2017)
12. Van Lint, J., Van Hinsbergen, C.: Short-term traffic and travel time prediction models. Artificial Intelligence Applications to Critical Transportation Issues **22**(1), 22–41 (2012)
13. Williams, B.M., Hoel, L.A.: Modeling and forecasting vehicular traffic flow as a seasonal arima process: Theoretical basis and empirical results. Journal of transportation engineering **129**(6), 664–672 (2003)
14. Xingjian, S., Chen, Z., Wang, H., Yeung, D.Y., Wong, W.K., Woo, W.c.: Convolutional lstm network: A machine learning approach for precipitation nowcasting. In: Advances in neural information processing systems. pp. 802–810 (2015)
15. Yao, H., Tang, X., Wei, H., Zheng, G., Li, Z.: Revisiting spatial-temporal similarity: A deep learning framework for traffic prediction. In: AAAI Conference on Artificial Intelligence (2019)
16. Yu, B., Yin, H., Zhu, Z.: Spatio-temporal graph convolutional networks: A deep learning framework for traffic forecasting. arXiv preprint arXiv:1709.04875 (2017)
17. Zhao, L., Song, Y., Zhang, C., Liu, Y., Wang, P., Lin, T., Deng, M., Li, H.:T-gcn: A temporal graph convolutional network for traffic prediction. IEEETransactions on Intelligent Transportation Systems (2019)

Depthwise Separable Convolutional Neural Network for Confidential Information Analysis

Yue Lu[1,2], Jianguo Jiang[1,2], Min Yu[1,2(✉)], Chao Liu[1], Chaochao Liu[1,2], Weiqing Huang[1], and Zhiqiang Lv[1]

[1] Institute of Information Engineering, Chinese Academy of Sciences, Beijing, China
yumin@iie.ac.cn
[2] School of Cyber Security, University of Chinese Academy of Sciences, Beijing, China

Abstract. Confidential information analysis can identify the text containing confidential information, thereby protecting organizations from the threat posed by leakage of confidential information. It is effective to build a confidential information analyzer based on a neural network. Most of the existing studies pursue high accuracy to design complex networks, ignoring speed and consumption. The optimal defense is to automatically analyze confidential information without compromising routine services. In this paper, we introduce a lightweight network, DSCNN, that can be adapted to low-resource devices. We also introduce two hyperparameters to balance accuracy and speed. Our motivation is to simplify convolutions by breaking them down because the space dimension and channel dimension are not closely related in the convolutions. Experimental results on real-world data from WikiLeaks show that our proposed DSCNN performs well for confidential information analysis.

Keywords: Depthwise Separable Convolutional Neural Network · Confidential information analysis · Information Security · Natural Language Processing

1 Introduction

Mobile devices generate a vast amount of data every minute. The data may contain confidential information that has not yet been marked. Such confidential information can be leaked without being noticed and pose a threat to national security, business trade, or personal life. Many organizations institute enforcement policies to protect confidential information [18]. These policies require every email sent by an employee to the Internet is reviewed by his manager. On the one hand, these policies limit the operating efficiency of the organization and waste a lot of manpower resources. On the other hand, it is ineffective if the employee's managers are not extremely well-versed in the scope of confidential matters for

G. Li et al. (Eds.): KSEM 2020, LNAI 12275, pp. 450–461, 2020.
https://doi.org/10.1007/978-3-030-55393-7_40

the entire organization. Therefore, constructing confidential information analyzers has become a trend. As it can help organizations identify confidential information, the need for high quality automated confidential information analyzers becomes much more profound.

Confidential information analysis is a cross-task of Information Security (IS) and Natural Language Processing (NLP). Its goal is to categorize text into different security levels (such as Confidential and Non-Confidential), or more fine-grained levels (such as Top-Secret, Secret, Confidential, and Unclassified) [1]. In similar areas of confidential information analysis, such as text mining [17], sentiment analysis [8], fake news detection [15], and confidential information detection [7],CNNs have attracted extensive attention because of their excellent performance. A large number of deep and complex CNNs have been designed for tasks related to text classification [14]. Conneau et al. used 19 convolution layers to build a Very Deep CNN (VDCNN) [3]. Johnson et al. built a Deep Pyramid CNN (DPCNN) with 15 layers [11]. These models can easily achieve high accuracy with sufficient computational resources and processing time. However, such models do not work well in low-resource and time-limited devices or applications. In real-world applications such as Data Leakage Prevention (DLP) [16] and Security Information and Event Management (SIEM) [21], confidential information analyzers require the ability to run in real-time on a computationally limited device to enforce the appropriate protection mechanism without degrading regular services.

In this paper, we present a lightweight model named **D**epthwise **S**eparable **C**onvolutional **N**eural **N**etwork (**DSCNN**) for confidential information analysis. Our motivation is that separating spaces and channels when convoluting text can reduce the computational complexity of convolutions. The space dimension and channel dimension are not closely related that it is preferable not to map them together. From the perspective of a model, channels are different pre-trained word embeddings without strict sequence. Additionally, we describe two hyper-parameters that efficiently balance accuracy and speed. These two hyper-parameters can be used when designing the appropriate size models for different devices. We conduct the comparison experiments of our proposed method and other popular methods. We also conduct extensive experiments of hyper-parameter sensitivity. The results show that our proposed method has a better performance in analyzing confidential information. The main contributions of this work include:

1) We propose a DSCNN for confidential information analysis. The DSCNN makes convolution easier by operating in space dimension and channel dimension respectively to reduce computational complexity.
2) We introduce two simple hyper-parameters, channel multiplier and space multiplier, to balance accuracy and speed. These two hyper-parameters can be used to further reduce the amount of calculation.
3) Extensive experiments using real-world data from WikiLeaks show that our proposed model not only achieves a high accuracy but also saves a lot of time and resources.

The rest of the paper is categorized as follows. The previous approaches are reviewed in Sect. 2. Our proposed method is presented in Sect. 3. Section 4 presents the results of the experiment. Finally, in Sect. 5, we conclude and give perspectives.

2   Preliminaries

Recently, CNNs have achieved strong performance for tasks related to text classification [2]. Dos Santos et al. designed a CNN to extract features at the character-level and word-level for sentiment analysis [4]. Kalchbrenner et al. introduced a global pooling named dynamic k-max pooling to build CNNs [12]. Kim used various pre-trained word embeddings to build CNNs for sentence classification [13]. Yin et al. presented a multi-channel CNN which accepts multiple word embeddings as the input [19]. Johnson et al. proposed a low-complexity CNN shaped like a pyramid for text categorization [11].

The CNN is a kind of neural network model, which relies on the layer called convolution layer for feature extraction. At the heart of this convolution layer is a learnable filter. This filter does convolution operation while scanning the vectorial text. The outputs of the convolution operation are the extracted feature maps. Suppose a learnable filter $\mathbf{w} \in \mathbb{R}^{h \times d}$ is scanning the text sequence $\mathbf{x} \in \mathbb{R}^{s \times d}$ The h represents the size of the filter $\mathbf{w}$, the s represents the length of the sequence $\mathbf{x}$, and the d represents the dimension of the word embeddings. A extracted feature is obtained by the convolution operation:

$$f_i = \sigma(\mathbf{w} \cdot \mathbf{x}_{i:i+h-1} + \mathbf{b}), \tag{1}$$

where the $\mathbf{x}_{i:i+h-1}$ representations a region of the above text sequence, the $\mathbf{w}$ represents the above filter, the $\mathbf{b}$ represents a bias, and the σ represents a non-linear function. The filter scans all the sequences $\{\mathbf{x}_{1:h}, \mathbf{x}_{2:h+1}, \ldots, \mathbf{x}_{s-h+1:s}\}$ to produce a feature map $\mathbf{f} = [f_1, f_2, \ldots, f_{s-h+1}]$. Typically models use a lot of filters (with different window sizes) to obtain a variety of feature maps.

The multi-channel CNN is an improvement on the single-channel CNN. Compared with single-channel CNN, the multi-channel CNN uses different pre-training word embedding vectors to initialize multiple channels as inputs. The multi-channel CNN brings the following advantages: On the one hand, multiple channels rather than one channel can bring more information available to a common word. On the other hand, one channel can be missing a rare word, other channels can supplement it. However, it is worth noting that the multi-channel model will bring lots of computation. In multi-channel CNN, the convolution layer attempts to learn the filter in three dimensions, which has two space dimensions (width and height) and one channel dimension. The convolution operation is a joint mapping of these dimensions. In single-channel CNN, the convolution operation is only a mapping of space dimensions. We think the space dimension and the channel dimension are not closely related. The channel dimension do not have the same strict order as the space dimension in the model, so we consider that joint mapping is unnecessary. To simplify the convolution operations, the standard convolution can be decomposed into a series of operations that convolving independently on spaces and channels. Our work is similar to Howard et al. [6] on image classification.

3 Methodology

In this section, we first present the network structure of DSCNN. We then detail the core of DSCNN – depthwise separable convolution. Finally, we describe two hyper-parameters – channel multiplier and space multiplier.

3.1 Network Structure of DSCNN

DPCNN [11] is a simple network structure with less computation and better performance. It is a pyramid-shaped network structure whose upper layer is half the size of the lower layer. We use the DPCNN network structure as the basic network of our DSCNN.

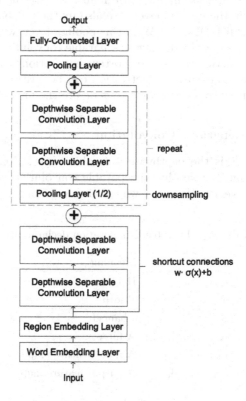

Fig. 1. Network structure of DSCNN.

The network structure of DSCNN is illustrated in Fig. 1. A DSCNN model consists of input, word embedding layer, region embedding layer, depthwise separable convolution layer, pooling layer, fully-connected layer, and output. In our DSCNN, the input is a text sequence and the output is its label. The first layer is a word embedding layer. The word embedding layer is used to convert text into vectors. In our DSCNN, we use a variety of pre-trained word embeddings to initialize the word embedding layer. Each channel of the layer corresponds to a pre-trained word embedding. So our DSCNN is also a multi-channel CNN. As mentioned in Sect. 2, multi-channel networks have more advantages than single-channel networks. The second layer is a region embedding layer. The region embedding layer works in a similar way to the N-gram model [9,10]. It is used to extract features of a small region in the width and height of space dimension. The following layers are the alternations of two depthwise separable convolution layers and one pooling layer. The depthwise separable convolution layer is used to extract features and model long distance dependencies. It extracts features in three dimensions, which has two space dimensions (width and height) and one channel dimension. After each depthwise separable convolution layer, there is a pooling layer used to downsample feature maps. We use the max-pooling with size 3 and stride 2 in the pooling layer. We also fix the number of feature maps in this pooling layer like DPCNN. With this pooling layer, we can model longer distance dependencies later by depthwise separable convolution layers. To enable the training of deep networks, we use shortcut connections with pre-activation. The shortcut connections with pre-activation can be written as $\mathbf{w}\sigma(\mathbf{x}) + b$ [5]. Finally, there is a fully-connected layer to generate the final classification.

3.2   Depthwise Separable Convolution

The core of DSCNN is the depthwise separable convolution. Compared to a standard convolution, the depthwise separable convolution can greatly reduce computational complexity. Hereafter we use the notation given in Table 1.

Table 1. This table lists notation declarations.

Symbol	Meaning
h	Height
w	Width
d	Channel
K_{h_K}, wK, d_F, d_G	A filter
F_{h_F}, wF, d_F	A input feature map
G_{h_G}, wG, d_G	A output feature map

As shown in Fig. 2(a), the standard convolution use d_G filters of size $h_K \times w_K \times d_F$ to extract feature maps. It extracts features in both the space dimension

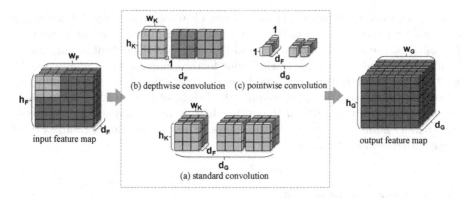

Fig. 2. The standard convolution extracts the space and channel features in one step (a), while the depthwise separable convolution extracts space features in the depthwise convolution (b) and extracts channel features in the pointwise convolution (c).

and the channel dimension. The h_K represents the height of the filter and the w_K represents the width of the filter in the space dimension. The d_F represents the channel number of the filter in the channel dimension. The channel number of the filter is the same as the channel number of the input feature maps because the filter convolves on a multi-channel input. The d_G represents the number of filters. The number of filters is the same as the channel number of the output feature maps.

As shown in Fig. 2(b), (c), the depthwise separable convolution contains two parts: a depthwise convolution and a pointwise convolution. The depthwise separable convolution is decomposed to operate separately in the space dimension and channel dimension. The depthwise convolution is operating in the space dimension, while the pointwise convolution is operating in the channel dimension. The depthwise convolution use d_F depthwise filters of size $h_K \times w_K \times 1$ to extract feature maps in the space dimension. The h_K represents the height of the depthwise filter, the w_K represents the width of the depthwise filter, and the 1 represents the channel number of the depthwise filter. The d_F represents the number of depthwise filters. The number of depthwise filters is the same as the input channel because one input channel corresponds to one depthwise filter. The pointwise convolution use d_G pointwise filters of size $1 \times 1 \times d_F$ to extract feature maps in the channel dimension. The 1×1 represent respectively the height and width of the pointwise filter. The d_F represents the channel number of the pointwise filter. The channel number of the pointwise filter is the same as the number of the depthwise filter. The d_G represents the number of pointwise filters. The number of pointwise filters is the same as the channel number of the output feature maps.

As discussed in Sect. 1, we think the space dimension and the channel dimension are not closely related. The standard convolution extracts the space and channel features in one step, while the depthwise separable convolution extracts space features in one step and extracts channel features in another step. Splitting one step into two steps can reduce computational complexity. Suppose we take F_{h_F, w_F, d_F} as the input feature maps and G_{h_G, w_G, d_G} as the output feature maps. The h_F represents the height of the input feature maps, the w_F represents the width of the input feature maps, and the d_F represents the channel number of input feature maps. The h_G represents the height of the output feature maps, the w_G represents the width of the output feature maps, and the d_G represents the channel number of output feature maps. The computational complexity of the standard convolution is:

$$h_K \cdot w_K \cdot d_F \cdot d_G \cdot h_F \cdot w_F. \tag{2}$$

The depthwise convolution costs:

$$h_K \cdot w_K \cdot d_F \cdot h_F \cdot w_F. \tag{3}$$

The pointwise convolution costs:

$$d_F \cdot d_G \cdot h_F \cdot w_F. \tag{4}$$

The depthwise separable convolution costs:

$$h_K \cdot w_K \cdot d_F \cdot h_F \cdot w_F + d_F \cdot d_G \cdot h_F \cdot w_F, \tag{5}$$

We get a reduction:

$$\frac{h_K \cdot w_K \cdot d_F \cdot h_F \cdot w_F + d_F \cdot d_G \cdot h_F \cdot w_F}{h_K \cdot w_K \cdot d_F \cdot d_G \cdot h_F \cdot w_F} = \frac{1}{d_G} + \frac{1}{h_K \cdot w_K}. \tag{6}$$

3.3   Channel Multiplier and Space Multiplier

To balance accuracy and speed for more applications, we introduce a simple parameter α called *channel multiplier*. It is used to set the channel number of feature maps and thin the depthwise separable convolution layer. The computational cost is:

$$h_K \cdot w_K \cdot \alpha d_F \cdot h_F \cdot w_F + \alpha d_F \cdot \alpha d_G \cdot h_F \cdot w_F, \tag{7}$$

where $\alpha \in (0, 1]$. $\alpha = 1$ is the baseline DSCNN and $\alpha < 1$ are reduced DSCNNs. We introduce the other parameter β called *space multiplier*. It is used to set the space size of feature maps and reduce the resolution of the input feature maps. The computational cost is:

$$h_K \cdot w_K \cdot d_F \cdot \beta h_F \cdot \beta w_F + d_F \cdot d_G \cdot \beta h_F \cdot \beta w_F, \tag{8}$$

where $\beta \in (0, 1]$. $\beta = 1$ is the baseline DSCNN and $\beta < 1$ are reduced DSCNNs.

4   Experiments and Discussion

We evaluated a variety of models on WikiLeaks Cable Dataset in this section. The purpose of these experiments is to clarify the influence of our proposed DSCNN for confidential information analysis.

4.1   Experiments Settings

Dataset: The WikiLeaks Cable Dataset consists of paragraphs extracted from Public Library of US Diplomacy (PlusD). We use white space as a delimiter, normalize punctuations, remove special characters, and convert the remaining characters to lowercase. After pre-processing, the details on the dataset are provided in Table 2. We randomly choose 80% of the original dataset for training and 10% of the original dataset for testing. The rest 10% of the original dataset to construct a validation set. We maintain a Secret/Confidential/Unclassified balance of the original dataset in each split and use 10-fold cross-validation.

Table 2. Statistics of wikiLeaks cable dataset.

Item	Content
Name	WikiLeaks cable dataset
Type	Sentence-level
#Classes	3
#Instances of secret	10,000
#Instances of confidential	10,000
#Instances of unclassified	10,000
Average length	145
Vocabulary size	125, 534
Test	10-fold CV

Hyper-parameters: We tune the hyper-parameters of our proposed model on the validation set.

- **Pre-trained Word Embeddings:** We initialize the word embedding layer with the following pre-trained word embeddings. We set the channel number of the word embedding layer as 4 and the dimension of vectors as 300. These pre-trained word embeddings are available on github[1]. We use the vectors of Word2Vec-GoogleNews, Word2VecModified-Wikipedia, GloVe-Crawl840B and GloVe-Wikipedia. The vectors of these word embedding do not fine-tune during training the classifiers. The Word2Vec-GoogleNews are trained on Google News through Word2Vec. The Word2VecModified-Wikipedia are trained on Wikipedia through modified Word2vec. The GloVe-Crawl840B are trained on Common Crawl through GloVe. The GloVe-Wikipedia are trained on Wikipedia through GloVe.
- **Hyper-Parameters in DSCNN:** We set the depth of DSCNN as 16, 14 convolution layers plus 2 embedding layers. We set the window size of the region embedding layer as 1, 3, 5 and the channel number of feature maps as 250. We train models with a mini-batch size of 64 and use Adam optimizer

---

[1] https://github.com/3Top/word2vec-api.

with the learning rate of 0.001. We use a 0.5 dropout rate on the fully-connected layer during training.

Evaluation. We use *accuracy* to measure these models because the dataset is balanced. We use *calculation* to evaluate the computational complexity of these model. With the same device configuration, the less computation is, the faster the speed is. We use *parameters* to evaluate the spatial complexity of these model. The fewer parameters, the less space.

4.2 Results and Discussion

Main Comparisions. First we show results for our proposed DSCNN based on the depthwise separable convolutions compared to other popular CNN models. The CharSCNN is a shallow network that extracts features from character-level to sentence-level. The TextCNN is a shallow network based on the word embedding. The MVCNN and MCCNN are multi-channel networks that have rich feature maps in convolution layers. The ConvNets and VDCNN are deep networks based character-level representation. The DPCNN is the state-of-the-art network for text classification. The Multi-Channel DPCNN is a network modified by us that use the diversity of different embedding to extract higher quality features. Compared the Multi-Channel DPCNN and our proposed DSCNN, the Multi-Channel DPCNN is based on the standard convolutions while the DSCNN is based on the depthwise separable convolutions.

Table 3. Results of our proposed DSCNN against other models.

Model	Type	Accuracy	Calculation	Parameters
CharSCNN [4]	shallow, char	64.51	–	–
TextCNN [13]	shallow, word	66.46	–	–
MVCNN [19]	shallow, word	68.17	–	–
MCCNN [2]	shallow, word	68.02	–	–
ConvNets [20]	deep, char	66.95	–	–
VDCNN [3]	deep, char	67.16	1503.36M	2.11M
DPCNN [11] (Baseline)	deep, word	68.85	1370.25M	2.63M
Multi-Channel DPCNN	deep, word	72.34	5481.00M	2.63M
DSCNN (Ours)	deep, word	**72.57**	**630.92M**	**0.89M**

From Table 3, we have the following observations: (1) As expected, our proposed DSCNN not only achieves a high accuracy but also saves a lot of time and resources. It costs less computation than the standard convolutions. The DSCNN uses $3 \times 3 \times 250 \times 250$ depthwise separable convolutions which use about 9 times less computation than the Multi-Channel DPCNN with the standard convolutions. Additionally, the DSCNN primarily focus on optimizing for calculation speed but also yield a small size network. (2) Deep models with multiple channels indeed give better performances. Single-channel networks do not outperform multi-channel networks. A single-channel network – DPCNN – achieves 68.85%, comparing to 72.34% of a multi-channel network – Multi-Channel DPCNN. It demonstrates the effectiveness of pre-trained word embeddings. The pre-trained word embedding vectors can introduce more useful external knowledge for short text.

Model Shrink. Table 4 shows a comparison between our proposed DSCNNs with different channel multipliers and the baseline DPCNN. We analyze the results from three aspects: accuracy, calculation, and parameters. The channel multiplier is used to set the channel number of feature maps. We observe that decreasing the channel multiplier α hurts the accuracy, but can reduce the calculation and parameters. Our proposed DSCNN with channel multiplier $\alpha = 0.75$ has 3 times less calculation and 5 times fewer parameters with the same accuracy as the baseline DPCNN.

Table 4. Results of channel multiplier (α).

Model	Accuracy	Calculation	Parameters
DPCNN (Baseline)	**68.85**	**1370.25M**	**2.63M**
1.00 DSCNN – 300	72.57	630.92M	0.89M
0.75 DSCNN - 300	**69.05**	**359.01M**	**0.50M**
0.50 DSCNN – 300	65.02	163.21M	0.22M
0.25 DSCNN – 300	60.17	43.54M	0.06M

Table 5. Results of space multiplier (β).

Model	Accuracy	Calculation	Parameters
DPCNN (Baseline)	**68.85**	**1370.25M**	**2.63M**
1.00 DSCNN – 300	72.57	630.92M	0.89M
1.00 DSCNN – 224	71.83	471.09M	0.89M
1.00 DSCNN – 192	70.66	403.79M	0.89M
1.00 DSCNN – 160	**68.81**	**336.49M**	**0.89M**
1.00 DSCNN – 128	65.74	269.19M	0.89M

Table 5 shows a comparison between our proposed DSCNNs with different space multipliers and the baseline DPCNN. We analyze the results from three aspects: accuracy, calculation, and parameters. The space multipliers is used to set the space size of feature maps. The accuracy decreases as the space size of feature maps decreases. The calculation reduces as the space size of feature maps decreases. The parameters remains the same because it is independent of the space size of feature maps. Our proposed DSCNN with space multiplier $\beta = 160$ has 4 times less calculation and 3 times fewer parameters with the same accuracy as the baseline DPCNN.

5 Conclusion

Confidential information analysis can protect organizations from the threat of confidential information leakage by identifying text that contains confidential information. In this paper, we proposed a lightweight model named DSCNN based on depthwise separable convolutions for improving the performance of confidential information analysis. The proposed method convolves in space and channel dimensions respectively, which can reduce the computational complexity of convolution operation. We then described the channel multiplier and space multiplier to balance accuracy and speed to fit different low-resource devices. We expect that separable convolution in depth will become the cornerstone of the design of CNNs in the future since they make the convolution easier and more efficient on multi-channel CNNs.

Acknowledgment. This work is supported by National Natural Science Foundation of China (No. 71871090).

References

1. Alzhrani, K.M.: Towards automating big texts security classification. Ph.D. thesis, University of Colorado Colorado Springs. Kraemer Family Library (2018)
2. Chen, K., Liang, B., Ke, W., Xu, B., Zeng, G.: Chinese micro-blog sentiment analysis based on multi-channels convolutional neural networks. J. Comput. Res. Dev. **55**(5), 945–957 (2018)
3. Conneau, A., Schwenk, H., Barrault, L., Lecun, Y.: Very deep convolutional networks for text classification. In: Proceedings of the 15th Conference of the European Chapter of the Association for Computational Linguistics (EACL 2017) (2017)
4. Dos Santos, C., Gatti, M.: Deep convolutional neural networks for sentiment analysis of short texts. In: Proceedings of COLING 2014, the 25th International Conference on Computational Linguistics: Technical Papers (COLING 2014), pp. 69–78 (2014)
5. He, K., Zhang, X., Ren, S., Sun, J.: Identity mappings in deep residual networks. In: Leibe, B., Matas, J., Sebe, N., Welling, M. (eds.) ECCV 2016. LNCS, vol. 9908, pp. 630–645. Springer, Cham (2016). https://doi.org/10.1007/978-3-319-46493-0_38
6. Howard, A.G., et al.: MobileNets: efficient convolutional neural networks for mobile vision applications. arXiv preprint arXiv:1704.04861 (2017)
7. Jiang, J., et al.: CIDetector: semi-supervised method for multi-topic confidential information detection. In: The 24th European Conference on Artificial Intelligence (ECAI 2020) (2013)

8. Jiang, J., et al.: Sentiment embedded semantic space for more accurate sentiment analysis. In: Liu, W., Giunchiglia, F., Yang, B. (eds.) KSEM 2018. LNCS (LNAI), vol. 11062, pp. 221–231. Springer, Cham (2018). https://doi.org/10.1007/978-3-319-99247-1_19

9. Johnson, R., Zhang, T.: Effective use of word order for text categorization with convolutional neural networks (2015)

10. Johnson, R., Zhang, T.: Semi-supervised convolutional neural networks for text categorization via region embedding. In: Advances in Neural Information Processing Systems (NIPS 2015), pp. 919–927 (2015)

11. Johnson, R., Zhang, T.: Deep pyramid convolutional neural networks for text categorization. In: Proceedings of the 55th Annual Meeting of the Association for Computational Linguistics (Volume 1: Long Papers) (ACL 2017), pp. 562–570. Association for Computational Linguistics (2017)

12. Kalchbrenner, N., Grefenstette, E., Blunsom, P.: A convolutional neural network for modelling sentences. In: Proceedings of the 52nd Annual Meeting of the Association for Computational Linguistics (ACL 2014), pp. 655–665. Association for Computational Linguistics (2014)

13. Kim, Y.: Convolutional neural networks for sentence classification. In: Proceedings of the 2014 Conference on Empirical Methods in Natural Language Processing (EMNLP 2014), pp. 1746–1751. Association for Computational Linguistics (2014)

14. Law, R., Li, G., Fong, D.K.C., Han, X.: Tourism demand forecasting: a deep learning approach. Ann. Tour. Res. **75**, 410–423 (2019)

15. Liu, C., et al.: A two-stage model based on BERT for short fake news detection. In: Douligeris, C., Karagiannis, D., Apostolou, D. (eds.) KSEM 2019. LNCS (LNAI), vol. 11776, pp. 172–183. Springer, Cham (2019). https://doi.org/10.1007/978-3-030-29563-9_17

16. Shvartzshnaider, Y., Pavlinovic, Z., Balashankar, A., Wies, T., Subramanian, L., Nissenbaum, H., Mittal, P.: Vaccine: using contextual integrity for data leakage detection. In: The World Wide Web Conference (WWW 2019), pp. 1702–1712. ACM (2019)

17. Vu, H.Q., Li, G., Law, R., Zhang, Y.: Exploring tourist dining preferences based on restaurant reviews. J. Travel Res. **58**(1), 149–167 (2019)

18. Yerazunis, W., Kato, M., Kori, M., Shibata, H., Hackenberg, K.: Keeping the good stuff. In: Confidential Information Firewalling with the CRM114 Spam Filter & Text Classifier. White Paper Black Hat USA (2010)

19. Yin, W., Schütze, H.: Multichannel variable-size convolution for sentence classification. In: Proceedings of the Conference on Computational Natural Language Learning (CoNLL 2016) (2016)

20. Zhang, X., Zhao, J., LeCun, Y.: Character-level convolutional networks for text classification. In: Advances in Neural Information Processing Systems (NIPS 2015), pp. 649–657 (2015)

21. Zhu, T., Li, G., Zhou, W., Yu, P.S.: Differential Privacy and Applications. Springer, Cham (2017). https://doi.org/10.1007/978-3-319-62004-6_9

The Short-Term Exit Traffic Prediction of a Toll Station Based on LSTM

Ying Lin[1], Runfang Wang[1] (ID), Rui Zhu[1]([X]), Tong Li[2], Zhan Wang[1], and Maoyu Chen[1]

[1] School of Software, Yunnan University, Kunming, Yunnan, China
rzhu@ynu.edu.cn
[2] School of Big Data, Yunnan Agricultural University, Kunming, Yunnan, China

Abstract. Short-term exit traffic flow prediction at a toll station is an important part of the intelligent traffic system. Accurate and real-time traffic exit flow forecast of toll stations can help people predict congestion situation in advance and then take corresponding measures. In this paper, we propose a traffic flow prediction model (LSTM_SPLSTM) based on the long short-term memory networks. This model predicts the exit traffic flow of toll stations by combining both the sequence characteristics of the exit traffic flow and the spatial-temporal characteristics with the associated stations. This LSTM_SPLSTM is experimentally verified by using real datasets which includes data collected from six toll stations. The MAEs of LSTM_SPLSTM are respectively 2.81, 4.52, 6.74, 7.27, 5.71, 7.89, while the RMSEs of LSTM_SPLSTM are respectively 3.96, 6.14, 8.77, 9.79, 8.20 10.45. The experimental results show that the proposed model has better prediction performance than many traditional machine models and models trained with just a single feature.

Keywords: Short-term exit traffic prediction · Sequence characteristics · Spatial-temporal characteristics · Long Short-term memory networks

1 Introduction

As we all know, toll stations are the main channel for vehicles to enter and exit high-way networks, and it has always been a bottleneck in traffic networks [1]. Short-term traffic flow of toll station is affected by many external factors such as time period, geographical location and spatial distance between highway network nodes. Due to these factors, the uncertainty of short term traffic flow forecasting is difficult to accurately predict by only using traditional forecasting methods. Therefore, studying how to obtain accurate and efficient traffic forecasts is necessary. High-quality predictions can not only relieve traffic pressure, but also make people travel more convenient.

At present, traffic prediction methods in the transportation fields are mainly based on statistics, machine learning, and deep learning. Among them, the statistical-based methods mainly include the autoregressive integrated moving average method (ARIMA) and the improved methods of ARIMA [2, 3]. The machine learning based methods mainly

G. Li et al. (Eds.): KSEM 2020, LNAI 12275, pp. 462–471, 2020.
https://doi.org/10.1007/978-3-030-55393-7_41

include Bayesian networks [4], support vector regression [5], gradient boosting decision tree (GBDT) [6], and neural networks [7]. The deep learning based methods include deep belief networks [8], stacked auto-encoders [9], and deep neural network models that extract the spatial-temporal characteristics of traffic data through combination of multiple models [10–12]. Due to the fact that Long Short Term Memory Network (LSTM) has long-term memory which is suitable for solving long-term dependence problems, it's widely used in natural language processing, weather prediction and other fields. Reference [13] first applied LSTM to the field of traffic flow prediction, and then more and more researchers began to improve LSTM and continue to exert the prediction effect of LSTM in the field of transportation [14–16]. In this paper, we propose a traffic prediction model based on LSTM to predict the exit traffic flow of toll stations. This model combines the sequence characteristics of the exit traffic flow with the spatial-temporal characteristics of the enter traffic flow of the associated stations. The main contributions of this paper as follows:

- We propose a model named LSTM_SPLSTM, which can respectively extract the sequence features of exit traffic flow of a target toll station and the spatial-temporal features of its associated toll stations. By combining the two features, it can accurately predict the short-term exit traffic flow of the target toll station;
- Considering the different impacts of associated stations on a target toll station, the Pearson correlation coefficient is used to measure the impacts and also used as the combined weight of the hidden layer of different associated toll stations in the spatial-temporal model;
- Experiments are performed on real toll station datasets. The experimental results show that the model we proposed has better prediction performance than many traditional machine models and models trained with a single feature.

2 Methodology

2.1 Problem Description

The structure of a road network is usually divided into many road sections with toll stations as dividing points. The stations on the same road segment are associated with each other and a toll station can be associated with multiple other toll stations. The Fig. 1 shows toll stations in a road network. The enter and exit traffic flows are illustrated in the Fig. 2. As shown in the Fig. 2, the road AB is divided by station A and B and the road AC is divided by station A and C. Additionally, station B and C are called the associated stations for station A. Cars entering the road AB or AC by station A belong to the enter flows of station A. On the contrary, cars leaving the road AB or AC through station A are called the exit flows of station A.

According to the charging records of a toll station, the traffic information of the station can be counted and expressed as the following:

$$X_{m\_in} = ((x_{in})_{m1}, (x_{in})_{m2}, \ldots, (x_{in})_{mt}) \tag{1}$$

$$X_{m\_out} = ((x_{out})_{m1}, (x_{out})_{m2}, \ldots, (x_{out})_{mt}) \tag{2}$$

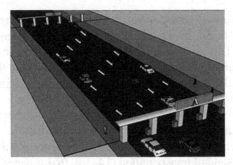

Fig. 1. Schematic of toll Stations in a road network

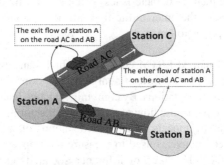

Fig. 2. The exit and enter flows of a station in road segments

Where m indicates a toll station, $(x_{in})_{mt}$ is the enter information of toll station m within time t, $(x_{out})_{mt}$ represents the exit information of toll station m within time t.

2.2   Model Design

Model Description. The traffic flow prediction of toll stations has typical spatial-temporal correlations and serial characteristics. Vehicles can drive into different roads from different toll stations and leave roads from the same toll station after a period of time. This shows that the enter traffic flow of associated toll stations at different moments will affect the exit traffic flow of target toll station in the future. Furthermore, the traffic flow records of a toll station are a series of statistical values within time intervals and those values reflect certain periodicity and trend over time. Therefore, a method combining sequence features and spatial-temporal features is more conducive to simulating and predicting traffic flow. As shown in the Fig. 3, the proposed traffic flow prediction model (LSTM_SPLSTM) is composed of two parts: one is a LSTM model that extracts sequence features from the exit traffic flow of target toll station; the other is SPLSTM model that extracts spatial-temporal features from the enter traffic flow of associated stations.

Sequence Feature Model. A toll station's traffic records are a series of statistical values within time intervals. The Fig. 5 illustrates the statistics of a toll station at different time

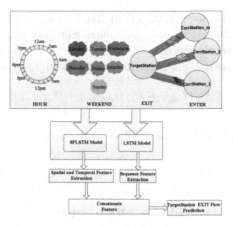

Fig. 3. The overall architecture of LSTM_SPLSTM

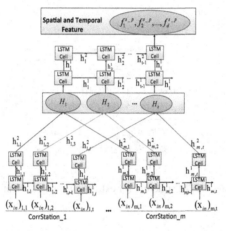

Fig. 4. SPLSTM structure for extracting spatial-temporal features

intervals. We can see that the toll station has a large traffic volume at noon and afternoon, and a smaller traffic volume in the early morning and night in a day. From the statistics of three weeks, the traffic flow at this toll station has certain periodicity.

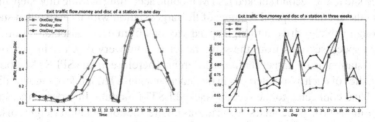

Fig. 5. A toll station's exit traffic, charge amount and mileages in one day and three weeks

The LSTM model is shown in the Fig. 6. The internal gate control mechanism of each LSTM neuron includes forgotten gate f, input gate i and output gate o. The neuron output of the hidden layer is denoted as $H = (h_1, h_2, \ldots, h_t)$, the weight is W, the bias is b, and the neuron state is c. g, φ, σ are three different sigmoid functions:

$$f_t = \sigma(W_{xf}(x_{out})_t + W_{hf}(x_{out})_{t-1} + W_{cf}c_{t-1} + b_f) \tag{3}$$

$$i_t = \sigma(W_{xi}(x_{out})_t + W_{hi}h_{t-1} + W_{ci}c_{t-1} + b_i) \tag{4}$$

$$c_t = f_t * c_{t-1} + i_t * g(W_{xc}(x_{out})_t + W_{hc}h_{t-1} + b_c) \tag{5}$$

$$o_t = \sigma(W_{xo}(x_{out})_t + W_{ho}h_{t-1} + W_{co}c_t + b_o) \tag{6}$$

$$h_t = o_t\varphi(c_t) \tag{7}$$

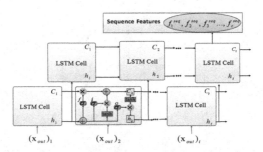

Fig. 6. LSTM structure for extracting sequence features

Spatial-Temporal Feature Model. The exit traffic flow of a toll station within a certain period of time is closely related to the entrance traffic flow of different historical moments of its associated toll stations. In order to obtain the impact of each associated site on the target site's exit traffic flow at different times, as shown in the Fig. 4, the SPLSTM model establishes an independent LSTM module for representing the enter traffic information of each associated site and connects the hidden layer information representing each associated site to the second stacked LSTM module according to same time step, thereby extracting the spatial-temporal features of the target station with its associated stations.

Although vehicles drive into high-speed sections from every associated station and leave high-speed sections from the same target station every day, the impact of each associated station on the target station is different. Therefore, in the SPLSTM model, the characteristics of each associated site in the independent LSTM model cannot be directly combined and connected to the second stacked LSTM module. In order to measure the difference impacts of associated sites on the target site, the combination weights of hidden layers in the spatial-temporal features model SPLSTM are represented according to the

Pearson correlation coefficient computed on historical data. The calculation formula is as follows:

$$per_{m\_k} = \left| \frac{T \sum\limits_{t=1}^{T} x_{m\_t} y_{k\_t} - \sum\limits_{t=1}^{T} x_{m\_t} \sum\limits_{t=1}^{T} y_{k\_t}}{\sqrt{T \sum\limits_{t=1}^{T} x_{m\_t}^2 - (\sum\limits_{t=1}^{T} x_{m\_t})^2} \sqrt{T \sum\limits_{t=1}^{T} y_{k\_t}^2 - (\sum\limits_{t=1}^{T} y_{k\_t})^2}} \right| \tag{8}$$

$$H_t = per_{m\_k} * h_{m,t} \tag{9}$$

Where x_{m_t} represents the enter traffic of associated station m at time t, y_{k_t} represents the exit traffic of the target station k at time t, and T represents the number of moments participating in the calculation. h_{m_t} represents the hidden layer output of the independent LSTM module of associated station m at time t.

3 Experiments and Results Analysis

3.1 Dataset Introduction and Hyper-parameter Setting

The datasets used in this paper are the charging records of toll stations in a certain area throughout August. After a series of preprocessing operations, we select six stations as forecast targets. Each target station has 2496 time-series data and has 11, 15, 36, 18, 19, and 58 associated sites respectively.

In the toll station exit traffic flow prediction model, for each target station and associated stations, the time step is 4 and the hidden layer size is 2, among which the number of hidden units is 64. The mean square error (MSE) is used as the loss function and the Adam optimization algorithm is used to optimize the network structure.

3.2 Experimental Evaluation Index

To evaluate the accuracy of the traffic prediction model, we mainly measure the error between the predicted value and the true value. The smaller the error, the closer the predicted value to the true value. We take root mean square error (RMSE) and average absolute error (MAE) as the evaluation indicators.

3.3 Experiment Analysis

In order to verify the effectiveness, this paper compares the performance of the proposed model (LSTM_SPLSTM) with the following 7 models: ARIMA, SVR, BP, GBDT, SAES, LSTM and SPLSTM. Table 1 shows the prediction results of different models on six toll stations' exit flow values for the next 15 min. It can be seen from Table 1 that when the prediction time is 15 min, The RMSEs and MAEs of LSTM_SPLSTM on different stations are lower than traditional machine learning methods. And it also performs better than LSTM, SAES and SPLSTM models.

Table 1. Performance comparison of 6 stations with a predicted duration of 15 min

Model	Index	S1	S2	S3	S4	S5	S6
ARIMA	RMSE	8.54	19.35	49.43	37.33	23.12	34.15
	MAE	7.43	16.27	40.75	31.53	19.48	29.53
SVR	RMSE	6.83	12.78	16.96	26.53	12.19	15.11
	MAE	5.37	10.58	12.97	21.81	9.25	11.64
BP	RMSE	6.75	14.74	26.59	24.00	17.07	25.16
	MAE	5.32	11.58	20.41	19.58	13.31	20.61
GBDT	RMSE	6.64	10.28	16.94	16.82	13.25	15.80
	MAE	4.84	7.71	12.49	12.72	9.69	12.27
SAES	RMSE	4.51	7.34	10.92	11.80	9.27	12.20
	MAE	3.27	5.44	8.33	8.99	6.82	9.24
LSTM	RMSE	4.55	7.30	11.66	11.89	9.32	11.77
	MAE	3.30	5.61	9.04	9.32	6.82	9.10
SPLSTM	RMSE	4.82	8.34	10.66	11.80	9.42	11.05
	MAE	3.40	5.80	7.87	8.70	6.55	8.32
LSTM_SPLSTM	RMSE	3.96	6.14	8.77	9.79	8.20	10.45
	MAE	2.81	4.52	6.74	7.27	5.71	7.89

For further proving the effectiveness of the model, this article predicts the exit traffic flow of target stations between 20 min and 60 min within 5 min spans. As shown in Figs. 7, 8, 9, 10, 11 and 12, with the increase of the prediction time, the prediction performance of all models shows an upward trend. But all machine learning methods have faster prediction errors increase than that of the deep learning methods. Among them, ARIMA, BP neural network and support vector regression have poor prediction performance. By combining the sequence characteristics and the spatial-temporal characteristics, the prediction effect of LSTM_SPLSTM still performs better as illustrated that the values of MAE and RMSE have the lowest rise with the increase of prediction time.

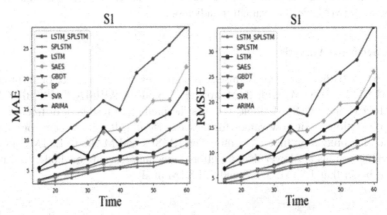

Fig. 7. Index performance of station S1 with different forecast intervals

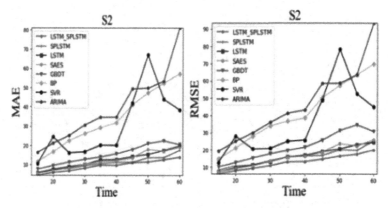

Fig. 8. Index performance of station S2 with different forecast intervals

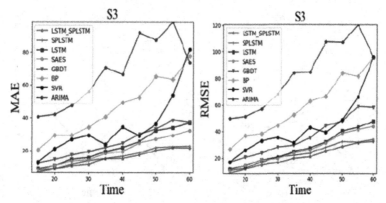

Fig. 9. Index performance of station S3 with different forecast intervals

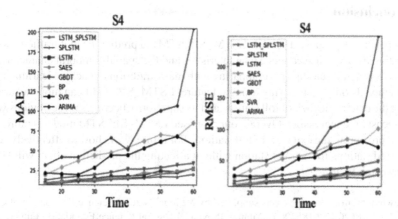

Fig. 10. Index performance of station S4 with different forecast intervals

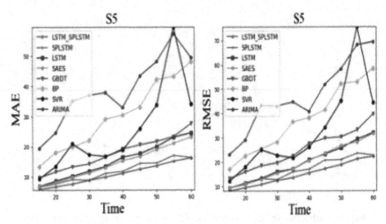

Fig. 11. Index performance of station S5 with different forecast intervals

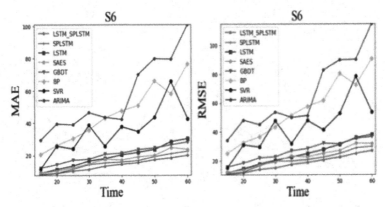

Fig. 12. Index performance of station S6 with different forecast intervals

4   Conclusion

In this paper, we propose a model (LSTM_SPLSTM) to predict the exit traffic flow of a toll station by using its sequence characteristics and the spatial-temporal characteristics with its associated stations. By comparing with many traditional machine learning models and models only considering a single feature, LSTM_SPLSTM can more accurately predict the exit traffic flow of toll station, and its superiority becomes more obvious when the forecast time increasing. Overall, our proposed LSTM_SPLSTM model is more suitable for predicting the exit flow of toll stations. For future work, how to effectively select associated stations for a target station without affecting the prediction effect will be our next research focus.

Acknowledgment. This work was supported by National Natural Science Foundation of China (Grant No. 61662085, 61862065); Yunnan Provincial Natural Science Foundation Fundamental Research Project (2019FB135); Yunnan University Data-Driven Software Engineering Provincial Science and Technology Innovation Team Project (2017HC012) and Yunnan University "Dong Lu Young-backbone Teacher" Training Program (C176220200).

References

1. Luo, Z.: Study on the capacity and ETC lane configuration of expressway toll station. East China Jiaotong University (2019)
2. Williams, B.M., Hoel, L.A.: Modeling and forecasting vehicular traffic flow as a seasonal ARIMA process: theoretical basis and empirical results. J. Transp. Eng. **129**(6), 664–672 (2003)
3. Ding, Q.Y., et al.: Forecasting traffic volume with space-time arima model. Adv. Mater. Res. **156**, 979–983 (2010)
4. Sun, S., Zhang, C., Yu, G.: A Bayesian network approach to traffic flow forecasting. IEEE Trans. Intell. Transp. Syst. **7**(1), 124–132 (2006)
5. Jeong, Y.S., et al.: Supervised weighting-online learning algorithm for short-term traffic flow prediction. IEEE Trans. Intell. Transp. Syst. **14**(4), 1700–1707 (2013)
6. Peiqun, L., Nannan, Z.: Short-term traffic flow forecast of toll station based on multi-feature GBDT model. J. Guangxi Univ. Natl. Sci. Ed. **3**, 39 (2018)
7. Karlaftis, M.G., Vlahogianni, E.I.: Statistical methods versus neural networks in transportation research: differences, similarities and some insights. Transp. Res. **19**(3), 387–399 (2011)
8. Huang, W., et al.: Deep architecture for traffic flow prediction: deep belief networks with multitask learning. Intell. Transp. Syst. IEEE Trans. **15**(5), 2191–2201 (2014)
9. Lv, Y., et al.: Traffic flow prediction with big data: a deep learning approach. Intell. Transp. Syst. IEEE Trans. **16**(2), 865–873 (2015)
10. Liu, Y., et al. Short-term traffic flow prediction with Conv-LSTM. In: 2017 9th International Conference on Wireless Communications and Signal Processing (WCSP) IEEE (2017)
11. Feng, N., Guo, S.N., Song, C., Zhu, Q.C., Wan, H.Y.: Multi-component spatial-temporal graph convolution networks for traffic flow forecasting. J. Softw. **30**(3), 759–769 (2019)
12. Guo, S., et al.: Attention based spatial-temporal graph convolutional networks for traffic flow forecasting. In: Proceedings of the AAAI Conference on Artificial Intelligence (2019)
13. Ma, X., et al.: Long short-term memory neural network for traffic speed prediction using remote microwave sensor data. Transp. Res. Part C Emerg. Technol. **54**, 187–197 (2015)
14. Kang, D., Lv, Y., Chen, Y.Y.: Short-term traffic flow prediction with LSTM recurrent neural network. In: 2017 IEEE 20th International Conference on Intelligent Transportation Systems (ITSC). IEEE (2017)
15. Wang, Z., Zhu, R., Li, T, et al.: A regularized LSTM network for short-term traffic flow prediction. In: The 6th International Conference on Information Science and Control Engineering, pp. 100–105. IEEE (2019)
16. Zhao, J., et al.: Traffic speed prediction under non-recurrent congestion: based on LSTM method and BeiDou navigation satellite system data. Intell. Transp. Syst. Mag. IEEE **11**(2), 70–81 (2019)

Long and Short Term Risk Control
for Online Portfolio Selection

Yizhe Bai⬤, Jianfei Yin$^{(\boxtimes)}$⬤, Shunda Ju⬤, Zhao Chen,
and Joshua Zhexue Huang⬤

College of Computer Science and Software Engineering, Shenzhen University,
Shenzhen, China
byz1994@yeah.net, {yjf,zx.huang}@szu.edu.cn,
jushunda2017@email.szu.edu.cn, rirnel@163.com

Abstract. Online portfolio selection is to allocate the capital among
a set of assets to maximize cumulative returns. Most of online portfolio selection algorithms focus on maximizing returns without effectively
controlling risk of loss in return. Further, many risk control algorithms
use the maximum drawdown, the Sharpe ratio, and others as risk indicators. However, these risk indicators are not sensitive to the short-term
of loss in return. This paper proposes the Long and Short Term Risk
(LSTR) control algorithm for online portfolio selection. LSTR achieves
high return and low risk by combining the effects of two parameters. The
first parameter learns the long-term risk of the market, and its posterior
probability changes slowly according to the mean reversion theory. The
second parameter senses the short-term risk of the market and makes a
quick response to changes in short-term returns. Through the multiplication of the two parameters, the risk control ability of online portfolio selection is effectively improved. The experimental results of the six
datasets demonstrate that the performance of LSTR is better than the
online portfolio selection algorithms with risk control and those without
risk control.

Keywords: Risk control · Long term learning · Short term control ·
Mean reversion theory

1   Introduction

The primary task of online portfolio select algorithms is to periodically adjust the
capital ratio among a set of risky assets to maximize the final return. In the past
few years, many online portfolio selection algorithms have been developed with
this design goal in mind, such as UP [2], EG [3], ONS [1], CFR-OGD [4] and so on.
However, these algorithms lack explicit risk control of loss in return, which leads
to serious losses in the market with large reversals [6, 12]. Therefore, people study
the definition of risk from the perspective of probability measurement of returns,

© Springer Nature Switzerland AG 2020
G. Li et al. (Eds.): KSEM 2020, LNAI 12275, pp. 472–480, 2020.
https://doi.org/10.1007/978-3-030-55393-7_42

such as variance, semi-variance, and the probability of adverse outcomes [5]. Any portfolio selection strategy that implements any of the three risk definitions requires a long trading period to collect enough observations of returns to make the empirical estimation of probability of risk related events. Thus, the returns generated by the algorithms based on the three risk definitions show stability in the long term but volatility in short term.

The improvement of the existing online portfolio selection algorithms in the aspect of risk is helpful to obtain effective trading algorithms [9,12]. One approach is to track the maximum drawdown of each portfolio vector, i.e., each expert and give more capital to the expert with a smaller maximum drawdown at each trading period [9]. However, the maximum drawdown is a long-term variable that describes the market and is not sensitive to the performance changes of portfolios in the short term. Another popular solution is to use reinforcement learning. In order to control the risk, the variables describing the risk are added to reward functions [8,10]. The Sharpe ratio is the most commonly used reward variable for reinforcement learning with risk control. It is a measure of the excess return per unit of risk for a trading strategy. This approach shows some advantages in terms of annual performance statistics. However, the Sharpe ratio needs to calculate the mean and standard deviation of the portfolio returns. The mean and standard deviation are long-term indicators that describe the market. Their short-term changes are small.

Based on the above analysis, we design an algorithm to control the long-term and short-term risks of loss in return by changing the proportion of the risk-free asset directly, called Long and Short Term Risk Control (LSTR). We summarize our main contributions as follows, i) we define a capital-ratio updating equation for online portfolio selection, where the multiplication effect of the long-term risk control parameter λ and the short-term risk control parameter η is found to be effective in the equation; ii) A risk indicator random variable $\mathcal{C}$ is defined to parameterize the probability of λ and control η ; iii) Based on the definition of $\mathcal{C}$, the learning and control algorithms for λ and η are proposed; iv) Based on λ and η , the LSTR algorithm is designed to improve the long and short term risk control ability of portfolios.

2   Related Work

The most popular methods of defining risk are variance, semi-variance, and the probability of adverse outcomes [5]. Taking the return as a random variable, the variance of the return is defined as the risk by the variance method. The semi-variance method computes the variance of the return only when the return is below the mean of the return. The probability of adverse outcomes, given by Roy's Safety First Criterion, defines a probability of the loss with respect to a target return. That is, $P\{(\phi - \zeta) \geq z\}$, where ϕ denotes target return, and $\phi - \zeta$ denotes the loss of the portfolio.

Mohr et al. proposed two risk control algorithms: RAPS and CRAPS [9] based on UP [2]. The algorithms track the experts with the lowest maximum drawdown. RAPS allocates higher ratio of capital to the lower maximum drawdown experts so far. CRAPS takes the maximum drawdown and the winner's capital into consideration to obtain the final expert capital ratio. Shen Weiwei et al. [10] used Bandit Learning to solve online portfolio problems with risk. They modeled the portfolio as multiple arms. The reward function for each arm is expressed as the Sharpe ratio. They then used the upper confidence bound to select the optimal arm. Liang Zhipeng et al. [8] implemented two reinforcement learning algorithms, DDPG and PPO in portfolio management. They used the Sharp ratio objective function to carry on the experiment, but the method is not very useful.

3   Problem Definition

Consider a financial market with n trading days and m assets. The closing price of an asset during the t period is represented by a price vector $\boldsymbol{p_t} \in \mathbb{R}_{++}^m$, and it's component p_t^i denotes the closing price of the ith asset on the tth trading day. The change in market price can be expressed by an m-dimensional relative price vector $\boldsymbol{x_t} = (x_t^1, x_t^2, ..., x_t^m) \in \mathbb{R}_+^m, t = 1, 2, ..., n$, where the element x_t^i denotes the ratio of the closing price of the ith asset in the tth period and the closing price of the $t-1$th period, expressed as $x_t^i = \frac{p_t^i}{p_{t-1}^i}$. Historical market sequence starts from period 1 to n, that is, $\boldsymbol{x_1^n} = \{\boldsymbol{x_1}, \boldsymbol{x_2}, ..., \boldsymbol{x_n}\}$. At the beginning of the tth period, the capital allocation can be represented by a portfolio vector $\boldsymbol{b_t} = (b_t^1, b_t^2, ..., b_t^m) \in \mathbb{R}_+^m, t = 1, 2, ..., n$. The element b_t^i represents the proportion of the capital of the ith asset in the tth period. $\boldsymbol{b_t} \in \Delta_m$, where $\Delta_m = \{\boldsymbol{b} : \boldsymbol{b} \succeq 0, \boldsymbol{b}^\top \mathbf{1} = 1\}$. For the tth period, $s_t = \boldsymbol{b_t}^\top \boldsymbol{x_t}$ is the daily return. The final cumulative capital after a sequence of n periods is: $S(X_1^n) = S_0 \Pi_{t=1}^n \boldsymbol{b_t}^\top \boldsymbol{x_t} = S_0 \Pi_{t=1}^n s_t$. In this paper, we define $S_0 = 1$.

4   Long and Short Term Risk Control Algorithm

In this selection, we first proposed a capital-ratio updating equation for online portfolio selection. Then we define a risk indicator random variable to parameterize the long and short term risk control parameters. Finally, the LSTR algorithm is derived based on the long and short term control parameters.

4.1   Risk Control Method

In order to control the risk of loss in return in any market situation, we can reasonably add a risk-free asset as the first asset in a given portfolio. This paper assumes that the risk-free asset is cash, and its daily return is 1. Thus, the portfolio vector $\boldsymbol{b_t}$ becomes an $m + 1$ dimensional vector, that is, $\boldsymbol{b_t} \in \mathbb{R}_+^{m+1}$. After many experiments, we found that $\boldsymbol{b_t}$ has excellent performance of risk control when it is defined by a capital-ratio updating equation:

$$\boldsymbol{b_t} = \lambda\eta\boldsymbol{e_1} + (1 - \lambda\eta)(1 - \boldsymbol{e_1}) \odot \boldsymbol{b_t}, \tag{1}$$

where $\odot$ denotes the element-wise product, the vector $\boldsymbol{e_1}$ is the unit vector with the first component is 1 and the others are 0. $\lambda \in (0,1)$ is the long-term risk control parameter. $\eta \in (0,1]$ is the short-term risk control parameter. λ and η are derived variables, which are parameterized by a risk indicator random variable $\mathcal{C}$ defined in the next section.

4.2   Long-Term Risk Control

The risk indicator random variable $\mathcal{C}$ is given as follows.

Definition 1. *Let s_t denote the daily return of the online portfolio selection on the tth trading day, ϕ denote the daily target return set by the investor, and $z \geq 0$ denote the loss that the investor can withstand per trading day:*

$$\mathcal{C}(s_t) = \begin{cases} 0, & \phi - s_t > z \\ 1, & \phi - s_t \leq z \end{cases} \tag{2}$$

where $\{\mathcal{C} = 0\}$ and $\{\mathcal{C} = 1\}$ represent event $\{\phi - s_t > z\}$ and event $\{\phi - s_t \leq z\}$, respectively.

In Definition 1, the event $\{\phi - s_t \leq z\}$ and the event $\{\phi - s_t > z\}$ respectively indicate whether the loss investor is acceptable. We count the number of times $\mathcal{C} = 0$ ($\mathcal{C} = 1$) to predict the probability q of $\mathcal{C} = 0$ ($\mathcal{C} = 1$) on a single trading day. We use the Beta distribution to describe the probability value q. The Beta distribution is characterized by two shape parameters, α and β: $P(q; \alpha, \beta) = \frac{1}{B(\alpha,\beta)} q^{\alpha-1}(1 - q)^{\beta-1}, 0 \leq q \leq 1, \alpha > 0, \beta > 0$, where $B(\alpha, \beta) = \int_0^1 q^{\alpha-1}(1 - q)^{\beta-1} dq$ denotes the Beta function. The long-term risk control parameter λ obeys the Beta distribution, i.e., $\lambda \sim P(q; \alpha, \beta)$.

In the Bernoulli trials of N times, ν denotes the number of occurrences of the random event $\mathcal{C} = 0$. We use the mean reversion theory [11] to describe the likelihood function $L(\nu|q)$, and the exponential of q is $N - \nu$ instead of ν: $L(\nu, N - \nu|q) = \binom{N}{\nu} q^{N-\nu}(1-q)^{\nu}$. Mean reversion theory points out that a stock's price will tend to move to the average price over time. The use of mean reversion strategies in the short term may be volatile, but it is very stable in the long-term run [11]. According to Bayes' theorem, we can get the following result: after a trading day, when $\mathcal{C} = 0$, the posterior probability of q becomes:

$$P(q|\nu, N - \nu) = P(q; \alpha, \beta + 1). \tag{3}$$

When $\mathcal{C} = 1$:

$$P(q|\nu, N - \nu) = P(q; 1 + \alpha, \beta). \tag{4}$$

λ can be estimated by three values λ^S, λ^E and λ^M. λ^S takes values randomly from the Beta distribution. λ^E and λ^M denote the mean and mode of the Beta distribution, respectively.

Algorithm 1. Calculation method of η.

Input:
    Adjust parameter τ;
Output:
    Proportion of cash in the portfolio η;
1: Initialize $\eta = 1$, $\kappa = 0$;
2: **for** $t = 1, 2, ..., n$ **do**
3:    **if** $\mathcal{C} = 1$ **then**
4:        $\kappa + +$;
5:        Update η using Eq.(5);
6:    **else**
7:        $\kappa = 0$, $\eta = 1$;
8:    **end if**
9: **end for**

4.3 Short-Term Risk Control

In Eq. (1), η denotes the proportion of cash in the portfolio. It is defined by:

$$\eta = \frac{1}{1 + exp\{\kappa + \tau\}}, \tag{5}$$

where κ denotes the number of consecutive events $\mathcal{C} = 1$, and τ is a constant that determines how much η falls when $\kappa = 1$. When the $\kappa + \tau \in [-4, 4]$ in Eq. (5), η decreases exponentially in the region $(1, 0)$. So when $\kappa = 1$, to have $\kappa + \tau \in [-4, 4]$, the constant τ must be in $[-5, 3]$. By Eq. (5), we know that the parameter η can quickly respond to the short-term returns of a portfolio.

The calculation method of η is shown in Algorithm 1. The initial cash is $\eta = 1$, indicating that it has not entered the market. The count variable κ is set to be 0 initially. As the number of consecutive occurrences of $\mathcal{C} = 1$ increases, κ also increases. So that, η drops to near 0 to ensure that capital can be used to invest quickly in other risky assets according to Eq. (1). Once the event $\mathcal{C} = 0$ happens, the continuous count parameter κ will be set to 0 immediately, and the proportion of capital for cash will be restored to 1 in order to avoid the risk of short-term loss of earnings.

Algorithm 2. LSTR algorithm.

Input:
    Historical market sequence x_1^n;
    The maximum daily loss z that the investor can bear;
    Adjust parameter τ;
    The target return ϕ set by the investor;
    An online portfolio selection A;
    $\lambda \in \{\lambda^S, \lambda^E, \lambda^M\}$;
Output:
    Final cumulative wealth S_n;
1: Initialize $S_0 = 1$, $b_1 = (\frac{1}{m+1}, ..., \frac{1}{m+1}) \in \mathbb{R}_+^{m+1}$, $\alpha = \beta = 2$, $\eta = 1$, $\kappa = 0$;
2: **for** $t = 1, 2, ..., n$ **do**
3:     Using online portfolio strategy A to compute a portfolio b_t;
4:     Calculate the daily return $s_t = b_t^T x_t$;
5:     Updates cumulative return $S_t = S_{t-1} \times s_t$;
6:     Through Eq.(2), get $\mathcal{C} = 0$ or $\mathcal{C} = 1$;
7:     **if** $\mathcal{C} = 1$ **then**
8:         $\kappa + +$;
9:         Update η using Eq.(5);
10:        Update α and β by Eq.(4);
11:     **else**
12:        $\kappa = 0$, $\eta = 1$;
13:        Update α and β by Eq.(3);
14:     **end if**
15:     Computes λ;
16:     Adjust b_t using Eq.(1);
17:     Portfolio strategy A updates the online portfolio selection rules;
18: **end for**

4.4 LSTR Algorithm

The LSTR algorithm is shown in Algorithm 2. First, the two shape parameters of the Beta distribution are initialized by $\alpha = \beta = 2$. When $\alpha > 1$ and $\beta > 1$, the Beta distribution has a unique mode, and λ^M works fine. LSTR obtains the portfolio vector b_t of the basic online portfolio selection algorithm through step (3). In step (7)–(14), LSTR automatically update the parameters through the risk indicator variable $\mathcal{C}$, including η, α, and β. By adjusting b_t in step (16), the adjusted b_t has a different proportion of cash than that of the portfolio selection A. Finally, the portfolio strategy is updated according to the rules of the online portfolio selection A.

5 Experimental Results

In Table 1, we show the six datasets used in this experiment. The datasets of this experiment have one more cash than the original datasets. The NYSE, DJIA, MSCI, and TSE datasets are all widely used[1] [7]. Two datasets, FSE and SP500,

---

[1] These datasets are downloaded from the https://github.com/OLPS/OLPS.

were collected at Yahoo Finance[2]. We choose six algorithms to compare with LSTR. UP [2], EG [3], ONS [1] are three online portfolio selection algorithms without risk control; RAPS and CRAPS [9] are two online portfolio selection algorithms with risk control.

Parameter Choices: For LSTR, $\phi = 1.003, z = 0.004$. In this experiment, let $\tau = 0$. This means that η directly drops from 1 to about $\frac{1}{1+e} \approx 0.268$. The parameters of other algorithms are set according to the suggestions of the authors. $LSTR^S$, $LSTR^E$, and $LSTR^M$ use three methods: λ^S, λ^E, and λ^M, respectively. We can see from Table 2, the results of $LSTR^S$ are slightly better than $LSTR^E$ and $LSTR^M$ in most cases. But $LSTR^E$ and $LSTR^M$ are more convenient and faster to calculate than $LSTR^S$.

Table 1. Summary of datasets

Dataset	Region	Time frame	Periods	Assets
NYSE	US	[1962.07.03–1984.12.31]	5651	37
DJIA	US	[2001.01.14–2003.01.14]	507	31
MSCI	Global	[2006.01.04–2010.03.31]	1043	25
FSE	GER	[2002.01.01–2003.06.16]	369	16
SP500	US	[2008.04.07–2011.01.14]	700	21
TSE	CA	[1994.04.01–1998.12.31]	1259	89

Cumulative Wealth (CW): In these six datasets, the cumulative wealth of $LSTR^M$ is superior to that of UBAH and the original algorithm. On NYSE, the cumulative wealth of $LSTR^M$ is about three times higher than that of UP and EG, and about seven times higher than that of ONS. The cumulative wealth of RAPS and CRAPS in six datasets is less than all LSTR.

Sensitivity of Risk: The Sharpe Ratio (SR) is an indicator that combines both return and risk. In Table 2, On the five datasets NYSE, MSCI, FSE, SP500, and TSE, the Sharpe ratios of LSTR are higher than those of the original algorithms. On the DJIA dataset, the Sharpe ratios of all algorithms are negative. The Maximum Drawdown (MDD) is the maximum observed loss of the portfolio from peak to valley before reaching a new peak. In these six datasets, the MDDs of LSTR are lower than that of UBAH and the original algorithms. Table 2 show that the MDDs and SRs of LSTR are better than those of RAPS and CRAPS on the six datasets. The results show that the risk sensitivity of LSTR is higher than that of RAPS and CRAPS.

---

[2] These datasets are collected from the https://finance.yahoo.com.

Table 2. Results of performance

Dataset	indices	UBAH	RAPS	CRAPS	UP	$LSTR_{UP}^{S}$	$LSTR_{UP}^{E}$	$LSTR_{UP}^{M}$	EG	$LSTR_{EG}^{S}$	$LSTR_{EG}^{E}$	$LSTR_{EG}^{M}$	ONS	$LSTR_{ONS}^{S}$	$LSTR_{ONS}^{E}$	$LSTR_{ONS}^{M}$
NYSE	CW	17.6211	24.5010	24.4553	24.5299	79.2908	79.9822	**80.1453**	24.4025	**78.0626**	77.8457	77.8463	17.2372	**122.0882**	122.0	121.9517
	SR	0.3080	1.0027	1.0062	1.0035	1.8397	1.8467	**1.8476**	1.0012	**1.8289**	1.8252	1.8254	1.0414	**1.3518**	1.3516	1.3516
	MDD	0.5350	0.1155	0.1164	0.1153	0.0654	0.0653	**0.0652**	0.1138	**0.0653**	0.0655	0.0655	0.1348	**0.1376**	0.1381	0.1381
DJIA	CW	0.7666	0.8382	0.8381	0.8374	0.8464	**0.8593**	0.8586	0.8127	0.8505	0.8528	**0.853**	1.5193	1.5131	1.53763	**1.5378**
	SR	-0.4755	-0.3091	-0.3031	-0.3091	-0.415	-0.359	-0.3615	-0.327	-0.4022	-0.3951	-0.3948	0.7608	0.9037	**0.9403**	**0.9403**
	MDD	0.3710	0.3673	0.365	0.3675	0.294	**0.2894**	0.2901	0.3673	**0.2822**	0.2894	0.2893	0.3362	0.2221	0.2221	**0.2220**
MSCI	CW	0.9010	0.9137	0.9032	0.9203	1.4072	**1.4254**	1.403	0.9246	**1.4233**	1.4154	1.4106	0.881	**1.4025**	1.3779	1.3834
	SR	-0.0878	-0.0521	-0.0601	-0.0502	0.4441	**0.4542**	0.4321	-0.0458	**0.4597**	0.4516	0.4472	-0.0233	**0.406**	0.3839	0.389
	MDD	0.6289	0.6242	0.6314	0.6279	0.3949	**0.3917**	0.3961	0.6274	**0.3893**	0.3934	0.3924	0.6612	**0.3989**	0.402	0.4009
FSE	CW	0.7316	0.7777	0.7706	0.7766	**1.0318**	1.0092	1.0069	0.8078	**1.0419**	1.0226	1.0219	1.8893	**1.9703**	1.9438	1.9434
	SR	-1.0232	-0.5613	-0.609	-0.5723	**0.0635**	-0.0335	-0.0407	-0.4799	0.093	0.0376	**0.0356**	0.7751	0.916	**0.9057**	0.9054
	MDD	0.4628	0.5007	0.4969	0.4994	**0.3631**	0.3698	0.3705	0.4994	**0.3559**	0.3703	0.3708	0.6706	0.3423	**0.339**	0.339
SP500	CW	1.0698	1.2357	1.2059	1.2164	**1.2711**	1.2604	1.2607	1.2227	**1.2685**	1.2638	1.2636	1.1943	**1.5815**	1.5697	1.5709
	SR	0.1503	0.3353	0.3065	0.3164	**0.4142**	0.3973	0.3979	0.3333	**0.4113**	0.4053	0.4051	0.2112	**0.5979**	0.5955	0.5958
	MDD	0.4993	0.5635	0.5607	0.5686	**0.4907**	0.4933	0.4932	0.5726	0.4966	0.4938	**0.4937**	0.9773	**0.8493**	0.8519	
TSE	CW	1.5691	1.5155	1.5286	1.546	2.3443	2.3601	**2.3692**	1.5396	2.3097	2.3566	**2.3591**	1.4561	**1.9003**	1.8843	1.8714
	SR	0.5903	0.5570	0.5732	0.582	1.726	1.7293	**1.7387**	0.5663	1.6839	1.7393	**1.7427**	0.3249	**0.5828**	0.575	0.5697
	MDD	0.2985	0.3372	0.3355	0.3334	0.1552	0.1553	**0.1548**	0.3337	0.1554	0.154	**0.1539**	0.5167	**0.3819**	0.3838	0.3832

6 Conclusions

In this paper, we propose the LSTR algorithm to control the long and short term risks for the online portfolio selection. We define a capital-ratio updating equation for online portfolio selection, where the multiplication effect of the long and short term risk control parameters is found to be effective in the equation. We define a risk indicator random variable $\mathcal{C}$ to parameterize the probability of the long-term risk parameter and the control of the short-term risk parameter. Based on the definition of $\mathcal{C}$, we developed the learning and control algorithms for the risk control parameters. The experimental results on the six datasets show that the combination effect of long and short term risk control parameters is better than the existing risk control algorithms.

Our future work is to optimize for transaction costs. Besides, we will try more different methods to learn long and short term control parameters.

Acknowledgement. We thank the editors and reviewers for their expert comments and constructive suggestions, and the support of Natural Science Foundation of China No. 61972261. Professor Jianfei Yin is the corresponding author for this work.

References

1. Agarwal, A., Hazan, E., Kale, S., Schapire, R.E.: Algorithms for portfolio management based on the Newton method. In: Proceedings of the 23rd International Conference on Machine learning, pp. 9–16. ACM (2006)
2. Cover, T.M.: Universal portfolios. Math. Financ. **1**(1), 1–29 (1991)
3. Helmbold, D.P., Schapire, R.E., Singer, Y., Warmuth, M.K.: On-line portfolio selection using multiplicative updates. Math. Financ. **8**(4), 325–347 (1998)
4. Huang, D., Yu, S., Li, B., Hoi, S.C., Zhou, S.: Combination forecasting reversion strategy for online portfolio selection. ACM Trans. Intell. Syst. Technol. (TIST) **9**(5), 58 (2018)
5. Huang, X.: Portfolio selection with a new definition of risk. Eur. J. Oper. Res. **186**(1), 351–357 (2008)

6. Li, B., Hoi, S.C.: Online portfolio selection: a survey. ACM Comput. Surv. (CSUR) **46**(3), 35 (2014)
7. Li, B., Sahoo, D., Hoi, S.C.: OLPS: a toolbox for on-line portfolio selection. J. Mach. Learn. Res. **17**(1), 1242–1246 (2016)
8. Liang, Z., Chen, H., Zhu, J., Jiang, K., Li, Y.: Adversarial deep reinforcement learning in portfolio management. arXiv preprint arXiv:1808.09940 (2018)
9. Mohr, E., Dochow, R.: Risk management strategies for finding universal portfolios. Ann. Oper. Res. **256**(1), 129–147 (2017). https://doi.org/10.1007/s10479-016-2176-6
10. Shen, W., Wang, J., Jiang, Y.G., Zha, H.: Portfolio choices with orthogonal bandit learning. In: Twenty-Fourth International Joint Conference on Artificial Intelligence (2015)
11. Summers, L.H.: Mean reversion in stock prices: evidence and implications. J. Financ. Econ. **22**(1), 27–59 (1987)
12. Uziel, G., El-Yaniv, R.: Online learning of portfolio ensembles with sector exposure regularization. arXiv preprint arXiv:1604.03266 (2016)

Author Index

Printed in the United States
By Bookmasters